Environmental Contamination Remediation and Management

Series Editors

Erin R. Bennett, School of the Environment, Trent University, Peterborough, Canada

Iraklis Panagiotakis, Environmental Engineer & Scientist, ENYDRON – Environmental Protection Services, Athens, Greece

Advisory Editors

Maria Chrysochoou, Department of Civil & Environmental Engineering, University of Connecticut, Storrs, CT, USA

Dimitris Dermatas, School of Civil Engineering, National Technical University of Athens, Zografou, Greece

Luca di Palma, Chemical Engineering Materials Environment, Sapienza University of Rome, Rome, Italy

Demetris Francis Lekkas, Environmental Engineering and Science, University of the Aegean, Mytilene, Greece

Mirta Menone, National University of Mar del Plata, Mar del Plata, Argentina

Chris Metcalfe, School of the Environment, Trent University, Peterborough, Canada

Matthew Moore, United States Department of Agriculture, National Sedimentation Laboratory, Oxford, MS, USA

There are many global environmental issues that are directly related to varying levels of contamination from both inorganic and organic contaminants. These affect the quality of drinking water, food, soil, aquatic ecosystems, urban systems, agricultural systems and natural habitats. This has led to the development of assessment methods and remediation strategies to identify, reduce, remove or contain contaminant loadings from these systems using various natural or engineered technologies. In most cases, these strategies utilize interdisciplinary approaches that rely on chemistry, ecology, toxicology, hydrology, modeling and engineering.

This book series provides an outlet to summarize environmental contamination related topics that provide a path forward in understanding the current state and mitigation, both regionally and globally.

Topic areas may include, but are not limited to, Environmental Fate and Effects, Environmental Effects Monitoring, Water Re-use, Waste Management, Food Safety, Ecological Restoration, Remediation of Contaminated Sites, Analytical Methodology, and Climate Change.

Vimal Chandra Pandey
Editor

Bio-Inspired Land Remediation

 Springer

Editor
Vimal Chandra Pandey ⓘ
Department of Environmental Science
Babasaheb Bhimrao Ambedkar University
Lucknow, India

ISSN 2522-5847　　　　　　ISSN 2522-5855　(electronic)
Environmental Contamination Remediation and Management
ISBN 978-3-031-04930-9　　　ISBN 978-3-031-04931-6　(eBook)
https://doi.org/10.1007/978-3-031-04931-6

© The Editor(s) (if applicable) and The Author(s), under exclusive license to Springer Nature Switzerland AG 2023
This work is subject to copyright. All rights are solely and exclusively licensed by the Publisher, whether the whole or part of the material is concerned, specifically the rights of translation, reprinting, reuse of illustrations, recitation, broadcasting, reproduction on microfilms or in any other physical way, and transmission or information storage and retrieval, electronic adaptation, computer software, or by similar or dissimilar methodology now known or hereafter developed.
The use of general descriptive names, registered names, trademarks, service marks, etc. in this publication does not imply, even in the absence of a specific statement, that such names are exempt from the relevant protective laws and regulations and therefore free for general use.
The publisher, the authors, and the editors are safe to assume that the advice and information in this book are believed to be true and accurate at the date of publication. Neither the publisher nor the authors or the editors give a warranty, expressed or implied, with respect to the material contained herein or for any errors or omissions that may have been made. The publisher remains neutral with regard to jurisdictional claims in published maps and institutional affiliations.

This Springer imprint is published by the registered company Springer Nature Switzerland AG
The registered company address is: Gewerbestrasse 11, 6330 Cham, Switzerland

Foreword

Land is a non-renewable resource and fundamental to human life. Land conservation and management are essential to protect food security and to promote sustainable livelihoods. Therefore, the degraded land restoration is vital for regaining biodiversity and ecosystems seruices and achieving Sustainable Development Goals (SDGs). Bio-inspired land remediation programs have been executed to secure upper soil surface during the United Nations Decade on Ecosystem Restoration (2021–2030). The book "Bio-Inspired Land Remediation" aims to cover a cutting-edge synthesis of scientific and experiential knowledge on bio-based remediation of degraded land. The book covers broad aspects of bio-inspired land remediation strategies including phytoremediation, bacterial remediation, fungal remediation, vermiremediation, and biochar remediation.

I am sure that this book will be a valuable reference for researchers, scientists, environmentalists, entrepreneurs, policymakers, and other stakeholders. I congratulate the editor and all the authors for their great efforts for compiling the book and hope that it will be beneficial for all the stakeholders and well-being of agriculture and environment systematically.

August 18, 2022

Himanshu Pathak
Director General, Indian Council
of Agricultural Research (ICAR)
New Delhi, India

Preface

Bio-Inspired Land Remediation offers cutting-edge knowledge on both basics and practical aspects of remediation of land pollution. Deterioration of the earth's land surfaces, at and below ground level due to natural causes or anthropogenic activities like urbanization, industrialization, and population growth, is the main cause of land pollution. Land pollution and land degradation are the most serious issues worldwide as land is a non-renewable resource that supports all human needs. Intentional or unintentional litter or waste product deposition containing heavy metals, metalloids, petroleum hydrocarbon, radioactive elements, polycyclic aromatic hydrocarbons, pesticides, herbicides, and other organic pollutants, urbanization or construction-related activities, mining, and agricultural activities including irrigation of agricultural land with polluted water are the reasons for land pollution.

The conservation and maintenance of the upper layer of land are of utmost essential to food security and sustainable livelihoods. Additionally, the land supports our biodiversity and helps to combat climate change adaptation and mitigation through carbon sequestration. Therefore, land restoration and management practices have become essential for reversing the trend of soil degradation and ensuring food security, thereby achieving United Nations Sustainable Development Goals. Bio-inspired land remediation refers to a number of biological remediation technologies for the treatment of soil using microorganisms, fungi, plants, earthworms, biochar, etc. However, several facts are still needed to do research into land remediation programs.

Biomanagement offers the unmatched potential to restore polluted lands to food security and sustainable livelihoods during the United Nations Decade on Ecosystem Restoration (2021–2030). Thus, bio-inspired remediation has a lot of potential to protect land quality toward nature sustainability. This book offers several potential approaches such as bacterial remediation, mycoremediation, vermiremediation, phytoremediation, biochar-based remediation, and others to provide cost-effective and eco-friendly ways to remediate the land. The book's seventeen chapters focus on different aspects of bio-inspired land remediation. This book is a notable asset, well-timed, and updated information for researchers, students, environmentalists, professors, ecological scientists, agriculturists, practitioners, policymakers, entrepreneurs, and other stakeholders.

Lucknow, India

Editor
Vimal Chandra Pandey

Acknowledgments

I sincerely wish to thank Erin Bennett (Series Editor) and Iraklis Panagiotakis (Co-series Editor), Ram Prasad R C (Project Coordinator), Madanagopal D (Production Project Manager), and Nel van der Werf (Publishing Editor) from Springer for their excellent support, guidance, and coordination during the production of this fascinating project. I thank all the contributors for their excellent chapter contributions. I also would like to thank all the reviewers for their time and expertise in reviewing the chapters for this book. I am really grateful to Dr. Himanshu Pathak, Secretary, Department of Agricultural Research & Education (DARE), and Director General, Indian Council of Agricultural Research (ICAR), New Delhi, India, for writing the foreword at short notice. And last but not least, I must thank my family for endless support and encouragement.

Contents

1	**Bioenergy Crop-Based Ecological Restoration of Degraded Land**	1
	Dragana Ranđelović and Vimal Chandra Pandey	
	1.1 Introduction	1
	1.2 Suitability of Bioenergy crops for Wide-Ranging Degraded Lands	4
	1.3 Plant Derived Bioenergy Sources	8
	1.4 Degraded Land Restoration by Energy Crops	10
	1.5 Livelihood Improvements	18
	1.6 Potential Challenges	19
	1.7 Growing Bioenergy Crops on Degraded Lands: Achieving UN-SDGs	20
	1.8 Conclusion	21
	References	22
2	**Understanding the Role of Ruderal Plant Species in Restoration of Degraded Lands**	31
	Dragana Ranđelović and Slobodan Jovanović	
	2.1 General Restoration Aspects of Degraded Lands	32
	2.2 Ruderal Plants—General Traits and Significance	34
	2.3 Annual Ruderal Plants	36
	2.4 Ruderal Perennials	40
	2.5 Woody Ruderals	46
	2.6 Invasive Ruderal Plants	50
	2.7 Conclusion Remarks	56
	References	58

3 Utilizing Polluted Land for Growing Crops 69
Shivakshi Jasrotia and Vimal Chandra Pandey
- 3.1 Introduction 69
- 3.2 Edible Crop Production from Polluted Land and Bio-fortification 70
- 3.3 Plants that Produce a Pollutant Free Edible Part 72
- 3.4 Strategies for Reducing Pollutants in Edible Parts 74
- 3.5 Connecting Phytoremediation and Food Production 76
- 3.6 Conclusions and Future Prospects 79
- References 80

4 Plant Assisted Bioremediation of Heavy Metal Polluted Soils 85
Sumita Chandel, Rouf Ahmad Dar, Dhanwinder Singh, Sapna Thakur, Ravneet Kaur, and Kuldip Singh
- 4.1 Introduction: Background of Heavy Metal Pollution 86
- 4.2 Sources of Heavy Metal Pollution 88
- 4.3 Plant Assisted Bioremediation: Techniques/Strategies 88
- 4.4 Significance of Plant Assisted Bioremediation of Heavy Metal in Agriculture 91
- 4.5 Role of Microflora and Flora in Plant Assisted Bioremediation 92
- 4.6 Mechanism of Plant Assisted Bioremediation 95
- 4.7 Conclusion and Future Prospects 105
- References 105

5 Cutting-Edge Tools to Assess Microbial Diversity and Their Function in Land Remediation 115
Indra Mani
- 5.1 Introduction 115
- 5.2 Culture-Dependent Techniques 116
- 5.3 Culture-Independent Techniques 118
- 5.4 Cutting-Edge High-Throughput Culture-Independent Approach for Microbial Diversity 120
- 5.5 Cutting-Edge High-Throughput Culture-Independent Approach for Microbial Function 123
- 5.6 Conclusion, Challenges, and Future Perspective 126
- References 128

6 Endophytic Microbes and Their Role in Land Remediation 133
Satinath Das, Pralay Shankar Gorai, Linee Goswami, and Narayan Chandra Mandal
- 6.1 Introduction 133
- 6.2 What Are Endophytic Microbes? 134
- 6.3 Effect of Endophytes in Soil Fertility Management 135
- 6.4 Nitrogen Fixation 135
- 6.5 Biofertilizer 137

	6.6	Pathogen Antagonism	137
	6.7	Siderophore Production	138
	6.8	Nutrient Cycling	138
	6.9	Plant–Endophytic Interaction and Their Role in Plant Growth Promotion	139
	6.10	Plant–Endophytic Interactions	139
	6.11	Plant Growth Promotion	141
	6.12	Identification of Endophytes and Their Utilization Against Persistent Organic Pollutants	142
	6.13	Effect of Endophytes Against Heavy Metal Contaminated Soil	150
	6.14	Phytoavailability	151
	6.15	Hyper Accumulation and Biosorption	152
	6.16	Toxicity Reduction	153
	6.17	Conclusion and Future Prospects	154
	References		155
7	**Fungal-Based Land Remediation**		165
	Soma Barman, Ratan Chowdhury, and Satya Sundar Bhattacharya		
	7.1	Introduction	166
	7.2	Fungal Organisms and Bioremediation	167
	7.3	Effectivity of Fungal Organism on Different Pollutants	169
	7.4	Mechanism of Action: Hypotheses and Evidence	175
	7.5	Conclusion and Recommendations	180
	References		181
8	**Microbial Detoxification of Contaminated Land**		189
	Nazneen Hussain, Linee Goswami, and Satya Sundar Bhattacharya		
	8.1	Introduction	189
	8.2	Environmental Impact of Pesticides	191
	8.3	Sources of POP and Their Current Status	195
	8.4	Factors Responsible for Microbial Degradation	197
	8.5	Microbial Enzymes Used in Pesticide Bioremediation	198
	8.6	Impact of Genetic Engineering in Pesticide Degradation	202
	8.7	Impact of Heavy Metal on Land Quality	203
	8.8	Future Prospects	210
	References		211
9	**Vermi-Remediation of Metal(loid)s Contaminated Surfaces**		221
	Linee Goswami, Subhasish Das, Nazneen Hussain, and Satya Sundar Bhattacharya		
	9.1	Introduction	221
	9.2	Factors Governing the Process of Vermicomposting	223
	9.3	Role of Earthworms in a Vermi-Reactor: Benefits and Limitations	224
	9.4	Vermi-Remediation and Solid Waste Management	225

	9.5	Vermicompost and Soil Health Management	226
	9.6	Detoxification of Soil Environment	229
	9.7	Future Prospects	233
	References		233

10 Fly Ash Management Through Vermiremediation 241
Sanat Kumar Dwibedi and Vimal Chandra Pandey
    10.1 Introduction 241
    10.2 Properties of FA 242
    10.3 Verms as Bioreactor 243
    10.4 Research Status on FA Use and Vermiremediation 243
    10.5 What is Vermiremediation? 247
    10.6 Biology of Earthworm and Its Functional Significance
         in Waste Degradation 250
    10.7 Process of Vermiremediation 252
    10.8 Strategies for Vermiremediation 252
    10.9 Conclusions and Prospects 254
    References 255

11 Management of Biomass Residues Using Vermicomposting Approach 261
Suman Kashyap, Seema Tharannum, V. Krishna Murthy,
and Radha D. Kale
    11.1 Introduction 262
    11.2 Vermicomposting 263
    11.3 Agents of Vermicompost 263
    11.4 Types of Earthworms Employed for Vermicomposting
         in India 264
    11.5 Vermicomposting on Ground Heaps 265
    11.6 Raw Materials for Degradation by Vermicomposting
         Process 266
    11.7 Animal Waste Biomass 269
    11.8 Forest Biomass 269
    11.9 Human Habitation Waste 270
    11.10 Municipal Waste 275
    11.11 Vermicomposting of Human Excreta 276
    11.12 Vermicomposting of Fly-Ash 276
    11.13 Vermiremediation of Contaminated Soils 276
    11.14 Properties of Vermicompost 277
    11.15 Quality of Vermicompost 278
    11.16 Advantages of Vermicompost 279
    11.17 Conclusion 281
    References 282

12	**Vermiremediation of Agrochemicals, PAHs, and Crude Oil Polluted Land**	**287**

Shivika Datta, Simranjeet Singh, Praveen C. Ramamurthy,
Dhriti Kapoor, Vaishali Dhaka, Deepika Bhatia, Savita Bhardwaj,
Parvarish Sharma, and Joginder Singh

12.1	Introduction	288
12.2	Agrochemicals: Classification, Effect on Environment, Health Hazards	289
12.3	PAHs (Classification, Effect on Environment, Health Hazards)	291
12.4	Crude Oil Polluted Land (Classification, Effect on Environment, Health Hazards)	293
12.5	Global Regulations on Use of Agrochemicals, PAHs, and Crude Oil	294
12.6	Strategies to Overcome the Harmful Effects of Agrochemicals, PAHs, Crude Oil	294
12.7	Vermicomposting in Bioremediation	300
12.8	Vermicompost: Mechanism	302
12.9	Conclusion	303
	References	304

13	**Biochar-Based Remediation of Heavy Metal Polluted Land**	**317**

Abhishek Kumar and Tanushree Bhattacharya

13.1	Introduction	318
13.2	Biochar and Its Production	318
13.3	Biochar Modification Methods	322
13.4	Properties of Biochar	323
13.5	Heavy Metals and Their Removal	327
13.6	Impact of Biochar on Mobility of Heavy Metal	331
13.7	Impact of Biochar on Bioavailability of Heavy Metal	333
13.8	Remediation of Polluted Sites by Application of Biochar	334
13.9	Applications of Biochar Other Than Heavy Metal Removal	335
13.10	Advantages and Risks Associated with Biochar Production and Application	337
13.11	Future Research	338
13.12	Conclusion	338
	References	339

14	**Soil Carbon Sequestration Strategies: Application of Biochar an Option to Combat Global Warming**	**353**

Shweta Yadav, Vikas Sonkar, and Sandeep K. Malyan

14.1	Introduction	353
14.2	Role of Carbon Dioxide (CO_2) in Global Warming	355
14.3	Soil Organic Carbon (SOC)	358
14.4	SOC Sequestration Strategies	359

14.5	Carbon Sequestration in Soil Through Biochar		364
14.6	Conclusions		368
References			369

15 Remediation of Pharmaceutical and Personal Care Products in Soil Using Biochar 375
Amita Shakya, Sonali Swain, and Tripti Agarwal
- 15.1 Introduction 376
- 15.2 Sources and Transport of PPCPs to the Soil Environment 377
- 15.3 Environmental and Health Risk of PPCPs 379
- 15.4 Fate and Occurrence of PPCPs in the Soil Environment 381
- 15.5 Strategies for Remediation of PPCPs from Soil 383
- 15.6 Biochar for PPCPs Removal from Soil 385
- 15.7 Factors Influencing the Removal of PPCPs Using Biochar 392
- 15.8 Mechanism of PPCPs Removal from Soil with Biochar 394
- 15.9 Possible Risk Factors Associated with Biochar Application for Removal of PPCPs from Soil 396
- 15.10 Conclusion and Future Approaches 396
- References 397

16 Biochar for Improvement of Soil Properties 403
Abhishek Kumar and Tanushree Bhattacharya
- 16.1 Introduction 403
- 16.2 Biochar Basics 406
- 16.3 Production of Biochar 406
- 16.4 Properties of Biochar 408
- 16.5 Impact of Biochar on Chemical Properties of Soil and Their Consequent Improvement 411
- 16.6 Impact of Biochar on Physical Properties of Soil and Their Consequent Improvement 417
- 16.7 Impact of Biochar on Other Properties of Soil and Their Consequent Improvement 421
- 16.8 Amendments and Modifications of Biochar for Enhancing the Impact on Soil Properties 426
- 16.9 Disadvantages of Biochar Application to Soil 427
- 16.10 Research Gaps 428
- 16.11 Conclusion 428
- References 429

17 Biochar Production and Its Impact on Sustainable Agriculture 445
Sanat Kumar Dwibedi, Basudev Behera, and Farid Khawajazada
- 17.1 Introduction 445
- 17.2 History of Biochar Production and Use 446
- 17.3 Benefits of Biochar Use 448
- 17.4 Procedure for Synthesis of Biochar 449
- 17.5 Methods of Preparation 451

17.6	Economic Feasibility of Biochar Production	456
17.7	Effects of Biochar on Agriculture	457
17.8	Future Prospects and Constraints in Biochar Systems	463
17.9	Conclusion	465
	References	467

Index 475

Editor and Contributors

About the Editor

Dr. Vimal Chandra Pandey featured in the world's top 2% scientists curated by Stanford University, USA. Dr. Pandey is leading Researcher in the field of environmental engineering, particularly phytomanagement of polluted sites. His research focuses mainly on the remediation and management of degraded lands, including heavy metal-polluted lands and post-industrial lands polluted with fly ash, red mud, mine spoil, and others, to regain ecosystem services and support a bio-based economy with phytoproducts through affordable green technology such as phytoremediation. Dr. Pandey's research interests also lie in exploring industrial crop-based phytoremediation to attain bioeconomy security and restoration, adaptive phytoremediation practices, phytoremediation-based biofortification, carbon sequestration in waste dumpsites, climate resilient phytoremediation, fostering bioremediation for utilizing polluted lands, and attaining United Nations Sustainable Development Goals. His phytoremediation work has led to the extension of phytoremediation beyond its traditional application. He is now engaged to explore profitable phytoremediation with least risk, low input, and minimum care. Dr. Pandey worked as CSIR-Pool Scientist and DS Kothari Postdoctoral Fellow at Babasaheb Bhimrao Ambedkar University, Lucknow; Consultant at the Council of Science and Technology, Uttar Pradesh; and DST-Young Scientist at CSIR-National Botanical Research Institute, Lucknow. He is Recipient of a number of awards/honors/fellowships and is Member of the National Academy of Sciences India. Dr. Pandey serves as a subject expert and panel member for the evaluation of research and professional activities in India and abroad for fostering nature sustainability. He has published over 100 scientific articles/book chapters in peer-reviewed journals/books. Dr. Pandey is also Author and Editor of several books published by Elsevier, with several more forthcoming. He is Associate Editor of *Land Degradation and Development* (Wiley); Editor of *Restoration Ecology* (Wiley); Associate Editor of *Environment, Development and Sustainability* (Springer); Associate Editor of *Ecological Processes* (Springer Nature); Academic Editor of *PLOS ONE* (PLOS); Advisory Board Member of *Ambio* (Springer); Editorial Board Member of *Environmental Management* (Springer); *Discover Sustainability* (Springer Nature) and *Bulletin of Environmental Contamination and Toxicology* (Springer). He also works/worked as Guest Editor for many reputed journals.

Contributors

Agarwal Tripti Department of Agriculture and Environmental Sciences, National Institute of Food Technology Entrepreneurship and Management, Kundli, Sonipat, Haryana, India

Barman Soma Soil and Agro-Bioengineering Laboratory, Department of Environmental Science, Tezpur University, Assam, India

Behera Basudev Institute of Agricultural Sciences, Siksha 'O' Anusandhan University (Deemed), Bhubaneswar, Odisha, India

Bhardwaj Savita Department of Botany, Lovely Professional University, Phagwara, India

Bhatia Deepika Department of Biotechnology, Baba Farid Group of Institutions, Deon, Bathinda, India

Bhattacharya Satya Sundar Department of Environmental Science, Tezpur University, Tezpur, Assam, India

Bhattacharya Tanushree Department of Civil and Environmental Engineering, Birla Institute of Technology, Mesra, Ranchi, Jharkhand, India

Chandel Sumita Department of Soil Science, Punjab Agricultural University, Ludhiana, Punjab, India

Chowdhury Ratan Department of Botany, Rangapara College, Amaribari, Rangapara, Assam, India

Dar Rouf Ahmad Department of Microbiology, Punjab Agricultural University, Ludhiana, Punjab, India;
Department of Industrial Microbiology, Sam Higginbottom University of Agriculture, Technology and Sciences, Prayagraj, India

Das Satinath Department of Botany, Mycology and Plant Pathology Laboratory, Santiniketan, West Bengal, India

Das Subhasish Department of Environmental Science, Mizoram University (PUC), Aizwal, Mizoram, India

Datta Shivika Department of Zoology, Doaba College, Jalandhar, India

Dhaka Vaishali Department of Microbiology, Lovely Professional University, Phagwara, India

Dwibedi Sanat Kumar Department of Agronomy, College of Agriculture, Odisha University of Agriculture and Technology, Bhubaneswar, India

Gorai Pralay Shankar Department of Botany, Mycology and Plant Pathology Laboratory, Santiniketan, West Bengal, India

Goswami Linee Department of Botany, Mycology and Plant Pathology Laboratory, Santiniketan, West Bengal, India;
Department of Biology, School of Science and Technology, Örebro University, Örebro, Sweden

Hussain Nazneen Department of Biosciences, Assam Don Bosco University, Guwahati, Assam, India

Jasrotia Shivakshi Department of Clinical Research, Delhi Institute of Pharmaceutical Sciences and Research, Government of N.C.T. of Delhi, Delhi, India

Jovanović Slobodan Faculty of Biology, Institute of Botany and Botanical Garden Jevremovac, University of Belgrade, Belgrade, Serbia

Kale Radha D. Mount Carmel College, Bengaluru, India

Kapoor Dhriti Department of Botany, Lovely Professional University, Phagwara, India

Kashyap Suman Biosciences, Dayanand Sagar University, Bengaluru, India

Kaur Ravneet Department of Botany, Punjab Agricultural University, Ludhiana, Punjab, India

Khawajazada Farid Department of Agronomy, College of Agriculture, Odisha University of Agriculture and Technology, Bhubaneswar, India

Krishna Murthy V. Department of Chemistry, Dayanand Sagar University, Bengaluru, India

Kumar Abhishek Department of Civil and Environmental Engineering, Birla Institute of Technology, Mesra, Ranchi, Jharkhand, India

Malyan Sandeep K. Research Management and Outreach Division, National Institute of Hydrology, Roorkee, India

Mandal Narayan Chandra Department of Botany, Mycology and Plant Pathology Laboratory, Santiniketan, West Bengal, India

Mani Indra Department of Microbiology, Gargi College, University of Delhi, New Delhi, India

Pandey Vimal Chandra Department of Environmental Science, Babasaheb Bhimrao Ambedkar University, Lucknow, Uttar Pradesh, India

Ramamurthy Praveen C. Interdisciplinary Centre for Water Research (ICWaR), Indian Institute of Sciences, Bangalore, India

Ranđelović Dragana Institute for Technology of Nuclear and Other Mineral Raw Materials, Belgrade, Serbia

Shakya Amita Department of Agriculture and Environmental Sciences, National Institute of Food Technology Entrepreneurship and Management, Kundli, Sonipat, Haryana, India

Sharma Parvarish School of Pharmaceutical Sciences, Lovely Professional University, Phagwara, India

Singh Dhanwinder Department of Soil Science, Punjab Agricultural University, Ludhiana, Punjab, India

Singh Joginder Department of Microbiology, Lovely Professional University, Phagwara, India

Singh Kuldip Department of Soil Science, Punjab Agricultural University, Ludhiana, Punjab, India

Singh Simranjeet Interdisciplinary Centre for Water Research (ICWaR), Indian Institute of Sciences, Bangalore, India

Sonkar Vikas Research Management and Outreach Division, National Institute of Hydrology, Roorkee, India

Swain Sonali Department of Agriculture and Environmental Sciences, National Institute of Food Technology Entrepreneurship and Management, Kundli, Sonipat, Haryana, India

Thakur Sapna Department of Forestry and Natural Resources, Punjab Agricultural University, Ludhiana, Punjab, India

Tharannum Seema Department of Biotechnology, PES University, Bengaluru, India

Yadav Shweta Research Management and Outreach Division, National Institute of Hydrology, Roorkee, India

Chapter 1
Bioenergy Crop-Based Ecological Restoration of Degraded Land

Dragana Ranđelović and Vimal Chandra Pandey

Abstract Increasing land degradation worldwide asks for restoration solutions that are often multi-purposed by nature. Establishment of Bioenergy crops, such as perennial grasses and short-rotation woody crops offers possibilities for both successful eco-restoration of various marginal lands and energy production. Besides many recognized benefits in terms of increased soil carbon stocks, reduction of GHG gasses and economical gains, there are still many potential challenges in bioenergy crop cultivation and production, particularly in terms of negative environmental implications. Comprehensive scientific studies are trying to recognize and overcome their existence and scope. Creation of sustainable bioenergy crops-based ecosystems on the various types of degraded lands through affordable restoration approach could pose a challenging task, but by its realization the fractional intentions of several UN-SDGs can be achieved.

Keywords Biofuel crops · Eco-restoration · Degraded soil · Polluted land · Waste dumpsites

1.1 Introduction

Land degradation presents one of the marked global issues of modern times. Not only that it impacts the environment, agricultural production, livelihoods and safety, but also causes a long-term effect on ecosystem services and human health. Land degradation is recognized as a complex phenomenon. However, owing to this complexity, there is still no unique definition of the term "land degradation", and interpretations vary according to the discipline there are oriented to and main factors taken into

---

D. Ranđelović
Institute for Technology of Nuclear and Other Mineral Raw Materials, Franchet d'Esperey Boulevard 86, 11000 Belgrade, Serbia

V. C. Pandey (✉)
Department of Environmental Science, Babasaheb Bhimrao Ambedkar University, Lucknow, India
e-mail: vimalcpandey@gmail.com

© The Author(s), under exclusive license to Springer Nature Switzerland AG 2023
V. C. Pandey (ed.), *Bio-Inspired Land Remediation*, Environmental Contamination Remediation and Management, https://doi.org/10.1007/978-3-031-04931-6_1

account. For instance, Olsson et al. (2019) refers to land degradation as "a negative trend in land condition, caused by direct or indirect human-induced processes including anthropogenic climate change, expressed as long-term reduction or loss of at least one of the following: biological productivity, ecological integrity or value to humans". Nevertheless, not every loss of productivity should be observed as land degradation, only the one characterized as persistent reduction of biological or economical productivity of land (Millennium Ecosystem Assessment 2005).

Other definitions use a more narrow approach. The United Nations Convention to Combat Desertification (UNCCD 1994) defines the term as "the reduction or loss, in arid, semi-arid and dry sub-humid areas, of the biological or economic productivity and complexity of rain fed cropland, irrigated cropland, or rangeland, pasture, forest and woodlands resulting from land uses or from a process or combination of processes, including processes arising from human activities and habitation patterns, such as: (a) soil erosion caused by wind and/or water; (b) deterioration of the physical, chemical and biological or economic properties of soil; and (c) long-term loss of natural vegetation", referring it especially to the degradation of the dry lands, known as desertification. Land degradation results in the reduction of the ecosystem services as a consequence of human activities or natural processes (ELD Initiative 2013). Common ground of more frequently used definitions is that the term refers to the long-term loss of functionality and productivity of all components of the land (considered as system comprising of soil, landscape, terrain, water, climate, biota etc.) (Eswaran et al. 2001). International efforts to standardize the terminology and develop universal, widely accepted definition of land degradation are still ongoing.

Global land degradation—Increasing nature of land degradation and its spreading among world biomes has called for the assessment of this problem on global scale. Global efforts to address land degradation arose in the 1980s, calling for common action of decision and policy-makers from local to global level. This resulted in international agreement named the United Nations Convention to Combat Desertification (UNCCD), established in 1994 with the aim towards the reduction of land degradation and desertification in all participant countries affected. Additionally, United Nations recognized the need for urgent halting and reverses the land degradation by compensation through land improvement, putting the Land Degradation Neutrality (LDN) concept in force by Rio 20+ outcome documents and establishing it as one of the targets in Sustainable Development Goals. Land Degradation Neutrality concept addresses the need to maintain and, where possible, restore land and soil quality aiming to achieve a land-degradation-neutral planet (Caspari et al. 2015). The effective implementation of these international policies into practice requires spatial information on degraded lands, supported by the recognition of causes and responses of natural and social surrounding.

As land degradation must be estimated taking into account its spatial, economic, environmental and cultural context, evaluations of such complex issue turned out to be a challenging task (Warren 2002). While certain land degradation assessments evaluate soil parameters, others use vegetation assessment or assessments of net primary productivity. Earlier assessments were based on extrapolation of local assessments, while modern approaches include usage of remote sensing technologies

(Dubovyk 2017). Estimation of global degradation range varies between 15 and 63%, but majority of assessments agglomerate around 25–30% of degraded land on global level (Safriel 2007). In order to overcome misinterpretations due to the differences in definitions of the term "land degradation", scenarios for the UNCCD's Global Land Outlook (that aim to predict changes in land use under alternative development scenarios up to 2050) were developed by using the concept of "land condition" and by quantifying changes in key trends of land use and ecosystem functions to determine anthropogenic impact in relation to the natural state (Van der Esch et al. 2017).

Reasons behind land degradation—Land degradation involves both natural ecosystem and the human social system, and changes in both biophysical natural ecosystem and socioeconomic conditions will affect the land degradation process (Millennium Ecosystem Assessment 2005). Certain causes were, however, identified as driving ones: biophysical causes (e.g. topography and climatic conditions) and non-sustainable land use practices (deforestation, urbanization, habitat fragmentation, improper agricultural practices and others) (Li et al. 2015). There are also indirect and not so obvious causes of land degradation, predominantly in the form of triggers for application of non-sustainable land use practices, such as: poverty, population density, migration, economic development, urbanization, agricultural extension etc. As the land degradation usually results from a complex effect of several causes, the clear separation between direct and indirect drivers can sometimes be difficult. Research by Song et al. (2018) on global land change in period 1982–2016 has shown that 60% of all land changes are associated with direct human activities, and 40% with indirect drivers. Network of Sustainable Land Management (SLM) specialists called World Overview of Conservation Approaches and Technologies (WOCAT) has defined six main categories of land degradation, according to the prevailing degradation process: soil erosion by water; soil erosion by wind; chemical soil deterioration; physical soil deterioration; biological deterioration; and water degradation (Harari et al. 2017). It has been generally recognized that key mechanisms behind land degradation include physical, chemical and biological processes. Some of the most important physical processes are erosion, compaction, sealing and crusting and certain types of environmental pollution. Important chemical processes are acidification, salinization, leaching, loss of fertility etc., while biological processes include reduction in total soil carbon and biodiversity loss (Eswaran et al. 2001). However, social, economic and political causes are often the main driving forces behind current land degradation processes.

Impact on ecosystem services and livelihoods—Land degradation is affecting ecosystem services in many areas of the world. Moreover, degradation could be considered as persistent reduction of ecosystem services. Such services are interconnected with changes in land use, and more research on clarifying the type and degree of that connection are needed (Hasan et al. 2020). However, it is clear that land use change is affecting main types of ecosystem services (as defined by Millennium Ecosystem Assessment 2005): supporting services (biomass and oxygen production, soil production, nutrient cycling etc.), provisioning services (food, fresh water, timber etc.), regulating services (climate regulation, carbon sequestration, waste decomposition, water purification etc.) and cultural services (recreation, visual effects, physical and mental health benefits, spiritual experiences etc.). Those services are dynamically interrelated, but a land use change that is orientated towards prioritizing certain ecosystem service may eventually result in decline of other, non-prioritized ecosystem services (Millennium Ecosystem Assessment 2005). Analyses of the cost of land degradation among the types of ecosystem services showed that 54% of the cost refers to the losses in supporting, regulating and cultural services, belonging to public goods. In addition, 42% of the world's poor population depends on services of degraded lands for providing food and income (Nkonya et al. 2016).

Land degradation followed by loss of ecosystem services impacts the livelihood security of people, including food and water security and climate change. These effects are especially pronounced among most vulnerable society groups, particularly those living in rural areas (IPBES 2018). Land degradation shows an asymmetric impact across the society, increasing poverty and deepening inequalities among various income groups. It is found that land degradation could have increased severe rural poverty rates by almost 10% between 2001 and 2015, if other factors held constant (Global Mechanism of the UNCCD 2019). As land resources influence livelihoods of population that depends on them, application of sustainable land management practices could be the way to avoid the land degradation, especially in more affected parts of the world (Gashu and Muchie 2018).

1.2 Suitability of Bioenergy crops for Wide-Ranging Degraded Lands

Land degradation contributes to the emission of greenhouse gases (GHG) and reduced carbon uptake by the land (Olsson et al. 2019). It is estimated that certain changes in land use, such as deforestation and expansion of agriculture contribute to approximately 15% of global emissions of GHG (United Nations Department of Economic and Social Affairs, United Nations Forum on Forests Secretariat 2021). One of the land management strategies that could contribute to combating the land degradation and simultaneously provide carbon sequestration is establishment of bioenergy crops (plants grown for the purpose of energy production), namely perennial grasses and short-rotation woody crops. Advantages of growing crops for bioenergy include absence of negative impact on the carbon dioxide balance in the atmosphere and reduction of GHG emissions. As the amount of quality land suitable for cultivation is a limited resource, marginal lands were recognized as viable option for growing Bioenergy crops. Such marginal or degraded lands that are unsuitable for food production include various erodible, acidic, saline and contaminated soils, reclaimed mine soils, urban marginal sites and abandoned or degraded former agricultural land. As stated by Shortall (2013), marginal land is the type of land that can be classified as unused, free, spare, abandoned, under-used, set aside, degraded, fallow, additional, appropriate or under-utilized land. Moreover, growing energy crops on such lands could even enhance ecosystem services, as they can reduce erosion processes, restore contaminated land and improve overall biodiversity of the area (Valcu-Lisman et al. 2016).

Biomass production on different categories of marginal lands is variable and depends on the characteristics of a particular site, applied land management practice and selection of suitable plant species for this purpose. Various research showed that biomass yields may range between 1 and 14 Mg ha^{-1} for perennial warm-season grasses and between 0.5 and 9.5 Mg ha^{-1} for short-rotation woody crops, while soil carbon sequestration rate may vary between 0.24 and 4 Mg C ha^{-1} yr^{-1} (according to Blanco-Canqui 2016).

Perennial grasses are characterized by higher yield potential in comparison to the annuals. Moreover, warm-seasonal C4 perennial grasses can provide higher annual biomass yield at the higher temperatures as they possess more efficient photosynthetic pathway than C3 plants. Due to the characteristics of their active underground organs, perennial plants are effective in recycling nutrients, therefore exhibiting a lower nutrient demand than the annuals (Santibáñez Varnero et al. 2018). Besides, perennial grasses are tolerant to many abiotic stresses (Ranđelović et al. 2018), adaptable to the range of habitats and suitable for multiple uses (Pandey and Singh 2020). Some of the most suitable perennial grasses for purpose of growing bioenergy crops over the globe are: switchgrass (*Panicum virgatum* L.), miscanthus (*Miscanthus × giganteus* Greef et Deuter), reed canary grass (*Phalaris arundinacea* L.), giant reed (*Arundo donax* L.), common reed (*Phragmites australis* (Cav.) Trin. ex Steud.) etc. (Sanderson and Adler 2008; Scordia and Cosentino 2019).

Woody Bioenergy crops, also called short-rotation woody crops, are fast-growing trees that can reach high yields, tolerate conditions of various soil types and require low inputs. In comparison to majority of annual crops, they have a lower impact on soil erosion and increased nutrient and organic matter input to the soil (Whittaker and Shield 2016). Although short-rotation woody crops have longer harvest rotations than the perennial crops, they compensate it by production of higher yields. Short-rotation woody crops are mainly represented with species such as poplars (*Populus* sp.), willows (*Salix* sp.), eucalyptus (*Eucalyptus* sp.), nettlespurge (*Jatropha curcas* L.), sycamore (*Plantanus occidentalis* L.), sweetgum (*Liquidambar styraciflua* L.) etc. (Lemus and Lal 2005; Pandey et al. 2012a; Pleguezuelo et al. 2015).

Results generated from investigation of growing energy crops on various types of marginal lands showed that the performance of various plant species were site-specific and species-specific (Blanco-Canqui 2016; Acharya et al. 2019). In order to reveal the potentials of bioenergy crops to grow on different categories of marginal lands a catalogue of crops suitable for growing conditions on different marginal lands in the territory of Europe was formed (SEEMLA 2016).

Perennial plants have demonstrated potential to be successfully grown on highly eroded lands, as they tend to form dense biomass cover in short time, and deep root system too. If used as conservation buffers in a landscape, such as hedges, filter strips or riparian buffers, these plants could successfully reduce wind or water erosion, therefore combining soil conservation practices with growing bioenergy crops (Kreig et al. 2019). Additionally, improved water quality in terms of reduced nitrogen and soil erosion rate in water can be generated by applying changes in cropping patterns and management practices of perennial crops (Valcu-Lisman et al. 2016). In addition, both wooden and perennial crops could be successfully established on steep slopes terrains and minimize soil erosion rates from such sites if good agricultural practices with minimal soil disturbances are applied (Jankauskas and Jankauskiene 2003).

Establishing energy crops on moderately polluted sites such as post-mining areas could be both economically viable and environmentally sound practice for simultaneously usage of biomass as energy source and improvement of the soil conditions. Coupling the phytoremediation with energy crops is another benefit that could be potentially gained on contaminated sites. Perennial grasses inhabiting post-mining sites are often recognized for their tolerance to metal toxicity as well as other characteristics of those sites, such as extreme pH values, sandy texture and low nutrient content (Ranđelović et al. 2014; Jakovljević et al. 2020). Similarly, certain tolerant woody species are also capable of growing at such sites (Migeon et al 2009; Shi et al. 2011). Naturally colonizing vegetation should preferably be used for this purpose in so-called sustainable phytoremediation approach (Pandey 2015), especially if it can contribute to the safe immobilization of the pollutants from contaminated sites. Selected plants should, however be preferably perennial, stress-tolerant, unsavoury to livestock, and able to generate both economic and ecological benefits for the site (Pandey 2017).

Formation of short rotation woody plants plantation on former mining sites is additional way to utilize biomass from mine lands. Performance of common energy crops that can be suitable for phytoremediation of various pollutants, such as *Miscanthus* × *giganteus*, *J. curcas*, *Salix* sp., *P. virgatum*, *A. donax* etc. was investigated for both purposes (Pandey et al. 2012a; Skousen et al. 2012; Jeżowski et al. 2017; Pandey 2017; Castaño-Díaz et al. 2018). The addition of soil amendments and microbial agents could additionally enhance the growth and yields of selected plants (Pogrzeba et al. 2017; Andrejić et al. 2019).

Saline soils, considered to be marginal lands of low productivity, are colonized by halophyte plant species that are able to thrive in saline conditions. Identification of suitable halophytes for biomass and energy production is currently in progress. Some plant species, such as *Desmostachya bipinnata* (L.) Stapf, *Kosteletzkya pentacarpos* (L.) Ledeb, *Salicornia bigelovii* Torr., *Tamarix jordanis* Boiss. etc. were recognized for their characteristics potentially suitable for energy production (Abideen et al. 2011; Bomani et al. 2011; Moser et al. 2013; Santi et al. 2014). Certain halophytes accumulate salts in their organs, which may generate problems during combustion or other biomass utilization processes, so halophytes with ability to exclude salts are generally considered to be a better choice for energy production (Sharma et al. 2016). However, before wider application of halophytes, hybridization and breeding should be conducted for domestication of wild species and their adaptation to agricultural management measures in order to obtain species with high yields and higher salinity thresholds, especially during the phase of germination and seedling emergence (Gul et al. 2013).

Wet and flood-prone marginal lands are also potentially suitable sites for growing dedicated energy crops. Das et al. (2018) investigated perennial bioenergy crops on wet marginal lands where soil properties and a biomass of switchgrass (*P. virgatum*) have been influenced by moisture gradient of the field. Barney et al. (2009) found that selection of adequate ecotypes of switchgrass for growing in excess soil moisture conditions could increase the range of environments suitable for growing Bioenergy crops. Similarly, short-rotation woody species such as willow or poplars that are naturally growing in floodplains and show ecological adjustments to the flooding conditions could be used as energy crops on such sites. However, although they are tolerant to wet conditions and can maintain efficient growth and productivity in such conditions, prolonged inundation could ultimately reduce their feedstock quality and increase the cost of the exploitation process (Bardhan and Jose 2012).

Using abandoned agriculture lands for bioenergy crops represents additional option for energy production on marginal lands. It has been estimated that bioenergy production on abandoned agricultural lands could satisfy approximately 8% of global energy demands (Campbell et al. 2008). Although there are still concerns about feasibility of using such sites and investigations showed that growing conventional crops on these lands as a bioenergy feedstock could potentially increase erosion rates and polluted runoff, field studies with low-input high-diversity mixtures of native perennial grasses grown for bioenergy purposes showed reduction of these impacts (Tilman et al. 2006). Projections of environmental implications on abandoned agricultural lands from production of bioenergy crops in subtropical region of Australia revealed that environmental improvements could be gained in open grazing areas, by using native woody perennial bioenergy crops under low management intensity, while other options did not produce favourable environmental outcomes (Miyake et al. 2015). However, there are indications that, if properly addressed, inclusion of perennial bioenergy crops on degraded parts of agricultural lands could create benefits in the landscape function and resilience and enhance ecosystem services such as wildlife habitat, soil and water quality (Blanco-Canqui 2016).

Additional research is required on the adequate utilization of various marginal lands for energy production with attention focused on selection of dedicated crops, such as extremophile energy crops. These crops would be adapted wild species or genetically modified existing crops that are capable of growing in extreme environments while retaining high productivity and low nutrient and water requirement (Bressan et al. 2011).

Growing energy crops on marginal lands becomes a field of intensive research and field trials. However, great care and careful planning are needed, as intensive management and exploitation measures on degraded lands could have negative impacts on soil, water and biodiversity conservation (Bonin and Lal 2012). Shifting marginal lands to bioenergy cultivation process should be carefully addressed in order not to cross certain thresholds by intensity of land use and compromise ecosystem services and biodiversity of such lands (Hennenberg et al. 2010). It is recognized that unsustainable bioenergy crop expansion could pose threat to biodiversity and habitats and could additionally degrade natural areas (Millennium Ecosystem Assessment 2005). Therefore, it should be secured that no land of conservation value or with significant carbon stocks is converted to biomass for energy production. European Directive 2009/28/EC on the promotion of the use of energy from renewable sources poses such requirements for sustainable biomass production, where bioenergy crops should not be obtained from land with high biodiversity value; land with high above ground or underground carbon stock and from peatlands. Biomass production should be environmentally responsible and any negative trade-offs for biodiversity, the environment and local communities should be avoided (Hennenberg et al. 2010).

1.3 Plant Derived Bioenergy Sources

Biomass is the renewable source of energy. Some sources of biomass are agricultural crops, algae, annual, perennial grasses or woody plants etc. Plants are producing biomass via photosynthesis, using sunlight energy to convert carbon-dioxide and water to carbohydrates and oxygen. Type and the amount of bioenergy that could be produced depend on the characteristic of biomass. Plants can be used for bioenergy production in two main ways: as energy crops (explicitly grown for that purpose) and as biomass residues (originating from plants grown for other purposes). Additionally, biomass can be converted to energy directly (by direct combustion) or indirectly (by conversion of row biomass material to fuels that are afterwards used for the energy production). Conversion of biomass to energy can be done thermochemically (by pyrolysis, combustion or gasification), biochemically (by using microorganisms and enzymes via technologies such as anaerobic digestion and fermentation) or chemically (use of chemicals to convert biomass to liquid fuels). These conversion technologies enable production of heat, power and biofuels. Besides, biomass it is the only renewable energy able to be processed into solid, liquid and gaseous fuels (World Energy Council 1994).

The production of heat is the leading modern bioenergy application throughout the world (WER 2013). Biomass efficiency for heating purposes depends on the plant chemical composition, especially the share of lignin (averagely 10–25 wt%), cellulose (40–50 wt%) and hemicellulose (20–40 wt%) (McKendry 2002). Relative proportion of cellulose and lignin is of particular importance for identification of plants suitable for energy crops, and their biomass is also known as lignocelluloses biomass. Some of the most important properties for biomass conversion process are calorific value, moisture and ash content, fixed carbon and alkali content. Biomass of perennial grasses generally shows higher contents of lignin and cellulose compared to the biomass of annual crops (Brown 2003). Generally, lignocellulosic biomass of woody species has higher contents of cellulose and lignin, while biomass of perennial grasses contains more hemicellulose and ash, making it less suitable for the combustion process (Scordia and Cosentino 2019). Among perennials, C3 plants have higher ash content in comparison to C4 plants (Zhao et al. 2012). Similarly, low moisture content woody and perennial species are more convenient for heating purposes, as higher water content has negative impact on biomass calorific value (SEEMLA 2016). Compacted forms of biomass such as wood pellets and briquettes can also be used for combustion. Short-rotation coppices of willow and poplar present the opportunity for sustainable source of biomass for such purpose. Moreover, high variation and presence of diverse cultivars among this species offer choices for optimizing the feedstock quality. However, the increased demand for heating sources is driving for more non-woody biomass resources (e.g. perennial grasses) to be used for this purpose (Santibáñez Varnero et al. 2018), and although their heating properties is usually lower than of woody biomass (Gami et al. 2011), they could provide sustainable amounts of feedstock due to their high biomass production.

Variety of liquid and gaseous fuels can also be produced from plant feedstock. Depending on the source of biomass, biofuels may belong to "first generation" (derived from food crops) and "second generation" (derived from lignocellulosic biomass of energy crops, including woody crops and perennial grasses). Most common liquid biofuel types are biodiesel and bioethanol. Among the gaseous fuels biogas (consisting of methane and carbon-dioxide) is the most commonly produced.

Bioethanol is considered to be an alternative to fossil fuels (especially petrol). Although technology for producing ethanol from food crops has been well developed and practically applied, competition with food sources has begun to be the issue of concern. Therefore, lignocelluloses biomass has recently gained attention as a source for bioethanol production. Research shows that net energy balance (energy in versus energy out) is generally lower in bioethanol gained from lignocelluloses materials in comparison to ethanol produced from sugar and starch-based feedstocks (Hayes 2008). There is ongoing research to identify plants suitable for bioethanol production, usually among ones with enhanced biomass production, such as *P. virgatum, P. arundinacea, Miscanthus × giganteus, A. donax* and others (Taiichiro and Shigenori 2010). Moreover, a significant portion of the research is dedicated to the use of woody species for production of bioethanol, especially fast-growing ones such as poplars and willows (Huang et al. 2009; Wang et al. 2011; Littlewood et al. 2014). Efficient and economically viable production of bioethanol from lignocellulose biomass depends primarily on the development of a suitable, simple and cost-effective pretreatment system for making cellulose from biomass accessible to the enzymes that break carbohydrate polymers into simple sugars available for further fermentation (Wi et al. 2015; Porth and El-Kassaby 2015). As the biomass composition of energy crops differ, individual approach in development and selection of suitable processing methods is needed for making bioethanol economically sustainable (Raud and Kikas 2020). Pilot plants established through the world demonstrated successful production of the bioethanol from agricultural waste, but the conversion of wood waste to bioethanol has turned out to be a challenging task (Johnson et al. 2009). Second-generation technologies using lignocellulose feedstock are still immature and need further development to demonstrate feasibility at commercial scale (Zhu et al. 2020).

An additional option for producing biofuels out of plant biomass is to extract the oils produced by plant seeds in a form of biodiesel. It is easy biodegradable fuel with potential to replace transportation fuels such as petroleum and diesel. Biodiesel is primarily generated by transesterification of plant oils. It is currently commercially produced from biomass of several species, such as canola, palm, rapeseed etc. Again, attention is paid to the potential use of lignocellulosic biomass for biodiesel production. *Jatropa curcas* was previously identified as one of the most promising species for biodiesel production, due to the stated high yields and 40–60% of oil content in seeds (Koh and Ghazi 2011), as well as the energy value of seed oil that was higher than in some types of coal (Wahyudi et al. 2019). However, grown in various field conditions, *J. curcas* generally did not meet the expectations due to the high fluctuation of yields, susceptibility to pests and diseases and toxicity of the seed cake (Moniruzzaman et al. 2017). Another non-edible plant potentially suitable for biodiesel production is *Pongamia pinnata*, whose seeds are found to contain 35% of oil, while fuel properties were found to be close to that of high-speed diesel (Ahmad et al. 2009). Technologies for production of biodiesel from second-generation crops are still at the beginning and need certain advances concerning seed production, management of plantations, biodiesel processing technology and supporting policies.

Biogas is a renewable energy resource produced during anaerobic bacterial degradation of biomass. Several second-generation crops showed potential for methane production. Perrenials *P. arundinacea* and *Elymus elongatus* cv. "Szarvasi-1" exceeded methane yields under favourable conditions in comparison to maize (Schmidt et al. 2018). Similarly, *A. donax* has been proposed as a suitable energy crop for biogas production. Although its production of methane was less than that of maize, the higher biomass production led to much higher biogas yield per hectare (Corno et al. 2015). Research on biogas production from other non-edible crops in terms of technology, economic benefit and environmental effects could contribute to the enhanced use of the renewable energy sources.

1.4 Degraded Land Restoration by Energy Crops

Degraded lands are inappropriate for agricultural crop cultivation due to low productivity (Gelfand et al. 2013). Generally, degraded lands include sodic land, saline land, nutrient poor land, urban marginal land, polluted land, waste dumpsites like fly ash dumps, mined land, red mud dumpsites, etc. These degraded lands have an uncertain and insignificant contribution to food security due to biotic and abiotic complications. Land degradation is phenomena of great concern because day by day it is increasing over the world. Hence, the transformation of degraded land in self-sustaining energy ecosystem is a current demand that will provide life-supporting services and support climate change mitigation (Hobbs et al. 2014). It depends mainly on the adaptation abilities of energy crops on degraded land. Many studies are available in terms of suitability of diverse energy crops to perform on various types of degraded lands (often under different watering and fertilization regimes), as well as their potential for production of various fuels (Table 1.1). Additionally, a high share of research conducted on field scale enables insights in performance of energy crops in real conditions of degraded sites.

As anthropogenic influence on land is growing and is often followed by environmental pollution, it is of particular importance to study the potential of bioenergy crops grown on different types of contaminated lands. Content and state of both organic and inorganic pollutants in soil influence not only the plant growth and

Table 1.1 Various research studies on energy crops grown on different types of degraded lands

Energy crop	Degraded land type	Experimental conditions	Research target	References
Arundo donax	Fertile and marginal soils	Field study	Environmental impact of bioenergy crop cultivated on fertile and marginal land via Life Cycle Assessment	Bosco et al. (2016)
Arundo donax	Reclaimed mine sites	Field study	Characterization of biomass, biochar, bio-oil and non-condensable gases generated from plant grown on mine sites	Oginni and Singh (2019)
Arundo donax, Miscanthus × giganteus	Moisture soils, inundate soils	Greenhouse conditions	Access moisture stress tolerance, physiological stress, and biomass yields	Mann et al. (2013)
Atriplex nitens, Suaeda paradoxa, Karelinia caspia	Saline soil	Field study	Biomass yield, chemical characteristics of biomass, biogas production	Akinshina et al. (2014)
Eucalyptus globulus	Fertile and non-fertile, irrigated and non-irrigated soils	Field study	Evaluating factors that affect the economic sustainability of Eucalyptus production (yields, prices and costs etc.) on marginal lands	Acuña et al. (2018)
Jatropha curcas	Marginal soils	Field study	Potential for cultivation of plant as bioenergy crop via propagation and growth tests	De Rossi et al. (2016)
Jatropha curcas	Abandoned agricultural land	Field study	Effects of irrigation systems with recycled wastewater on morphometric characteristics, plant growth and productivity, soil fertility status	Dorta-Santos et al. (2015)

(continued)

Table 1.1 (continued)

Energy crop	Degraded land type	Experimental conditions	Research target	References
Manihot esculenta	Contaminated site	Field study	Plant growth and remediation potential, bioethanol production	Shen et al. (2020)
Miscanthus × giganteus	Degraded coal mine soil	Field study	Effect of sewage sludge and sewage sludge with mineral fertilizer on plant height and biomass yield, changes of soil conditions	Jeżowski et al. (2017)
Miscanthus × giganteus	Saline soil	Control environmental glasshouse	Biomass yield and production, stress tolerance level	Stavridou et al. (2017)
Miscanthus × giganteus	Contaminated agricultural soil, post-military soil, petroleum contaminated soil	Field study	Calorific values of biomass	Nebeská et al. (2019)
Miscanthus × giganteus, Phalaris arundinacea, Salix schwerinii × Salix viminalis	Marginal land, brownfield sites, landfills	Field study	Biomass yield and contamination, fuel composition	Lord (2015)
Panicum virgatum	Marginal soil (podzolic)	Field study	Effects of cultivation technology and different types of cultivation systems on biomass yield	Taranenko et al. (2019)
Panicum virgatum, Populus × hybrid	Marginal land	Greenhouse, field study	Effect of soil microbes and seaweed extract on plants productivity	Fei et al. (2017)
Panicum virgatum, var. Shawnee and Carthage	Reclaimed mine sites	Field study	Effects of different fertilizer systems on biomass yield of selected plant varieties	Brown et al. (2015)
Phalaris arundinacea, Panicum virgatum	Wet marginal soils	Field study	Influence of moisture gradient on above-ground biomass yields	Das et al. (2018)

(continued)

Table 1.1 (continued)

Energy crop	Degraded land type	Experimental conditions	Research target	References
Pennisetum americanum × *P. purpureum*	Saline soil	Field study	Effects of mulching, plant density, and organic/inorganic fertilizers on biomass yield, plant height and soil microorganisms	Wang et al. (2014)
Populus nigra × *Populus maximowiczii* Henry cv Max 5, *Robinia pseudoacacia*, *Salix viminalis*	Marginal degraded soil	Field study	Assessment of survival rate, plant morphological traits and biomass yields by using different soil amendments	Stolarski et al. (2014)
Salix alba, *Salix viminalis*	Peat soil, alluvial soil, heavy clay soil	Field study	Accessing plant morphological traits and biomass yields	Stolarski et al. (2019)

biomass production, but also the quality of various derived energy sources. Many research studies have been dedicated to access the degree of plant contaminant uptake, or concentration of contaminants in final or by-products, such as oil, ash, or wood chips.

Some potential and perennial bioenergy grasses such as *Arundo donax* L., *Miscanthus* × *giganteus*, *Panicum virgatum* L. have been identified especially from Australia, Europe, and the United States for enhancing the contribution of bioenergy production at global level (Patel and Pandey 2020; Praveen and Pandey 2020; Alexopoulou 2018). For instance, *A. donax* and *Miscanthus* genotypes (*M.* × *giganteus*, *M. sinensis*, and *M. floridulus*) were tested on heavy metal contaminated soils, and results showed that the presence of trace elements reduced biomass production of investigated plants, while *M.* × *giganteus* kept the highest biomass production under conditions of Zn-contaminated soils (Barbosa et al. 2015). Additionally, analysis of percolated waters showed that *A. donax* promoted Phytostabilization of Cr, Zn and Pb in soil, and *Miscanthus* genotypes similarly prevented leaching of Zn in water, thus contributing to the overall remediation of the environment. Multiple studies confirmed tendency of *M.* × *giganteus* for retaining the majority of accumulated metals in its underground parts (Korzeniowska and Stanislawsk 2015; Pidlisnyuk et al. 2019; Andrejić et al. 2019), so low concentration of metals accumulated in above-ground organs should not be obstacle for its use as bioenergy crop. However, noted elevated contents of potassium in plant biomass associated with regulation of the metal exclusion as adaptive plant response could cause problems with

Fig. 1.1 Field trials with *Miscanthus* × *giganteus* on Pb–Zn–Cu flotation tailing site Rudnik in Serbia (photo by courtesy of Mr. Dželetović Željko)

fouling and slagging during combustion, so regulation of potassium content should be further investigated for making the combustion process more efficient (Laval-Gilly et al. 2017). Additionally, fuel characterizations of *M.* × *giganteus* biomass from a phytoremediation sites located in Poland and Germany showed differences in the thermal decomposition of biomass, possibly due to the differences in pH value and heavy metal content of the investigated soils (Werle et al. 2019). Moreover, performances of *M.* × *giganteus*, *M. sinensis*, and *M. sacchariflorus* were analyzed in different environments across Europe, showing significant influence of the environment on composition and quality of plant biomass (Van der Weijde et al. 2017). As such, environmental influence should be acknowledged when deciding the end-use of *Miscanthus* feedstock, as well as during development of novel varieties with improved biomass quality for biofuel production. *Miscanthus* spp. can generally be used for combustion, biofuel production and Phytostabilization (Figs. 1.1 and 1.2), bridging the environmental remediation and renewable energy production, so further investigations in terms of its utilization for such combined purpose are needed.

Similarly, *P. virgatum* was considered as model perennial energy crop, while at the same time its tolerance or capacity for removal of inorganic and organic contaminants from soils and water was recognized (Guo et al. 2019; Phouthavong-Murphy et al. 2020). Ability of *P. virgatum* to extract metals from contaminated sites was modelled by Chen et al. (2012), who developed different models between plant metal content and biomass yield for predicting the amount of Cd, Cr and Zn potentially extracted by plant. Obtained results suggested its use for Phytoremediation purposes, while acknowledging that the biomass yield is significantly correlated with uptake of metals. Cultivation of *P. virgatum* on Pb-contaminated soil for accessing its remediation efficiency and applying two conversion routes (enzymatic hydrolysis and fast pyrolysis processes) for biofuel production was implemented by Balsamo et al. (2015). Lead was mainly retained in the roots of *P. virgatum*, and the uptake rate increased with the Pb concentration in soil. However, Pb present in the biomass of *P. virgatum* from contaminated site had minimal or no effect on the fast pyrolysis

Fig. 1.2 Field trials with *Miscanthus* × *giganteus* on fly ash deposits of thermal power plant "Kolubara" in Veliki Crljeni, Serbia (photo by courtesy of Mr. Dželetović Željko)

processes and the following bio-oil products distribution in comparison to the plant biomass from control non-polluted site. Enzymatic hydrolysis with fungal cultures additionally showed that production of sugar by selected cultures was not adversely affected by the Pb content in *P. virgatum* biomass.

Besides those already mentioned, neglected and underutilized perennial grasses *Saccharum spontaneum* L. and *S. munja* Roxb. were also noticed for their potential to revegetate, remediate and restore fly ash dumpsites (Pandey et al. 2012b, 2015a; Pandey and Singh 2014; Pandey 2015, 2017; Pandey and Singh 2020), sponge iron solid waste dumps (Kullu and Behera 2011), coal-mined lands (Maiti et al. 2013) and rock phosphate mine restoration (Bhatt 1990). Thus, *Saccharum* spp. has been noticed as a potential bioenergy grass, but to date it is neglected and underutilized, and requires proper attention for exploitation of its unique characteristics for land restoration and bioenergy production (Fig. 1.3). Likewise, some other grasses such as *Arundo donax* L., *Desmostachya bipinnata* L. Stapf., *Panicum antidotale* Retz., *Saccharum* species, *Vetiveria zizanioides* L. are broadly dispersed over India and have abilities to grow naturally on degraded lands without outer inputs.

Various tree species were tested for their capacity to produce useful biomass and remediate contaminated sites. This is especially the case with short-rotation coppice crops of willows and poplars, but also the species such as *J. curcas*, *Alnus glutinosa*, *Eucalyptus* sp., *Robinia pseudoacacia* etc. Potential of willow species and their clones to accumulate metals such as Cd, Cu, Pb and Zn from polluted soils was recorded by various researchers (Tlustoš et al. 2007; Algreen et al. 2014; Yang et al. 2014; Lebrun et al. 2017). *Salix* genus is often used in short-rotation coppice system for energy production, showing fast growth and high biomass yields. A number of

Fig. 1.3 *Saccharum* spp. on fly ash disposal area of Renusagar thermal power plant, Renukoot, Sonbhadra district, Uttar Pradesh, India (photo by courtesy of Dr. Vimal Chandra Pandey)

clones and cultivars with improved traits enable wider use of willows for simultaneous Phytoremediation and bioenergy production. Positive results in removal of hazardous substances from various landfill leachates were reported in short-rotation willow coppice Phytoremediation systems used on large-scale in Sweden (Dimitriou and Aronsson 2005). However, to reach environmental and economic benefits, both biological and technical approach should be optimized. When biomass of *Salix viminalis* L. grown on contaminated dredged sediment disposal site was gasified in order to determine the fate of accumulated trace elements (namely Cd, Cr, Cu, Ni and Pb) upon the biomass conversion, gasification results showed that concentration of Cd and Zn in bottom and cyclone ash fractions exceeds thresholds for using the ash as soil fertilizer; therefore the ash originating from this process should be landfilled (Vervaeke et al. 2006). Optimization of gasification process would contribute towards concentrating trace elements in small ash fraction, so that the more voluminous fractions could be utilized in forms of fertilizer. The fate of the metals (Cd, Zn, Cu, Pb) present in the *Populus maximowiczii* × *P. trichocarpa* cultivar Skado grown on contaminated soil was studied in the end-products of the torrefaction and pyrolysis processes by Bert et al. (2017). Although concentration of accumulated metals in above-ground biomass was low, they were eventually concentrated in end-products. Similarly, content of metals would be a limiting factor in case of valorization of bio-oils from torrefaction and pyrolysis. Biomass of poplar from Phytoremediation site with contaminated soil was subjected to gasification experiments, where higher ash content and significantly lighter hydrocarbons in comparison to poplar from natural site were obtained, possibly due to the increased content of Ca and Mg that could act as catalyst in the tar (Aghaalikhani et al. 2017).

One of the recognized energy crops, *J. curcas,* was also investigated for its Phytoremediation ability. García Martín et al. (2020) found that *J. curcas* accumulates Fe, Cr, Cu, Mn, Ni and Zn in the aerial parts in higher concentration than in underground parts, thus exhibiting significant potential for metal Phytoextraction. Álvarez-Mateos et al. (2019) found reduction of 30–70% of Cr, Ni, Cu, Zn and Pb from mining site

Fig. 1.4 *Jatropha* spp. on fly ash disposal area of Tanda thermal power plant, Uttar Pradesh, India (photo by courtesy of Dr. Vimal Chandra Pandey)

soils, coupled with their higher transfer factors to shoots. Aggangan et al. (2017) tried to modify high transfer rate for Cu and Zn in *J. curcas* with mycorrhizal treatments. Upon the root colonization with investigated mycorrhizal inoculums Translocation of Cu and Zn to the aerial plant parts was inhibited and majority of elements were retained in roots of *J. curcas*. Moreover, concentrations of Cu and Zn in fruits and seeds remained below the detection limit, which enabled their further use for the production of biofuels. Similar biotechnological approaches coupled with selection of varieties with high seed oil content and yield could offer new possibilities for application of *J. curcas* in Phytoremediation of degraded lands. Besides accumulation of contaminants, additional reclamation benefits could be achieved, such as improved soil or water quality, increase of soil microorganism content and overall biodiversity at the site. *J. curcas* has been recognized for remediation and biofuel production (Fig. 1.4).

Eventually, to avoid environmental and health risks, end-products might be purified, or technology improved, allowing the use of metal-enriched plant biomass for efficient energy production. Distribution of metals in the end-products depends on the conversion process used, as well as from the optimization of process parameters. More research is needed in this sense in order to find viable solutions for conversion of biomass from remediated sites to bioenergy.

Leguminous plant-based biofuel production has also been reported and found suitable for land restoration because of their potential to enhance soil productivity owing to their connection with N_2-fixing bacteria, which is particularly suitable for various types of degraded sites and marginal lands. Suitable plants could be *Acacia mangium, Galega* sp., *Medicago sativa, Onobrychis viciifolia* and others (Singh et al. 2019). It has also been reported by da Costa et al. (2015) that energy yield of *A. mangium* grown in Amazon biome was two times larger than *Acacia auriculiformis,* including differences in biomass distribution. Density of this species as well as their calorific value in investigated area was increased in comparison to native species, revealing their potential for establishing energy forests. Additionally,

P. pinnata is recognized as nitrogen-fixing tree species that produce oilseed and could simultaneously contribute to restoration of degraded lands (Leksono et al. 2018). *Galega* sp., *M. sativa* and *O. viciifolia* were found to store the higher N_2 content in their dry biomass that enhances the *C:N* ratio. The higher *C:N* ratio based leguminous plants are suitable for higher biogas production and may be used as an indicator to identify potential legume plant for more biogas production (Slepetys et al. 2012). It also assists to control Soil organic carbon by removal of residues, soil cultivation and land use change. However, a proper management of crop residues is needed to ultimately help to increase soil carbon and nitrogen stocks (Wu et al. 2018; Martani et al. 2020).

Environmental footprint of bioenergy crops depends on several factors, including the type of crop, land use and soil type. Important issue in bioenergy crop cultivation is their influence on soil properties, both short- and long-term, especially on degraded or marginal lands. In this sense, improvement of soil conditions in order to support sustainable yields and ecosystem services should be one of the main tasks in successful management for bioenergy. For example, the impact of introduced perennial bioenergy crops on soil quality showed positive effects on soil carbon pools, microbial and enzymatic activities, as well as activities of soil fauna (Emmerling et al. 2017). Similarly, bioenergy crops on contaminated agricultural sites showed increased diversity of soil fauna (Chauvat et al. 2014). Contrary to that, non-favourable changes in soil physical properties, such as decline of porosity and water infiltration rate, followed the conversion of reclaimed mine soil to bioenergy crop production site (Guzman et al. 2019). Quantifying change of soil parameters during land use change to bioenergy crops in different environments presents important task in which further research on small and large scales are needed.

1.5 Livelihood Improvements

Establishment of multiple energy cropping systems on degraded lands such as nutrient poor lands, polluted lands and waste dumpsites is a current demand to improve livelihood and to reduce environmental problems. Generally, the integrated assessments of bioenergy deployment should consider social dimension and livelihoods in more detail, as they are particularly important for practical implementation of bioenergy production. Government policy on biofuels should be intended to utilize degraded lands together with farmers, unemployed villagers, practitioners, companies, entrepreneurs, self-help groups, etc. Energy crop cultivation on degraded lands and especially nutrient poor land will help poor villagers by providing more ecological-resilient cash energy crops than traditional crops (Scordia et al. 2018). However, particular care should be given to promotion of programmes and communication with small-scale farmers on bioenergy production issues in order to avoid misunderstandings, particularly in the terms of expected profits and access to resources. The cultivation of energy cropping systems on degraded lands is labour-intensive and will provide job opportunities to local people including both skilled and non-skilled workers by various steps such as planting, harvesting, collecting, baling, densification, carrying, energy production (Mckendry 2002) and decentralized bioenergy systems such as processing and distribution (Valentine et al. 2012). On the other hand, agricultural residue of this system can be used for making compost and biochar. Besides the employment generation, these energy systems on degraded land will also help in climate change mitigation, promotion of local tourism and cultural activities. In India, wide-ranging driven policies (Bioenergy Policy, Green India Mission, MNREGA, Ethanol Blended Petrol Program, National Biodiesel Mission, Biodiesel Blending Program, etc.) can be linked with energy crop-based degraded land restoration. If energy crops should be used for restoration and remediation of marginal and contaminated lands, besides providing transparent and established values on cost-effectiveness of the process in order to create more realistic expectations, it is also important to simultaneously advocate remediation of human relationship with the land.

1.6 Potential Challenges

Although bioenergy cropping systems on degraded lands offer multiple benefits including ecological and socioeconomic aspects, there can be many potential challenges in their cultivation and production. First, the potential challenges of restoration of degraded lands are hostile conditions such as physicochemical and biological characteristics (i.e. low or high pH, heavy metals, metalloids, poor microbial activities, poor soil-nutrient status, higher temperature, water scarcity, etc.) that may not be suitable for growing bioenergy crops. Such scarce conditions may cause diverse effects, like yield reduction, accumulation of pollutants in plant tissues, plant metabolic disorders, reduction of vitality, occurrence of pest and diseases etc. However, as arising from previous examples, presence of contaminants in biomass of plants from contaminated sites may reduce the range of their final uses as energy crops. Therefore, combating impeding issues for each specific case of marginal land and optimizing the cropping systems in order to be most adapted to the site conditions is one of the future tasks and challenges in wider application of bioenergy crops.

The degraded land restoration is among the most important tasks for achieving UN-Sustainable Development Goals (UN-SDSGs). The challenge of degraded land's restoration could be solved through sound ecological restoration technologies implemented by skilled practitioners. In recent years, ecological restoration technologies are popularized as affordable and effective (Pandey et al. 2015b; Pandey 2017, 2021) for remediation and management of degraded land. They should particularly take into account plant species selection, climate changes, plant and soil carbon storage potential, planting density and technology, and application of maintenance measurements. If not done in a proper way by using optimal management practices, bioenergy production on degraded land could cause negative effects such as deterioration of water quality, soil erosion, nutrient depletion and increase in greenhouse gas emissions (Wu et al. 2018). It has been recognized that growing woody plants with or in rotation with herbaceous bioenergy crops could additionally improve both the soil properties and the crop yields (Schrama et al. 2014). Similarly, intercropping of perennials and legume plants could lead to the sustainable biomass production (Nabel et al. 2018).

However, bioenergy monocultures are certainly able to restore degraded lands, but on the other hand, they may decrease the biodiversity (Pandey et al. 2016). Second, it is of vital importance to correctly manage the bioenergy cropping systems; otherwise, their potential invasiveness can shift the native species (Pandey et al. 2016). For example, invasiveness of *Prosopis juliflora* in process of restoration of degraded lands in India showed reduction of local flora to much lower number of species in comparison to non-invaded areas (Edrisi et al. 2020). Monocultures of bioenergy crops, especially those growing in conditions of reduced landscape heterogeneity increase the potential for future invasion of non-native species, as in case of *P. virgatum* (Hartman et al. 2011). However, if the concept of bioenergy multi-cropping system is applied and diverse mixtures of suitable species are used, the biodiversity may be increased alongside with enhanced biomass production and ecosystem services (Awasthi et al. 2017).

More comprehensive scientific research concerning various species with potential of producing bioenergy in different locations and climatic conditions, as well as under various management practices are needed in order to gain better insights in real pros and cons of bioenergy production. So far, there are a limited number of Life Cycle Assessments conducted evaluating environmental costs in bioenergy production chain (Wu et al. 2018), which is of vital importance in addressing the environmental influence of such energy production type and its potential consequences.

1.7 Growing Bioenergy Crops on Degraded Lands: Achieving UN-SDGs

The utilization of degraded lands for good health and well-being of human is urgently required as well as the attaining UN-SDGs. However, the restoration of degraded lands is still the ambitious and tough task for the soil–plant scientists and practitioners. But recent advances in restoration technologies have significantly contributed to our ability to restore lost ecosystem services from degraded lands. Growing bioenergy crops on degraded land is gaining high importance in global land restoration programmes. UN Decade on Ecosystem Restoration (2021–2030) emphasizes such possibilities, coupled with sustainable ecosystem restoration practices.

Therefore, it is required to explore affordable and sustainable restoration technologies to achieve maximum UN-SDGs. Bioenergy crop-based restoration is gaining importance as affordable and sustainable approach in restoration programmes. Hence, it is vital to manage bioenergy multi-cropping system in such a way that the energy plant biodiversity can deliver life-supporting services towards the intentions of international policies (such as GHG emissions and soil organic matter sequestration). Otherwise, results may be deteriorating towards ecosystem functions, food security and biodiversity loss.

In developing countries like India, local villagers can be employed in bioenergy crops-based restoration programmes for accelerating the recovery of the degraded lands, thus increasing per capita income and reducing poverty. Plantation of multiple cropping systems of bioenergy crops on degraded lands significantly increases biodiversity that offers habitat to a surplus of flora and fauna species, provides ecosystem services, mitigates climate change, and reduces CO_2 emission and environmental pollution. Therefore, creating a sustainable bioenergy crops-based multifunctional ecosystem on the degraded lands through affordable restoration approach can achieve the fractional intentions of several UN-SDGs, especially those that address poverty (SDG 1), good health and well-being of people and societies (SDG 3), affordable and clean energy (SGD 7), decent work and economic growth (SDG 8), climate action (SDG 13), life below water (SDG 14), and life on land (SDG 15). Scientific contribution towards these goals is vital, as investigations of biomass sources and bioenergy products continue, coupled with field research concerning issues such as land use transition, soil and water quality, biodiversity and socio-economic effects.

1.8 Conclusion

Land degradation requires wider global attention, as well as the efforts to restore ecological functions of such lands. As bioenergy production is taking higher share in global energy consumption, matching these approaches in a way that could reach maximum UN-SGDs represents challenging scientific, societal and economic task.

During conversion of marginal and degraded lands into sites for growing bioenergy crops, many issues, such as land use changes, soil carbon and nitrogen content, GHG emissions, biodiversity, Water use efficiency, erosion rate, livelihood improvements and economical values of products should be considered. During efforts to reach sustainable bioenergy production it is of paramount importance to carefully address environmental issues and avoid unsustainable crop expansion in order not to cause deteriorate effects on soil and water quality, GHG emissions, biodiversity, erosion etc. Sustainable practices, such as using dedicated bioenergy crops, mixed cultures, rotation and intercropping and application of optimized agronomic practices can be beneficial to biodiversity, soil carbon and nitrogen content and GHG mitigation.

Various types of degraded lands require different approaches and management methods for overcoming obstacles in bioenergy crop production. Field studies revealed that performances vary depending on plant species used as bioenergy crop and site-specific conditions and limitations. Beside already recognized and widely studied bioenergy crops, there is a need to search for other, preferably multipurpose plant species, which could be additionally used for bioenergy production. Site limitations could also play crucial role in determining the success of bioenergy crops. For example, during restoration of contaminated marginal land it is important to access not only the growth parameters, but also the contamination level of final products in order to make them safe for bioenergy consumption.

Using Life Cycle Assessment tool for developing site-specific designs and applying sustainable management practices for producing bioenergy crops should aid in recognition of environmental footprints and bottlenecks for bioenergy production process on various types of degraded lands. Future scientific research should reveal still unknown mechanisms, behaviours and connections between various parameters in soil–plant-water systems of diverse degraded sites, especially in terms of their restoration under growing anthropogenic pressures in climate change conditions. Finally, wider social acceptance of bioenergy production through creating adequate policies and livelihood improvement that could be brought to producers, especially to local people in developing countries, is of great importance for implementing bioenergy production and should be responsibly promoted taking into account both advantages and potential disadvantages of this process.

References

Abideen Z, Ansari R, Khan MA (2011) Halophytes: potential source of ligno-cellulosic biomass for ethanol production. Biomass Bioenergy 35(5):1818–1822

Acharya BS, Blanco-Canqui H, Mitchell RB, Cruse R, Laird D (2019) Dedicated bioenergy crops and water erosion. J Environ Qual 48(2):485–492

Acuña E, Rubilar R, Cancino J, Albaugh TJ, Maier CA (2018) Economic assessment of *Eucalyptus globulus* short rotation energy crops under contrasting silvicultural intensities on marginal agricultural land. Land Use Policy 76:329–337

Aggangan N, Cadiz N, Llamado A, Raymundo A (2017) *Jatropha curcas* for bioenergy and bioremediation in mine tailing area in Mogpog, Marinduque, Philippines. Energy Procedia 110:471–478

Aghaalikhani A, Savuto E, Di Carlo A, Borello D (2017) Poplar from phytoremediation as a renewable energy source: gasification properties and pollution analysis. Energy Procedia 142:924–931

Ahmad M, Zafar M, Khan A, Sultana S (2009) Biodiesel from *Pongamia pinnata* L. oil: a promising alternative bioenergy source. Energy Sources, Part A: Recov Util Environ Effects 31(16):1436–1442

Akinshina N, Toderich K, Azizov A, Saito L, Ismail S (2014) Halophyte biomass—a promising source of renewable energy. J Arid Land Stud 24–1:231–323

Alexopoulou E (2018) Perennial grasses for bioenergy and bioproducts. Academic Press, Elsevier, USA, p 2018

Algreen M, Trapp S, Rein A (2014) Phytoscreening and phytoextraction of heavy metals at Danish polluted sites using willow and poplar trees. Environ Sci Pollut Res 21:8992–9001

Álvarez-Mateos P, Alés-Álvarez FJ, García-Martín JF (2019) Phytoremediation of highly contaminated mining soils by *Jatropha curcas* L. and production of catalytic carbons from the generated biomass. J Environ Manag 231:886–895

Andrejić G, Šinžar-Sekulić J, Prica M, Dželetović Ž, Rakić T (2019) Phytoremediation potential and physiological response of *Miscanthus × giganteus* cultivated on fertilized and non-fertilized flotation tailings. Environ Sci Pollut Res 26:34658–34669

Awasthi A, Singh K, Singh RP (2017) A concept of diverse perennial cropping systems for integrated bioenergy production and ecological restoration of marginal lands in India. Ecol Eng 105:58–65

Balsamo R, Kelly W, Satrio J, Ruiz-Felix N, Fetterman M, Wynn R, Hagel K (2015) Utilization of grasses for potential biofuel production and phytoremediation of heavy metal contaminated soils. Int J Phytoremediation 17(5):448–455

Barbosa B, Boléo S, Sidella S, Costa J, Duarte MP, Mendes B, Cosentino S, Fernando AL (2015) Phytoremediation of heavy metal-contaminated soils using the perennial energy crops Miscanthus spp. and *Arundo donax* L. Bioenergy Res 8(4):1500–1511

Bardhan S, Jose S (2012) The potential for floodplains to sustain biomass feedstock production systems. Biofuels 3(5):575–588

Barney J, Mann J, Kyser G, Blumwald E, Van Deynze A, DiTomaso J (2009) Tolerance of switchgrass to extreme soil moisture stress: ecological implications. Plant Sci 177(6):724–732

Bert V, Allemon J, Sajet P, Dieu S, Papin A, Collet S, Gaucher R, Chalot M, Michiels B, Raventos C (2017) Torrefaction and pyrolysis of metal-enriched poplars from phytotechnologies: effect of temperature and biomass chlorine content on metal distribution in end-products and valorization options. Biomass Bioenerg 96:1–11

Bhatt V (1990) Biocoenological succession in reclaimed rock phosphate mine of Doon Valley. Ph.D. thesis HN Bahuguna Garhwal University, Srinagar UK

Blanco-Canqui H (2016) Growing dedicated energy crops on marginal lands and ecosystem services. Soil Sci Soc Am J 80:845–858

Bomani B, Hendricks R, Elbluk M, Okon M, Lee E, Gigante B (2011) NASA's GreenLab research facility—a guide for a self-sustainable renewable energy ecosystem. Technical Publication. https://ntrs.nasa.gov/citations/20120001794

Bonin C, Lal R (2012) Agronomic and ecological implications of biofuels. In: Sparks D (ed) Advances in agronomy, vol 117, pp 1–50

Bosco S, Nassi o Di Nasso N, Roncucci N, Mazzoncini M, Bonari E (2016) Environmental performances of giant reed (Arundo donax L.) cultivated in fertile and marginal lands: a case study in the Mediterranean. Eur J Agron 78:20–31

Bressan R, Reddy M, Chung S, Yun D, Hardin L, Bohnert H (2011) Stress-adapted extremophiles provide energy without interference with food production. Food Security 3:93–105

Brown R (2003) Biorenewable resources: engineering new products from agriculture, 1st edn. Iowa State Press, Ames, Iowa, p 286

Brown C, Griggs T, Keene T, Marra M, Skousen J (2015) Switchgrass biofuel production on reclaimed surface mines: I. Soil Quality and Dry Matter Yield. Bioenergy Res 9(1):31–39

Campbell JE, Lobell DB, Genova RC, Field CB (2008) The global potential of bioenergy on abandoned agriculture lands. Environ Sci Technol 42(15):5791–5794. https://doi.org/10.1021/es800052w

Caspari T, Lynden G, Bai Z (2015) Land degradation neutrality: an evaluation of methods. ISRIC-World Soil Information, Wageningen, Netherlands, p 53

Castaño-Díaz M, Barrio-Anta M, Afif-Khouri E, Cámara-Obregón A (2018) Willow short rotation coppice trial in a former mining area in Northern Spain: effects of clone fertilization and planting density on yield after five years. Forests 9:154

Chauvat M, Perez G, Hedde M, Lamy I (2014) Establishment of bioenergy crops on metal contaminated soils stimulates belowground fauna. Biomass Bioenerg 62:207–211

Chen BC, Lai HY, Juang KW (2012) Model evaluation of plant metal content and biomass yield for the phytoextraction of heavy metals by switchgrass. Ecotoxicol Environ Saf 80:393–400

Corno L, Pilu R, Tambone F, Scaglia B, Adani F (2015) New energy crop giant cane (*Arundo donax* L.) can substitute traditional energy crops increasing biogas yield and reducing costs. Bioresour Technol 191:197–204

da Costa L, Lima R, de Ferreira MJ (2015) Biomass and energy yield of leguminous trees cultivated in Amazonas. Floresta 45:705–712

Das S, Teuffer K, Stoof CR, Walter MF, Walter MT, Steenhuis TS, Richards BK (2018) Perennial grass bioenergy cropping on wet marginal land: impacts on soil properties, soil organic carbon, and biomass during initial establishment. Bioenergy Res 11(2):262–276

De Rossi A, Vescio R, Russo D, Macrì G (2016) Potential use of Jatropha Curcas L. on marginal lands of Southern Italy. Procedia Soc Behav Sci 223:770–775

Dimitriou I, Aronsson P (2005) Willows for energy and phytoremediation in Sweden. Unasylva 221(56):47–50

Dorta-Santos M, Tejedor M, Jiménez C, Hernández-Moreno JM, Palacios-Díaz MP, Díaz FJ (2015) Evaluating the sustainability of subsurface drip irrigation using recycled wastewater for a bioenergy crop on abandoned arid agricultural land. Ecol Eng 79:60–68

Dubovyk O (2017) The role of remote sensing in land degradation assessments: opportunities and challenges. Eur J Remote Sens 50(1):601–613

Edrisi SA, El-Keblawy A, Abhilash PC (2020) Sustainability analysis of *Prosopis juliflora* (Sw.) DC based restoration of degraded land in North India. Land 9:59

ELD Initiative (2013) The rewards of investing in sustainable land management. Interim report for the economics of land degradation initiative: a global strategy for sustainable land management. www.eld-initiative.org/

Emmerling C, Schmidt A, Ruf T, von Francken-Welz H, Thielen S (2017) Impact of newly introduced perennial bioenergy crops on soil quality parameters at three different locations in W-Germany. J Plant Nutr Soil Sci 180(6):759–767

Eswaran H, Lal R, Reich PF (2001) Land degradation: an overview. In: Bridges EM, Hannam ID, Oldeman LR, Pening de Vries FWT, Scherr SJ, Sompatpanit S (eds) Responses to land degradation; Proceedings of 2nd international conference on land degradation and desertification, Khon Kaen, Thailand. Oxford Press, New Delhi, India

Fei H, Crouse M, Papadopoulos Y, Vessey JK (2017) Enhancing the productivity of hybrid poplar (*Populus × hybrid*) and switchgrass (*Panicum virgatum* L.) by the application of beneficial soil microbes and a seaweed extract. Biomass Bioenerg 107:122–134

Gami B, Limbachiya R, Parmar R, Bhimani H, Patel B (2011) An evaluation of different non-woody and woody biomass of Gujarat, India for preparation of pellets—a solid biofuel. Energy Sources, Part A: Recov Util Environ Effects 33(22):2078–2088

García Martín JF, González Caro MC, López Barrera MC, Torres García M, Barbin D, Álvarez Mateos P (2020) Metal accumulation by *Jatropha curcas* L. Adult plants grown on heavy metal-contaminated soil. Plants 9:418

Gashu K, Muchie Y (2018) Rethink the interlink between land degradation and livelihood of rural communities in Chilga district, Northwest Ethiopia. J Ecol Environ 42:17

Gelfand I, Sahajpal R, Zhang X, Izaurralde RC, Gross KL, Robertson GP (2013) Sustainable bioenergy production from marginal lands in the US Midwest. Nature 493:514–517

Global Mechanism of the UNCCD (2019) Land degradation, poverty and inequality. Conservation international, DIE, Bonn, Germany

Gul B, Abideen Z, Ansari R, Khan MA (2013) Halophytic biofuels revisited. Biofuels 4(6):575–577

Guo Z, Gao Y, Cao X, Jiang W, Liu X, Liu Q, Chen Z, Zhou W, Cui J, Wang Q (2019) Phytoremediation of Cd and Pb interactive polluted soils by switchgrass (*Panicum virgatum* L.). Int J Phytoremediation 21(14):1486–1496

Guzman JG, Ussiri DAN, Lal R (2019) Soil physical properties following conversion of a reclaimed minesoil to bioenergy crop production. CATENA 176:289–295

Harari N, Gavilano A, Liniger HP (2017) Where people and their land are safer: a compendium of good practices in disaster risk reduction. Centre for Development and Environment (CDE), University of Bern, and Swiss NGO Disaster Risk Reduction (DRR) Platform, Bern and Lucerne, Switzerland

Hasan SS, Zhen L, Miah MG, Ahamed T, Samie A (2020) Impact of land use change on ecosystem services: a review. Environ Dev 34. https://doi.org/10.1016/j.envdev.2020.100527

Hartman JC, Nippert JB, Orozco RA, Springer CJ (2011) Potential ecological impacts of switchgrass (*Panicum virgatum* L.) biofuel cultivation in the Central Great Plains, USA. Biomass Bioenerg 35(8):3415–3421

Hayes D (2008) An examination of biorefining processes, catalysts and challenges. Catal Today 145(1–2):138–151

Hennenberg KJ, Dragišić C, Haye S, Hewson J, Semroc B, Savy C, Wiegmann K, Fehrenbach H, Fritsche UR (2010) The power of bioenergy-related standards to protect biodiversity. Conserv Biol 24(2):412–423

Hobbs RJ, Higgs E, Hall CM, Bridgewater P, Chapin FS et al (2014) Managing the whole landscape: historical, hybrid, and novel ecosystems. Front Ecol Environ 12:557–564

Huang H, Ramaswamy S, Al-Dajani W, Tschirner U, Cairncross R (2009) Effect of biomass species and plant size on cellulosic ethanol: a comparative process and economic analysis. Biomass Bioenergy 33:234–246

IPBES (2018) The IPBES assessment report on land degradation and restoration. In: Montanarella L, Scholes R, Brainich A (eds) Secretariat of the intergovernmental science-policy platform on biodiversity and ecosystem services, Bonn, Germany, pp 744

Jakovljević K, Mišljenović T, Savović J, Ranković D, Ranđelović D, Mihailović N, Jovanović S (2020) Accumulation of trace elements in *Tussilago farfara* colonizing post-flotation tailing sites in Serbia. Environ Sci Pollut Res 27(4):4089–4103

Jankauskas B, Jankauskiene G (2003) Erosion-preventive crop rotations for landscape ecological stability in upland regions of Lithuania. Agric Ecosyst Environ 95:129–142

Jeżowski S, Mos M, Buckby S, Cerazy-Waliszewska J, Owczarzak W, Mocek A, Kaczmarek Z, McCalmont JP (2017) Establishment, growth, and yield potential of the perennial grass *Miscanthus* × *giganteus* on degraded coal mine soils. Front Plant Sci 8:726

Johnson T, Johnson B, Scott-Kerr C, Kiviaho J (2009) Bioethanol—status report on bioethanol production from wood and other lignocellulosic feedstocks. In: 63rd Appita annual conference and exhibition, Melbourne, 19–22 April, p 3

Koh M, Ghazi T (2011) A review of biodiesel production *from Jatropha curcas* L. oil. Renew Sustain Energy Rev 15(5):2240–2251

Korzeniowska J, Stanislawska-Glubiak E (2015) Phytoremediation potential of *Miscanthus* × *giganteus* and *Spartina pectinata* in soil contaminated with heavy metals. Environ Sci Pollut Res 22:11648–11657

Kreig J, Ssegane H, Chaubey I, Negri M, Jager H (2019) Designing bioenergy landscapes to protect water quality. Biomass Bioenergy 128:105327

Kullu B, Behera N (2011) Vegetational succession on different age series sponge iron solid waste dumps with respect to topsoil application. Res J Environ Earth Sci 3:38–45

Laval-Gilly P, Henry S, Mazziotti M, Bonnefoy A, Comel A, Falla J (2017) *Miscanthus* x *giganteus* composition in metals and potassium after culture on polluted soil and its use as biofuel. BioEnergy Res 10(3):846–852

Lebrun M, Macri C, Miard F, Hattab-Hambli N, Motelica Heino M, Morabito D, Bourgerie S (2017) Effect of biochar amendments on As and Pb mobility and phytoavailability in contami-nated mine technosols phytoremediated by Salix. J Geochem Explor 182:149–156

Leksono B, Rahman S, Purbaya D, Samsudin Y, Lee S, Maimunah S, Maulana A, Wohono J, Baral H (2018) Pongamia (*Pongamia pinnata*): a sustainable alternative for biofuel production and land restoration in Indonesia. Preprints, 2018110604.https://doi.org/10.20944/preprints201811.0604.v1

Lemus R, Lal R (2005) Bioenergy crops and carbon sequestration. Crit Rev Plant Sci 24(1):1–21

Li Z, Deng X, Yin F, Yang C (2015) Analysis of climate and land use changes impacts on land degradation in the North China Plain. Adv Meteorol 2015. Article ID 976370

Littlewood J, Guo M, Boerjan W, Murphy R (2014) Bioethanol from poplar: a commercially viable alternative to fossil fuel in the European Union. Biotechnol Biofuels 7(1):113–125

Lord RA (2015) Reed canarygrass (*Phalaris arundinacea*) outperforms miscanthus or willow on marginal soils, brownfield and non-agricultural sites for local, sustainable energy crop production. Biomass Bioenergy 78:110–125

Maiti SK (2013) Establishment of grass and legume cover. In: ecorestoration of the coalmine degraded lands. Springer, India, pp 151–161

Mann J, Barney J, Kyser G, Ditomaso J (2013) *Miscanthus* x *giganteus* and *Arundo donax* shoot and rhizome tolerance of extreme moisture stress. GCB Bioenergy 5:693–700

Martani E, Ferrarini A, Serra P, Pilla M, Marcone A, Amaducci S (2020) Belowground biomass C outweighs soil organic C of perennial energy crops: insights from a long-term multispecies trial. CBC Bioenergy 13(3):459–472

McKendry P (2002) Energy production from biomass (part 1): overview of biomass. Bioresour Technol 83:37–46

Migeon A, Richaud P, Guinet F, Chalot M, Blaudez D (2009) Metal accumulation by woody species on contaminated sites in the North of France. Water Air Soil Pollut 204(1–4):89–101

Millennium Ecosystem Assessment (2005) Ecosystems and human well-being: desertification synthesis. World Resources Institute, Washington, DC

Miyake S, Smith C, Peterson A, McAlpine C, Renouf M, Waters D (2015) Environmental implications of using "underutilised agricultural land" for future bioenergy crop production. Agric Syst 139:180–195

Moniruzzaman M, Yaakob Z, Shahinuzzaman M, Khatun R, Islam A (2017) Jatropha biofuel industry: the challenges. In: Jacob-Lopes E, Zepka L (eds) Frontiers in bioenergy and biofuels. https://www.intechopen.com/books/frontiers-in-bioenergy-and-biofuels/jatropha-biofuel-industry-the-challenges

Moser BR, Dien BS, Seliskar DM, Gallagher JL (2013) Seashore mallow (Kosteletzkya pentacarpos) as a salt-tolerant feedstock for production of biodiesel and ethanol. Renew Energy 50:833–839

Nabel M, Schrey SD, Temperton VM, Harrison L, Jablonowski ND (2018) Legume intercropping with the bioenergy crop *Sida hermaphrodita* on marginal soil. Front Plant Sci 9

Nebeská D, Trögl J, Žofková D, Voslařová A, Štojdl J, Pidlisnyuk V (2019) Calorific values of *Miscanthus* x *giganteus* biomass cultivated under suboptimal conditions in marginal soils. Studia Oecologica 13(1):61–67

Nkonya E, Mirzabaev A, von Braun J (2016) Economics of Land degradation and improvement—a global assessment for sustainable development. Springer International Publishing, pp 686

Oginni O, Singh K (2019) Pyrolysis characteristics of *Arundo donax* harvested from a reclaimed mine land. Ind Crops Prod 133:44–53

Olsson L, Barbosa H, Bhadwal S, Cowie A, Delusca K, Flores-Renteria D, Hermans K, Jobbagy E, Kurz W, Li D, Sonwa DJ, Stringer L (2019) Land degradation. In: Shukla PR, Skea J, Calvo Buendia E, Masson-Delmotte V, Pörtner H-O, Roberts DC, Zhai P, Slade R, Connors S, van Diemen R, Ferrat M, Haughey E, Luz S, Neogi S, Pathak M, Petzold J, Portugal Pereira J, Vyas P, Huntley E, Kissick K, Belkacemi M, Malley J (eds) Climate change and land: an IPCC special report on climate change, desertification, land degradation, sustainable land management, food security, and greenhouse gas fluxes in terrestrial ecosystems. Intergovernmental Panel on Climate Change (IPCC)

Pandey VC (2015) Assisted phytoremediation of fly ash dumps through naturally colonized plants. Ecol Eng 82:1–5

Pandey VC (2017) Managing waste dumpsites through energy plantations. In: Bauddh K, Singh B, Korstad J (eds) Phytoremediation potential of bioenergy plants. Springer, Singapore, pp 371–386

Pandey VC (2021) Direct seeding offers affordable restoration for fly ash deposits. Energy, Ecol Environ. https://doi.org/10.1007/s40974-021-00212-7

Pandey VC, Singh N (2014) Fast green capping on coal fly ash basins through ecological engineering. Ecol Eng 73:671–675

Pandey VC, Singh DP (2020) Phytoremediation potential of perennial grasses. Elsevier, pp 371

Pandey VC, Singh K, Singh JS, Kumar A, Singh B, Singh R (2012a) *Jatropha curcas*: a potential biofuel plant for sustainable environmental development. Renew Sustain Energy Rev 16:2870–2883

Pandey VC, Singh K, Singh RP, Singh B (2012b) Naturally growing *Saccharum munja* on the fly ash lagoons: a potential ecological engineer for the revegetation and stabilization. Ecol Eng 40:95–99

Pandey VC, Bajpai O, Pandey DN, Singh N (2015a) *Saccharum spontaneum*: an underutilized tall grass for revegetation and restoration programs. Genet Resour Crop Evol 62:443–450

Pandey VC, Pandey DN, Singh N (2015b) Sustainable phytoremediation based on naturally colonizing and economically valuable plants. J Clean Prod 86:37–39

Pandey VC, Bajpai O, Singh N (2016) Energy crops in sustainable phytoremediation. Renew Sustain Energy Rev 54:58–73

Patel D, Pandey VC (2020) Switchgrass–an asset for phytoremediation and bioenergy production. In: Pandey VC, Singh DP (Authored book with contributors) Phytoremediation potential of perennial grasses. Elsevier, Amsterdam, pp 179–193. https://doi.org/10.1016/B978-0-12-817 732-7.00008-0

Phouthavong-Murphy JC, Merrill AK, Zamule S, Giacherio M, Brown B, Roote C, Das P (2020) Phytoremediation potential of switchgrass (*Panicum virgatum*), two United States native varieties, to remove bisphenol-A (BPA) from aqueous media. Sci Rep 10:835

Pidlisnyuk V, Erickson L, Stefanovska T, Popelka J, Hettiarachchi G, Davis L, Trögl J (2019) Potential phytomenegment of military polluted sites and biomass production using biofuel crop *Miscanthus* x *giganteus*. Environ Pollut 249:330–337

Pleguezuelo CR, Zuazo V, Bielders C, Bocanegra J, Torres F, Martinez J (2015) Bioenergy farming using woody crops: a review. Agron Sustain Dev 35:95–119

Pogrzeba M, Rusinowski S, Sitko K, Krzyżak J, Skalska A, Małkowski E, Ciszek D, Werle S, McCalmont JP, Mos M, Kalaji HM (2017) Relationships between soil parameters and physiological status of Miscanthus x giganteus cultivated on soil contaminated with trace elements under NPK fertilisation vs. microbial inoculation. Environ Pollut 225:163–174

Porth I, El-Kassaby Y (2015) Using *Populus* as a lignocellulosic feedstock for bioethanol. Biotechnol J 10(4):510–524

Praveen A, Pandey VC (2020) Miscanthus–a perennial energy grass in phytoremediation. In: Pandey VC, Singh DP (Authored book with contributors) Phytoremediation potential of perennial grasses. Elsevier, Amsterdam, pp 79–95. https://doi.org/10.1016/B978-0-12-817732-7.00004-3

Ranđelović D, Cvetković V, Mihailović N, Jovanović S (2014) Relation between edaphic factors and vegetation development on copper minewastes: a case study from Bor (Serbia, SE Europe). Environ Manage 53(4):800–812

Ranđelović D, Jakovljević K, Mihailović N, Jovanović S (2018) Metal accumulation in populations of *Calamagrostis epigejos* (L.) Roth from diverse anthropogenically degraded sites (SE Europe, Serbia). Environ Monit Assess 190:183

Raud M, Kikas T (2020) Perennial grasses as a substrate for bioethanol production. Environ Climate Technol 24(2):32–40

Safriel UN (2007) The assessment of global trends in land degradation. In: Sivakumar MVK, Ndiang'ui N (eds) Climate and land degradation. Environmental Science and Engineering (Environmental Science). Springer, Berlin, Heidelberg, pp 1–38

Sanderson M, Adler P (2008) Perennial forages as second generation bioenergy crops. Int J Mol Sci 9(5):768–788

Santi G, D'Annibale A, Eshel A, Zilberstein A, Crognale S, Ruzzi M, Valentini R, Moresi M, Petruccioli M (2014) Ethanol production from xerophilic and salt-resistant Tamarix jordanis biomass. Biomass Bioenergy 61:73–81

Santibáñez Varnero C, Urrutia MV, Ibaceta SV (2018) Bioenergy from perennial grasses. In: Nageswara-Rao M, Soneyi J (eds) Advances in biofuels and bioenergy. IntechOpen. https://doi.org/10.5772/intechopen.74014

Schmidt A, Lemaigre S, Delfosse P, von Francken-Welz H, Emmerling C (2018) Biochemical methane potential (BMP) of six perennial energy crops cultivated at three different locations in W-Germany. Biomass Convers Bior 8:873–888

Schrama M, Vandecasteele B, Carvalho S, Muylle H, van der Putten WH (2014) Effects of first- and second-generation bioenergy crops on soil processes and legacy effects on a subsequent crop. GCB Bioenergy 8(1):136–147

Scordia D, Testa G, van Dam JE, van den Berg D (2018) Suitability of perennial grasses for energy and non-energy products. In: Perennial grasses for bioenergy and bioproducts, pp 217–244

Scordia D, Cosentino SL (2019) Perennial energy grasses: resilient crops in a changing European agriculture. Agriculture 9(8):1–19

SEEMLA (2016) Sustainable exploitation of biomass for bioenergy from marginal lands in Europe: catalogue for bioenergy crops and their suitability in the categories of MagLs, Deliverable D 2.2. of the SEEMLA project funded under the European Union's Horizon 2020 research and innovation programme GA No: 691874. https://www.seemla.eu/catalogue-for-bioenergy-crops-and-their-suitability-in-the-categories-of-magls/

Sharma R, Wungrampha S, Singh V, Pareek A, Sharma MK (2016) Halophytes as bioenergy crops. Front Plant Sci 7:1372

Singh S, Jaiswal DK, Krishna R, Mukherjee A, Verma JP (2019) Restoration of degraded lands through bioenergy plantations. Restor Ecol 28(2):263–266

Shen S, Chen J, Chang J, Xia B (2020) Using bioenergy crop cassava (*Manihot esculenta*) for reclamation of heavily metal-contaminated land. Int J Phytoremediation 22(12):1313–1320

Shi X, Zhang X, Chen G, Chen Y, Wang L, Shan X (2011) Seedling growth and metal accumulation of selected woody species in copper and lead/zinc mine tailings. J Environ Sci 23(2):266–274

Shortall O (2013) "Marginal land" for energy crops: exploring definitions and embedded assumptions. Energy Policy 62(C):19–27

Skousen J, Keene T, Marra M, Gutta B (2012) Reclamation of mined land with switchgrass, miscanthus, and arundo for biofuel production. J Amer Soc Mining Reclam 2(1):177–191

Slepetys J, Kadziuliene Z, Sarunaite L, Tilvikiene V, Kryzeviciene A (2012) Biomass of plants grown for bioenergy production. Renew Energy Energ Effi 11:66–72

Song XP, Hansen MC, Stehman SV, Potapov P, Tyukavina A, Vermote E, Townshend JR (2018) Global land change from 1982 to 2016. Nature 560:639–643

Stavridou E, Hastings A, Webster R, Robson P (2017) The impact of soil salinity on the yield, composition and physiology of the bioenergy grass *Miscanthus* x *giganteus*. GCB Bioenergy 9:92–104

Stolarski MJ, Krzyzaniak M, Szczukowski S, Tworkowski J, Bieniek A (2014) Short rotation woody crops grown on marginal soil for biomass energy. Pol J Environ Stud 23(5):1727–1739

Stolarski MJ, Szczukowski S, Tworkowski J, Krzyżaniak M (2019) Extensive willow biomass production on marginal land. Pol J Environ Stud 28(6):4359–4367

Taiichiro H, Shigenori M (2010) Energy crops for sustainable bioethanol production; which, where and how? Plant Prod Sci 13(3):221–234

Taranenko A, Kulyk M, Galytska M, Taranenko S. (2019) Effect of cultivation technology on switchgrass (*Panicum virgatum* L.) productivity in marginal lands in Ukraine. Acta Agrobot 72(3):1786

Tilman D, Hill J, Lehman C (2006) Carbon-negative biofuels from low-input high-diversity grassland biomass. Science 314(5805):1598–1600

Tlustoš P, Száková J, Vysloužilová M, Pavlíková D, Weger J, Javorská H (2007) Variation in the uptake of arsenic, cadmium, lead, and zinc by different species of willows Salix spp. grown in contaminated soils. Cent Eur J Biol 2:254–275

United Nations Department of Economic and Social Affairs, United Nations Forum on Forests Secretariat (2021) The Global Forest Goals Report 2021

UNCCD (1994) United Nations Convention to Combat Desertification in countries experiencing serious drought and/or desertification, particularly in Africa. UN Doc A/AC.241/27 Paris

Valcu-Lisman A, Kling C, Gassman P (2016) The optimality of using marginal land for bioenergy crops: tradeoffs between food, fuel, and environmental services. Agric Econ Res Rev 45:217–245

Valentine J, Clifton-Brown J, Hastings A, Robson P, Allison G, Smith P (2012) Food vs. fuel: the use of land for lignocellulosic 'next generation' energy crops that minimize competition with primary food production. GCB Bioenergy 4:1–19

Van der Esch S, ten Brink B, Stehfest E, Bakkenes M, Sewell A, Bouwman A, Meijer J, Westhoek H, van den Berg M (2017) Exploring future changes in land use and land condition and the impacts on food, water, climate change and biodiversity: scenarios for the global land outlook. PBL Netherlands Environmental Assessment Agency, The Hague

Van der Weijde T, Dolstra O, Visser RGF, Trindade LM (2017) Stability of cell wall composition and saccharification efficiency in miscanthus across diverse environments. Front Plant Sci 7

Vervaeke P, Tack FMG, Navez F, Martin J, Verloo MG, Lust N (2006) Fate of heavy metals during fixed bed downdraft gasification of willow wood harvested from contaminated sites. Biomass Bioenerg 30(1):58–65

Wahyudi W, Nadjib M, Bari F, Permana F (2019) Increasing of quality biodiesel of Jatropha seed oil with biodiesel mixture of waste cooking oil. J Biotech Res 10:183–189

Wang D, Guo JR, Liu XJ, Song J, Chen M, Wang BS (2014) Effects of cultivation strategies on hybrid pennisetum yield in saline soil. Crop Sci 54(6):2772

Wang M, Wang J, Tan JX (2011) Lignocellulosic bioethanol: status and prospects. Energy Sources Part A 33(7):612–619

Warren A (2002) Land degradation is contextual. Land Degrad Dev 13:449–459

WER (2013) World energy resources survey. World Energy Council, London

Werle S, Tran K, Magdziarz A, Sobek S, Pogrzeba M, Lovas T (2019) Energy crops for sustainable phytoremediation—fuel characterization. Energy Procedia 158:867–872

Whittaker C, Shield I (2016) Short rotation woody energy crop supply chains. In: Ehiaze J, Ehimen A (eds) Biomass supply chains for bioenergy and biorefining. Elsevier, pp 217–248

Wi S, Cho E, Lee D, Lee S, Lee Y, Bae H (2015) Lignocellulose conversion for biofuel: a new pretreatment greatly improves downstream biocatalytic hydrolysis of various lignocellulosic materials. Biotechnol Biofuels 8:228

World Energy Council (1994) New renewable energy resources: a guide to the future. Kogan Page, London, p 391

Wu Y, Zhao F, Liu S, Wang L, Qiu L, Alexandrov G, Jothiprakash V (2018) Bioenergy production and environmental impacts. Geosci Lett 5(1):14

Yang W, Wang Y, Zhao F, Ding Z, Zhang X, Zhu Z, Yang X (2014) Variation in copper and zinc tolerance and accumulation in 12 willow clones: implications for phytoextraction. J Zhejiang Univ Sci B 15(9):788–800

Zhao X, Zhang L, Liu D (2012) Biomass recalcitrance. Part I: the chemical compositions and physical structures affecting the enzymatic hydrolysis of lignocellulose. Biofuel Bioprod Biorefin 6:465–482

Zhu P, Abdelaziz O, Hulteberg C, Riisager A (2020) New synthetic approaches to biofuels from lignocellulosic biomass. Curr Opin Green Sustain Chem 21:16–21

Chapter 2
Understanding the Role of Ruderal Plant Species in Restoration of Degraded Lands

Dragana Ranđelović and Slobodan Jovanović

Abstract Ruderal plants are dynamic functional group characterized by their resistance to changing conditions of areas under anthropogenic influence. Opportunistic character enables them wide and fast spreading on growing number of degraded sites, including polluted ones. Not only are ruderal plants able to easily colonize open, degraded areas of land, but they can also be observed as a biological signal that the degraded land is able to recover. Therefore, many ruderal plants are studied for their capacity to cope with various inorganic and organic contaminants in the environment and recognized for their potential in remediation of various degraded lands. Many annual, perennial and woody species are investigated on range of contaminated sites where they showed potential to be applied in phytoremediation technologies. Degraded habitats represent favorable areas for introduction and spreading of invasive species, and some ruderal species. Those ruderal species, owing traits such as high seed production rate, efficient vegetative spreading and rapid nutrient uptake, could rapidly spread in such areas out of their native range, becoming invasive. Common traits of ruderal plants are mostly matching the desirable traits of plant suitable for phytoremediation purposes. As research in the area of phytoremediation continues to develop toward increasing the remediation efficiency, ruderal plants are gradually being subjected to various experiments and applications aiming to improve element bioavailability, plant tolerance or accumulation capacity. Overall, ruderal plants show high potential for remediation of degraded lands, and their importance will grow over time with increasing rate of anthropogenic disturbances and climate changes.

Keywords Ruderal · Phytoremediation · Land degradation · Pollution

---

D. Ranđelović (✉)
Institute for Technology of Nuclear and Other Mineral Raw Materials, Boulevar Franchet d'Esperey 86, 11000 Belgrade, Serbia
e-mail: d.randjelovic@itnms.ac.rs

S. Jovanović
Faculty of Biology, Institute of Botany and Botanical Garden Jevremovac, University of Belgrade, Takovska 43, 11000 Belgrade, Serbia

2.1 General Restoration Aspects of Degraded Lands

Land degradation is recognized as one of the major modern global environmental problems (Jie et al. 2002; Gisladottir and Stocking 2005; Dwibedi et al. 2021). However, there are many definitions of this term that vary according to their scope and focus. Land degradation is defined as "the loss of actual or potential productivity or utility as a result of natural or anthropic factors; it is the decline in land quality or reduction in its productivity" (Eswaran et al. 2001), lowering of the land productive capacity (UNEP 1992), or "negative trend in land condition, caused by direct or indirect human-induced processes including anthropogenic climate change, expressed as long-term reduction or loss of at least one of the following: biological productivity, ecological integrity or value to humans" (Olsson et al. 2019). Similarly, there are different views concerning mapping and quantification of land degradation worldwide, although many research studies ultimately refer this process to 25–30% of total world lands (IUCN 2015).

Mechanisms that affect land degradation include various physical, chemical and biological aspects and are regional and site-specific. Marked physical processes are soil compaction, water and wind erosion, desertification, pollution and unsustainable land use, while significant chemical processes are salinization, acidification, leaching and fertility loss (Lal 2012; Baumhardt et al. 2015). Biological processes consist of reduction of soil carbon and biodiversity loss (Eswaran et al. 2001; Montanarella 2007).

As one of the global actions to combat land degradation, United Nations included the concept of land degradation neutrality (LDN) as one of the targets of sustainable development goals. LDN concept was defined as "the state whereby the amount and quality of land resources necessary to support ecosystem functions and services and enhance food security remain stable or increase within specified temporal and spatial scales and ecosystems" (UNCCD 2015). This concept is used to address degree and type of land degradation and enhance actions for avoiding, halting or combating this process. Various restoration and rehabilitation efforts may be applied in this sense. While both land remediation and land restoration aim at implementation of strategies in order to reverse negative impacts, there are certain differences between these two categories. While land restoration is defined as "the act of restoring to a former state or position", or bringing it back to the original state (Bradshaw 1997; Mentis 2020), the term remediation is considered as a way of cleaning and revitalizing the land, removing the contaminants while focusing on processes rather than returning the land to a previous state (Bradshaw 2002). Restoration of degraded land may utilize various approaches depending on type and intensity of degradation. However, certain forms of severe land degradation, such as extreme pollution, can be very costly to restore, in which case the remediation approach could be an acceptable alternative for obtaining transition from degraded to recovered state (Bradshaw 1997). Certain countries exhibiting issues with industrial heritage and development focused their actions on remediation of hazardous mining and industrial sites in this sense (Van Liedekerke et al. 2014). Nevertheless, in some cases, land remediation can be followed by reclamation activities upon the removal of the pollution that was threatening to the ecosystem and biota health, resulting in return of the land to the previous land use type.

There are various remediation techniques for reducing the contamination of soil and restoring its vital functions. Soil degradation caused by contamination is an important part of the industrial expansion-related problems, and it may often pose risk to the environment and biota. Majority of soil pollution cases worldwide are caused by anthropogenic activities, such as industrial activities, mining, waste disposal, atmospheric deposition and application of fertilizers (Cachada et al. 2018). Two major groups of soil pollutants may be marked as inorganic and organic pollutants. Either way, soil represents a major environmental sink for a range of these pollutants. While organic contaminants are prone to degradation by various biotic and abiotic processes, inorganic elements (such as metal(loid)s and radionuclides) are not readily degradable, and they may persist in soils for a long time or accumulate in tissues of biota and cause toxic effects (Adriano 2003). Therefore, remediation of inorganic contaminants represents an important issue, as they are long-term contaminants in ecosystems and the environment. General technologies for remediation of metal(loid)s and radionuclides from contaminated soils are based on various physical, chemical and biological methods, including immobilization, toxicity reduction, extraction, etc. (Pandey and Singh 2019). Moreover, these remediation technologies can be applied ex-situ or in-situ. Selection of the most appropriate remediation method depends on the site characteristics, pollution type and severity and future land use. Each of the techniques poses specific applicability, advantages and limitations. Certain remediation technologies are used at full scale, while others are still in different phases of development.

Biological or bio-based remediation methods are based on use of living organisms, mainly microorganisms and plants, for removal or neutralization of contaminants. These methods are generally recognized as low-cost, non-invasive and sustainable solution for remediation that improves both physicochemical and biological quality of degraded lands (Liu et al. 2018). Phytoremediation is based on use of plants and Microorganisms associated with them for removal, degradation or isolation of contaminants from the environment (Prasad 2004). Several categories of phytoremediation, based on mechanism of contaminant removal, can be singled out: phytoextraction, phytostabilization, phytodegradation, phytovolatilization, rhizofiltration and rhizodegradation (Pandey and Bajpai 2019). Phytoextraction focuses on using plants for removal of contaminants from the environment via their uptake into harvestable plant parts, while phytostabilization refers to immobilization of contaminants from soil via adsorption, precipitation processes, sorption or complexation processes in the rhizosphere zone (Grzegórska et al. 2020).

Phytodegradation represents breakdown of environmental contaminants through plant metabolic processes within the plant or by the effect of compounds that are produced by the plant, while rhizodegradation is the breakdown of contaminants in the rhizosphere through microbial activity that is enhanced by the present of plant roots (EPA 2000). Phytovolatilization represents direct or indirect process for remediation of volatile compounds via plant steams or leaves, resulting from the activities of the root zone, respectively (Limmer and Burken 2016). Rhizofiltration is a phytoremediation technique that uses plant roots to absorb, concentrate and precipitate contaminants from the rhizosphere (Dushenkov et al. 1995; Pandey et al. 2014). Phytoremediation research increased in the 1970s, while commercial adoption of the technology started in 1980s (Liu et al. 2018; Pandey and Souza-Alonso 2019). Although it is still considered to be a technology in developing phase, it is considered mostly suitable for large areas with low to moderate level of contamination, where it shows clear advantages over the other remediation methods (Chaney and Baklanov 2017).

2.2 Ruderal Plants—General Traits and Significance

Urban, industrial and agricultural sprawl are a major driver of contemporary landscape changes, and the spontaneously growing ruderal flora and their vegetation in such areas represent specific examples of interactions between natural processes and the multilateral impact of human activities (Goddard et al. 2010; Shochat et al. 2010; Ramalho and Hobbs 2012).

Ruderal plants are a dynamic group characterized by their resistance to ever-changing conditions mainly in urban and industrial areas. Different anthropogenic activities and consequently presence of degraded and polluted areas are the reason of their existence (Jovanović 1994).

Unlike the typical weed species in the narrower sense (i.e., segetal weeds in agroecosystems), ruderal plants are developed and maintained in habitats that are under constant or occasional human influence, but not in order to create productive areas. Such habitats are usually found along roads, in neglected yards, on walls and roofs, in tree lines, on ruins, construction sites, landfills, along railway, road and defensive embankments, on wet and nitrified river banks near human settlements, on cemeteries, around village pens, on degraded pastures, forest clearings. Hence the name of this specific group of plants, which comes from the Latin words *rudus, ruderis* = ruins or rubble, because ruins and similar neglected and polluted places around the settlement are their most typical habitats. Ruderal plants are adapted to these specific, often extremely unfavorable habitat conditions in terms of hydro-thermal regime and type of substrate, as well as in terms of mechanical influences such as trampling, mowing, grazing, and burning, often appearing as pioneer species. Later, through various succession stages, the more stable cenotic relations are established, which are conditioned by the type and intensity of various anthropogenic influences (Jovanović 1994; Prach et al. 2001; Tamakhina et al. 2019). According to Yalcinalp and Meral (2019), as other plant categories are likely to inhabit the areas previously colonized by ruderal species, existence of ruderal plants in areas where maintenance is not necessary implies the overall increase in biodiversity. The ruderal flora in urban areas is often surprisingly rich, combining plants of contrasting habitat requirements and geographical distributions. Moreover, ruderal plants can also be viewed as a specific indicator that an area has a potential to host certain biota, while area inhabitable for even ruderal species signifies that there is no or there is very limited capacity of habitat to support life (Yalcinalp and Meral 2019).

Based on Grime's theory of the plant life-history strategies (Grime 2001), ruderals represent a special category of plants that colonize low stress habitats with high disturbance regimes, allocate majority of resources and energy to the seed production and often belong to annuals or short-lived perennials (r-strategists). Common characteristics of ruderal species include high relative growth rate and flowering, short-lived leaves and short statured plants with minimal lateral expansion (Grime 2001). However, many ruderals can be categorized into the transition group between typical ruderal strategy and stress tolerates, especially in the presence of different toxic elements in soils of polluted areas and habitats. This position in Grime's CSR triangle suggests that many ruderal plants can live successfully under conditions of a certain amount of stress and high disturbance, but they are not competitively strong (i.e., they cannot withstand strong competition with other types of plants).

According to Jovanović (1994), some of the most important bioecological properties of ruderal plants may be generally and briefly summarized in several main points: (1) mostly equivalent species; (2) wide range of distribution (often cosmopolitan species); (3) domination of annual plants (therophytes) that reproduce exclusively by seed; (4) significant proportion of biennial and perennial plants (mainly hemicryptophytes); (5) massive seed production; (6) phenological variability and flexibility as adaptive advantage in wide range of environmental conditions; (7) increased presence of polyploid forms with higher adaptability in disturbed habitats; and (8) anthropochoric—as one of the most important ways of dissemination that causing their predominantly allochthonous (alien) character.

Positive aspects of ruderal plants, ruderal vegetation in urban and industrial areas, are numerous, and multiple compensate some negative effects such as sources of some allergen plants, reservoirs of weeds, as well reservoirs of some virus diseases which can be transferred to cultivated crops (Hadač 1978; Jovanović 1994). The most important positive feature of ruderal plants is manifested in the fact that they are able to quickly colonize open soil and prevent or mitigate erosion processes. At the same time, ruderals produce oxygen and biomass, absorb the carbon dioxide, and many of them accumulate heavy metals as well as other toxic elements whose concentrations in cities and various technogenic areas are usually above the permitted values (Jovanović 1994). Also, ruderals are mostly nitrophilous and capture a considerable amount of nitrates from soil and water. This is a valuable process due to its contribution to the inhibition of eutrophication in the water basins (Hadač 1978). Ruderal vegetation also contributes to urban ecosystem services and enhances the well-being of citizens (Maller et al. 2009; Bratman et al. 2012).

2.3 Annual Ruderal Plants

Many research studies accessed the presence of ruderal plants on the anthropogenically disturbed lands, especially contaminated urban, industrial and mining sites (Liu et al. 2014; Ranđelović et al. 2014; Salinitro et al. 2019). They are usually recognized as an early successional stage in vegetation colonization of these degraded lands. However, successional directions of early, ruderal-dominated stages may considerably vary. Investigations of Prach et al. (2001) on spontaneous vegetation succession in various human-altered habitats showed that in agricultural, industrial or urban landscapes under anthropogenic influence succession starts with ruderal annuals, being followed by ruderal perennials. Although the ability of certain ruderal species developed on mine-influenced sites to arrest the succession process was recognized (Prach and Řehounková 2006; Zajac and Zarzycki 2012), a long-term monitoring of contaminated mine waste fluvial deposits showed that some key pioneer species (many of which belong to the ruderal species) also act as promoters of colonization processes (Nikolić 2020). Tamakhina et al. (2019) found that on the tailing dumps of tungsten molybdenum plant primary succession develops toward gradual replacement of ruderal plants with natural flora, followed by the increase of species diversity. Moreover, research by Chan et al. (2015) stated that significant decrease of the barren ground areas was enabled due to enhanced colonization of ruderal plants on mine waste dumps in Germany. Additionally, Mudrák et al. (2016) assumed, based on their research, that ruderal plant *Calamagrostis epigejos* may facilitate other plants during high-stress conditions in early successional phases, while later, upon stabilization of conditions and development of more competitive interactions, it may arrest the later succession phase. Not only are ruderal plants able to easily colonize open, degraded areas of land, but they can also be utilized as a biological signal that the degraded land is able to recover. Moreover, investigation of the various contaminated sites and their flora recognized the capacity of certain ruderal species to accumulate pollutants, revealing their potential to be used in phytoremediation technologies.

Early-stage colonization of degraded lands is often characterized by increased presence of annual ruderal species (Munoz et al. 2016). Early successional traits of these first colonizers (such as fast establishment, annual life cycle, seed reproduction, intensive seed production and rapid nutrient uptake) are important for initial stabilization of ground material, decrease of the erosion processes and the increase of the soil organic matter content. Some of these species are also able to immobilize bioavailable fractions of contaminants in the environments they thrive, making them less available for spreading in surrounding soils and waters (Ranđelović et al. 2019).

Ruderal species *Poa annua* L., a widespread annual weed grass, was found to be suitable indicator of Ni soil pollution, exhibiting significant correlation with soil bioavailable Ni pool and content of Ni in aerial plant parts (Salinitro et al. 2019). Similarly, increased concentrations of Pb were found by Tamas and Kovacs (2005) in shoots and roots of *P. annua* on Pb/Zn mining site in Hungary (273 ± 172 mg kg^{-1} and 497 ± 613 mg kg^{-1}, respectively). Experiments with *P. annua* for mobilization of Hg from contaminated industrial soils by using various mobilizing agents showed that their addition promoted the uptake of Hg in aerial part of *P. annua* several hundred times in comparison with the control (up to 380 mg kg^{-1} in shoots), simultaneously increasing transfer of Hg from roots to shoots and increasing the potential of plant for phytoextraction of this element (Pedron et al. 2011).

Chenopodium album L. is annual ruderal characterized with fast growth, large biomass, drought tolerance and universal adaptability (including the extremely harsh environments), that contributed to its worldwide distribution. Therefore, it has become the species of interest in studies for environmental remediation purposes (Hu et al. 2012). Zulfiqar et al. (2012) noticed that *C. album* could accumulate significant concentrations of Cd in its shoots in comparison with *Chenopodium murale* L. and found that *C. album* may act as efficient phytoremediator of marginally Cd-contaminated soils. An initial study by Liang et al. (2016) on pioneer vegetation of metal-smelting zone in Hebei, China, showed that overall ability of metal accumulation was highest in *C. album*. Gupta and Sinha (2008) found that *C. album* thriving on the fly ash dykes of the thermal power plant in Uttar Pradesh (India) had high bioaccumulation factors (BAC) for elements Cd (30.9) and Pb (26.1) from soil DTPA-extractable fraction to the roots. However, as the species showed relatively low transfer factors from root to shoot (TF) for both Cd (0.72) and Pb (1.45), authors noticed that the species may potentially be used in phytostabilization of fly ash dykes. The same authors conducted research on *C. album* grown on soil amended with tannery sludge, showing that the species is able to accumulate high quantities of Cr in the aerial parts of the plant (up to 272.6 mg kg^{-1} in leaves, when grown on soil amended with 10% of tannery sludge), and as such could be used for phytoextraction of Cr from tannery waste contaminated sites (Gupta and Sinha 2007). Research of Ebrahimi (2016) showed that the assisted phytoremediation (by using ethylenediaminetetraacetic acid (EDTA) synthetic chelate) enhanced capacity of *C. album* for uptake of Pb from contaminated soils in its shoots, by using multiply doses of EDTA and thus reducing risk of contaminant leaching.

Arabidopsis thaliana (L.) Heynh. is widespread annual plant regularly found on disturbed habitats. Although it is considered to be a typical ruderal species, *A. thaliana* has a genotypic variation that exhibits the whole stress tolerance ruderal strategy spectrum in relation to local adaptation of the plant to the different environments (Takou et al. 2019). Moreover, *A. thaliana* is known as model organism for various ranges of research, including genomic studies. Analyses of *Arabidopsis* genome have significantly contributed toward understanding of sequestration and detoxification of contaminants in plants, while around 700 genes found in *A. thaliana* encode proteins which have the capacity to contribute the phytoremediation process (Cobbett and Meagher 2002).

Increased concentration of Zn, Pb, Cd and Cu were found in leaves of wild *A. thaliana* (201 mg kg^{-1}, 52.1 mg kg^{-1}, 0.89 mg kg^{-1} and 16.6 mg kg^{-1}, respectively) in urban soils of Botanical Garden Park in St. Peterburgh, Russia. Results showed increased concentrations of metal in comparison with background values (Drozdova et al. 2019). *A. thaliana* tested on soils from mine tailing sites in the northeast of China fortified with Cd, Pb and Mn confirmed increased level of accumulated metals in plant with increasing metal content in soil, as well as the limited effect of Cd, Pb and Mn on plant growth and physiology indicating a strong metal-tolerance ability of this species (Zhang et al. 2018). In contrast with its wild type, whose tolerance for metals has been documented, various modifications of *A. thaliana* genome gave more valuable insights in enhanced metal accumulation and potential application of this species for phytoremediation purposes. Dominguez-Solis et al. (2004) created *A. thaliana* lines of various capabilities to provide cysteine under metal-stress conditions. Increased availability of cysteine has enabled growth of *A. thaliana* under high concentration of Cd in growth medium (up to 400 μM of Cd), where significant concentration of Cd was accumulated in leaves (up to 670 mg kg^{-1}), thereby showing no visible signs of phytotoxicity. By applying this approach, accumulation of Cd in previously non-accumulating species was markedly enhanced, revealing the potential of transgenic plants for phytoremediation. Similarly, transgenic *A. thaliana* was created in order to express bacterial MerE gene responsible for methylmercury accumulation. Engineered *A. thaliana* was able to transport and accumulate significantly higher content of methylmercury in comparison with non-engineered, wild plant, showing potential of this method to be used in phytoremediation of methylmercury pollution (Sone et al. 2013). Application of plant transgenic approach in remediation of mercury contamination by using *A. thaliana* engineered with bacterial mercuric reductase and organomercurial lyase (merA and merB) genes resulted in conversion of methylmercury and Hg(II) to much less toxic Hg(0) state that was able to be released from leaves in volatile form (Heaton et al. 1998). Moreover, transgenic plants were able to volatilize 3 to 4 times more Hg(0) from plant tissue than their wild types, indicating potential of using genetic engineering to develop plants with enhanced capacity for use in phytovolatilization. Opposite to that, when gene from *Escherichia coli* named *ZntA*, responsible for encoding a Pb(II), Cd(II) and Zn(II) pump was applied to *A. thaliana*, resistance of transgenic plants for Pb (II) and Cd (II) was noticed (Lee et al. 2003). Results showed decreased content of Pb and Cd in transgenic *A. thaliana* in comparison with wild type, demonstrating use of bacterial genes for developing plants with reduced uptake of heavy metals.

Cosmopolitan ruderal species *Polygonum aviculare* L. is adaptable to a range of man-made habitats that differ by the degree of disturbance. Research by Vasilyeva et al. (2019) showed that on various urban soils of Orenburg city, Russia, *P. aviculare* actively absorbs Pb and Cd, where measured concentrations in plants significantly correlate with total content of Pb and to the water-available content of Cd. Similarly, Polechońska et al. (2013) showed that *P. aviculare* that inhabits urban sites along the highway accumulates increased concentration of Cd, Cu, Pb and Zn in shoots, while significantly greater accumulation of Cd and Pb occurs in roots of this plant, suggesting limited mobility and sequestration in roots as a mechanism of metal tolerance. During assessment of metal accumulation in wild plants surrounding Ag, Au and Zn mining areas at the Zacatecas, Mexico, González and González-Chávez (2006) discovered that *P. aviculare* accumulated 9236 mg kg^{-1} of Zn in its aerial parts (value that is near the threshold for hyperaccumulator plant) at mine waste site with lower pH, while it accumulated somewhat lower concentrations of 925 mg kg^{-1} in aerial parts at slightly alkaline slag heap site. Similarly, *P. aviculare* growing on a multi-metal contamination industrial site in Italy accumulated 3.5–11.5 mg kg^{-1}

Hg in roots and 0.3–10.5 mg kg^{-1} Hg in shoots, while accumulation factors (BCF) for Hg varied from 64.6 to 2032, showing the capacity for hyperaccumulation of Hg (Massa et al. 2010).

Fast-growing annual *Solanum nigrum* L. is native to Europe and Asia and was introduced to America, Australia and South Africa. It is highly adapted to a broad range of habitats and environmental conditions including wastelands and industrial sites, where it shows enhanced tolerance for elevated content of environmental contaminants (Rehman et al. 2017). Investigation of Liu et al. (2014) in Mn mining area of Guangxi, South China, showed that *S. nigrum* exhibits increased abilities for metal accumulation, especially of Mn and Cd, and can be potentially used in remediation of metal-contaminated soils. Wei et al. (2005) found that *S. nigrum* exhibits hyperaccumulating abilities in case of Cd, as in concentration gradient experiments its leaves and stems accumulated 124.6 mg/kg and 103.8 mg/kg of Cd (respectively) from soil spiked with 25 mg/kg Cd. Moreover, transfer factor (TF) from root to shoot was >1, satisfying one of the criteria for successful accumulation of metals.

This species was subject of different research that included enhanced remediation by application of various amendments. Gao et al. (2012) studied co-application of chelators (citric acid) and metal resistant Microorganisms (*Paecilomyces lilacinus* and *Hypocrea virens*) on phytoextraction and growth ability of *S. nigrum* in the presence of Cd and Pb soil contamination. Results showed that co-application of chelates, and metal resistant strains exhibit synergistic effect that improved plant biomass (up to 50%) and enhanced Cd accumulation (up to 35%) in plant. Marques et al. (2008) studied application of EDDS and EDTA chelates for promoted Zn accumulation by *S. nigrum* inoculated with arbuscular mycorrhizal fungi (AMF). The addition of EDTA increased concentration of accumulated Zn up to 231%, 93% and 81% in the leaves, stems and roots of *S. nigra*, respectively, while the application of EDDS increased the accumulation in leaves, stems and roots up to 140, 124 and 104%, respectively. Application of synthetic chelate EDTA, in association with AMF, appears to be the treatment able to enhance concentration of Zn taken up from the soil by *S. nigrum*, thus markedly reducing the time required for successful remediation of Zn-contaminated soils. Similarly, Li et al. (2019) investigated technique for multi-metal-contaminated mine tailings by using *S. nigrum* and biochar/attapulgite as soil amendments. Uptake of metals in plant roots was significantly increased with the addition of amendments, suggesting the enhancement of phytostabilization process. With the application of soil amendments, the removal rates of metals were significantly increased by 29.6–148% in the 10% attapulgite amendment treatment and 9.69–20.8% in the 10% biochar amendment treatment in comparison with the control. Moreover, role of *S. nigrum* was additionally recognized in remediation of organic xenobiotics from soil, where *S. nigrum* hairy root clone SNC-9O showed capability to efficiently degrade polychlorinated biphenyls (PCBs). Namely, after 30 days of incubation the residual PCBs were 40% in comparison with the controls (Mackova et al. 1997).

2.4 Ruderal Perennials

Ruderal perennials could usually be found accompanied with annual forms or as one of the later stages in early and middle-succession of disturbed lands. Vegetation types where ruderal perennials are commonly present are prone to repeated stresses appearing each year so more competitive perennial species cannot be established. For example, in certain habitats where the effect of disturbance is causing production of discontinuous vegetative cover, species of ruderal perennial type can provide new spatial distribution of shoots each growing season, therefore representing efficient colonizers of these temporary gaps (Grime 2001). Additionally, studies on perennial plants in stressful environments show that certain stress-tolerant traits may evolve in order for adjustment to be made between resources allocated for reproduction in one season and storage for survival and growth within the next season (Stanton et al. 2000). The largest number of perennials in ruderal flora and vegetation belongs to the life form of hemicryptophytes, whereas the geophytes and hamephytes are much less represented (Jovanović 1994; Jovanović et al. 2013; Rat et al. 2017).

One of the species commonly found on varieties of anthropogenically degraded sites is *Calamagrostis epigejos* (L.) Roth., rhizomatous perennial native to Europe, but also distributed in Africa, Asia and North America. It tolerates wide range of conditions, such as those of low-nutrient or high-nutrient ability, dry to moisture sites or open to shady places. This species is considered to be a strong competitor and successful colonizer of disturbed sites, due to its growth and nutrient-use strategy.

Investigations by Mudrák et al. (2010) revealed that *C. epigejos* is a dominant species of post-mining sites over Central Europe in the early stages of succession. Similarly, Prach and Pyšek (1994) noticed that clonal plants such as *C. epigejos* represent the dominant component of the vegetation cover during initial 10 years of succession within human-made habitats in Central Europe. It is recognized that in some cases it can arrest the succession on lignite post-mining sites (Baasch et al. 2011). However, it is considered that in extreme environmental conditions this species may provide positive conditions for establishment of other plants due to its organic litter that provides nutrients and favorable microclimate conditions for plant development (Štefanek 2015). Investigations of Ranđelović et al. (2018) on metal accumulation in *C. epigejos* from different types of anthropogenically degraded sites indicated that the species uptakes a significant portion of the available fraction of heavy metals (Fe, Mn, Cd, Cu, Pb and Zn) from the soil and stores it in the roots, showing a potential for metal phytostabilization. However, as the proportion of available metal fraction was relatively small in comparison with total content of metals in investigated soils, *C. epigejos* could not be recommended as single remediation option. Similarly, *C. epigejos* colonizing fly ash deposit site of thermoelectric plant in Obrenovac, Serbia, showed pronounced tolerance to metals and metalloids, as well as low root-to-shoot transfer factors for As, Cu, Zn and Mo, according to Mitrović et al. (2008). Additionally, Pietrzykowski and Likus-Cieslik (2018) indicated that *C. epigejos* is also tolerant to high S concentrations in the soil. Existent of exclusion mechanism as metal-tolerance strategy of this species was confirmed by investigation of Lehmann and Rebele (2004) that accessed the potential of *C. epigejos* for phytoremediation of Cd-contaminated soils. Authors found that, during three growing seasons, root-to-leaf translocation factor showed no significant relation with increase of soil Cd contamination. Only 4–7% out of total plant content of Cd was allocated to aerial organs of the plant. Having in mind widespread colonization of anthropogenically devastated sites such as areas of mining wastes, areas near the roadsides or neglected agricultural areas, *C. epigejos* could efficiently contribute to the contaminant immobilization and their naturally assisted remediation.

As application of amendments is one of the means for enhanced phytoremediation, Bert et al. (2012) applied Thomas Basic Slag (5% dry weight) soil amendment to a sediment contaminated with Zn, Cd, Cu, Pb and As and planted with *C. epigejos*. Results showed efficacy of the treatment, as Cd and Zn concentration in shoots of *C. epigejos* were decreased, as well as the $Ca(NO_3)_2$- extractable Cd and Zn fractions in amended sediment. Nowińska et al. (2012) concluded that *C. epigejos* can also be used as a potential energy crop due to its high calorific value and suitable chemical composition for unhampered combusting process. Linking phytostabilization ability of this species with its potential use as energy crop could open a new direction for utilization of *C. epigeios* in remediation of degraded lands (Ranđelović et al. 2020a).

Tussilago farfara L. (Fig. 2.1) is ruderal rhizomatous plant native to Europe, western Asia and north Africa, but nowadays spreaded worldwide. It is known to colonize various habitats including waste lands and roadsides. This plant is often used for medicinal purposes, especially for relieving inflammatory conditions and

Fig. 2.1 *Tussilago farfara* L. on urban ruderal site

Fig. 2.2 *T. farfara* colonizing slag dump of copper mining industrial area (Bor, Serbia)

infectious diseases (Chen et al. 2020). Sometimes it appears as a pioneer species of various degraded sites (Fig. 2.2).

Jakovljević et al. (2020) examined accumulation of trace elements (Fe, Al, Pb, Zn, Cu, Cd, Mn, As, Sb, Ag, Ti and Sr) in *T. farfara* colonizing various postflotation tailing sites associated with Pb–Zn–Ag–Sb–W metallogeny in Serbia. Results suggested that two strategies of metal accumulation could exist in *T. farfara*: the sequestration of toxic elements in roots, and detoxification through leaf loss. Despite the active absorption of Pb, Zn, Cu, Cd, As, Sb, concentrations in shoots of *T. farfara* was below the hyperaccumulation thresholds, suggesting limited capability of this species to be used for phytoextraction. Occurrence of *T. farfara* on range of As-contaminated soils showed that this species could be used for their remediation, as it showed increased uptake and immobilization of As in roots in comparison with the shoots (Chyck et al. 2012). Contrary to that, Wechtler et al. (2019) found that on industrial brownfield site characterized with high content of Ni, Pb and Zn *T. farfara* was able to accumulate high concentrations of Zn in its shoots (401.25 ± 107.15 mg kg^{-1}). Soil-to-shoot bioconcentration factor for Zn was 3.069, suggesting that *T. farfara* is promising species for accumulation of Zn from contaminated soils. Similarly, Gałuszka et al. (2020) recorded extremely increased concentration of S (up to 4.17%) in the aboveground parts of *T. farfara* from acid mine drainage areas in Poland. Investigations of Robinson et al. (2008) showed that, on shooting range soil contaminated with Pb (concentrations ranging 111,000–158,000 mg kg^{-1}), *T. farfara* was able to hyperaccumulate 1100–2280 mg kg^{-1} of Pb in its leaves. However, at lower Pb concentrations, this plant species did not exhibit hyperaccumulating abilities, but rather just increased toxic concentrations of Pb in its leaves. Therefore, passive lead absorption mechanism of *T. farfara* could eventually be utilized in remediation of highly Pb-contaminated soils, while at the same time, there should be awareness of risk from this exposure pathway of Pb from entering the food chains. Therefore, due to the tendency of plant to accumulate certain contaminants, use of *T. farfara* originating from contaminated sites for medicinal purposes could additionally pose a risk to human health and should be avoided.

Fig. 2.3 *Plantago major* L. developing on asbestos mining dump site (Stragari, Serbia)

Plantago major L. is synanthropic ruderal species distributed worldwide from native Eurasia range. The species is associated with various human activities and able to tolerate range of climate and soil types, included degraded ones (Fig. 2.3). Together with the ruderal annuals such as *P. aviculare* and *C. album*, *Plantago major* is one of the most frequent perennial species of urban ruderal habitats in Europe, especially ones affected by trampling (Jovanović 1994). Plant is also well known for its healing properties, and it is often used for treatment of injuries and skin diseases, respiratory diseases, digestion problems, blood circulation, etc. (Samuelsen 2000).

Investigations by Malizia et al. (2012) on sites with different level of anthropogenic pollution in Rome, Italy, showed that *P. major* consistently accumulates higher concentration of Cu, Mn, Zn, Pb, Cr and Pd in roots than in shoots. Additionally, results also revealed seasonal variation in metal concentration of *P. major*. Bekteshi and Bara (2013) conducted research on uptake of heavy metals in *P. major* from various contaminated locations in city of Durres, Albania, where good correlation between soil and leaf content of Cu, Zn, Mn and Ni was observed. Species showed ability to accumulate Pb from contaminated soil and water in its roots. Results of Romeh et al. (2015) showed that after 25 days roots of *P. major* accumulated 9284.66 mg kg^{-1} of Pb, compared to only 25.29 mg kg^{-1} in leaves, displaying medium-to-root BCF factor of 380.93 and clear rhizofiltration ability. In Pb-contaminated soils after 20 days roots and shoots accumulated 77.12 mg kg^{-1} and 30.4 mg kg^{-1}, respectively, showing the potential for phytostabilization of this contaminant. Investigations of Filipović-Trajković et al. (2012) on plants inhabiting industrial area of Pb/Zn smelting plant showed that *P. major* accumulated 660 mg kg^{-1} of Pb, 2500 mg kg^{-1} of Zn and 33.25 mg kg^{-1} of Cd in its roots, while considerably lower concentrations were accumulated in its leaves and fruits, confirming the plant's capacity for phytostabilization.

The ability of *P. major* to metabolize diverse organic pesticides in the phytoremediation processes was additionally recognized. In the investigation of the phytoremediation of soil contaminated with imidacoprid conducted by Romeh (2020), *P. major* was inoculated with effective microorganisms (EM1) and applied to the peat moss-amended soil in laboratory condition and soil in field condition. Concentration of imidacloprid in *P. major* roots at the end of experiment was 48.61 mg kg^{-1}, while in the P. major roots amended with EM1 plus peat moss was 21.25 mg kg^{-1}. However, *P. major* inoculated with effective microorganisms (EM1) with addition of peat moss showed as the most effective combination in enhancing the degradation rate of imidacloprid from soil. This suggests an interaction of plant roots where EM1 increases the degradation process and peat moss stimulates enumeration of microorganisms. Similarly, Romeh (2016) determined efficiency of 0.05% *Rumex dentatus* L. leaves extract in enhancing phytoremediation of *P. major* on soils contaminated with carbosulfan insecticide. This combination showed significant effect on uptake of carbosulfan into plant roots (1.31 to 2.39-fold in first 4 days of experiment in comparison with *P. major* alone), where it was rapidly degraded to carbofuran. Carbofuran translocated into the leaves of *P. major* amended with leaves extract of *R. dentatus* reached the maximum (10.43 mg kg^{-1}) after two weeks of exposure in comparison with *P. major* alone. Results showed that phytoremediation of carbosulfan from the contaminated soil can be enhanced by using *P. major* amended with *R. dentatus* leaves extract. Surfactant-enhanced phytoremediation of soils contaminated with systemic fungicide azoxystrobin was evaluated using *P. major* by Romeh (2015). Surfactant Tween 80 was used for enhancing the availability and uptake of azoxystrobin by *P. major*. The results showed synergistic effect on translocation and uptake of azoxystrobin in roots and leaves (25.72 mg/kg and 18.0 mg/kg, respectively) in comparison with *P. major* alone (20.62 mg kg^{-1} and 15.03 mg kg^{-1} in roots and leaves, respectively), suggesting that Tween 80 could be convenient agent for enhanced phytoremediation of fungicide.

Verbascum thapsus L. is widespread perennial plant, originated from Euroasia and northern Africa and introduced in temperate areas worldwide. It is often found as colonizer of coarse grounds, bare and disturbed soils (Fig. 2.4), on wastelands, clearing sites or near the roads. It is also well known for its use in treatment of various medical conditions, such as pulmonary diseases, heart diseases, neuralgia or topical injuries (Riaz et al. 2013).

Investigations of Turnau et al. (2010) on wild plants grown on mixture of soil and Zn–Pb industrial waste showed that *V. thapsus* was able to accumulate from 5370 to 8780 mg kg^{-1} of Zn and 1140–2440 mg kg^{-1} Pb in leaves, showing potential for phytoaccumulation of these elements. Despite of present non-favorable conditions, plant was able to produce highly vital seeds for further reproduction. Sasmaz et al. (2016) found that *V. thapsus* accumulates high rates of Tl in its roots and shoots (up to 2979 mg kg^{-1} and 1879 mg kg^{-1}, respectively) on Ag mine deposit sites, showing preferable Tl deposition in roots at their higher concentrations in deposits. Similarly, Sasmaz et al. (2015) revealed that on polymetallic ore deposits in the

Fig. 2.4 *Verbascum thapsus* L. growing on industrial copper smelter site (Bor, Serbia)

Gumuskoy mining area, Turkey, *V. thapsus* was able to absorb Hg from polluted soil, indicating the ability to transfer it from roots to shoots and showing potential for phytoremediation of such sites. Čudić et al. (2016) conducted five-year research on wild plants inhabiting highly contaminated landfill of a zinc processing factory in Šabac, Serbia. Measured accumulation of metal(loid)s in *V. thapsus* was initially high and afterward was descending during years two to five (e.g., concentration of metals in aerial parts of plant was decreasing from 1840.8 to 138.4 mg kg^{-1} for Pb, from 141.9 to 0.1 mg kg^{-1} for Cd and 7807.3 to 474.2 mg kg^{-1} for Zn), while measured concentration of elements followed the order Zn > Pb > Cu > Cd > As > Ni > Cr. However, *V. thapsus* showed increased translocation of Pb, Cu, Ni, Cr and As in aerial parts, showing potential for accumulation of these elements. Moreover, this research accessed potential use of plants biomass as an energy source. *V. thapsus* gross calorific value was 19,735 kJ kg^{-1}, that, coupled with low ash content, makes this plant potentially useful for biomass energy conversion process. As *V. thapsus* showed increased accumulation of Pb and Zn in aboveground parts, there is potential for additional metal recovery from fly ash remained from the combustion process, as a way for biomass utilization after phytoextraction.

2.5 Woody Ruderals

Woody life forms as ruderals are not so typical for disturbed habitats. Thus, they can be considered only as an optional or facultative category of ruderal plants. Namely, in habitats with a pronounced anthropogenic influence, ruderal phanerophytes are mostly present singly or in the form of a small number of seedlings, juveniles and other lower age categories (Jovanović 1994). This is especially the case in the early succession stages of vegetation development with a dominant presence of annual ruderal colonizers. However, in the middle- and later-succession stages of disturbed lands, the share of woody species gradually increases together with perennial ruderals, but still considerably less (Prach and Pyšek 1994; Prach et al. 2001). The share of phanerophytes in ruderal flora and vegetation is generally low, especially in the case of native tree and shrub species (Rat et al. 2017). Mainly, they are a species of different ecology (from hygrophytes to xerophytes), which are associated with resistance to various pollutions in the air, soil or water in urban or ruderal habitats prone to human influence (e.g., certain species of the genera *Populus, Betula, Sambucus, Acer, Pistacia, Prunus, Crataegus*, etc.). Some of them have been investigated in more detail for potential use in phytoremediation.

Pistacia lentiscus L. is small evergreen tree or shrub native to Mediterranean region of southern Europe, northern Africa and western Asia. It grows in dry and rocky areas, open woods, coastlines and along roadsides. It is known as pioneer species and can be often found on man-made habitats. The plant is sometimes harvested from the wild, as its resin, known as mastic, owns a wide range of medical and culinary uses.

Metal tolerance of *P. lentiscus* was found in research of Concas et al. (2015), who accessed the content of Pb and Zn in plants growing on Pb/Zn mining site in Italy. The soil-to-plant concentration factors were consistently low (≤ 0.05), revealing the exclusion strategy of *P. lentiscus*. Metal contents in roots and shoots showed a rough relationship to respective soil contents, showing ability of plant to be an indicator. Moreover, on pyritic mine soils shoots of *P. lentiscus* showed a positive significant correlation with Cu and Zn exchangeable soil fraction ($r = 0.77$ and $r = 0.72$, for Cu and Zn, respectively), according to Parra et al. (2016). Performances of *P. lentiscus* on amended Pb/Zn mine site showed that the root-to-leaves transfer remained lower than 1 (on average 0.56–0.88 for Pb and 0.54–0.74 for Zn in the different plots), while bioavailability of soil contaminants Pb and Zn was decreased due to application of various soil amendments (Bacchetta et al. 2012). Additionally, while accumulation rate of metals in plant grown on mine polluted soils consistently followed the order Zn > Pb > Zn regardless of metal bioavailability in soil, survival of *P. lentiscus* after 6 month on contaminated soil showed high rate (77–100%) in comparison with the rate of *Phragmites australis* (25–58%), as stated by Bacchetta et al. (2015). Overall research data generally confirm the ability of *P. lentiscus* for revegetation and phytostabilization of metal-contaminated areas of Mediterranean region, due to its resistance to metals and preserved biomass production.

Fig. 2.5 *Populus* sp. colonizing degraded coal separation basin site (Piskanja, Serbia)

Populus nigra L. is facultative ruderal deciduous tree distributed throughout Europe, northern Africa and in central and west Asia, predominantly in riparian woodlands. It is considered as one of the key riparian species, as it is able to tolerate flood disturbance and high water levels. However, it is currently considered close to extinction in certain parts of its European range, due to habitat degradation and disturbed genetic integrity from intercrossing with other poplar species and cultivated hybrids (Jelić et al. 2015). Generally, poplars, their varieties and cultivars are known for the abilities to remove contaminants from soil and water. Extensive root system, high water transpiration and fast growth make them as species of choice for application of phytoremediation technologies on various categories of degraded sites (Fig. 2.5).

Populus nigra was thoroughly tested to tolerance and accumulation of various elements. Uptake of heavy metals (Cd, Mn, No, Pb and Zn) in *P. nigra* in pot experiment with artificially polluted chernozem soil was studied by Biró and Takács (2007). After 6–8 months, Pb and Ni were found to be preferably accumulated in roots, while Cd, Mn and Zn had higher concentrations in leaves of *P. nigra*. Within 8 months, content of Cd and Zn in poplar exhibited 5–6% of their available content in soil. Similarly, El-Mahrouk et al. (2020) tested phytoremediation ability of *P. nigra* in pot experiment with various concentrations of Cd, Cu and Pb. They found these elements predominantly localized in roots in comparison with leaves and steams, coupled with bioconcentration factor (BCF) soil-to-root of <1. Moreover, 1, 5-4 fold increase in transfer factor (TF) was noticed for Cd and Pb with increase of their concentration in soil. Vuksanović et al. (2017) examined five genotypes of *P. nigra* in order to select those with preferable copper tolerance and accumulation for phytoremediation purposes. All the examined genotypes differed in their copper tolerance and accumulation, whereas *P. nigra* genotype DN3 showed the best performances based on morphological parameters, biomass accumulation and photosynthetic pigment contents, narrowing the selection of candidate genotypes for copper phytoextraction purposes. Investigations of Iori et al. (2015) reveal genetic architecture of Cd accumulation and tolerance in *P. nigra*, mapping certain candidate genes related to Cd transport and detoxification. The ability of plant to accumulate Cd in roots and restrict its translocation to the aboveground organs as an adaptive response for coping with increased level of Cd pollution in long-term studies of Jakovljević et al. (2014) was additionally recognized.

Poplar hybrids were recognized for their enhanced remediation abilities of both inorganic and organic contaminants. Potential of five hybrid poplars for phytoremediation of Cu-contaminated soils accessed by Cornejo et al. (2017) showed that genotype *P. deltoides* × *P. nigra* had highest Cu accumulation capacity (1321.1 ± 108.1 mg kg^{-1}, 169.8% more than average accumulation of other tested genotypes) in its roots, showing the potential for phytostabilization of Cu-contaminated land. Similarly, *Populus deltoides* × *P. nigra* I-214 clone was tested in greenhouse on two ranges of Zn-contaminated soils, containing 13.10 and 131 mg kg^{-1} of extractable Zn. Clone I-214 had ecophysiological responses to increased Zn concentrations, including modifications of leaf area, decrease of Chlb content, increased concentration of Zn in older leaves, etc., and revealed its potential for remediation of Zn-contaminated sites (Di Baccio et al. 2003). Hybrid poplar *P. deltoides* × *P. nigra* DN-34 was studied for hydroponics and soil removal of organic contaminant 1,4-dioxane by Aitchison et al. (2000). After 15 days, 18.8 ± 7.9% of the initial dioxane concentration remained in soil, in comparison with 72.0 ± 7.7% remaining in unplanted soil. Within 9 days, more than 50% of initial dioxane concentration was removed from hydroponics. Total of 76 and 83% of the dioxane from hydroponic and soil experiment, respectively, was taken by the poplar clone and transpired in atmosphere, showing a great potential of this hybrid for phytovolatilization. Similarly, Doty et al. (2017) studied endophyte-assisted phytoremediation of trichloroethylene (TCE) by hybrid poplars. Endophyte strain *Enterobacter* sp. PDN3 was isolated from *P. deltoides* × *P. nigra* and used for inoculation of hybrid poplars in remediation experiment. The inoculated poplar trees exhibited increased growth and reduced phytotoxic effects, while at the same time, they excreted 50% more chloride ion into the rhizosphere in comparison with the control, which indicated increased TCE metabolism in plants. The combination of native pollutant-degrading endophytic bacteria and fast-growing poplar trees may be acceptable solution for phytoremediation of certain organic chemicals. However, matching appropriate genotypes with contaminants could help to support the ecosystem services resulted from phytoremediation activity, while phytoremediation success may be increased by using genotypes that tolerate wide range and type of contaminants (known as generalists), e.g., *P. deltoides* × *P. nigra* DN 34 or *P. nigra* × *P. maximowiczi* NM6 (Zalesny et al. 2019). When comparing these genotypes, clone DN34 generally showed higher elemental pollutant concentrations, but NM60 exhibited 3.4 times greater biomass productivity and carbon storage than DN34, and its stand-level annual uptake was 28–657% greater than on DN34, eventually pointing out the phytoremediation superiority of clone NM60 (Zalesny et al. 2020).

Betula pendula Roth. is a fast-growing deciduous tree distributed throughout Eurasia and introduced to North America and Canada. This pioneer species has the ability to grow on diverse ahthropogenic, preferably sandy sites, where it plays important role in colonization process (Rebele 1992). *B. pendula* can be regularly found as ruderal species of various barren soils, industrial wastelands and mining dumps (Fig. 2.6).

Fig. 2.6 Afforestation test plot with *Betula pendula* Roth. on copper mine waste piles (Bor, Serbia)

Different studies recognized *B. pendula* as good accumulator of trace elements in its leaves, finding it a good monitoring species of environmental pollutants. Investigation of element content in leaves of *B. pendula* from Pb–Zn mine flotation tailing dumps showed increased accumulation of Pb and Zn in leaves (up to 69–530 mg kg^{-1} and 1660–3100 mg kg^{-1}, respectively), exhibiting values 8-67-fold (for Pb) and 7-14-fold (for Zn) higher than in control plants growing on unpolluted sites (Marguí et al. 2007). However, authors didn't find significant correlation between pseudo-total and BCR-extractable phases with content of Pb and Zn in leaves of *B. pendula*. Contrary to that, researches of Dmuchowski et al. (2014) found positive correlation of Zn content in soil and Zn content in birch leaves ($r = 0.88$) on metallurgical waste heaps near Warsaw, Poland. While content of Zn in leaves was up to 482 mg kg^{-1} on metallurgical waste comparing to 180 mg kg^{-1} on unpolluted site, Dmuchowski et al. (2013) determined even sharper differences between Zn content in leaves of *B. pendula* from mining metallurgical complex in Bukowno, Poland, and non-contaminated sites (up to 2352 mg kg^{-1} and up to 285 mg kg^{-1}, respectively). Similarly, Pavlović et al. (2017) found that soil chemistry can explain 82.99% of element (B, Cu, Sr and Zn) variability in *B. pendula* leaves and only 27.6% of element variability in bark of trees growing on urban localities with different levels of pollution. When Soudek et al. (2007) studied uptake of ^{226}Ra into the trees growing at a mill tailing dump at a former uranium mill in South Bohemia, Czech Republic, they found that maximal radium activity was obtained in the leaves of birch (0.41 Bq ^{226}Ra/g dry weight) and that concentration ratio soil-to-leaves was 0.084, showing generally low ability for radium uptake. Accumulation of uranium in leaves and twigs of *B. pendula* was also studied in former uranium mining and milling complex in Germany (Brackhage and Gert 2002). *B. pendula* exhibited very low concentration of U in aboveground parts, therefore posing low environmental risk for transfer and accumulation of radionuclides in food chains.

Betula pendula was also tested for efficiency in phytoremediation of soils contaminated with organic pollutants. Sipila et al. (2008) showed that *B. pendula* clone Wales

W008 was able to enhance polyaromatic hydrocarbon (PAH) degradation on soils with higher PAH contamination (300 mg kg^{-1}), while on soils with lower PAH level (50 mg kg^{-1}) this clone did not enhance the degradation process. Various levels of root exudates noted at high and low PAH levels could be responsible for such concentration dependence of birch phytoremediation effect. Authors hypotethized that this could be the effect of diversification of aerobic aromatic ring-cleavage bacterial populations in rhizosphere-associated soil of B. *pendula*. Similarly, performance of *B. pendula* and *B. pubescens* clones and associated rhizoidal bacteria was tested for degradation of PAH from metal-contaminated soils of different sand: peat ratio (Tervahauta et al. 2009). PAH degradation significantly differed between the two clones and soil type and showed the highest values for sandy soil in the presence of birches. Hence, *Betula* ssp. can be employed as additional tool in remediation of organic contaminants from polluted sandy soils.

2.6 Invasive Ruderal Plants

Due to the features typical for ruderal strategy (such as fast growth, high reproductive rate, seed dispersal ability and adaptation to a wide range of environmental conditions), invasive species own competitive advantage over native ones in the absence of natural enemies and specific herbivores from their native range, and may outcompete native species in the environments they are introduced to (Sakai et al. 2001). Moreover, high-stressed habitats on various degraded lands represent favorable areas for introduction and spreading of invasive species (Rendeková et al. 2019). As many non-native plants are applying ruderal strategy, numerous empirical evidence supports the idea that disturbed habitats have more invasive species (Crooks and Suarez 2006; Chiuffo et al. 2018). Research conducted alongside of roads revealed that the number of invasive species decreases with the distance from the road (Tyser and Worley 1992; Gelbard and Belnap 2003). As mainly pioneer species that characterize the early stages of succession on disturbed habitats, invasive species grow best in conditions of minimal competition. However, the share of invasive species in the plant community usually decreases with its age, i.e., their number decreases as the succession progresses (Fridley 2011).

Invasive plants can be annual, perennial or woody species that grow fast and spread aggressively, thereby affecting the functioning and properties of invaded habitats (Hejda et al. 2009). Some of the induced changes are impact on the biodiversity, alteration of hydrology and nutrient cycling and change of soil properties. Being characterized with opportunistic traits that enable them rapid colonizing of the large areas, invasive plants could pose considerable risk in face of climate changes, as they have potential to respond to a shifting of niches much faster than some native species (Crossman et al. 2011). Annual invasive species (such as *Ambrosia artemisiifolia, Bromus tectorum, Artemisia annua, Impatiens glandulifera,* etc.) are known to rapidly occupy degraded or bare sites and, due to their rapid seed producing, can even arrest the succession process in its primary phase, halting the development of more complex, perennial plant communities. Additionally, annual invasive plants are more resistant to herbivores than perennials (Vesk et al. 2004) because perennial herbaceous plants have a longer period of vulnerability that precedes reproduction. On the other hand, a common trait of many invasive perennial plants is a coupled vegetative spreading and seed dispersal that allows them increased population growth and recovery rate (Rejmanek and Richardson 1996). Therefore, invasive perennial ruderals (such as *Reynoutria japonica*, *Cirsium arvense*, *Senecio jacobaea*, etc.) are more persistent and could be harder to remove from invaded ecosystem in comparison with annuals. Contrary to annual and perennial invaders as well as to native woody ruderals, non-indigenous invasive woody species (e.g., *Amorpha fruticosa*, *Leucaena leucocephala*, *Ailanthus altissima*, *Acer negundo*, *Robinia pseudoacacia*, etc.) may spontaneously to form own or mixed dense populations, i.e., the alien forest communities, presenting the last stage in vegetation succession on disturbed and polluted habitats. It is a case of succession being deflected toward dominance by the introduced species (Vitousek et al. 1996; Glišić et al. 2014; Batanjski et al. 2015; Titus and Tsuyuzaki 2020). Despite to fact that only 0.5–0.7%, i.e., more than 620 tree and shrub species in the world are invasive (Richardson and Rejmanek 2011), woody plant invasions are rapidly increasing in importance around the world. Regardless of the dangers of invasive tree and shrub species to natural vegetation, the possibility of their use in phytoremediation of disturbed areas is increasingly explored.

Ambrosia artemisiifolia L. is an annual ruderal plant native to Central and Northern America, from where it has been introduced worldwide. In introduced areas, *A. artemisiifolia* colonizes disturbed grounds producing a large number of seeds that can survive in a dormant state for more than 40 years (King 1966). Species is competitive to native plants, and it is considered as highly invasive. Moreover, pollen of *A. artemisiifolia* is important cause of human allergy, evoking reactions such as allergenic rhinitis, asthma or dermatitis. Sustainable methods for control of *A. artemisiifolia* growth and spreading must be undertaken in different parts of its introduction range.

Among other, *A. artemisiifolia* is recognized for tolerating high level of heavy metals in soil. Successful establishment of *A. artemisiifolia* along contaminated roadside edges was found to be in connection with its greater tolerance of heavy metals (Fig. 2.7). Seedling survival rates of *A. artemisiifolia* under Ni and Cu concentrations in soil reaching 100 mg kg^{-1} were 73% and 87%, respectively, in comparison with less than 20% seedling survival rates for *Lotus corniculatus*, *Coronilla varia* and *Trifolium arvense* at Ni and Cu concentrations of 50 mg kg^{-1} in soil (Bae et al. 2016), indicating tolerance to metals even during early growth phases. Ranđelović et al. (2020b) studied accumulation patterns of As, B, Ba, Cd, Co, Cu, Fe, Mn, Mo, Ni, Pb, Sb, Sr and Zn in *A. artemisiifolia* on diverse anthropogenically modified, non-polluted to polluted sites. Trend of shoot accumulation was generally observed for majority of elements, with significant accumulation of B in shoots regardless of its concentration in soil and significant correlation of Ba, Pb and Zn in shoots with their content in soil. Similarly, Chaplygin et al. (2018) found that *A. artemisifolia* predominantly accumulates Cd, Pb, Zn, Cu, Cr, Ni and Mn in aboveground parts

Fig. 2.7 *Ambrosia artemisiifolia* L. on roadside edge

on soils under increased technogenic load. Accumulation of metals in shoots of *A. artemisiifolia* may pose certain environmental risk, as accumulated elements could easily be transferred to other components of the environment. Additionally, Cloutier-Hurteau et al. (2014) found that certain part of accumulated metals could be even disseminated via pollen grains, therefore increasing risk of further contamination of invaded sites.

Since mowing is one of the dominant strategies for the control of this invasive species, certain care should be taken in terms of storage, treatment and utilization of *A. artemisiifolia* biomass. Thermal conversion of metal-accumulating *A. artemisiifolia* to biochar material showed that stable form of potentially toxic elements increased with the increase of pyrolytic temperature and that temperature range 500–600 °C with retention time up to 1 h 30 min and with smaller particle sizes is optimal for producing biochar with low environmental risk (Yousaf et al. 2018). Furthermore, Lian et al. (2020) tested the capacity of Cd(II) and Pb(II) adsorption in aqueous solution with biochar produced from *A. artemisiifolia*. biochar pyrolyzed on temperature of 450 °C showed the highest adsorption capacity for both Cd(II) and Pb(II), with precipitation, ion exchange and complexation with functional groups as main mechanisms of metal adsorption. This research suggests that *A. artemisiifolia* derived biochar has the potential to be applied as cost-effective adsorbent for removing pollutants from wastewater.

Reynoutria japonica Houtt. (syn. *Fallopia japonica* Houtt.) is herbaceous perennial plant native to East Asia and introduced worldwide. Plant is listed on the IUCN list of the Worlds' 100 worst invasive species (Lowe et al. 2000) and certain countries developed control measures and suppression programs regarding this invasive species. It is present in various riparian areas, man-made environments, degraded lands and roadsides. Further distribution of this species is going to be influenced by anthropogenic modification of habitats and climate changes, especially in riparian sites, and specific management plans should be implemented to prevent her spreading, primarily in the areas of high conservation interest (Jovanović et al. 2018), but also on range of ruderal sites.

The ability of *R. japonica* to accumulate heavy metals (such as Pb, Cu, Cd and Cr) in its leaves under field conditions was shown by Rahmonov et al. (2014), while investigation of Sołtysiak (2020) proved that *R. japonica* is tolerant to presence of Cd, Cr, Cu, Pb and Zn in soil, being able to successfully regenerate, grow and develop on soils containing various ranges of these elements. Moreover, in the case of soils polluted with 100 mg kg^{-1} of Cd leaves of *R. japonica* accumulated up to 700 mg kg^{-1} Cd, even 630 times more than the control, showing potential for cadmium phytoaccumulation. Similar ability of *R. japonica* leaves was previously recognized by Hulina and Đumija (1999a) for accumulation of Zn and Cd in urban areas of Croatia. Contrary to that, Berchová-Bímová et al. (2014) found increased concentration of Cd, Fe and Pb in underground parts of invasive *Reynoutria* species. The metal accumulation in underground tissues was most pronounced in the case of Cd. The mean concentration of Cd was 10 times higher in plant underground tissues (11.25 ± 1.013 mg kg^{-1}) than in soils (0.90 ± 0.109 mg kg^{-1}) and 2 times higher in comparison with its content in leaves (5.01 ± 0.463 mg kg^{-1}). Additionally, more invasive hybrid of *Reynoutria japonica* var. *japonica* and *R. sachalinensis* (= *Reynoutria* × *bohemica* Chrtek et Chrtková) was found to accumulate Zn, Cu, Cr, Cd and Pb in its leaves in small concentrations, while Mn and Fe tend to be retained in roots, as a part of metal exclusion strategy on different antropogenically modified sites in Serbia (Hlavati Širka et al. 2016). Although higher ability for accumulation of metals in plant parts indicates certain potential of *R. japonica* for application in phytoremediation technologies, due to high invasiveness of this species, its use in such programs is not considered to be safe environmental solution.

Being an invasive species that requires periodical removal, there is a growing need for adequate utilization of *R. japonica* biomass. Namely, this species was studied by Koutník et al. (2020) as source of activated carbon for removal of xenobiotics (diclofenac and paracetamol) from water. Results showed that carbon adsorbent impregnated with H_3PO_4 exhibited highest adsorption capacity for both diclofenac (87.09 mg g^{-1}) and paracetamol (136.61 mg g^{-1}). Similarly, biomass of *R. japonica* showed potential to be used as biosorbent for removal of Zn^{2+} ions from aqueous solutions (Melčáková and Horvathova 2010). Kinetics of zinc biosorption by inactive biomass was fast; with a biomass concentration of 10 g/l metal was adsorbed within 10 min. The maximum removal efficiency of 99.4% (or the highest sorption capacity for Zn^{2+} being $q_{max} = 17$ mg/g) was achieved using the biomass of leaves.

Amorpha fruticosa L. is fast-growing semi-aquatic deciduous shrub native to central and eastern part of North America, but nowadays introduced across Asia and Europe. It is considered to be among the most invasive alien shrub species of Europe (Radovanović et al. 2017). *A. fruticosa* prefers humidity in both native and invaded range, inhabiting mainly riparian and periodically floodplain habitats regardless of the level of their degradation (Pedashenko et al. 2012). Once introduced, it can easily spread to various disturbed wet areas (Fig. 2.8) where, due to its high reproductive capacity (both by seeds and vegetative) forms dense thickets, outcompetes native flora and changes vegetation structure and successional patterns (Weber 2005).

Fig. 2.8 *Amorpha fruticosa* L. colonizing flotation tailing dam at metalliferous mine area (Bor, Serbia)

Tests of various plants seedlings growth on copper and lead/zinc mine tailings from China showed that *A. fruticosa* exhibited the highest tolerance to metal stress by maintaining normal root development and biomass production in comparison with other plants (Shi et al. 2011). Moreover, *A. fruticosa* accumulated higher concentration of elements in roots than in shoots (e.g., 161.86 ± 57.0 mg kg^{-1} and 38.79 ± 10.9 mg kg^{-1} of Zn, respectively; 60.35 ± 37.8 mg kg^{-1} and 4.65 ± 1.2 mg kg^{-1} of Cu, respectively), showing potential for phytostabilization of investigated mine tailings sites. Similarly, seedling growth of *A. fruticosa* on contaminated spoils from two abandoned mines accessed by Seo et al. (2008) showed great surviving rate (75–100% after 18 months). The addition of organic fertilizer resulted in the highest concentrations of As, Pb and Zn in *A. fruticosa* roots (42.9 mg kg^{-1} of Cd, 226.9 mg kg^{-1} of Cu, 356.4 mg kg^{-1} of Pb and 1056 mg kg^{-1} of Zn), significantly increasing heavy metal uptake by *A. fruticosa* compared to unfertilized control. Results of Sikdar et al. (2020) corroborate this findings, as their investigation on assisted phytostabilization of Pb/Zn mine tailings with organic amendments and triple superphosphate confirmed that *A. fruticosa* accumulated higher concentrations of metals in roots (up to 945 mg kg^{-1} of Pb, 4555 mg kg^{-1} of Zn, 64.65 mg kg^{-1} of Cd and 609 mg kg^{-1} of Cu) exhibiting low translocation to aboveground tissues and confirming its phytostabilization potential on multi-metal polluted sites. However, due to the increase invasiveness of the species, it should be considered for such purposes only on degraded sites within its native range.

Leucaena leucocephala (Lam.) de Wit is small fast-growing tree or shrub native to Mexico and nowadays characterized with pantropical range of distribution. It is known as one of the Worlds' 100 worst invasive species according to IUCN list (Lowe et al. 2000), being an aggressive colonizer of many ruderal sites in tropical and subtropical region. Plant forms dense, homogenous thickets that are difficult to control, and once established, it poses significant threat to native biodiversity. Plant is also known as one of the highest quality fodder trees of the tropics.

Increased accumulations of Fe, Mn, Zn and Cu in roots of *L. leucocephala* growing on fly ash were noticed by Gupta et al. (2000). Similarly, in pot and field phytoremediation trials on Pb-contaminated tailings (total Pb content > 9850 mg kg^{-1}), plant showed similar trends (Meeinkuirt et al. 2012). *L. leucocephala* absorbed 2803.3 ± 77.2 mg kg^{-1} of Pb in roots and 667.8 ± 142.6 mg kg^{-1} of Pb in shoots (transfer factor root-to-shoot 0.24) on mine tailing soil with addition of cow manure, while in field trials on Pb mine tailing amended with Osmocote and organic fertilizer it accumulated 1695.3 ± 247.6 mg kg^{-1} pf Pb in roots and 1007.5 ± 129.4 mg kg^{-1} of Pb in shoots (transfer factor root-to-shoot 0.59). During eventual use of this species in remediation programs with assistance of organic materials, special attention should be paid to sites that can be easily accessed by animals, which can cause entering of metals into the food chains. Additional caution measures should be taken if the species is used for these purposes outside of its native range, due to its invasiveness.

On the other hand, the conversion of *L. leucocephala* biomass to biochar revealed an option for developing particular amendment for soil reclamation. *L. leucocephala* bark has been modified to biochar through slow pyrolysis process on temperature range 300–600 °C and time 35–205 min (Anupam et al. 2015). Characterization of such biochar showed high resistivity toward degradation and suitability for soil amelioration. Additionally, Jha et al. (2016) found that *Leucaena* biochar has ameliorating effect on acid soils in India, increasing exchangeable base cations and pH of soil solution, while simultaneously increasing the process of nitrification.

Ailanthus altissima (Mill.) Swingle is rapid-growing tree native to northern and central China that is nowadays introduced worldwide. In Europe and USA it is considered as highly invasive species. Tree also became invasive outside its natural climate zone, invading territories from cool temperate to tropical climate (Kowarik and Säumel 2007). *A. altissima* grows on a wide range of habitats, most of which are subject to a higher level of human disturbance. Moreover, allelochemical toxins produced in the bark and leaves tend to accumulate in soil and inhibit the growth of neighboring plants, negatively influencing the plant community (Lawrence et al. 1991).

Assessment of the heavy metal effects on germination and early seedlings growth of species *A. altissima* and *Acer negundo* L. showed that *A. altissima* exhibited higher germination capacity of seeds treated with Cd and Pb nitrate 90 µM solution (88.66 and 94.67%, respectively) and that seedlings had 4-7 fold higher biomass production than seedlings of *A. negundo* (Djukić et al. 2013). Experiments on in vitro micropropagation of *A. altissima* showed that cultures exposed to Cu, Zn and Mn tolerate multiple-metal pollution (Gatti 2008), confirming that this species can be used in phytoremediation of contaminated sites. Similarly was noticed by Yang et al. (2015) who found that *A. altissima* growing on Pb/Zn mining area showed phytostabilization potential through possibility to concentrate Pb and Cd primarily in its roots. At the same research, investigated plants accumulated >300 mg kg^{-1} of Zn in their leaves, exhibiting soil-to-plant concentration factor for leaves >1 and showing certain capacity for phytoextraction of Zn. Addition of fertilizer Osmocote on the As–Pb mine Technosol greatly improved the growth of *A. altissima* plants, revealing different mechanisms of tolerance to As and Pb: While majority of As was absorbed inside the root system with poor translocation to the aboveground parts, adsorption of Pb was mainly restricted to the root surface, with smaller contents entering the root system (Lebrun et al. 2020). Thereunto, carbon-derived materials from *A. altissima* showed certain application potential in combating water pollution. Lignin modified from wood of *A. altissima* showed as effective adsorbent for Co(II) and Hg(II) from water, showing maximum adsorption capacities of 7.1–7.7 and 4.4–5.3 mg g^{-1} of the modified lignin, respectively (Demirbas 2007). Additionally, Bangash and Alam (2009) showed that carbon prepared from the wood of *A. altissima* at 800 °C can effectively remove acid blue 1 dye from aqueous solutions, thereby showing potential in remediation of industrial wastewaters.

Ailanthus altissima exhibits range of roles in the environment, compromising biodiversity of natural habitats where invasive, but also providing a number of human-related services where properly controlled. Therefore, eventual utilization of this species should be restricted mainly to areas that are not suitable for its natural establishment (Sladonja et al. 2015), whereas its use in natural and close-to-natural environments should be avoided.

2.7 Conclusion Remarks

Ruderal plants are characterized by wide adaptability to high disturbance and lower intensity of stress, which makes them successful colonizers of many anthropogenically degraded areas. Some of the traits that enable them such success are high fecundity, phenotypical plasticity as a response to resource availability variation and prevailing equivalence to the various ecological factors. Increased rate of anthropogenic disturbances within last decades has favored this functional group enabling them the occupation of unfilled or newly created niches, which consequently lead to their expansion in many regions of the world.

Not only that ruderal species are able to thrive on disturbed sites, but they also often represent their first inhabitants, playing a role of certain biological signal for ecosystem recovery processes. Posing such threats, ruderal species have become recognized for exhibiting additional potential for biological remediation of various degraded lands. Increased level of anthropogenic disturbance during last decades has brought continual pollution to many areas, including urban and industrial sites. Therefore, many ruderal plants were studied for their capacity to cope with presence of various inorganic and organic contaminants in the environment. Researches confirmed that some species are able to accumulate significant concentration of targeted chemical elements (metals, metalloids or radionuclides) in their roots, showing potential for phytostabilization on polluted industrial and urban sites. Investigations also revealed, though more rarely, potential of certain ruderal species for accumulation or even hyperaccumulation of selected chemical elements in plants shoots, presenting those species as potential candidates for phytoextraction process on contaminated sites. Certain plant species were also able to degrade or metabolize organic pollutants, such as xenobiotics or polycyclic aromatic hydrocarbons.

As phytoremediation research continues to develop and modify toward enhancement of the remediation efficiency, ruderal plants are gradually being subjected to diverse innovative experiments and applications developed to improve element bioavailability, plant tolerance or accumulation capacity. Some of them include plant breeding, hybridization or creation of genetically engineered plants in order to improve or develop suitable plant traits. This is particularly valuable when it comes to increasing biomass of high accumulating plants or enhancing the accumulation abilities of existing high biomass species (e.g., poplar trees). Other approaches are orientated toward biologically assisted remediation, using rhizosphere bacteria and arbuscular mycorrhizal fungi to affect the availability of elements in rhizosphere zone. Another practice is known as chemically assisted remediation that applies synthetic or natural chelating agents for altering the metal mobilization processes and enhancing their removal from contaminated soils. Generally, addition of different soil amendments, especially ones that contain certain form of organic matter to the contaminated soil, usually increases the growth and establishment of ruderal plants (as nutrient availability at the site becomes improved), which in certain cases may result in increased ability of the plant for phytostabilization or phytoextraction of targeted elements. Similarly, there are possibilities of using certain high-biomass ruderal species for integrated phytoremediation approach, where land remediation is coupled with energy production or other types of plant biomass valorization. Opportunistic character of ruderal species enables them wide and fast spreading on growing number of degraded sites, so some of them become invasive starting to threat the native plant communities. However, there are growing possibilities of using biomass of invasive species for remediation purposes, such as using it as raw or modified material for sorption of pollutants from wastewaters, or in thermochemically converted state (as biochar) for remediation of polluted soils.

Further developments in the field of phytoremediation will undoubtedly involve ruderal species as some of their common traits are matching desirable traits of phytoremediation plant candidates. In conclusion, ruderal plants are representing functional plant group with high potential for application in remediation of degraded lands, whose importance will grow with increasing rate of existing and novel anthropogenic disturbances and ongoing climate changes.

Dedication This chapter is dedicated to the late professor Slobodan Jovanović, whose enthusiasm, knowledge and committed teaching about the ecology of ruderal and invasive plants inspired many generations of biologists and environmentalists.

References

Adriano D (2003) Trace elements in terrestrial environments: biogeochemistry, bioavailability and risks of metals, 2nd ed. Springer, New York, NY, USA

Aitchison E, Kelley S, Alvarez P, Schnoor J (2000) Phytoremediation of 1,4-dioxane by hybrid poplar trees. Water Environ Res 72(3):313–321

Anupam K, Swaroop V, Deepika Lal PS, Bist V (2015) Turning Leucaena leucocephala bark to biochar for soil application via statistical modelling and optimization technique. Ecol Eng 82:26–39

Baasch A, Kirmer A, Tischew S (2011) Nine years of vegetation development in a postmining site: effects of spontaneous and assisted site recovery. J Appl Ecol 49(1):251–260

Bacchetta G, Cao A, Cappai G, Carucci A, Casti M, Fercia ML, Lonis R, Mola F (2012) A field experiment on the use of Pistacia lentiscus L. and Scrophularia canina L. subsp.bicolor (Sibth. et Sm.) Greuter for the phytoremediation of abandoned mining areas. Plant Biosystems—Int J Dealing with All Aspects of Plant Biology 146(4):1054–1063

Bacchetta G, Cappai G, Carucci A, Tamburini E (2015) Use of native plants for the remediation of abandoned mine sites in Mediterranean semiarid environments. Bull Environ Contam Toxicol 94(3):326–333

Bae J, Benoit DL, Watson AK (2016) Effect of heavy metals on seed germination and seedling growth of common ragweed and roadside ground cover legumes. Environ Pollut 213:112–118

Bangash F, Alam S (2009) Adsorption of acid blue 1 on activated carbon produced from the wood of Ailanthus altissima. Braz J Chem Eng 26(2)

Batanjski V, Kabaš E, Kuzmanović N, Vukojičić S, Lakušić D, Jovanović S (2015) New invasive forest communities in the riparian fragile habitats—a case study from Ramsar site Carska bara (Vojvodina, Serbia). Šumarski List 3–4:155–169

Baumhardt R, Stewart B, Sainju U (2015) North American soil degradation: processes, practices, and mitigating strategies. Sustainability 7:2936–2960

Bekteshi A, Bara G (2013) Uptake of heavy metals from plantago major in the region of Durrës, Albania. Polish J Environ Stud 22(6):1881–1885

Berchová-Bímová K, Soltysiak J, Vach M (2014) Role of different taxa and cytotypes in heavy metals absorption in knotweeds (Fallopia). Sci Agric Bohem 45(1):11–18

Bert V, Lors C, Ponge JF, Caron L, Biaz A, Dazy M, Masfaraud JF (2012) Metal immobilization and soil amendment efficiency at a contaminated sediment landfill site: a field study focusing on plants, springtails, and bacteria. Environ Pollut 169:1–11

Biró I, Takács T (2007) Study of heavy metal uptake of Populus nigra in relation to phytoremediation. Cereal Res Commun 35(2):265–268

Brackhage C, Gert DE (2002) Long-term differences in transfer and accumulation of potentially toxic trace elements and radionuclides in trees on uranium mining dumps (Erzgebirge, Germany). In: Merkel BJ, Planer-Friedrich B, Wolkersdorfer C (eds) Uranium in the aquatic environment. Springer, Berlin, Heidelberg, pp 471–478

Bradshaw A (2002) Introduction and philosophy. In: Perrow M, Davy A (eds) Handbook of ecological restoration, vol 1: principles of restoration. Cambridge University Press, Cambridge, pp 3–9

Bratman GN, Hamilton JP, Daily GC (2012) The impacts of nature experience on human cognitive function and mental health. Ann New York Acad Sci 1249:118–136

Bradshaw A (1997) What do we mean by restoration? In: Urbanska K, Webb NR, Edwards PJ (eds) Restoration ecology and sustainable development. Cambridge University Press, pp 8–14

Cachada A, Rocha-Santos T, Duarte A (2018) Soil and pollution: an introduction to the main issues. In: Duarte A, Cachada A, Rocha-Santos T (eds) Soil pollution: from monitoring to remediation, 1st ed. Academic Press, Elsevier, pp 1–28

Chan B, Dudeney A, Meyer S (2015) Surface regeneration of coal tips: 15 years of mine rehabilitation in a former coal mining region in Southwest Germany. Legislation, technology and practice of mine land reclamation, proceedings of the Beijing international symposium land reclamation and ecological restoration. LRER, Beijing, China, pp 617–624

Chaney R, Baklanov I (2017) Phytoremediation and phytomining: status and promise. Adv Bot Res 83:189–221

Chaplygin V, Minkina T, Mandzhieva S, Burachevskaya M, Sushkova S, Poluektov E, Antonenko E, Kumacheva V (2018) The effect of technogenic emissions on the heavy metals accumulation by herbaceous plants. Environ Monit Assess 190(3)

Chen S, Dong L, Quan H, Zhou X, Ma J, Xia W, Zhou H, Fu X (2020) A review of the ethnobotanical value, phytochemistry, pharmacology, toxicity and quality control of Tussilago farfara L. (coltsfoot). J Ethnopharmacol. https://doi.org/10.1016/j.jep.2020.113478

Chiuffo M, Cock M, Prina A, Hierro J (2018) Response of native and non-native ruderals to natural and human disturbance. Biol Invasions 20:2915–2925

Chyc M, Krzyżewska I, Tyński P (2012) Occurrence of coltsfoot (Tussilago farfara L.) on arsenic contaminated area [in Polish]. Prace Naukowe Gig Górnictwo I Środowisko 4:73–84

Cloutier-Hurteau B, Gauthier S, Turmel M, Comtois P, Courchesne F (2014) Trace elements in the pollen of Ambrosia artemisiifolia: What is the effect of soil concentrations? Chemosphere 95:541–549

Cobbett C, Meagher R (2002) Arabidopsis and the genetic potential for the phytoremediation of toxic elemental and organic pollutants. The Arabidopsis Book 1(2002):e0032

Concas S, Lattanzi P, Bacchetta G, Barbafieri M, Vacca A (2015) Zn, Pb and Hg contents of Pistacia lentiscus L. grown on heavy metal-rich soils: implications for phytostabilization. Water, Air Soil Pollut 226(10):340

Cornejo J, Tapia J, Guerra F, Yáñez M, Baettig R, Guajardo J, Alarcón E, Vidal G (2017) Variation in copper accumulation at the tissue level of five hybrid poplars subjected to copper stress. Water, Air Soil Pollut 228(6)

Crooks J, Suarez A (2006) Hyperconnectivity, invasive species, and the breakdown of barriers to dispersal. In Crooks KR, Sanjayan M (eds) Conservation biology—connectivity conservation, vol 14. Cambridge University Press, Cambridge, UK, pp 451–478

Crossman ND, Bryan BA, Cooke DA (2011) An invasive plant and climate change threat index for weed risk management: Integrating habitat distribution pattern and dispersal process. Ecol Ind 11(1):183–198

Čudić V, Stojiljković D, Jovović A (2016) Phytoremediation potential of wild plants growing on soil contaminated with heavy metals. Arch Ind Hyg Toxicol 67(3):229–239

Demirbas A (2007) Adsorption of Co(II) and Hg(II) from water and wastewater onto modified lignin. Energy Sources, Part A: Recov Util Environ Effects 29(2):117–123

Di Baccio D, Tognetti R, Sebastiani L, Vitagliano C (2003) Responses of Populus deltoides x Populus nigra (Populus x euramericana) clone I-214 to high zinc concentrations. New Phytol 159(2):443–452

Djukić M, Bojović DD, Grbić M, Skočajić D, Obratov-Petković D, Bjedov I (2013) Effect of Cd and Pb on Ailanthus altissima and Acer negundo seed germination and early seedling growth. Fresenius Environ Bull 22(2a):524–530

Dmuchowski W, Gozdowski D, Baczewska AH, Bragoszewska P (2013) Evaluation of various bioindication methods used for measuring zinc environmental pollution. Int J Environ Pollut 51:238–254

Dmuchowski W, Gozdowski D, Brągoszewska P, Baczewska AH, Suwara I (2014) Phytoremediation of zinc contaminated soils using silver birch (Betula pendula Roth). Ecol Eng 71:32–35

Domínguez-Solís J, López-Martín C, Ager F, Ynsa D, Romero L, Gotor C (2004) Increased cysteine availability is essential for cadmium tolerance and accumulation in Arabidopsis thaliana. Plant Biotechnol J 2:469–476

Doty S, Freeman J, Cohu C, Burken J, Firrincieli A, Simon A, Khan Z, Isebrands J, Lukas J, Blaylock MJ (2017) Enhanced degradation of TCE on a superfund site using endophyte-assisted poplar tree phytoremediation. Environ Sci Technol 51(17):10050–10058

Drozdova I, Alekseeva-Popova N, Dorofeyev V, Bech J, Belyaeva A, Roca N (2019) A comparative study of the accumulation of trace elements in Brassicaceae plant species with phytoremediation potential. Appl Geochem 108:104377

Dushenkov V, Kumar N, Motto H, Raskin I (1995) Rhizofiltration: the use of plants to remove heavy metals from aqueous streams. Environ Sci Technol 29(5):1239–1245

Dwibedi SK, Pandey VC, Divyasree D, Bajpai O (2021) Biochar-based land development. Land Degrad Dev. https://doi.org/10.1002/ldr.4185

Ebrahimi M (2016) Enhanced phytoremediation capacity of chenopodium album L. grown on Pb-contaminated soils using EDTA and reduction of leaching risk. Soil Sediment Contamination: Int J 25(6):652–667

El-Mahrouk E, Eisa E, Ali H, Hegazy M, Abd El-Gayed M (2020) Populus nigra as a phytoremediator for Cd, Cu, and Pb in contaminated soil. BioResources 15(1):869–893

EPA (2000) Introduction to remediation. Report EPA/600/R-99/107, US Environmental Protection Agency, Ohio, USA

Eswaran H, Lal R, Reich P (2001) Land degradation: an overview. In: Bridges EM, Hannam ID, Oldeman LR, Pening de Vries FWT, Scherr SJ, Sompatpanit S (eds) Responses to land degradation. In: Proceedings 2nd. international conference on land degradation and desertification, Khon Kaen, Thailand. Oxford Press, New Delhi, India

Filipović-Trajković R, Ilić Z, Šunić LJ, Andjelković S (2012) The potential of different plant species for heavy metals accumulation and distribution. J Food Agric Environ 10(1):959–964

Fridley J (2011) Invasibility, of communities and ecosystems. In: Simberloff D, Rejmánek M (eds) Encyclopedia of biological invasions. University of California Press, Berkeley and Los Angeles, pp 356–360

Gałuszka A, Migaszewski ZM, Pelc A, Trembaczowski A, Dołęgowska S, Michalik A (2020) Trace elements and stable sulfur isotopes in plants of acid mine drainage area: implications for revegetation of degraded land. J Environ Sci 94:128–136

Gao Y, Miao C, Wang Y, Xia J, Zhou P (2012) Metal-resistant microorganisms and metal chelators synergistically enhance the phytoremediation efficiency of Solanum nigrum L in Cd- and Pb-contaminated soil. Environ Technol 33(12):1383–1389

Gatti E (2008) Micropropagation of Ailanthus altissima and in vitro heavy metal tolerance. Biol Plant 52(1):146–148

Gelbard JL, Belnap J (2003) Roads as conduits for exotic plant invasions in a semiarid landscape. Conserv Biol 17:420–432

Gisladottir G, Stocking M (2005) Land degradation control and its global environmental benefits. Land Degrad Dev 16(2):99–112

Glišić M, Lakušić D, Šinžar-Sekulić J, Jovanović S (2014) GIS analysis of spatial distribution of invasive tree species in the protected natural area of Mt. Avala (Serbia). Botanica Serbica 38(1):131–138

Goddard MA, Dougill AJ, Benton TG (2010) Scaling up from gardens: biodiversity conservation in urban environments. Trends Ecol Evol 25:90–98

González R, González-Chávez M (2006) Metal accumulation in wild plants surrounding mining wastes. Environ Pollut 144(1):84–92

Grime P (2001) Plant strategies, vegetation processes, and ecosystem properties. John Wiley & Sons, New York, USA, p 417

Grzegórska A, Rybarczyk P, Rogala A, Zabrocki D (2020) (2020): Phytoremediation—from environment cleaning to energy generation—current status and future perspectives. Energies 13:2905. https://doi.org/10.3390/en13112905

Gupta A, Sinha S (2007) Phytoextraction capacity of the Chenopodium album L. grown on soil amended with tannery sludge. Bioresource Technol 98(2):442–446

Gupta A, Sinha S (2008) Decontamination and/or revegetation of fly ash dykes through naturally growing plants. J Hazard Mater 153:1078–1087

Gupta M, Kumari A, Yunus M (2000) Effect of fly-ash on metal composition and physiological responses in Leucaena leucocephala (lamk.) de. wit. Environ Monit Assess 61:399–406

Hadač E (1978) Ruderal vegetation of the Broumov basin, NE Bohemia. Folia Geobotanica et Phytotaxonomica 13(2):129–163

Heaton A, Rugh C, Wang N, Meagher R (1998) Phytoremediation of mercury- and methylmercury-polluted soils using genetically engineered plants. J Soil Contam 7(4):497–509

Hejda M, Pyšek P, Jarošík V (2009) Impact of invasive plants on the species richness, diversity and composition of invaded communities. J Ecol 97(3):393–403

Hlavati Širka V, Jakovljević K, Mihailović N, Jovanović S (2016) Heavy metal accumulation in invasive Reynoutria × bohemica Chrtek & Chrtková in polluted areas. Environ Earth Sci 75:951

Hu R, Sun K, Su X, Pan Y, Zhang Y, Wang X (2012) Physiological responses and tolerance mechanisms to Pb in two xerophils: Salsola passerina Bunge and Chenopodium album L. J Hazard Mater 205–206:131–138

Hulina N, Đumija L (1999) Ability of Reynoutria japonica Houtt. (Polygonaceae) to accumulate heavy metals. Periodicum Biologorum 101(3):233–235

Iori V, Gaudet M, Fabbrini F, Pietrini F, Beritognolo I, Zaina G, Scarascia Mugnozza G, Zacchini M, Massacci A, Sabatti M (2015) Physiology and genetic architecture of traits associated with cadmium tolerance and accumulation in Populus nigra L. Trees, 30(1):125–139

IUCN (2015) Land degradation neutrality: implications and opportunities for conservation. Technical brief, 2nd Ed., November 2015. IUCN, Nairobi, p 19p

Jakovljević T, Bubalo MC, Orlović S, Sedak M, Bilandžić N, Brozinčević I, Redovniković IR (2014) Adaptive response of poplar (Populus nigra L.) after prolonged Cd exposure period. Environ Sci Pollut Res 21(5):3792–3802

Jakovljević K, Mišljenović T, Savović J, Ranković D, Ranđelović D, Mihailović N, Jovanović S (2020) Accumulation of trace elements in Tussilago farfara colonizing post-flotation tailing sites in Serbia. Environ Sci Pollut Res 27(4):4089–4103

Jelić M, Patenković A, Skorić M, Mišić D, Kurbalija Novičić Z, Bordács S, Várhidi F, Vasić I, Benke A, Frank G, Šiler B (2015) Indigenous forests of European black poplar along the Danube River: genetic structure and reliable detection of introgression. Tree Genet Genom 11(5)

Jha P, Neenu S, Rashmi I, Meena BP, Jatav RC, Lakaria BL, Biswas AK, Singh M, Patra AK (2016) Ameliorating effects of Leucaena biochar on soil acidity and exchangeable ions. Commun Soil Sci Plant Anal 47(10):1252–1262

Jie C, Jing-zhang C, Man-zhi T, Zi-tong G (2002) Soil degradation: a global problem endangering sustainable development. J Geog Sci 12:243–252

Jovanović S (1994) Ekološka studija ruderalne flore i vegetacije Beograda [Ecological study of ruderal flora and vegetation in the city of Belgrade]. Monograph publication, University of Belgrade, Faculty of Biology, pp 222. ISBN 86-7087-001-1

Jovanović S, Jakovljević K, Đorđević V, Vukojičić S (2013) Ruderal flora and vegetation of the town of Žabljak (Montenegro)—an overview for the period 1990–1998. Botanica Serbica 37(1):55–69

Jovanović S, Hlavati-Širka V, Lakušić D, Jogan N, Nikolić T, Anastasiu P, Vladimirov V, Šinžar-Sekulić J (2018) Reynoutria niche modelling and protected area prioritization for restoration and protection from invasion: a Southeastern Europe case study. J Nat Conserv 41:1–15

King LJ (1966) Weeds of the world. In: Biology and control. Interscience Publication, New York, USA

Koutník I, Vráblová M, Bednárek J (2020) Reynoutria japonica, an invasive herb as a source of activated carbon for the removal of xenobiotics from water. Bioresource Technol:123315. https://doi.org/10.1016/j.biortech.2020.123315

Kowarik I, Säumel I (2007) Biological flora of Central Europe: Ailanthus altissima (Mill.) Swingle. Persp Plant Ecol Evol Systemat 8(4):207–237

Lal R (2012) Climate change and soil degradation mitigation by sustainable management of soils and other natural resources. Agric Res 1:199–212

Lawrence J, Colwell A, Sexton O (1991) The ecological impact of allelopathy in Ailanthus altissima (Simaroubaceae). Am J Bot 78(7):948–958

Lebrun M, Alidou Arzika I, Miard F, Nandillon R, Bayçu G, Bourgerie S, Morabito D (2020) Effect of fertilization of a biochar and compost amended technosol: consequence on Ailanthus altissima growth and As- and Pb-specific root sorption. Soil Use Manag. https://doi.org/10.1111/sum.12646

Lee J, Bae H, Jeong J, Lee J, Yang Y, Hwang I, Martinoia E, Lee Y (2003) Functional expression of a bacterial heavy metal transporter in Arabidopsis enhances resistance to and decreases uptake of heavy metals. Plant Physiol 133:589–596

Lehmann C, Rebele F (2004) Assessing the potential for cadmium phytoremediation with Calamagrostis epigejos: a pot experiment. Int J Phytorem 6(2):169–183

Li X, Zhang X, Wang X, Cui Z (2019) Phytoremediation of multi-metal contaminated mine tailings with Solanum nigrum L. and biochar/attapulgite amendments. Ecotoxicol Environ Saf 180:517–525

Lian W, Yang L, Joseph S, Shi W, Bian R, Zheng J, Li L, Shan S, Pan G (2020) Utilization of biochar produced from invasive plant species to efficiently adsorb Cd (II) and Pb (II). Bioresource Technol 124011. https://doi.org/10.1016/j.biortech.2020.124011

Liang S, Gao N, Li Z, Shen S, Li J (2016) Investigation of correlativity between heavy metals concentration in indigenous plants and combined pollution soils. Chem Ecol 32(9):872–883

Liedekerke V, Prokop G, Rabl-Berger S, Kibblewhite M, Louwagie G (2014) Progress in the management of contaminated sites in Europe. Report EUR 26376 EN, European Commission, pp 70

Limmer M, Burken J (2016) Phytovolatilization of organic contaminants. Environ Sci Technol 50(13):6632–6643

Liu J, Zhang X, Li T, Wu Q, Jin Z (2014) Soil characteristics and heavy metal accumulation by native plants in a Mn mining area of Guangxi, South China. Environ Monit Assess 186(4):2269–2279

Liu L, Li W, Song W, Guo M (2018) Remediation techniques for heavy metal-contaminated soils: principles and applicability. Sci Total Environ 633:206–219

Lowe S, Browne M, Boudjelas S, De Poorter M (2000) 100 of the world's worst invasive alien species: a selection from the global invasive species database, Invasive Species Specialist Group (ISSG) a specialist group of the Species Survival Commission (SSC) of the World Conservation Union (IUCN)

Mackova M, Macek T, Kucerova P, Burkhard J, Pazlarova J, Demnerova K (1997) Degradation of polychlorinated biphenyls by hairy root culture of Solanum nigrum. Biotech Lett 19(8):787–790

Malizia D, Giuliano A, Ortaggi G, Masotti A (2012) Common plants as alternative analytical tools tomonitor heavy metals in soil. Chem Cent J 6(Suppl 2):S6

Maller CJ, Henderson-Wilson C, Townsend M (2009) Rediscovering nature in everyday settings: or how to create healthy environments and healthy people. EcoHealth 6:553–556

Marguí E, Queralt I, Carvalho ML, Hidalgo M (2007) Assessment of metal availability to vegetation (Betula pendula) in Pb-Zn ore concentrate residues with different features. Environ Pollut 145(1):179–184

Marques A, Oliveira R, Samardjieva K, Pissarra J, Rangel A, Castro P (2008) EDDS and EDTA-enhanced zinc accumulation by Solanum nigrum inoculated with arbuscular mycorrhizal fungi grown in contaminated soil. Chemosphere 70(6):1002–1014

Massa N, Andreucci F, Poli M, Aceto M, Barbato R, Berta G (2010) Screening for heavy metal accumulators amongst autochtonous plants in a polluted site in Italy. Ecotoxicol Environ Saf 73(8):1988–1997

Meeinkuirt W, Pokethitiyook P, Kruatrachue M, Tanhan P, Chaiyarat R (2012) Phytostabilization of a Pb-contaminated mine tailing by various tree species in pot and field trial experiments. Int J Phytorem 14(9):925–938

Melčáková I, Horvathova H (2010) Study of biomass of Reynoutria japonica as a novel biosorbent for removal of metals from aqueous solutions. GeoScience Eng L VI(1):55–70

Mentis M (2020) Environmental rehabilitation of damaged land. Forest Ecosyst 7(19). https://doi.org/10.1186/s40663-020-00233-4

Mitrović M, Pavlović P, Lakušić D, Djurdjević L, Stevanović B, Kostić O, Gajić G (2008) The potential of Festuca rubra and Calamagrostis epigejos for the revegetation of fly ash deposits. Sci Total Environ 407(1):338–347

Montanarella L (2007) Trends in land degradation in Europe. In: Sivakumar MVK, Ndiang'ui N (eds) Climate and land degradation. Environmental Science and Engineering (Environmental Science). Springer, Berlin, Heidelberg, pp 83–104

Mudrák O, Frouz J, Velichova V (2010) Understory vegetation in reclaimed and unreclaimed post-mining forest stands. Ecol Eng 36:783–790

Mudrák O, Doležal J, Frouz J (2016) Initial species composition predicts the progress in the spontaneous succession on post-mining sites. Ecol Eng 95:665–670

Munoz F, Violle C, Cheptou P (2016) CSR ecological strategies and plant mating systems: outcrossing increases with competitiveness but stress-tolerance is related to mixed mating. Oikos 125(9):1296–1303

Nikolić N (2020) highly patterned primary succession after fluvial deposition of mining waste. Nat Sci 10(1):1–5

Nowińska K, Kokowska-Pawłowska M, Patrzałek A (2012) Metale w Calamagrostis epigejos i Solidago sp. ze zrekultywowanych nieuŻyt-ków poprzemysłowych. Infrastruktura i Ekologia Terenów Wiejskich/ / Infrastructure and Ecology of Rural Areas. Nr 2012/ 03. [in Polish]

Olsson L, Barbosa H, Bhadwal S, Cowie A, Delusca K, Flores-Renteria D, Hermans K, Jobbagy E, Kurz W, Li D, Sonwa DJ, Stringer L (2019) Land degradation. In: Shukla PR, Skea J, Calvo Buendia E, Masson-Delmotte V, Pörtner HO, Roberts DC, Zhai P, Slade R, Connors S, van Diemen R, Ferrat M, Haughey E, Luz S, Neogi S, Pathak M, Petzold J, Portugal Pereira J, Vyas P, Huntley E, Kissick K, Belkacemi M, Malley J (eds) Climate change and land: an IPCC special report on climate change, desertification, land degradation, sustainable land management, food security, and greenhouse gas fluxes in terrestrial ecosystems. Intergovernmental Panel on Climate Change (IPCC)

Pandey VC, Bajpai O (2019) Phytoremediation: from theory towards practice. In: Pandey VC, Bauddh K (eds) Phytomanagement of polluted sites. Elsevier, Amsterdam, pp 1–49. https://doi.org/10.1016/B978-0-12-813912-7.00001-6

Pandey VC, Singh V (2019) Exploring the potential and opportunities of recent tools for removal of hazardous materials from environments. In: Pandey VC, Bauddh K (eds) Phytomanagement of polluted sites. Elsevier, Amsterdam, pp 501–516. https://doi.org/10.1016/B978-0-12-813912-7.00020-X

Pandey VC, Souza-Alonso P (2019) Market opportunities in sustainable phytoremediation. In: Pandey VC, Bauddh K (eds) Phytomanagement of polluted sites. Elsevier, Amsterdam, pp 51–82. https://doi.org/10.1016/B978-0-12-813912-7.00002-8

Pandey VC, Singh N, Singh RP, Singh DP (2014) Rhizoremediation potential of spontaneously grown Typha latifolia on fly ash basins: study from the field. Ecol Eng 71:722–727. https://doi.org/10.1016/j.ecoleng.2014.08.002

Parra A, Zornoza R, Conesa E, Gómez-López MD, Faz A (2016) Evaluation of the suitability of three Mediterranean shrub species for phytostabilization of pyritic mine soils. CATENA 136:59–65

Pavlović D, Pavlović M, Marković M, Karadžić B, Kostić O, Jarić S, Mitrović M, Gržetić I, Pavlović P (2017) Possibilities of assessing trace metal pollution using Betula pendula Roth. leaf and bark—experience in Serbia. J Serbian Chem Soc 82(6):723–737

Pedashenko H, Apostolova I, Vassilev K (2012) Amorpha fruticosa invasibility of different habitats in lower Danube. Phytologia Balcanica 18(3):285–291

Pedron F, Petruzzelli G, Barbafieri M, Tassi E, Ambrosini P, Patata L (2011) Mercury mobilization in a contaminated industrial soil for phytoremediation. Commun Soil Sci Plant Anal 42:2767–2777

Pietrzykowski M, Likus-Cieslik J (2018) Comprehensive study of reclaimed soil, plant, and water chemistry relationships in highly s-contaminated post sulfur mine site Jeziórko (Southern Poland). Sustainability 10:2442

Polechońska M, Zawadzki K, Samecka-Cymerman A, Kolon K, Klink A, Krawczyk J, Kempers AJ (2013) Evaluation of the bioindicator suitability of Polygonum aviculare in urban areas. Ecol Ind 24:552–556

Prach K, Pyšek P (1994) Spontaneous establishment of woody plants in Central European derelict sites and their potential for reclamation. Restor Ecol 2(3):190–197

Prach K, Řehounková K (2006) Vegetation succession over broad geographical scales: which factors determine the patterns? Preslia 78:469–480

Prach K, Pyšek P, Bastl M (2001) Spontaneous vegetation succession in human-disturbed habitats: a pattern across seres. Appl Veg Sci 4(1):83–88

Prasad M (2004) Phytoremediation of metals and radionuclides in the environment: the case for natural hyperaccumulators, metal transporters, soil-amending chelators and transgenic plants. In: Prasad MNV (ed) Heavy metal stress in plants: from biomolecules to ecosystems, 2nd edn. Springer, Berlin, pp 345–391

Radovanović N, Kuzmanović N, Vukojičić S, Lakušić D, Jovanović S (2017) Floristic diversity, composition and invasibility of riparian habitats with Amorpha fruticosa: a case study from Belgrade (Southeast Europe). Urban Forest Urban Green 24:101–108

Rahmonov O, Czylok A, Orczewska A, Majgier L, Parusel T (2014) Chemical composition of the leaves of Reynoutria japonica Houtt. and soil features in polluted areas. Open Life Sci 9(3)

Ramalho CE, Hobbs RJ (2012) Time for a change: dynamic urban ecology. Trends Ecol Evol 27:179–187

Ranđelović D, Cvetković V, Mihailović N, Jovanović S (2014) Relation between edaphic factors and vegetation development on copper mine wastes: a case study from Bor (Serbia, SE Europe). Environ Manage 53(4):800–812

Ranđelović D, Jakovljević K, Mihailović N, Jovanović S (2018) Metal accumulation in populations of Calamagrostis epigejos (L.) Roth from diverse anthropogenically degraded sites (SE Europe, Serbia). Environ Monit Assess 190(4):183

Ranđelović D, Mihailović N, Jovanović S (2019) Potential of equisetum ramosissimum Desf. for remediation of antimony flotation tailings: a case study. Int J Phytoremediation 21(91):1–7

Ranđelović D, Jakovljević K, Jovanović S (2020a) The application of Calamagrostis epigejos (L.) Roth. in phytoremediation technologies. In: Pandey VC, Singh DP (eds) Phytoremediation potential of perennial grasses. Elsevier, pp 259–282

Ranđelović D, Jakovljević K, Mišljenović T, Savović J, Kuzmanović M, Mihailović N, Jovanović S (2020b) Accumulation of potentially toxic elements in invasive Ambrosia artemisiifolia on sites with different levels of anthropogenic pollution. Water Air Soil Pollut 231:272. https://doi.org/10.1007/s11270-020-04655-2

Rat M, Gavrilović M, Radak B, Bokić B, Jovanović S, Božin B, Boža P, Anačkov G (2017) Urban flora in the Southeast Europe and its correlation with urbanization. Urban Ecosyst 20:811–822

Rebele F (1992) Colonization and early succession on anthropogenic soils. J Vegetation Sci 3(2):201–208

Rehman M, Rizwan M, Ali S, Ok Y, Ishaque W, Saifullah S, Nawaz M, Akmal F, Waqar M (2017) Remediation of heavy metal contaminated soils by using Solanum nigrum: a review. Ecotoxicol Environ Saf 143:236–248

Rejmanek M, Richardson DM (1996) What attributes make some plant species more invasive? Ecology 77:1655–1661

Rendeková A, Mičieta K, Hrabovský M, Eliašová M, Miškovic J (2019) Effects of invasive plant species on species diversity: implications on ruderal vegetation in Bratislava City, Slovakia, Central Europe. Acta Societatis Botanicorum Poloniae 88(2)

Riaz M, Zia-Ul-Haq M, Jaafar H (2013) Common mullein, pharmacological and chemical aspects. Rev Bras 23:6

Richardson DM, Rejmanek M (2011) Trees and shrubs as invasive alien species—a global review. Divers Distrib 17:788–809

Robinson BH, Bischofberger S, Stoll A, Schroer D, Furrer G, Roulier S, Gruenwald A, Attinger W, Schulin R (2008) Plant uptake of trace elements on a Swiss military shooting range: uptake pathways and land management implications. Environ Pollut 153(3):668–676

Romeh A (2015) Evaluation of the phytoremediation potential of three plant species for azoxystrobin-contaminated soil. Int J Environ Sci Technol 12:3509–3518

Romeh A (2016) Efficiency of Rumex dentatus L. leaves extract for enhancing phytoremediation of Plantago major L. in soil contaminated by carbosulfan. Soil Sediment Contamination: Int J 25(8):941–956

Romeh A (2020) Synergistic use of Plantago major and effective microorganisms, EM1 to clean up the soil polluted with imidacloprid under laboratory and field condition. Int J Phytorem 22(14):1515–1523

Romeh A, Khamis M, Metwally S (2015) Potential of Plantago major L. for phytoremediation of lead-contaminated soil and water. Water, Air Soil Pollut 227(1)

Sakai AK, Allendorf FW, Holt JS, Lodge DM, Molofsky J, With KA, Baughman S, Cabin RJ, Cohen JE, Ellstrand NC, McCauley DE, O'Neil P, Parker IM, Thompson JN, Weller SG (2001) The population biology of invasive species. Annu Rev Ecol Evol Syst 32:305–332

Salinitro, M., Tassoni, A., Casolari, S., Laurentiis, F., Zappi, A., Melucci, D. (2019): Heavy Metals Bioindication Potential of the Common Weeds Senecio vulgaris L., Polygonum aviculare L. and Poa annua L., Molecules, 24: 2813.

Samuelsen A (2000) The traditional uses, chemical constituents and biological activities of Plantago major L: a review. J Ethnopharmacol 71(1–2):1–21

Sasmaz M, Akgül B, Yıldırım D, Sasmaz A (2015) Mercury uptake and phytotoxicity in terrestrial plants grown naturally in the Gumuskoy (Kutahya) mining area, Turkey. Int J Phytoremediation 18(1):69–76

Sasmaz M, Akgul B, Yıldırım D, Sasmaz A (2016) Bioaccumulation of thallium by the wild plants grown in soils of mining area. Int J Phytoremediation 18(11)

Seo KW, Son Y, Rhoades CC, Noh NJ, Koo JW, Kim J-G (2008) Seedling growth and heavy metal accumulation of candidate woody species for revegetating Korean mine spoils. Restor Ecol 16(4):702–712

Shi X, Zhang X, Chen G, Chen Y, Wang L, Shan X (2011) Seedling growth and metal accumulation of selected woody species in copper and lead/zinc mine tailings. J Environ Sci 23(2):266–274

Shochat E, Lerman SB, Anderies JM, Warren PS, Faeth SH, Nilon CH (2010) Invasion, competition, and biodiversity loss in urban ecosystems. Bioscience 60:199–208

Sikdar A, Wang J, Hasanuzzaman M, Liu X, Feng S, Roy R, Ali Sial T, Lahori AS, Jeyasundar P, Wang X (2020) Phytostabilization of Pb-Zn mine tailings with Amorpha fruticosa aided by organic amendments and triple superphosphate. Molecules 25(7):1617

Sipila T, Keskinen A, Åkerman M, Fortelius C, Haahtela K, Yrjala K (2008) High aromatic ring-cleavage diversity in birch rhizosphere: PAH treatment-specific changes of I.E.3 group extradiol dioxygenases and 16S rRNA bacterial communities in soil. ISME J 2:968–981

Sladonja B, Sušek M, Guillermic J (2015) Review on invasive tree of heaven (Ailanthus altissima (Mill.) Swingle) conflicting values: assessment of its ecosystem services and potential biological threat. Environ Manage 56(4):1009–1034

Sołtysiak J (2020) Heavy metals tolerance in an invasive weed (Fallopia japonica) under different levels of soils contamination. J Ecol Eng 21(7):81–91

Sone Y, Nakamura R, Pan-Hou H, Sato MH, Itoh T, Kiyono M (2013) Increase methylmercury accumulation in Arabidopsis thaliana expressing bacterial broad-spectrum mercury transporter MerE. AMB Express 3(1):52

Soudek P, Petrová Š, Benešová D, Tykva R, Vaňková R, Vaněk T (2007) Comparison of 226Ra nuclide from soil by three woody species Betula pendula, Sambucus nigra and Alnus glutinosa during the vegetation period. J Environ Radioact 97(1):76–82

Stanton M, Roy B, Thiede D (2000) Evolution in stressful environments. I. Phenotypic variability, phenotypic selection, and response to selection in five distinct environmental stresses. Evolution 54(1):93–111

Štefanek M (2015) Role of plant dominants on abandoned tailings containment from manganese-ore maining in Chvaletice, Eastern Bohemia, Czech Republic (overview of long-term case studies). J Landsc Ecol 8:3

Takou M, Wieters B, Kopriva S, Coupland G, Linstädter A, De Meaux J (2019) Linking genes with ecological strategies in Arabidopsis thaliana. J Exp Bot 70(4):1141–1151

Tamakhina YA, Ahkubekova A, Gadieva A, Tiev R (2019): Primary succession of plants of technogenic dumps of the Kabardino Balkarian Republic. IOP Conf Series: Mater Sci Eng 663(2019):012052

Tamas J, Kovacs E (2005) Vegetation pattern and heavy metal accumulation at a mine tailing at Gyöngyösoroszi, Hungary. Zeitschrift Fur Naturforschung C 60(3–4):362–367

Tervahauta AI, Fortelius C, Tuomainen M, Åkerman M-L, Rantalainen K, Sipilä T, Lehesranta S, Koistinen K, Karenlampi S, Yrjälä K (2009) Effect of birch (Betula spp.) and associated rhizoidal bacteria on the degradation of soil polyaromatic hydrocarbons, PAH-induced changes in birch proteome and bacterial community. Environ Pollut 157(1):341–346

Titus JH, Tsuyuzaki S (2020) Influence of a non-native invasive tree on primary succession at Mt. Koma, Hokkaido, Japan. Plant Ecol 169:307–315

Turnau K, Ostachowicz B, Wojtczak G, Anielska T, Sobczyk L (2010) Metal uptake by xerothermic plants introduced into Zn-Pb industrial wastes. Plant Soil 337:299–311

Tyser R, Worley C (1992) Alien flora in grasslands along road and trail corridors in Glacier National Park, USA. Conserv Biol 6:253–262

UNCCD (2015) Report of the intergovernmental working group on the follow-up to the outcomes of the United Nations conference on sustainable development (Rio+20)

UNEP (1992) World atlas of desertification. Edward Arnold, London, UK

Vasilyeva T, Galaktionova L, Lebedev S (2019) Assessment of remediation potential of flora of the Southern Urals, Conference on Innovations in Agricultural and Rural development IOP Conf Series: Earth Environ Sci 341(2019):012037

Vesk P, Leishman M, Westoby M (2004) Simple traits do not predict grazing response in Australian dry shrublands and woodlands. J Appl Ecol 41:22–31

Vitousek PM, D'Antonio CM, Loope LL, Westbrooks R (1996) Biological invasions as global environmental change. Am Sci 84:468–478

Vuksanović V, Kovačević B, Katanić M, Orlović S, Miladinović D (2017) In vitro evaluation of copper tolerance and accumulation in populus nigra. Arch Biol Sci 69(4):679–687

Weber E (2005) Invasive plant species of the world. Geobotanical Institute, Swiss Federal Institute of Technology, CABI Publishing, Oxon

Wechtler L, Laval-Gilly P, Bianconi O, Walderdorff L, Bonnefoy A, Falla-Angel J, Henry S (2019) Trace metal uptake by native plants growing on a brownfield in France: zinc accumulation by Tussilago farfara L. Environ Sci Pollut Res 26:36055–36062

Wei S, Zhou Q, Wang X, Zhang K, Guo L (2005) A newly found Cd-hyperaccumulator Solanum nigrum L. Chinese Sci Bull 50:33

Yacinalp E, Meral A (2019) Ruderal plants in urban and sub-urban walls and roofs. Ege Üniversitesi Ziraat Fakültesi Dergisi 56(2):81–90

Yang Y, Liang Y, Ghosh A, Song Y, Chen H, Tang M (2015) Assessment of arbuscular mycorrhizal fungi status and heavy metal accumulation characteristics of tree species in a lead–zinc mine area: potential applications for phytoremediation. Environ Sci Pollut Res 22(17):13179–13193

Yousaf B, Liu G, Abbas Q, Ubaid Ali M, Wang R, Ahmed R, Wang C, Al-Wabel M, Usman A (2018) Operational control on environmental safety of potentially toxic elements during thermal conversion of metal-accumulator invasive ragweed to biochar. J Clean Prod 195:458–469

Zajac E, Zarzycki J (2012) Revegetation of reclaimed soda waste dumps: effects of topsoil parameters. J Elementol 17:525–536

Zalesny R, Headlee W, Gopalakrishnan G, Bauer E, Hall R, Hazel D, Isebrands J, Licht L, Negri C, Nichols E, Rockwood D, Wiese AH (2019) Ecosystem services of poplar at long-term phytoremediation sites in the Midwest and Southeast, United States. Wiley Interdiscip Rev: Energy Environ:e349. https://doi.org/10.1002/wene.349

Zalesny R, Zhu JY, Headlee WL, Gleisner R, Pilipović A, Acker JV, Bauer E, Birr B, Wiese AH (2020) Ecosystem services, physiology, and biofuels recalcitrance of poplars grown for landfill phytoremediation. Plants 9(10):1357. https://doi.org/10.3390/plants9101357

Zhang X, Li M, Yang H, Li X, Cui Z (2018) Physiological responses of Suaeda glauca and Arabidopsis thaliana in phytoremediation of heavy metals. J Environ Manage 223:132–139

Zulfiqar S, Wahid A, Farooq M, Makbool N, Arfan M (2012) Phytoremediation of soil cadmium using Chenopodium species. Pak J Agric Sci 49(4):435–445

Chapter 3
Utilizing Polluted Land for Growing Crops

Shivakshi Jasrotia and Vimal Chandra Pandey

Abstract The world's population is growing at an expeditious rate, which is a global problem specific in context to the directly proportional need of edible food supply. Of late in research the term 'Sustainable food production' has caught the interest of many researchers worldwide. This kind of agriculture practice focuses on usage of land other than normal land which could be wastelands/polluted lands to produce edible food crops. What makes this current choice of polluted land usage difficult is the related complications and on field agricultural techniques in contrast to the conventional farming practices. For such land usage to be brought into practice, there are additional challenges and much needed suitable agro-technological interventions; the outcome of these would ensure a safe and sustainable crop production system. Issues like cost–benefit analysis, investigating the related entry of pollutants in phytoproducts associated with further labelling, certification and its marketing are few points to be considered for achieving a positive large-scale exploitation of polluted lands. The key to success lies in identification of such phytoaccumulators which can survive in contaminated land, absorb pollutants from soil which further avoid the transfer to the edible part of the same plant for safe consumption.

Keywords Edible crops · Polluted land · Contaminated biomass · Threats · Bio-fortification

3.1 Introduction

One of the most diverse and relatively complex ecosystems of the world is soil. It is the existence of soil that caters to many services related to mankind apart from just providing good food. Services like carbon storage, greenhouse gas regulation,

S. Jasrotia
Department of Clinical Research, Delhi Institute of Pharmaceutical Sciences and Research, Government of N.C.T. of Delhi, Delhi, India

V. C. Pandey (✉)
Department of Environmental Science, Babasaheb Bhimrao Ambedkar University, Lucknow, Uttar Pradesh, India
e-mail: vimalcpandey@gmail.com

Fig. 3.1 Food security pillars

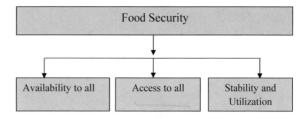

flood mitigation etc. are associated with the existence of soil (Kopittke et al. 2019). Currently, more than 98% of the world's food is being produced by soil. Rapid population growth is an unprecedented pressure on the soil, and current intensive agricultural and farming practices are slowly making an irreversible and unsustainable impact on our soils. By the years 2021–22 there would be around 70–80% increase in the need of edible food, feed and related fibre production for the enormous expected outgrowth of global population (Montanarella and Vargas 2012). This escalating rate of upcoming global pressure of food demand, directly points us to a rapid conclusion of finding alternate ways to the conventional food supply practices that have been followed. To meet this additional land area of 2.7–4.9 Mha yr^{-1} would be needed further (Abhilash et al. 2016). The challenge is not just limited to growing and supplying food by utilizing polluted land space, but the mechanism to be followed must also be sustainable in the long run.

Figure 3.1 represents that there are several factors of concern related to such an attempt of growing food in pre-polluted land. Factors of concern can range from (a) Food safety (b) Leaching of pollutants in plant cells (c) Sustainable farming practices (d) Plant species identification (should be tolerant to native pollution in soil) (e) Climate change impact and environmental concerns.

The study or research should be in accordance with the global climate change concerns and negative impact of extensive farming practices. This thought variedly channels researchers to study and investigate such plant species that are non-generous in generating hazardous bio waste but should be able to easily grow on polluted land space. In short, such species need to be grown and cultivated that generate less harmful impact on the environment.

3.2 Edible Crop Production from Polluted Land and Bio-fortification

Globally, several agricultural lands are polluted with rich contaminants that are further classified as essential micronutrients (e.g., Fe, Mg, Cu, Zn and Se). In this regard, bio-fortification of edible plants is one optional way forward for the treatment (Vamerali et al. 2009). Biofortification of edible plants with essential and important micronutrients (i.e. Zn, Fe, Cu, Mg and Se) can be targeted and further achieved on polluted land via cropping (Zhu et al. 2009; Vamerali et al. 2009).

An important and super essential dietary micronutrient for both species of animals and plants is Selenium (Madejón et al. 2011). When talking about food components, several grains like wheat (*Triticum aestivum* L.), rye (*Secale cereale* L.) and barley (*Hordeum vulgare* L.) have Selenomethionine (SeMet) as the major chemical species of Se present in them. This presence generates a contribution of nearly 60–80% of the total Se content (Stadlober et al. 2001). In addition to SeMet another variant of Se is also found known as Selenomethylcysteine (SeMeSeCys). For few rice samples collected from the Se-contaminated regions of south-central China, this information has been confirmed in research by X-Ray absorption technique (Williams et al. 2009). SeMeSeCys and SeMet, both are known to provide extra health benefits over the readily available inorganic Se, with SeMeSeCys also believed to have anti-carcinogenic properties (Rayman 2008). In general the level of selenium in rice grains is reported to range between 33–50%, because of this high range accumulation in rice as well as locally contaminated soil, Se-accumulating crops with soil can further be used for bio-fortification studies (Beilstein et al. 1991). Studies have shown the presence of Se can further aid in lowering the uptake of lead in rice varieties, therefore cultivation of rice grains in Se-contaminated soils can on the contrary reduce the uptake of other pollutants/metals too (Yu et al. 2014). It was also observed in studies that if linseed (*Linum usitatissimum* L.) was grown on contaminated and polluted soils with higher ranges of metals like Fe, Cu and Zn, there was definite increase in the height as well as increase in the capsules per plant (Rastogi et al. 2014). Therefore, if linseed is cultivated on metal-contaminated soil, it has a possibility that it could enhance the nutrient density in the seeds as these metals are important and essential micronutrients. The bio-fortification as well as remediation potential of maize (*Zea mays* L.) and radish that was grown in pyrite waste dump at Torviscosa (Udine), Italy were studied by Vamerali et al. (2009). The studies revealed that in maize grains (in mg kg^{-1}), presence of heavy metals like Cr (0.12), Cd (0.001), Cu (3.28), Co (0.002), Mn (6.17), Zn (40.2), Pb (0.001) and Ni (0.41) was found to be much lower. But, in radish the concentrations of both Cd (2.34) and Pb (4.20) were recorded to be above the permissible limit of the European Union. Some more studies have also proved that plants easily growing and flourishing in contaminated soils can have the potential and capability to fall in regulatory limits with respect to the accumulation & collection of toxic metals in edible parts of plants. For example, the Cd, Pb and Zn metal concentrations and its accumulation in maize grain (Meers et al. 2010), further as accumulation in beet root and lettuce (Warren et al. 2003) and then Ni concentration in carrot and onion (Stasinos and Zabetakis 2013) were detected below the limit. It is hence reported in various research studies that healthy crop production/cultivation on polluted soils is possible and is being investigated broadly. The studies have also summed up that such cultivation may still be in regulatory limits giving scope to use such contaminated lands.

3.3 Plants that Produce a Pollutant Free Edible Part

Intercropping/co-cropping (which involves one hyperaccumulator + pollutant-free food crop) is an alternative to both producing/cultivating food with simultaneous remediation of contaminated soil. This should enhance crop growth, reduce translocation of harmful pollutants to edible parts and act as an attractive mitigation method for remediation. This offers less financial losses to farm owners with quick remediation (Haller et al. 2018). But, this also brings to notice for being one of the most hazardous strategies which accounts to a planned systematic mapping of translocation patterns of relevant and essential pollutants in various species. This is to make sure that the pollutant doesn't enter the food chain. Studies like phytoremediation and its process, could not be very successful due to our lack of understanding of transport and further related tolerance mechanism in plants with single genus and its varieties (Becker 2000; Feist and Parker 2001; Dickinson et al. 2009; Haller et al. 2018).

Current and previous studies (Dickinson et al. 2009; Mahar et al. 2016) have laid their focus much on firstly identifying and investigating the hyperaccumulators namely as *Pteris vittata* (As), *Berkheyacoddii* (Ni), *Alyssum* spp. (Ni and Co), and *Noccaea caerulescens* (Cd, Ni, Pb, Zn). But, the studies of Linger et al. (2002) have brought out a new scope in CPFP i.e. combined phytoremediation and food production that even less efficient accumulators, can work correctly provided they have the potential to provide food/fibres crops. Very crucial for this strategy is the correct identification of such plant species that have ability to absorb contaminants in its tissue without further translocating them to the edible parts or combinations of hyperaccumulators/excluders (Singh et al. 2011; Haller et al. 2017) but in edible plants this important knowledge about the translocation patterns is presently limited and may also be inconsistent. It is out of the interest of current research, to focus on plants which are phytoextracting in nature, but totally avoid accumulating pollutants in the edible part. But, with time few promising studies have reported few plant species may translocate heavy metals and further also be able to release the organochlorines to roots, stems and leaves with much less concentration in edible parts (Liu et al. 2012; Haller et al. 2017; Pandey and Mishra 2018; Wang et al. 2019). Relevance to this was found in a Northern India study on fly ash dump which had contamination of metals like Cr, Cd, Cu, Co, Mn, Zn, Mo, Ni, As, Pb, Se etc. It was found that the fruit tree named *Ziziphus mauritiana* serves as an accumulator for few metals but majority of the metal concentration was far below the set WHO standards in its edible parts (Pandey and Mishra 2018). Another study focussed on field experiments in North-Eastern part of China where it was found *Apium graveolens* had the potential to significantly enhance the remediation process of PAHs with removal efficiency in the range of 31 and 50% post three months. Yet, the concentration of active PAHs in the edible, above ground part was observed to be lower than the Chinese Standard of limits which is 5 mg kg^{-1} in food (GB2762-2017) (Wang et al. 2019). In the field studies of Nicaragua, three cultivars of *Amaranthus* spp. were found highly responsive and active to soils contaminated with toxaphene and pesticide metabolites. The congeners of toxaphene, a-HCH, b-HCH, g-HCH and dieldrin were found to be bioaccumulated in the leaves, roots and/or stem. The concentrations were untraceable or were below the set EU Maximum Residue Level (MRL) in the cultivars *A. cruentus* 'Don Le_on' and *A. caudatus* 'CAC48 Perú'(Haller et al. 2017). A vice versa to the same situations has been observed where the concentrations of metal were hyper accumulated in seeds of the plants which were further considered unsafe to consume further (Haller et al. 2017). In Bulgaria for Cd, Cu, Zn & Pb polluted soils, few more organs of plants like peanuts (*Arachis hypogaea* L.), rapeseed (*Brassica napus* L.), sunflower (*Helianthus annuus* L.), and sesame (*Sesamum indicum* L.) were inspected. Least harmful concentration was observed in seeds/shells of the fruit whereas active locations like the leaves, the roots and stems observed the highest concentrations of locked and accumulated metals (Angelova et al. 2005). The study also made a striking remark on the presence of Pb and Cu to exceed twice the permissible levels in sunflower oil making it unsafe for public consumption. Experiments in southern China showed successful cases of a co-cropping system as an experiment in which Cu and Zn were successfully separated from sewage sludge using (*Sedum alfredii*) hyperaccumulator plant. The harvested crops like *Zea mays* and *Alocasia macrorrhizos* were successfully used as animal feeds under Chinese standards (Xiaomei et al. 2005; Wu et al. 2007). It is a known fact that the associations between selected plant species used in co-cropping are super-complex and also important. This could be due to alterations in shared rhizospheres (due to presence of various plants) and thereby alter and also affect the availability of pollutants to the surrounding plants (Tang et al. 2012). Research of Whiting et al. (2001) proved that such a co-cropping system can be successful. In the co-cropping research experiments between a hyperaccumulator (*N. caerulescens*) and a non-accumulator (*Thlaspi arvense*), the latter increased much in growth. There was also reduction in zinc uptake in the non-accumulator whereas in contrast the hyperaccumulator had an increased zinc uptake. In various food crops there was an increased uptake of Pb, Cu and Cd due to co-cropping with *Kummerowia striata* (Liu et al. 2012). As many factors including the pH of soil, with the genetics of plant, actively present plant-associated microorganisms and the co-cropping etc. are responsible to affect the process of bioaccumulation or even the translocation patterns. Therefore it is potentially hazardous to infer some crucial information about translocation from one case study to another. These few marked research studies hereby indicated that there needs to be strict monitoring in the bioaccumulation of pollutants at specific sites and identify the safe edible part. The level of stored pollutant in every part can be different with no equal distribution in edible parts. For example: crops such as *Ipomea batata*, *Colocasia esculenta*, *Cucurbita moschata* and *Dioscorea bulbifera* had 3 to 40 times extra organochlorine contamination in the pulp, making peeling and rinsing a crucial step before consumption (Cabidoche and Lesueur-Jannoyer 2012; Clostre et al. 2014).

3.4 Strategies for Reducing Pollutants in Edible Parts

Most noteworthy concern to deal with here is to firstly grow edible plants and crops on contaminated soils with absorbed accumulated pollutant inside the plant and that too is not in regulatory limits (Ye-Tao et al. 2012). In an attempt to mass scale exploitation of polluted lands for crop production the major challenge is to prevent the related potential health risks. In this process discussed above the complete removal of the accumulated pollutant does not take place in the plant ever (Eapen et al. 2007). This can give rise to biomagnification of the pollutant in the food chain further (Köhler and Triebskorn 2013). Also, the growth/yield of plants can be easily affected by the toxic elements in soil, lack of nutrients and lack of necessary beneficial organisms (Abhilash et al. 2013). Such, on ground conditions give rise to the need of both particular agronomic practices coupled with agro-technological interventions to aid the yield and restrict the flow of pollutant transfer to further phytoproducts (Tripathi et al. 2015). Based on Ye-Tao et al. (2012), research the strategy to cope this can be summed as (1) selecting and breeding for low-accumulating cultivars (phytoexcluders) for polluted lands, (2) reducing the bioavailability of pollutants in the soil and (3) restricting the uptake and translocation of pollutants to edible parts. The ensuing sections briefly highlight various strategies that can be employed to achieve these endpoints.

Using low-accumulating cultivars—Many researchers have brought forward the fact that it is the plant species, the cultivar with relevant species-specific traits on which the accumulation of pollutants depends on a plant. For example the studies of Ye-Tao et al. (2012) had extensively worked on various cultivars of rice, maize; wheat and soybean (*Glycine max* L.) with respect to study review the differences in uptake of heavy metals. To use polluted land, it is therefore important to screen the suitable species of cultivars having reduced contamination. Once identification of such a cultivar is done, both site and crops specific agronomic practices can be further optimized. This should aid and accelerate in plant–microbe interactions with increase in the nutrient. Making the phytoavailability of the pollutants, there should be a reduction in toxicity and also hold more efficiency for fertilizer (Gilbert 2013).

Reducing the bioavailability of pollutants in the soil—Certain soil corrective measures like as addition of lime, addition of phosphate and few silicon-based materials, or even adding some adsorption agents (e.g., iron oxides, zeolites, manganese oxides and clay minerals) can aid cost-effective chemical immobilization for reduction in heavy metal uptake in plants (Ye-Tao et al. 2012; Kashem et al. 2010). There can be few more organic healthy bio-corrections like addition of biochar, manure, sludge, peat, agricultural residues, compost or vermicompost. Not only can these bio-corrections be favoured due to their potential to reduce pollutant availability to the plants but also function to provide nutrition to plants. These amendments may further aid to degrade organic pollutants by supporting microbial consortia. For example, biochar in 10% was found successful when added to heavy metal-polluted soil, to enhance and accelerate the production of rape seed and sideways even brought down the heavy metal concentrations of Cd, Zn and Pb by 71, 87 and 92%, respectively

(Houben et al. 2013). In radish growing on polluted land PAH concentrations as well their uptake with further accumulation was found much reduced post addition of activated carbon, charcoal or compost in soil (Marchal et al. 2014). Similar results were found with experiments that used humic acid for bio-fortification (Vamerali et al. 2009), on the contrary chelating agents have been effectively found useful to reduce toxicity of metals. Few more applications like crop rotation, intercropping, drip irrigation system and also the inoculation of plant growth-promoting rhizobacteria (PGPR) and endophyte with application of microbial enzymes have also been trusted to amplify the process of bioremediation with respect to contaminants present in soil, and further aid in plant growth and reduce accumulation of the pollutant in focused edible parts (Karigar and Rao 2011; Rao et al. 2010; Segura and Ramos 2013; Alvareza et al. 2012). These healthy agro-based practices which enhance the plant–microbe interactions are very much needed for sustainable agriculture on lands affected by pollution and pollutants.

Minimizing the entry of toxic pollutants into the plant parts—To increase the fertility of contaminated soils and attempt to degrade the pollutants in the root zone itself, another approach like Rhizosphere engineering is practised (Dubey and Kumar 2022). This brings changes and manipulations in the microbial community structure of the soil (Hur et al. 2011), AMF colonization (Gao et al. 2012) with endophytic kind of microbial association (Germaine et al. 2009). Biotechnological and microbiological approaches like metatranscriptomics and metaproteomics can be used to extract and also recover the novel microbial strains and identify from the polluted system new and current degradation pathways (Machado et al. 2012; Junttila and Rudd 2012). With the advancements made in genetic studies, for a variety of agricultural traits, one can easily identify the quantitative trait loci (QTLs). These QTLs can offer and give far better links to identify certain responsive traits which could possibly be further used to enhance and even accelerate the growth, as well as the yield and control the stress tolerance of crops to be grown in polluted soils. Another promising avenue is root genetics which can show very good potential for the modification in root architecture, rhizoremediation of the pollutants, with increase in the water use efficiency and also focus on better uptake of the nutrient, help in translocation and use efficiency (Meister et al. 2014; Villordon et al. 2014; Tian et al. 2014; Schmidt 2014). On polluted sites, enhancing the process of nanoremediation (i.e. degradation of pollutants) by nanotechnology will offer much promising & a better approach for minimizing and limiting the entry of toxic pollutants into various plant parts (Karn et al. 2009). In light of this there exists certain nanoparticles (NPs) like the nZVI, ZnO, TiO_2, carbon nanotubes, fullerenes and bimetallic nanometals which are perfect for remediation of the soil (Karn et al. 2009). The role played by NPs is very crucial to firstly, be able to immobilize heavy metals (Cr(VI), Pb(II), As(III) and Cd) present in soil and secondly also help to reduce the concentration of heavy metals in leachates to a permissible level (Mallampati et al. 2013). In tannery waste contaminated soil, NPs also have shown the tendency of redox reaction to interconvert heavy metals like Cr(VI) to the less toxic forms like Cr(III). Post addition of NPs in a Pb-contaminated fire range soil, it was investigated that the TCLP-leachable Pb fraction decreased from 66 to 10% (Singh et al. 2012; Liu and Zhao 2013). NPs are also being used for degradation of organic pollutants such as carbamates, chlorinated organic solvents, DDT and PCBs, (Zhang 2003; El-Temsah 2013). Post remediation by nanoparticles, the treated contaminated land could be further used for agricultural production. Close examination of pros and cons of nanotechnology need to be monitored further to make it more developed, approachable for successful use for contaminated land remediation.

3.5 Connecting Phytoremediation and Food Production

Due to the extensive use of mining and urban civilization, soil pollution is a major environmental threat to nature and human lives. Against the previously used alternatives to treat soil pollution like incineration and soil washing etc., phytoremediation has been one of the most suitable & sustainable for remediation (Gomes 2012). It is a green remediation and also one of the most ecologically responsible alternatives. It is currently one of the safest alternatives to produce safe crops in contaminated zones of soil and water (Jasrotia et al. 2017). A combination of phytoremediation and food production (CPFP) can hence produce safe food for consumption (Haller and Jonsson 2020).

3.5.1 Benefits and Limitations

Once a hyper accumulator plant is identified for CPFP what makes CPFP a success is the sheer deep knowledge related to the plant's physiological as well as the biochemical mechanisms. This identification and characteristic knowledge is important to develop an efficient CPFP model/programmes and further prevent the exposure of toxic levels of soil pollutants/leaching/exposure to hazards to the consumers. Once plant species is identified any of the five named plant remediation from the following can be applied (Fig. 3.2). Table 3.1 further signifies targeted benefits of CPFP (adapted from Haller and Jonsson 2020).

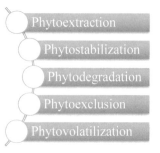

Fig. 3.2 Types of phytoremediation alternatives for CPFP

Table 3.1 Depicts the characteristics of three main strategies that may be adopted to minimize human exposure to pollutants in CPFP projects (modified from Haller and Jonsson 2020)

Selection of strategy	Strategy level 1	Strategy level 2	Strategy level 3
Description of strategy	Either it is removed from the soil followed with degradation of the pollutant prior to food production (Strategy 1)	Before remediation try to save the food chain from contamination (Strategy 2)	During the production of a pollutant-free edible part, steps like extraction/degradation of pollutants should be done (Strategy 3)
Type of phytotechnology	Phytoextraction of organic pollutants metals	Phytoexclusion Phytostabilization Phytovolatilization Succession based cropping	Intercropping and Phytoextraction with plants that avoid accumulation of pollutants in the edible parts
Disadvantages	Expected financial loss (Income) until remediation is carried out	There is no remediation of present inorganic pollutants and only natural attenuation happens of organic pollutant hence, is an interim solution. It poses a risk if mismanaged as the pollutants could enter the food chain	Poses a risk if mismanaged, also the pollutants have high chances to enter the food chain
Advantages	Safe	Low cost Easy to adopt	Process has immediate implementation with no delay. But, during this period incomes are certainly diminished
Sustainability quotient	Effective	Moderate	Highly sustainable

3.5.2 Utilizing Harvested Plant Material as a Resource

Once extracted the pollutant needs quick dispersal solution to avoid leaching in soil which is a common hurdle for successful phytoremediation projects on site. Based on the study and result findings of few researchers like Mitsch and Jørgensen (2004), Song and Park (2017), and Song et al. (2016) harvesting can be concluded as an important yet expensive solution. This is again followed by pitching and identifying proper eco-safe disposal sites/facilities and legal policy framework of the location for disposal standards. What we should look forward to is solutions where financial returns can be gained from leftover plant biomass post harvesting from remediation sites. Figure 3.3 explains a schematic for a successful phytoremediation system that

Fig. 3.3 Schematic for expected successful phytoremediation project

encapsules a higher financial gain from post-harvest treatment of extractant from plant residual biomass.

3.5.3 Policy Implementation

Post the inception of a new idea/technology, its production and post-release of a new product in market after its launch has to pass through a series of hurdles to reach its final goal. Genetically modified crops (GM Crops like cotton, soybeans) faced a series of challenges to reach the plates from farms, on a similar note, the 2 major identified hurdles for CPFP projects are (a) Involved legal legislations and (b) Public acceptance. For local environmental based policy and its implementations, one needs to study the local area of application with a set government framework. The goal should be to use degraded land appropriately and to set standards for disposal of waste (biomass post-harvest). For public acceptance for food grown on degraded land, post treatment needs a thorough process of PPP (public private partnership) with stakeholders. This should be in an attempt to have a flow of transparency of the process/related information, with correct labelling of food (from CPFP), disclosure of site and public reviews and their engagement.

3.5.4 Viewpoint and Study Needs

We on a global level currently need a healthy food chain and related supply to address the growing food availability issue. From cultivable farms, the focus now is shifted to using lands which are degradable in nature or polluted in form. Immediate strategies are to be researched and investigated for this goal, which at the same time should be able to prevent pollutants from entering the food chain, and be able to have sufficient healthy food supply. Reasonable management of hazardous waste could ensure a

3 Utilizing Polluted Land for Growing Crops

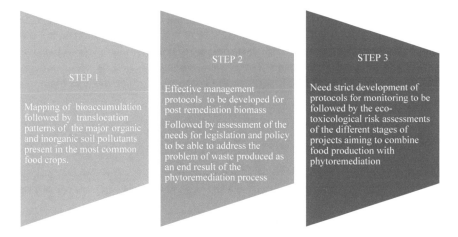

Fig. 3.4 Schematic of protocol for active implementation

better future with a decreasing rate of global food problems. One of the serious limitations lies in the uniqueness of every site specifically speaking, in terms of the spatial distribution of contaminants etc. limits the one-fits-all solutions and makes it impossible further (Haller and Jonsson 2020). Also, another challenge is lack of funds for investment on phytoremediation projects to have commercial success (Mench et al. 2010). More research aimed at funding should be released with the aim to have many researchers and scientists collaborate for betterment. Figure 3.4 gives a schematic of the protocols to be followed (Haller and Jonsson 2020).

3.6 Conclusions and Future Prospects

This study assessed both the risks and related opportunities for collaborating and also combining the two processes of phytoremediation of both the elemental and organic pollutants with food production (CPFP). There has been enough evidence that such combination studies work well for most pollutants (in context to climatic or socioeconomic variables), but still a number of challenges exist, and need to be taken care of. The challenges could be either very simple remediation-technological issues, reason being as one-fits-all solutions can barely address the variation in heterogeneous and complex nature of the different polluted sites globally. A second major challenge is post-harvest technology to be used for the contaminated biomass which is recovered from the phytoextraction process. This further has a capacity and high potential to raise the operation cost. The challenges for successful remediation food production projects can further be fuelled by inadequate and inappropriate soil governance.

This can be concluded that CPFP (combined phytoremediation and food production) yet hasn't reached technological maturity. But, monitoring pollution pathways, with perfect combinations of soil type (needs study and analysis), with related species of plants and agronomic practices can still allow production of safe food on polluted land. This would also allow restrictions on passing of pollutants in the food chain with reduction in soil pool of pollutants. This is a safe strategy until edible biomass has no harmful concentrations, with proper care and proper disposal of inedible biomass.

References

Abhilash PC, Dubey RK, Tripathi V, Srivastava P, Verma JP, Singh HB (2013) Remediation and management of POPs-contaminated soils in a warming climate: challenges and perspectives. Environ Sci Pollut Res 20:5879–5885

Abhilash PC, Tripathi V, Edrisi SA, Dubey RK, Bakshi M, Dubey PK, Singh HB, Ebbs SD (2016) Sustainability of crop production from polluted lands. Energy Ecol Environ 1:54–65. https://doi.org/10.1007/s40974-016-0007-x

Alvarez A, Yañez ML, Benimeli CS, Amoroso MJ (2012) Maize plants (Zea mays) root exudates enhance lindane removal by native Streptomyces strains. Int Biodeterior Biodegrad 66:14–18

Angelova V, Ivanova R, Ivanov K (2005) Heavy metal accumulation and distribution in oil crops. Commun Soil Sci Plant Anal 35:2551–2566. https://doi.org/10.1081/css-200030368

Becker H (2000) Phytoremediation using plants to clean up soils (brief article). Agric Res 48(6):4

Beilstein MA, Whanger PD, Yang GQ (1991) Chemical forms of selenium in corn and rice grown in a high selenium area of China. Biomed Environ Sci 4(4):392–398

Cabidoche YM, Lesueur-Jannoyer M (2012) Contamination of harvested organs in root crops grown on chlordecone-polluted soils. Pedosphere 22(4):562–571. https://doi.org/10.1016/s1002-0160(12)60041-1

Clostre F, Letourmy P, Thuriès L, Lesueur-Jannoyer M (2014) Effect of home food processing on chlordecone (organochlorine) content in vegetables. Sci Total Environ 490:1044–1050. https://doi.org/10.1016/j.scitotenv.2014.05.082

Dickinson NM, Baker AJM, Doronila A, Laidlaw S, Reeves RD (2009) Phytoremediation of inorganics: realism and synergies. Int J Phytoremediation 11:97–114. https://doi.org/10.1080/15226510802378368

Dubey RC, Kumar P (2022) Rhizosphere Engineering. Academic Press. United States, ISBN: 9780323899734

Eapen S, Singh S, D'Souza SF (2007) Advances in development of transgenic plants for remediation of xenobiotic pollutants. Biotechnol Adv 25:442–451. https://doi.org/10.1016/j.biotechadv.2007.05.001

El-Temsah YS (2013) Effects of nano-sized zero-valent iron (nZVI) on DDT degradation in soil and its toxicity to collembola and ostracods. Chemosphere 92:131–137

Feist LJ, Parker DR (2001) Ecotypic variation in selenium accumulation among populations of Stanleya pinnata. New Phytol 149(1):61–69. https://doi.org/10.1046/j.1469-8137.2001.00004.x

Gao X, Lu X, Wu M, Zhang H, Pan R, Tian J, Li S, Liao H (2012) Co-inoculation with rhizobia and AMF inhibited soybean red crown rot: from field study to plant defense-related gene expression analysis. PLoS ONE 7:e33977

Germaine KJ, Keogh E, Ryan D, Dowling DN (2009) Bacterial endophyte-mediated naphthalene phytoprotection and phytoremediation. FEMS Microbiol Lett 296:226–234

Gilbert N (2013) Case studies: a hard look at GM crops. Nature 497:24–26. https://doi.org/10.1038/497024a

Gomes HI (2012) Phytoremediation for bioenergy: challenges and opportunities. Environ Technol Rev 1(1): 59–66. https://doi.org/10.1080/09593330.2012.696715

Haller H, Jonsson A (2020) Growing food in polluted soils: a review of risks and opportunities associated with combined phytoremediation and food production (CPFP). Chemosphere 126826. https://doi.org/10.1016/j.chemosphere.2020.126826

Haller H, Jonsson A, Lacayo Romero M, Jarquín Pascua M (2017) Bioaccumulation and translocation of field-weathered toxaphene and other persistent organic pollutants in three cultivars of amaranth (A. Cruentus 'R127 Mexico', A. Cruentus 'Don Leon' Y A. Caudatus 'CAC 48 Perú') e A field study from former cotton fields in Chinandega, Nicaragua. Ecological Engineering

Haller H, Jonsson A, Fröling M (2018) Application of ecological engineering within the framework for strategic sustainable development for design of appropriate soil bioremediation technologies in marginalized regions. J Clean Prod 172:2415–2424

Houben D, Evrard L, Sonnet P (2013) Beneficial effects of Biochar application to contaminated soils on the bioavailability of Cd, Pb and Zn and the biomass production of rapeseed (Brassica napus L.). Biomass Bioenergy 57:196–204

Hur M, Kim Y, Song H-R, Kim JM, Choi YI, Yi H (2011) Effect of genetically modified poplars on soil microbial communities during the phytoremediation of waste mine tailings. Appl Environ Microbiol 77(21):7611–7619. https://doi.org/10.1128/AEM.06102-11

Jasrotia S, Kansal A, Mehra A (2017) Performance of aquatic plant species for phytoremediation of arsenic-contaminated water. Appl Water Sci 7:889–896. https://doi.org/10.1007/s13201-015-0300-4

Junttila S, Rudd S (2012) Characterization of a transcriptome from a non-model organism, Cladonia rangiferina, the grey reindeer lichen, using high-throughput next generation sequencing and EST sequence data. BMC Genom 13:575

Karigar CS, Rao SS (2011) Role of microbial enzymes in the bioremediation of pollutants: a review. Enzyme Res. https://doi.org/10.4061/2011/805187

Karn B, Kuiken T, Otto M (2009) Nanotechnology and in situ remediation: a review of the benefits and potential risks. Environ Health Perspect 117:1823–1831

Kashem MA, Kawai S, Kikuchi N, Takahashi H, Sugawara R, Singh BR (2010) Effect of Lherzolite on chemical fractions of Cd and Zn and their uptake by plants in contaminated soil. Water Air Soil Pollut 207:241–251. https://doi.org/10.1007/s11270-009-0132-7

Köhler HR, Triebskorn R (2013) Wildlife ecotoxicology of pesticides: can we track effects to the population level and beyond? Science 341:759–765. https://doi.org/10.1126/science.1237591

Kopittke PM, Menzies NW, Wang P, McKenna BA, Lombi E (2019) Soil and the intensification of agriculture for global food security. Environ Int 132:105078. https://doi.org/10.1016/j.envint.2019.105078

Linger P, Mussing J, Fisher H, Kobert J (2002) Industrial hemp (Cannabis sativa L.) growing on heavy metal contaminated soil fibre quality and phytoremediation potential. Ind Crop Prod 16(1):33–42. https://doi.org/10.1016/S0926-6690(02)00005-5

Liu R, Zhao D (2013) Synthesis and characterization of a new class of stabilized apatite nanoparticles and applying the particles to in situ Pb immobilization in a fire-range soil. Chemosphere 91:594–601

Liu L, Hu L, Tang J, Li Y, Zhang Q, Chen X (2012) Food safety assessment of planting patterns of four vegetable-type crops grown in soil contaminated by electronic waste activities. J Environ Manage 93(1):22–30. https://doi.org/10.1016/j.jenvman.2011.08.021

Machado A, Magalhães C, Mucha AP, Almeida CM, Bordalo AA (2012) Microbial communities within saltmarsh sediments: composition, abundance and pollution constraints. Estuar Coast Shelf Sci 99:145–152

Madejón P, Barba-Brioso C, Lepp NW, Fernández-Caliani JC (2011) Traditional agricultural practices enable sustainable remediation of highly polluted soils in Southern Spain for cultivation of food crops. J Environ Manag 92:1828–2183

Mahar A, Wang P, Ali A, Awasthi MK, Lahori AH, Wang Q, Zhang Z (2016) Challenges and opportunities in the phytoremediation of heavy metals contaminated soils: a review. Ecotoxicol Environ Saf 126:111–121. https://doi.org/10.1016/j.ecoenv.2015.12.023

Mallampati SR, Mitoma Y, Okuda T, Sakita S, Kakeda M (2013) Total immobilization of soil heavy metals with nano-Fe/Ca/CaO dispersion mixtures. Environ Chem Lett 11:119–125

Marchal G, Smith KE, Mayer P, de Jonge LW, Karlson UG (2014) Impact of soil amendments and the plant rhizosphere on PAH behaviour in soil. Environ Pollut 188:124–131

Meers E, Van Slycken S, Adriaensen K, Ruttens A, Vangronsveld J, Laing GD, Witters N, Thewys T, Tack FMG (2010) The use of bio-energy crops (Zea mays) for "phytoattenuation" of heavy metals on moderately contaminated soils: a field experiment. Chemosphere 78(1):35–41. https://doi.org/10.1016/j.chemosphere.2009.08.01

Meister R, Rajani MS, Ruzicka D, Schachtman DP (2014) Challenges of modifying root traits in crops for agriculture. Trends Plant Sci 19:779–788

Mench M, Lepp N, Bert V, Schwitzguebel J-P, Gawronski SW, Schröder P, Vangronsveld J (2010) Successes and limitations of phytotechnologies at field scale: outcomes, assessment and outlook from COST action 859. J Soils Sediments 10(6):1039–1070

Mitsch WJ, Jørgensen SE (2004) Ecological engineering and ecosystem restoration. Wiley, Hoboken, N.J

Montanarella L, Vargas R (2012) Global governance of soil resources as a necessary condition for sustainable development. Curr Opin Environ Sustain 4(5):559–564. https://doi.org/10.1016/j.cosust.2012.06.007

Pandey VC, Mishra T (2018) Assessment of Ziziphus mauritiana grown on fly ash dumps: Prospects for phytoremediation but concerns with the use of edible fruit. Int J Phytoremediation 20:1250–1256. https://doi.org/10.1080/15226514.2016.1267703

Rao MA, Scelza R, Scotti R, Gianfreda L (2010) Role of enzymes in the remediation of polluted environments. J Soil Sci Plant Nutr 10:333–353

Rastogi A, Mishra BK, Singh M, Mishra R, Shukla S (2014) Role of micronutrients on quantitative traits and prospects of its accumulation in linseed (Linum usitatissimum L.). Arch Agron Soil Sci 60:1389–1409

Rayman MP (2008) Food-chain selenium and human health: emphasis on intake. Br J Nutr 100:254–268. https://doi.org/10.1017/s0007114508939830

Schmidt W (2014) Root systems biology. Front Plant Sci 5:1–2

Segura A, Ramos JL (2013) Plant–bacteria interactions in the removal of pollutants. Curr Opin Biotechnol 24:467–473

Singh B, Gupta S, Azaizeh H, Shilev S, Sudre D, Song W, Martinoia E, Mench M (2011) Safety of food crops on land contaminated with trace elements. J Sci Food Agric 91:1349–1366

Singh R, Misra V, Singh RP (2012) Removal of Cr(VI) by nanoscale zero-valent iron (nZVI) from soil contaminated with tannery wastes. Bull Environ Contam Toxicol 88:210–214

Song U, Park H (2017) Importance of biomass management acts and policies after phytoremediation. J Ecol Environ 41(1):13

Song U, Kim DW, Waldman B, Lee EJ (2016) From phytoaccumulation to postharvest use of water fern for landfill management. J Environ Manag 182:13–20

Stadlober M, Irgolic SM, KJ, (2001) Effects of selenate supplemented fertilisation on the selenium level of cereals—Identification and quantification of selenium compounds by HPLC-ICP-MS. Food Chem 73:357–366

Stasinos S, Zabetakis I (2013) The uptake of nickel and chromium from irrigation water by potatoes, carrots and onions. Ecotoxicol Environ Saf 91:122–128. https://doi.org/10.1016/j.ecoenv.2013.01.023

Tang Y-T, Deng T-H-B, Wu Q-H, Wang S-Z, Qiu R-L, Wei Z-B, Morel JL (2012) Designing cropping systems for metal-contaminated sites: a review. Pedosphere 22(4):470–488. https://doi.org/10.1016/s1002-0160(12)60032-0

Tian YL, Zhang HY, Guo W, Wei XF (2014) Morphological responses, biomass yield and bioenergy potential of sweet sorghum cultivated in cadmium—contaminated soil for biofuel. Int J Green Energy 12:577–584

Tripathi V, Fraceto LF, Abhilash PC (2015) Sustainable clean-up technologies for soils contaminated with multiple pollutants: plant–microbe–pollutant and climate nexus. Ecol Eng 82:330–335

Vamerali T, Bandiera M, Mosca G (2009) Field crops for phytoremediation of metal-contaminated land. Env Chem Lett 8:1–17

Villordon AQ, Ginzberg I, Firon N (2014) Root architecture and root and tuber crop productivity. Trends Plant Sci 19:419–425

Wang H, Zhao Y, Muhammad A, Liu C, Luo Q, Wu H, Wang X, Zheng X, Wang K, Du Y (2019) Influence of celery on the remediation of PAHs contaminated farm soil. Soil Sediment Contam Int J 28(2):1–13

Warren GP, Alloway BJ, Lepp NW, Singh B, Bochereau FJM, Penny C (2003) Field trials to assess the uptake of arsenic by vegetables from contaminated soils and soil remediation with iron oxides. Sci Total Environ 311:19–33. https://doi.org/10.1016/S0048-9697(03)00096-2

Whiting SN, Leake JR, McGrath SP, Baker AJM (2001) Assessment of Zn mobilization in the rhizosphere of Thlaspi caerulescens by bioassay with non-accumulator plants and soil extraction. Plant Soil 237(1):147–156. https://doi.org/10.1023/a:1013365617841

WilliamsPN LE, Sun GX, Scheckel K, Zhu YG, Feng X, Zhu J, Carey AM, Adomako E, Lawgali Y, Deacon C, Meharg AA (2009) Selenium characterisation in the global rice supply chain. Environ Sci Technol 43(15):6024–6030. https://doi.org/10.1021/es900671m

Wu Q-T, Hei L, Wong JWC, Schwartz C, Morel J-L (2007) Co-cropping for phyto-separation of zinc and potassium from sewage sludge. Chemosphere 68:1954–1960

Xiaomei L, Qitang W, Banks MK (2005) Effect of simultaneous effect of simultaneous establishment of sedum alfredii and zea mays on heavy metal accumulation in plants. Int J Phytoremediation 7(1):43–53. https://doi.org/10.1080/16226510590915800

Ye-Tao TA, Teng-Hao-Bo DE, Qi-Hang WU (2012) Designing cropping systems for metal-contaminated sites: a review. Pedosphere 22:470–488

Yu L, Zhu J, Huang Q, Su D, Jiang R, Li H (2014) Application of a rotation system to oilseed rape and rice fields in Cd-contaminated agricultural land to ensure food safety. Ecotoxicol Environ Saf 108:287–293

Zhang WX (2003) Nanoscale iron particles for environmental remediation: an overview. J Nanoparticle Res 5:323–332

Zhu YG, Pilon-Smits EAH, Zhao FJ, Williams PN, Meharg AA (2009) Selenium in higher plants: understanding mechanisms for biofortification and phytoremediation. Trends Plant Sci 14(8):436–442. https://doi.org/10.1016/j.tplants.2009.06.006

Chapter 4
Plant Assisted Bioremediation of Heavy Metal Polluted Soils

Sumita Chandel, Rouf Ahmad Dar, Dhanwinder Singh, Sapna Thakur, Ravneet Kaur, and Kuldip Singh

Abstract Industrial and anthropogenic activities are the major reason for heavy metal pollution. To date, thousands of hectares of farmland globally and in India specifically have been contaminated by heavy metals. This has adversely affected the crop productivity, soil microbial diversity and eventually deteriorated the soil quality. Soil quality is closely associated with crop quality, human health and welfare. Therefore, the remediation of these metal-polluted soils becomes imperative. Conventional remediation methods like precipitation, oxidation/reduction, filtration, evaporation and adsorption etc. are energy demanding or require a large number of chemical reagents and are associated with possible production of secondary pollutants. Fortunately, some microorganisms with the capability to induce resistance to heavy metals, and reduce or adsorb them in non-toxic form can be used for possible bioremediation

S. Chandel (✉) · D. Singh · K. Singh
Department of Soil Science, Punjab Agricultural University, Ludhiana, Punjab, India
e-mail: Sumita-coasoil@pau.edu

D. Singh
e-mail: dhanwinder@pau.edu

K. Singh
e-mail: kuldip@pau.edu

R. A. Dar
Department of Microbiology, Punjab Agricultural University, Ludhiana, Punjab, India
e-mail: roufdar-mb@pau.edu

S. Thakur
Department of Forestry and Natural Resources, Punjab Agricultural University, Ludhiana, Punjab, India
e-mail: sapnathakur@pau.edu

R. Kaur
Department of Botany, Punjab Agricultural University, Ludhiana, Punjab, India
e-mail: ravneet-bot@pau.edu

R. A. Dar
Department of Industrial Microbiology, Sam Higginbottom University of Agriculture, Technology and Sciences, Prayagraj, India

© The Author(s), under exclusive license to Springer Nature Switzerland AG 2023
V. C. Pandey (ed.), *Bio-Inspired Land Remediation*, Environmental Contamination Remediation and Management, https://doi.org/10.1007/978-3-031-04931-6_4

of polluted soils, thus representing an economical and environment-friendly remediation method. These microbes detoxify the heavy metals, clean up the environment and increase the soil fertility, but, the adsorbed or converted metal still remains in the soil is the problem associated with it. Phytoremediation can be another option for detoxification of heavy metal polluted soils. However, phytoremediation alone has its limitations. Hence, the most effective way of remediation of heavy metal polluted soils is an integrated approach that involves both plants and microbes. Understanding the whole mechanism of plant assisted bioremediation along with bioavailability, uptake, translocation, sequestration and different defence mechanisms will help to develop heavy metal stress-resistant cultivars and highly efficient plant species for phytoremediation in harmony with microflora through genetic engineering technologies. Hence, this chapter will provide an understanding of plant assisted bioremediation, the fate of heavy metals in plant and soil, different plant defence mechanisms and potential microflora for plant assisted bioremediation.

Keywords Bioremediation · Heavy metal pollution · Microflora · Phytoextraction · Soil fertility

4.1 Introduction: Background of Heavy Metal Pollution

Heavy metals being toxic and bioaccumulative in nature, are environmental pollutants with prolonged persistence in the environment, thus leading to detrimental effects on floral wealth and human health (Rzymski et al. 2014). They are metals possessing a specific density of more than 5 g cm^{-3} and have adverse impacts on the life and environment (Järup 2003). Some metals known as micronutrients, (copper, iron, manganese, molybdenum, nickel and zinc) play a vital role in the normal functioning of plant cells such as biosynthesis of nucleic acids, chlorophyll, carbohydrates, secondary metabolites, stress resistance and maintenance of biological membranes as well as overall growth of the plants (Rengel 2004). However, when their internal concentration transcends a certain threshold limit, they negatively influence plant growth and become toxic, forming a bell-shaped dose–response relationship (Marschner 1995). Moreover, the concentration of heavy metals is generally location-specific, subjected to the source of individual pollutants.

As per the World Health Organization (WHO), the common toxic 'heavy metals' of public health concern are arsenic (As), cadmium (Cd), cobalt (Co), chromium (Cr), mercury (Hg), nickel (Ni), lead (Pb), selenium (Se), manganese (Mn), copper (Cu) and molybdenum (Mo). The standards for heavy metals in soil, plant and water as per Bureau of Indian standards (BIS) and WHO have been presented in Table 4.1.

Table 4.1 Normal and critical range of heavy metals in soil, plant and water

Heavy metal (s)	Normal range in soil (mg kg^{-1})	Critical soil total concs (mg kg^{-1})	Normal range in plant (mg kg^{-1})	Critical concentration in plants	Permissible limit in water (mg L^{-1})
Arsenic	0.1–40	20–50	0.02–7	5–20	0.01
Cadmium	0.01–2.0	3–8	0.1–2.4	5–3	0.003
Cobalt	0.5–65	25–50	0.02–1	15–50	0.05
Chromium	5–1500	75–100	0.03–14	5–30	0.05
Mercury	0.01–0.5	0.3–15	0.005–0.17	1–3	0.001
Nickel	2–750	100	0.02–5	10–100	0.02
Lead	2–300	100–400	0.2–20	30–300	0.01
Selenium	0.1–5	5–10	0.0001–0.2	5–30	0.01
Manganese	20–10,000	1500–3000	20–1000	300–500	0.3
Copper	2–250	60–125	5–20	20–100	1.5
Molybdenum	0.1–40	2–10	0.03–5	10–50	0.07

Data from Bowen (1979), Kabata-Pendias and Pendias (1984), BIS (2012)

4.1.1 World Status

Rapid industrialization and exponential increase in the human population has increased the discharge of massive loads of heavy metal pollutants in the environment (Zhang et al. 2020). Globally, around 500 M ha of our land resources are facing the problem of soil contamination ended up with higher concentrations of heavy metals compared to the regulatory levels (Liu et al. 2018). Industries and other human activities discharge approximately 2 million tons/day of sewage and effluents into the water bodies making them unfit for various agricultural and other activities. Fly ash dumping sites of coal-based thermal power stations are also a major source of heavy metal pollution around the world (Pandey and Singh 2010; Pandey 2020). In developing nations, the situation is more critical where about 90% of sewage and 70% of industrial wastes (generally untreated/partially treated) are being discharged into surface water resources (Anonymous 2010). Over the past few years, the annual global release of heavy metals has surpassed 0.2 lacs MT for Cadmium, 9.3 lacs MT for Copper, 7.83 lacs MT for lead and 1.35 lacs MT for Zinc (Thambavani and Prathipa 2012). Further, heavy metal poisoning has become a universal public health concern. Heavy metal pollution in soils also has tremendously impacted the global economy, which has annually been estimated to be beyond US$10 billion.

4.1.2 Indian Status

The data available on the nature and extent of metal pollution and its impact assessment on the plant, soil and human health is not very conclusive in India. However, as per Indian central pollution control board (CPCB 2009), approximately 38,254 megaliters per day (MLD) of sewage and 25,000 MLD of untreated industrial wastewater generated from urban areas are released into the surface water bodies, wreaking degradation of the quality of water resources. Bhardwaj (2005) has estimated that by 2050, wastewater generation in India is going to be around 1,22,000 MLD. The utilization of such wastewater loaded surface water sources for irrigation purposes in agricultural fields has magnified the heavy metals concentrations in soils of agricultural fields particularly those situated in the vicinity of urban areas (Saha and Panwar 2013). However, the heavy metal accumulation in the soils will vary depending upon the source, concentration and duration of application (Rattan 2005). Usage of sewage water as irrigation for 20 years successively may result in significant accumulation of zinc (2.1 times), copper (1.7 times), iron (1.7 times), nickel (63.1%) and lead (29%) in the soils as compared to soils irrigated with tube well water (Simmons 2006). Such unrestricted transfer of heavy metals in arable land through wastewater irrigation will trigger more metal uptake by crops and will enter the food chain (Rattan 2002).

4.2 Sources of Heavy Metal Pollution

Geogenic and anthropogenic activities are mainly responsible for heavy metal pollution in the environment. Geogenic processes such as biogenic, terrestrial, volcanic processes, erosion, leaching and meteoric are the main sources of heavy metals in the environment (Muradoglu et al. 2015). While, industrialization, urbanization and modernization of the agricultural sector are substantially contributed to the release of heavy metal pollutants into the surrounding which gets deposited on the soil through natural processes of sedimentation and precipitation. In addition, anthropogenic processes such as irrigation with sewage and industrial wastewater, mining activities, fly ash disposal, excessive application of pesticides and fertilizers, have disturbed the natural balance of geochemical cycles, which in turn has resulted in the entry of heavy metals into the soil (Zhang et al. 2011; Dixit et al. 2015). The major contributors of heavy metals in the environment are listed in Table 4.2.

4.3 Plant Assisted Bioremediation: Techniques/Strategies

Plant assisted bioremediation involves the symbiotic relationship between rhizospheric microorganisms and the plant roots (Kumar et al. 2017). The symbiotic relationship intensifies bioavailability of the heavy metals and stimulates absorption capacity of the roots. Remediation of metal-polluted soil by soil microbes especially the rhizospheric population is known as rhizoremediation (Kuiper et al. 2004). Rhizoremediation involving plant growth-promoting rhizobia, mycorrhiza and other microorganisms is very efficient in promoting plant biomass and thus its efficiency to stabilize and remediate metal-polluted soil (Jing et al. 2007). Plant roots release exudates, may be enzymatic or non-enzymatic that modify the soil environment and habitat to numerous microorganisms. Rhizosphere plays a great role in the remediation of metal polluted soil. Heavy metals can only be transformed via several processes such as sorption, methylation, complexation or change in valence oxidation state, affecting their mobility and bioavailability. Microbes have an important role in the processes like carbon sequestration, plant growth, productivity and phytoremediation. Microorganisms (bacteria, fungi and microalgae) along with plants are the potential agents of bioremediation. They enhance the plant growth through different enzymatic activities, nitrogen fixation and reducing the ethylene production (Pandey and Singh 2019). Bacteria respond to the heavy metals and the molecules generated through oxidative stress in different ways. These are entrapped in the capsules, transported through heavy metals by the cell membrane, absorbed on the cell walls, precipitated or oxidized/reduced (Singh et al. 2010). The microbial response to heavy metals is important in harnessing them as potential candidates for remediation of metal polluted soils (Hemambika et al. 2011). Plant growth-promoting bacteria (PGPB) also known as growth-promoting agents are now assessed for their metal detoxifying potential in remediating metal-polluted soils (Ahemad 2014). Fungi are important as

Table 4.2 Major sources of heavy metals

Heavy metal (s)	Contributors of heavy metals in the environment
Arsenic	Volcanic eruptions, semiconductors, smelting coal mines, power plants, petroleum refining, metal adhesives, ammunition, wood preservatives, pesticides and herbicides, animal feed additives
Copper	Biosolids electroplating, mining activities, petroleum refining and smelting operations
Cadmium	Geogenic sources, metal smelting and refining process, combustion of fossil fuels, fertilizers, sewage sludge
Chromium	Sewage sludge, solid wastes, electroplating, tanning industries
Lead	Mining and smelting of metalliferous ores, leaded gasoline combustion, sewage and industrial waste, paints
Mercury	Volcanic eruptions, wild forest fires, emissions from industries producing caustic soda, combustion of coal, peat and wood
Selenium	Coal mining, oil refineries, fossil fuels, glass manufacturing industry, varnish and pigment formulation
Nickel	Volcanic eruptions, forest fire, landfilling operations, oceanic gaseous exchange, weathering of soils and geological processes
Zinc	Smelting and refining industries, mining operations, electroplating industry, bio solids

Source Lone et al. (2008)

these augment the phytoremediation by changing the bioavailability of metal through different ways like modifying the pH of the soil, production of different chelators, and controlling the redox reaction etc. (Ma et al. 2011a, b). Also, a high surface-to-volume ratio make bacteria a potential biosorbing agents. While, the plants absorb these metals and translocate them to various plant tissues and organs. Plants remediate the heavy metal polluted soil by adopting different techniques/strategies such as phytoextraction, phytostabilization, phytovolatilization, phytostimulation (Pandey and Bajpai 2019; Pathak et al. 2020). These are described as:

Phytoextraction: Phytoextraction is the process of the uptaking and storing of heavy metals from the soil by the plants (McGrath 1998). There are two fundamental ways of phytoextraction:

- **Natural**: The natural way of removal of heavy metals by the plants, also known as unassisted phytoremediation.
- **Assisted**: Microbes, plant hormones and chelating agents assist the plant in the remediation of heavy metal polluted soils (Malik et al. 2022).

Natural phytoremediation can be accomplished by either (1) hyperaccumulator plants or (2) genetic engineering of the plant with certain characteristics of hyperaccumulators for the accomplishment of phytoextraction (Chaney et al. 2005). The hyperaccumulator plants are the plants whose tissues can contain certain heavy metals from 1000 to 10,000 mg kg^{-1} (Black 1995). They can collect and concentrate the heavy metals in the harvestable tissues, biomass without affecting the plant growth. The heavy metal concentration in the hyperaccumulator plants is approximately 100 times higher compared to the ordinary plants. It is approximately 1000 mg kg^{-1} for arsenic and nickel, 100 mg kg^{-1} for cadmium and 10,000 mg kg^{-1} for zinc and manganese. The most prominent examples of hyperaccumulator plants are *Arabidopsis, Alyssum, Noccaea and the members of* Brassicaceae family.

Phytostabilization: Phytostabilization involves complexation, precipitation, sorption or metal reduction (Ghosh and Singh 2005). Plants restrict the movement of the metals in the roots by the assimilation, aggregation, adsorption and precipitation. They also help to avoid movement of the metals through water, wind, drainage and dispersion of soil (USEPA 1999). The phytostabilization stabilizes the metal contaminant rather than translocating it to the edible parts, that in turn can reach human beings (Prasad and Freitas 2003). In this, there is the aggregation of metal by roots or root exudates that immobilize and lower the accessibility of the soil pollutants. Chromium and lead are toxic metals that are remediated by phytostabilization. The proficiency of the phytostabilization is increased by the addition of nutrients to soil viz. lime and phosphate. *Brassica juncea* has been reported to be an efficient Phytostabilizer as it accumulates chromium in the roots (Bluskov and Arocena 2005).

Phytovolatilization: The release of metal pollutants to the atmosphere by the plants in altered or unaltered form after metabolic and transpirational pull is called phytovolatilization (USEPA 1999). Selenium, arsenic and mercury are the main metal pollutants that can be remediated through phytovolatilization (Dietz and Schnoor 2001).

4.4 Significance of Plant Assisted Bioremediation of Heavy Metal in Agriculture

Agriculture is the main backbone of the Indian economy and socio-political stability. With approximately 7% of the growth, Indian economy is the 7th largest in the world. The contribution of agriculture and its allied sectors was 51.81% during 1950–51, which declined to 18.20% by 2013–2014 and now it is approximately 14.39% (2018–19). These figures are still higher than most of the countries. Soil quality is one of the main contributing factors for sustainable agriculture. Sustainability is defined as the living within the regenerative capacity of the biosphere (Wackernagel et al. 2002). Inappropriate agricultural management practices, excessive use of a large number of chemicals, insecticides, pesticides, sludge and manure are attributing declination of soil quality. Developmental activities such as industrialization, urbanization and transportation are competing with natural resources, and impacting soil quality and biodiversity (Godfray et al. 2010). Despite several adverse effects of industries on the environment, they are considered important because of the unlimited human desires. The dependency of agriculture and industries on each other and their impact on the environment and human life is depicted in Fig. 4.1. Different anthropogenic activities have become the main reason for the deterioration of natural resources (soil + water). In the long run, polluted soil and water will not be suitable to grow the food which will directly or indirectly impact the socio-economic condition of the country (Saha et al. 2017). In the agroecosystem, agriculture and industries are the main reason for

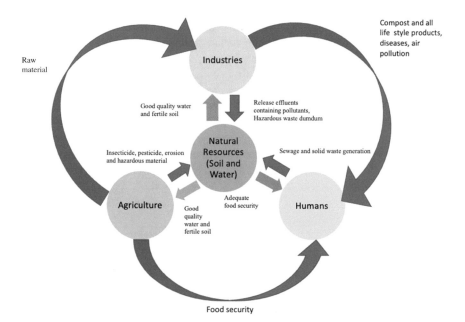

Fig. 4.1 Relationship of agriculture and industrial activities affecting natural resources and humans

soil pollution by heavy metals such as cadmium, chromium, lead, arsenic, nickel and mercury. Introduction of the above metals in the soil environment and their impact on soil is quite alarming, as:

- Entry of these metals in the agroecosystem results in degradation of soil structure and affects moisture retention in the soil profile.
- Heavy metals interact with soil components, microorganisms (nitrogen transformation, mineralization/immobilization etc.), root cells and affect the transformation of soil nutrients and their uptake.
- Build-up of salinity problem in the polluted soil along with the effect on water and nutrient uptake.
- Heavy metals lead to alkalinity development, result in more ammonia volatilization losses.
- The agronomic efficiency and partial factor productivity of polluted soil are normally lower than the unpolluted soil.
- The shelf life of the crops irrigated with industrial wastewater is lower than irrigated with fresh water.
- Heavy metal pollution results in huge ecological disturbances

4.5 Role of Microflora and Flora in Plant Assisted Bioremediation

Plant assisted bioremediation is an eco-friendly approach, encompassing the complex phenomenon of interaction between the microbes and plant genotype with its biotic and abiotic environment. The most important components of plant assisted bioremediation consist the in situ selection of genotypes and symbiotic microorganisms in the rhizosphere (having the capability of degrading the organic contaminants completely). Further, the identification of candidate genes and alleles linked with biochemical and physiological processes also has a key role in the development of a potential plant assisted bioremediation strategy. High levels of metal extraction and translocation to shoots and organic degradation are keys to develop an efficient phytoremediation measure. A most promising approach to substitute the costly remediation technologies is the use of plants assisted by microbes to clean up heavy metal polluted soils and water (Malik et al. 2022). Therefore, the selection of appropriate microbes and plant species is a prerequisite for effective remediation of heavy metal pollution.

4.5.1 Potential Microbes Involved in Bioremediation

Microorganisms help in the uptake of heavy metals through both active (bioaccumulation) and passive (adsorption) modes. Microbes (bacteria, fungi and algae) have been utilized to remediate the contaminated sites. The high surface to volume ratio, ubiquitous nature, the capability of growing in extreme conditions and active chemisorption sites make bacteria a potential candidate for bioremediation (Srivastava et al. 2015; Mosa et al. 2016). Higher absorption, uptake and recovery capacity of fungi make them suitable biosorbents for the remediation of toxic metals (Fu et al. 2012). Also, algae compared to other biosorbents produce high biomass. The high sorption capacity of algae and the presence of various metal binding chemical groups like hydroxyl, carboxyl, phosphate and amide make them a suitable contender for remediation of heavy metals (Abbas et al. 2014). The microorganisms involved in bioremediation of heavy metals are summarized in Table 4.3.

Numerous researchers have reported bacterial accumulation and sorption along with other plant growth promoting features responsible for the enhanced plant growth in polluted soils (Ma et al. 2011a, b; Kumar et al. 2009). Higher accumulation of heavy metals in plants without having any phytotoxicity is due to decreased internal availability of metals or metalloids and higher rhizospheric plant bioavailability (Deng et al. 2013a, b; Weyens et al. 2010). Nickle uptake by *Alyssum murale* was significantly enhanced by *Sphingomonas macrogoltabidus*, *Microbacterium liquefaciens* and *Microbacterium arabinogalactanolyticum* inoculation compared to the un-inoculated control (Abou-Shanab et al. 2003). Correspondingly, the inoculation of *Phaseolus vulgaris* with *Pseudomonas putida* KNP9 protected it from metal toxicity (lead and cadmium) and improved its growth with respect to controls (Tripathi et al. 2005). Therefore, the application of metal remediating plant growth-promoting bacteria (PGPB) along with plant growth promoting activities makes the remediation process more effective and efficient (Glick 2012). The utilization of the mining sites with the higher concentration of heavy metals is a global challenge to environmental sustainability (Ahirwal and Pandey 2021). In this direction, researchers demonstrated that the *Pseudomonas aeruginosa*-HMR1 removes heavy metals and exhibits plant growth-promoting attributes. Thus, the *P. aeruginosa*-HMR1 can be used for the restoration of mining lands for forestry, ornamental plants and agricultural purposes (Bhojiya et al. 2021).

Fungi have been found to have more tolerance to metals than bacteria (Deng and Cao 2017; Deng et al. 2013a). Fungi easily reach the microsites that are not accessible to the plant roots and thus can compete with other microbes for food and metal uptake. These protect the plant roots from directly interacting with metals and increase the soil hydrophobicity, thus hindering metal transport. Moreover, the extended mycelia formation by fungi also makes them suitable for bioremediation. In metal-polluted soils, various fungi like *Aspergillus, Trichoderma* and the arbuscular mycorrhizae (AM) have demonstrated the capacity to improve the phytoremediation process (Deng et al. 2011, 2013a). These fungi have high capability of immobilization of toxic/heavy metals by forming either the insoluble compounds, chelation or through biosorption. The fungal species and ecotype greatly affect phytoremediation efficiency. Some examples of bioremediation of heavy metals are given in Table 4.4.

Table 4.3 Potential microbes for remediation of heavy metal pollution

Microorganisms	Metal	Metal concentration (Initial) (mg L^{-1})	Efficiency (%) or Sorption capacity (mg g^{-1})	References
Bacteria				
Bacillus laterosporus	Cd	1000	159.5	Zouboulis et al. (2004)
Bacillus licheniformis	Cd	1000	142.7	Zouboulis et al. (2004)
Desulfovibrio desulfuricans	Cu	100	98.2	Kim et al. (2015)
Acinetobacter sp.	Cr	16	87%	Bhattacharya et al. (2019)
Bacillus subtilis	Cr	0.57	99.60%	Kim et al. (2015)
Methylobacterium organophilum	Pb		18%	Kim et al. (1996)
Cellulosimicrobium sp. (KX710177)	Pb	50	99.33%	Bharagava and Mishra (2018)
Staphylococcus sp.	Cu	100	98.20%	Kim et al. (2015)
Flavobacterium sp.	Cu	1.194	20.30%	Kumaran et al. (2011)
Micrococcus sp.		100	65.00%	Jafari et al. (2015)
Enterobacter cloacae	Pb	7.2	2.3	Kang et al. (2005)
Pseudomonas aeruginosa	Co	58.93	8.92	Kang et al. (2005)
	Ni	58.69	8.26	
Pseudomonas sp.	Cu	300	5.52	Rajkumar et al. (2008)
	Zn	275	3.66	
Fungi				
Aspergillus niger	Pb	100	34.4	Dursun et al. (2003)
	Cr(VI)	50	6.6	
Phanerochae techrysosporium	Pb	100	88.16	Iqbal and Edyvean (2004)
	Zn	100	39.62	
Rhizopus oryzae	Cu	100	34	Fu et al. (2012)
Sphaerotilus natans	Cr	200	60%	Kumar et al. (2017)
Saccharomyces cerevisiae (Y)	Cr	570.25	95%	Benazir et al. (2010)
Aspergillus versicolor	Cu	50	29.06%	Tas et al. (2010)
Algae				

(continued)

Table 4.3 (continued)

Microorganisms	Metal	Metal concentration (Initial) (mg L^{-1})	Efficiency (%) or Sorption capacity (mg g^{-1})	References
Codium vermilara	Ni	147	13.2	Romera et al. (2007)
Lessonia nigrescens	Ar(V)	200	45.2	Hansen et al. (2006)
Sargassum muticum	Sb	10	5.5	Ungureanu et al. (2015)
Spirogyra sp.	Pb	200	140	Gupta and Rastogi (2008)
Chlorella vulgaris	Cu	50 mg dm^{-3}	97.70%	Goher et al. (2016)
Spirulina sp.	Cu	5	81.20%	Mane and Bhosle (2012)
Nostoc sp.	Pb	1	99.60%	Kumaran et al. (2011)
	Cd	1	95.40%	Kumaran et al. (2011)
	Ni	1	88.23%	Kumaran et al. (2011)

4.5.2 Potential Plants Involved in Bioremediation

The potential phytoextractive plant species has the ability to accumulate the high content of the metals into the aboveground biomass without showing any toxicity symptoms. Their potential of phytoextraction can be enhanced by the use of fast-growing hyperaccumulator tree species with extensive root systems, thus ensuring its economic and environmental feasibility. Remarkable genetic variability has been reported to exist among plants of Salicaceae species adapted to soil of varying level metal contaminants (Dickinson and Pulford 2005; Puschenreiter et al. 2010; Marmiroli et al. 2011; Yang et al. 2015). In addition to this, adoption of native fast growing tree species may provide us with a better possible solution. Some of the plants suitable for plant assisted bioremediation are given in Table 4.5.

4.6 Mechanism of Plant Assisted Bioremediation

The mechanism for the plant assisted bioremediation involves bioavailability, uptake, translocation, sequestration and different defence mechanism that can help to develop heavy metal stress-resistant cultivars and highly efficient plant species

Table 4.4 Plant assisted microflora involved in bioremediation of heavy metals

Microorganisms	Plant	Heavy metals	Role/effect	Reference
Rhizobium sp. strains E20-8 and NII-1	*Pisum sativum* L	Cd	Remediates Cd pollution through various mechanisms like cytoplasmic sequestration, periplasmic allocation, extracellular immobilization and biotransformation	Cardoso et al. (2018)
Microbacterium sp. NE1R5, *Curtobacterium* sp. NM1R1	*Brassica nigra*	As, Zn, Cu, Pb	Enhancement of seed germination and root development	Román-Ponce et al. (2017)
Thiobacillus thiooxidans	*Gladiolus grandiflorus* L	Cd, Pb	Uptake and accumulation of Cd and Pb increases along with enhanced root length, plant height and dry biomass	Mani et al. (2016)
Pseudomonas brassicacearum, Rhizobium leguminosarum	*Brassica juncea*	Zn	Attenuates metal toxicity and promotes metal chelation	Adediran et al. (2015)
Pseudomonas sp.	Soybean, mungbean, wheat	Ni, Cd, Cr	Promotion of growth	Gupta et al. (2002)
Sinorhizobium sp. Pb002	*Brassica juncea*	Pb	Lead phytoextraction efficiency is enhanced by *B. juncea* plants	Di Gregorio et al. (2006)
Pseudomonas sp. A3R3	*Alyssum serpyllifolium, Brassica juncea*	Ni	More biomass (*B. juncea*) and Ni content (*A. serpyllifolium*) in plants grown in Ni-stressed conditions	Ma et al. (2011a, b)
Bradyrhizobium sp. 750, *Pseudomonas cytisi* sp., *Ochrobactrum*	*Lupinus luteus*	Cu, Cd, Pb	Enhanced the accumulation of metals	Dary et al. (2010)
Pseudomonas aeruginosa, Pseudomonas fluorescens, Ralstonia metallidurans	Maize	Cr, Pb	Accelerated soil metal mobilization and increased uptake of Cr and Pb	Braud et al. (2009)
Bacillus weihenstephanensis strain SM3	*Helianthus annuus*	Ni, Cu, Zn	Accelerated Cu and Zn accumulation in plants, also increased the water soluble Ni, Cu and Zn concentrations in soil with their metal mobilizing potential	Rajkumar et al. (2008)
Achromobacter xylosoxidans strain Ax10	*Brassica juncea*	Cu	Enhanced Cu uptake by plants	Ma et al. (2009)

(continued)

Table 4.4 (continued)

Microorganisms	Plant	Heavy metals	Role/effect	Reference
Glomerales species *Rhizophagus Funneliformis Claroideoglomus*	*Lactuca sativa Daucus carota*	Sb	Increased its uptake and accumulation in plants particularly in roots	Pierart et al. (2018)
Funneliformis mosseae, Rhizophagus irregularis, Claroideoglomus lamellosum	*Ricinus communis*	Cr(III), Cr(VI)	Reduction of Cr(VI) concentration in soils	Gil-Cardeza et al. (2018)
AM fungi	*Solanum melongena*	Pb, As, Cd	AM increased metal (loids) uptake and biomass	Chaturvedi et al. (2018)
Glomus mosseae	*Cajanus cajan*	Cd, Pb	Lead distribution pattern seems to be changed by fungal symbiont in extra radical hyphae of fungi, roots and shoots. Inoculation of fungal cultures in pigeon pea demonstrated the bioremediation potential by assisting it to grow in heavily metal-contaminated soils	Garg and Aggarwal (2011)
Scleroderma citrinum	*Pinus sylvestris* L	Zn, Cd, Pb	Reduction in translocation of Zn, Cd, or Pb from roots to shoots in pine seedlings	Krupa and Kozdrój (2007)
Bacillus thuringiensis GDB-1	*Alnus firma*	Cd, Ni, As, Cu, Pb, Zn	Promoted accumulation of metal(loid)s (As, Cu, Pb, Ni and Zn)	Babu et al. (2013)

for phytoremediation in harmony with microflora through genetic engineering technologies.

4.6.1 Bioavailability

It is defined as a part of the total elemental concentration available to plants that determines the uptake and accumulation of heavy metal ions in plants. Heavy metals exist in soils with several degrees of fractions i.e. soil solution form, soluble metal complexes and free metal ions forms. Several factors that determine the bioavailability of heavy metal elements are environmental conditions (moisture, temperature and oxidation state), soil properties (pH and organic matter) and enhanced biological activity by microbes (Yang et al. 2012; Bravin et al. 2012). These factors regulate the release of heavy metals into the soil and influence the plant uptake from soils. Environmental factors like high temperature enhance the physical, chemical and biological activities in soil–plant system, while precipitation and rainfall are known

Table 4.5 Potential plants for remediation of heavy metal pollution from soil

Plant species	Metal	Reference
Alyssum spp., Phyllanthus serpentines, Isatispinnatiloba, Berkheyacoddii	Nickel (Ni)	Li et al. (2003), Bani et al. (2010), Chaney et al. (2010), Mesjasz-Przybyłowicz et al. (2004), Altinözlü et al. (2012)
Azolla pinnata, Solanum photeinocarpum, Thlaspicaerulescens, Rorippaglobosa, Turnip landraces, Prosopis laevigata	Cadmium (Cd)	Rai (2008), Zhang et al. (2011), Lombi et al. (2001), Wei et al. (2008), Li et al. (2016), Buendía-González et al. (2010)
Pteris spp., Corrigiolatelephiifolia, Eleocharis acicularis, Azolla carobiniana	Arsenic (As)	Srivastava et al. (2006), Sakakibara et al. (2011), Garcia-Salgado et al. (2012)
Eleocharis acicularis, Aeolanthusbiformifolius, Ipomoea alpine, Haumaniastrumkatangense, Pteris vittata	Copper (Cu)	Sakakibara et al. (2011), Chaney et al. (2010), Mitch (2002), Sheoran et al. (2009), Wang et al. (2012)
Pteris vittata	Chromium (Cr)	Kalve et al. (2011)
Thlaspicaerulescens, Eleocharis acicularis, Thlaspicalaminare, Deschampsiacespitosa	Zinc (Zn)	Cunningham and Ow (1996), Sakakibara et al. (2011), Sheoran et al. (2009), Kucharski et al. (2005)
Medicago sativa, Brassica spp.,Thlaspirotundifolium Helianthus annuus, Euphorbia cheiradenia, Betula occidentalis, Deschampsiacespitosa	Lead (Pb)	Koptsik (2014), Cunningham and Ow (1996), Chehregani and Malayeri (2007), Kucharski et al. (2005)
Lecythisollaria, Astragalus racemosus	Selenium (Se)	Marques et al. (2009)
Schima superba, Macadamia neurophyllaMaytenusbureavianaAlyxiarubricaulis	Manganese (Mn)	Yang et al. (2012), Sheoran et al. (2009), Marques et al. (2009), Chaney et al. (2010)
Achillea millefolium, Marrubium vulgare, Rumex induratus, Hordeum spp., Festuca rubra, Helianthus tuberosus, Poa pratensis, Armoracia lapathifolia, Brassica juncea	Mercury (Hg)	Wang et al. (2012), Rodriguez et al. (2003), Sas-Nowosielska et al. (2008)

to accelerate plant growth and development. High soil moisture content regulates the movement of water-soluble trace elements during bioremediation. Soil properties, viz., pH, organic matter/organic carbon and cation exchange capacity (CEC) are the important factors that control the bioavailability of cations in soil. Soils with higher organic matter and high pH will form complex with heavy metals more firmly and become less available to plants for uptake and accumulation. Acidification of the rhizosphere is considered to increase the metal accumulation potential of plants raised on heavy metal contaminated soils. At acidic pH, heavy metals are found in free ionic forms and are more bioavailable, but at the basic pH metals form insoluble metal complexes with phosphates and carbonates (Sandarin and Hoffman 2007; Rensing and Maier 2003). Biological activities within the soil–plant system alter the bioavailability of metal elements. Microbes in the rhizosphere can produce chelating compounds, enhance the key nutrient uptake and also the availability of soil heavy metals (Rajkumar et al. 2012). Some plants secrete the organic components that form soluble complexes with heavy metal ions in soils. These soluble complex formations promote the mobility of heavy metals in soils. Yang et al. (2012) reported that root exudates include various organic acids and amino acids viz., oxalic acid, citric acid, tartaric acid, succinic acid, aspartic acid and glutamic acid, that form heavy metal soluble complexes and increase the mobility of Cd, Cu, Zn and Pb in soils (Fig. 4.2).

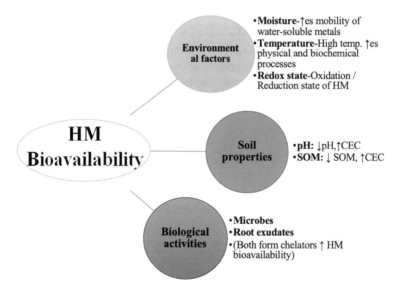

Fig. 4.2 Bioavailability of the heavy metals (HM) to plants

4.6.2 Plant Uptake

The movement of heavy metals in soils depends upon precipitation, redox potential, absorption/ adsorption and its complexation/methylation responses mediated by microbes along with plants (Kumar et al. 2017). The mechanism of plant metal uptake, rejection, translocation and sequestration is specific and highly variable within the plant varieties (Lone et al. 2008). Plants adopt two main strategies to combat heavy metal stress by either reduce metal uptake or increase vacuolar sequestration. The heavy metal is bioactivated by the root microbe's interaction first which leads to root absorption and further compartmentalization.

(i)   *Bioactivation of metals by root-microbe interaction*

Several studies depicted the positive interaction of microorganisms with plant species in the rhizosphere (Dakora and Phillips 2002; Kuiper et al 2004). Plant growth promoting rhizobacteria increase the bioavailability of metal ions by dissolving them via changing the chemical properties (pH, redox state, organic matter) of soils in the rhizosphere and modify the heavy metal speciation in the root zone (Jing et al. 2007). They solubilize the ions like phosphate, siderophore and increase acid production (Kumar et al. 2017). During heavy metal stress, mycorrhizae release natural acids that enhance zinc solubility and its mobility, ultimately playing a significant role to strengthen plant survival rate (Giasson et al. 2008).

Bacterial endophytes are considered to be beneficial for host plants usually during stress conditions, because they regulate the plant growth promoting mechanisms like phytohormone production by activating enzymes viz., 1-aminocyclopropane-1-carboxylic acid (ACC) deaminase, ethylene and Indole acetic acid (IAA) (Hardoim et al. 2008; Rajkumar et al. 2012). Endophytes also known to enhance nitrogen fixation and phosphate availability in rhizosphere, hence helps to recover the plant during heavy metal (HM) stress conditions (Kuklinsky-Sobral et al. 2004).

(ii) *Root absorption and compartmentalization*

The transport of nutrients and heavy metals from soils to plant roots occurs via symplastic and apoplastic transport. In symplastic transport heavy metals enter the root cells through the plasma membrane of the endodermis of the root. While in apoplastic transport, it enters the root apoplast via spacing within the cells. Generally, heavy metals and nutrient ions cross the membranes only with the aid of naturally occurring membrane transport proteins (Fig. 4.3). The abundance of these proteins depends upon tissue type and environmental conditions. If a small amount of nutrients is present in soils, then the plant requires high-affinity transporters for uptake; whereas if the nutrients in the soil are present in high concentrations (e.g. agricultural soils with fertilizers), then low-affinity transporters would be more useful for plant uptake (Cailliatte et al. 2010).

Several transporter families have been reported in plants such as heavy metal ATPase (HMA), natural resistance and macrophage proteins (NRAMP), Zrt, Irt-like proteins (ZIP) etc. (Table 4.6). In the cytosol, toxic metals rapidly bind to chelators

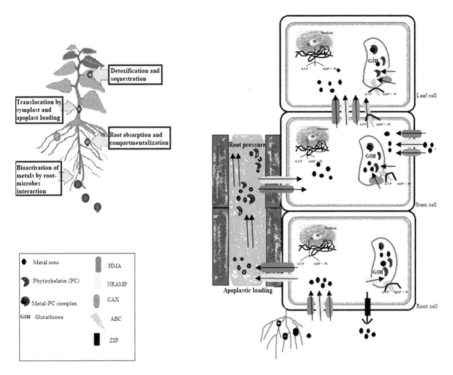

Fig. 4.3 Uptake of heavy metal and its compartmentalization in various plant part through different transporter proteins

and are transferred to the vacuole for sequestration. Ingle et al. (2005) observed that histidine is involved in Ni-chelation in root cells and helps plant to tolerate Ni toxicity. Cr (III) in root cells is chelated with acetate and sequestered in the vacuole (Bluskov and Arocena 2005).

4.6.3 Translocation

Heavy metal transporters are required for translocation of metallic ions from root symplast to xylem apoplast due to endodermal barrier (casparian strips) in the root. The translocation of heavy metal ions depends on two factors: root pressure and leaf transpiration (Kumar et al. 2017).

(i) *Root symplast to apoplast through xylem tissues*

Xylem loading of metals from root symplast is an important phenomenon making the plant to tolerate heavy metal toxicity instead of promoting its accumulation in root cells that would inactivate the enzymes involved in metabolic processes. Cation

Table 4.6 Metal transporters involved in heavy metal uptake, transport and sequestration during phytoremediation (Bhargava et al. 2012)

Transporter family	Transporter gene	Plant species	Metal transported	References
Natural resistance-associated macrophage proteins (NRAMP)	nramp1 nramp1-3 nramp4	Malus baccata Lycopersicon esculentum Thlaspi japonicum	Fe (Iron) Fe Iron) Fe (Iron)	Xiao et al. (2008), Bereczky et al. (2003), Mizuno et al. (2005)
Fe-regulated transporter (IRT)	irt1 irt1-2 irt1-2	A. thaliana T. caerulescens L. esculentum	Fe (Iron) Fe (Iron) Fe (Iron)	Kerkeb et al. (2008), Schikora et al. (2006), Plaza et al. (2007), Bereczky et al. (2003)
Zn-regulated transporter (ZRT)	Zip zip1-12 zip4 znt1-2	Medicago truncatula A. thaliana O. sativa T. caerulescens	Zn (Zinc) Zn (Zinc) Zn (Zinc) Zn (Zinc)	Lopez-Millan et al (2004), Roosens et al. (2008), Ishimaru et al. (2005), Van de Mortel et al (2006)
P-type ATPase	hma9 hma8 hma3 hma4	Oryza sativa Glycine max A. thaliana A. halleri	Cu (Copper), Zn (Zinc), Cd (Cadmium), Cu (Copper) Co (Copper), Zn (Zinc), Cd (Cadmium), Pb (Lead), Cd (Cadmium)	Lee et al. (2007), Bernal et al. (2007), Morel et al. (2008), Courbot et al. (2007)
Copper transporter	copt1	A. thaliana	Cu (Copper)	Sancenon et al. (2004), Andres-Colas et al. (2010)
Yellow stripe-like (YSL)	Ysl3 Ysl2	T. caerulescens A. thaliana	Fe (Iron), Ni (Nickel) Fe (Iron), Cu (Copper)	Gendre et al. (2006), DiDonato et al. (2004)
Cation diffusion facilitator (CDF)	mtp1 mtp1 mtp1 mtp1	Thlaspi goesingense A. thaliana A. halleri Nicotiana tabacum	Zn (Zinc), Ni (Nickel) Zn (Zinc) Zn (Zinc) Zn (Zinc), Co (Cobalt)	Kim et al. (2004), Kawachi et al. (2008), Willems et al. (2007), Shingu et al. (2005)

diffusion facilitator (CDF) type of proteins conveys a broad array of metal divalent ions from cytoplasm toward the outer cell parts and even within the subcellular compartments (Hanikenne et al. 2005). HMA2 proteins are energy-dependent transporters, despite having selective nature they also get activated by analogue metal ions. Hussain et al. (2004) isolated HMA2 and HMA4 transporters in *Arabidopsis* for Zn transportation within cellular compartments and homeostasis. Milner and Kochain (2008) deciphered the importance of HMA2 and HMA4 genes in metal loading into the xylem.

(ii) *Root apoplast to aerial (stem and leaves) tissues*

Hyper accumulator plants rapidly translocate the absorbed metal ions from the root to the above-ground parts, while non-accumulators accumulate heavy metals only in their root portions. Heavy metals can be stored in root vacuoles. Due to the limited space and high heavy metal concentration in the soil matrix, it gets translocated to shoot tissue where sequestration and detoxification rate is comparatively high (Kumar et al. 2017). Generally, metals are stored in only chelated form but are transported from one cellular compartment to other in free ionic state according to the selectivity of transporter proteins (Ortiz et al. 1995). Research experiments showed that hyperaccumulator plants accumulate high concentration of heavy metals in stem and leaf vacuoles than the root tissues. In the leaf tissues, high amount of metals accumulates in epidermal tissues compared to the cortical and vascular tissues (Kupper et al. 2001; Kumar et al. 2017).

4.6.4 *Sequestration/Detoxification*

To cope up with heavy metal stress, plants adapt different survival strategies like compartmentalization, exclusion, complexation and synthesis of binding proteins (metallothioneins and phytochelatins). Heavy metal toxicity inside the plant cell gets detoxified by complex formation and compartmentalization to make them less available to metabolic active sites. Organic acids, glutathione precursor of phytochelatins and metallothiones play a significant role in detoxification/sequestration. Phytochelatins (PC) have an imperative role to detox cadmium in fungi and plants through conjugation. Glutathione enhances the PC synthesis and thus more PC-metal complex formation in the vacuole which ultimately enhances cadmium tolerance in plants (Lee et al. 2003).

In plants, different heavy metal ions (Cu, Hg, Zn, Pb, Cd) stimulate the enzyme, γ-glutamyl-cysteinyl dipeptidyl transpeptidase (PC synthase) for phytochelatin synthesis which results in glutathione conversion (GSH) to phytochelatin (Fig. 4.4). These phytochelatins are produced from glutathione (GSH) through oxidation and reduction reactions. The metal ion binds to cysteine sulfhydryl residues of phytochelatins and its sequestration occurs inside the vacuole (Zhu et al 2004; Kumar et al. 2017). In hyperaccumulator plants, toxic effects of Ni were overcome by enhancing GSH-dependent antioxidant mechanism that protects the plant from

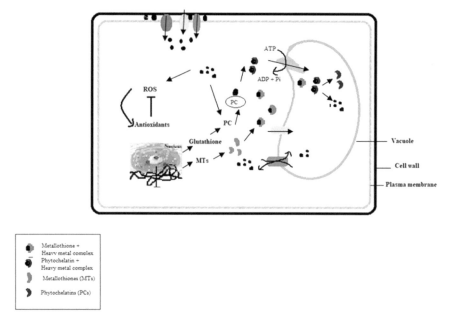

Fig. 4.4 Sequestration and detoxification of heavy metals (HM) in plant cell

oxidative damage (Freeman et al. 2005). Metallothiones are metal-binding proteins that modulate the concentration of metals inside the cell by binding heavy metal ions to cysteine and thiol groups (Khan et al. 2004). Mn^{2+} metal detoxification involves uptake of ions from the plasma membrane, binding with malate and transportation through tonoplast to vacuole where Mn unbinds from malate and form complex with oxalate (Memon et al. 2001).

Heavy metal toxicity hindered the functional group of important molecules that disrupt the metabolic enzyme activity and consequently inhibit or suppress photosynthetic rate, respiration rate and all physiological and biochemical processes of plants (Gupta et al. 2015; Ali et al. 2013). Naturally, plants develop various defense mechanisms against heavy metal stress inside the plant body which include compartmentalization reduction, suppression of high-affinity phosphate transport system, sequestration and translocation (Zhao et al. 2009). When metal ions cross enter into plant tissues by crossing these barriers then various cellular defense mechanisms (as a second line of defense viz., ROS production, antioxidants) are initiated to detox the adverse effect of noxious heavy metals (Silva and Matos 2016).

4.7 Conclusion and Future Prospects

The pollution due to heavy metals is of great concern because of its potential impact on human and animal health. It is imperative to protect the natural resources and biodiversity, by using cheaper and effective technologies. In phytoremediation, the plants have to retain the pollutant in their root or other parts by producing large biomass and microbes converting toxic forms of heavy metals to non-toxic forms. But till now no plant is known to fulfil both these criteria. At the heavily contaminated sites with both organic and inorganic pollutants, there is a limitation of plant growth and microbial activity, thus having reduced plant assisted bioremediation efficiency. Recent progress in molecular, biochemical and plant physiology fields provides a strong scientific base for achieving this goal. During the last decade, substantial efforts have been made by the researchers to identify plant hyperaccumulators, bioremediators for heavy metals and their mechanism of uptake, translocation. There is a huge genetic variation in different plant species, even among the cultivars of the same species. So, research must be carried out to study the mechanism of metal uptake, accumulation, exclusion, translocation and compartmentation for each species as they play a specific role in phytoremediation. Further, research is needed to study metal uptake at the cellular level including influx and efflux of different metals by different cell organelles and membranes.

- There is a need of microbial profiling of rhizosphere under controlled and field conditions to examine the antagonistic and synergistic effects of different metal ions in soil and polluted waters.
- Selected essential rhizosphere microorganisms and microbial strains able to degrade toxic pollutants can be studied in the natural habitat. Molecular techniques will further help in elucidating the fate and effect of these selected strains in the soil environment.
- Standardization of the methods for heavy metal recovery from the hyperaccumulator plants will allow the detection of the new strains of the micro-organisms who can degrade or reduce the toxic metals to non-toxic metals as well as improve the fertility status of the soil.

References

Abbas SH, Ismail IM, Mostafa TM, Sulaymon AH (2014) Biosorption of heavy metals: a review. J Chem Sci Technol 3(4):74–79

Abou-Shanab RA, Angle JS, Delorme TA, Chaney RL, van Berkum P, Moawad H, Ghanem K, Ghozlan HA (2003) Rhizobacterial effects on nickel extraction from soil and uptake by Alyssum murale. New Phytol 158:219–224

Adediran GA, Ngwenyaa BT, MosselmansJFW Heal KV, Harvie BA (2015) Mechanisms behind bacteria induced plant growth promotion and Zn accumulation in Brassica juncea. J Hazard Mater 283:490–499

Ahemad M (2014) Phosphate solubilizing bacteria-assisted phytoremediation of metalliferous soils: a review. 3 Biotech. https://doi.org/10.1007/s13205-014-0206-0

Ahirwal J, Pandey VC (2021) Restoration of mine-degraded land for sustainable environmental development. Restor Ecol 29(4):e13268. https://doi.org/10.1111/rec.13268

Ali H, Khan E, Sajad MA (2013) Phytoremediation of heavy metals—concepts and applications. Chemosphere 91:869–881

Altinözlü H, Karagöz A, Polat T, Ünver I (2012) Nickel hyperaccumulation by natural plants in Turkish serpentine soils. Turk J Bot 36:269–280. https://doi.org/10.3906/bot-1101-10

Andres-Colas N, Perea-Garcia A, Puig S, Penarrubia L (2010) Deregulated copper transport affects Arabidopsis development especially in the absence of environmental cycles. Plant Physiol 153:170–184

Anonymous (2010) World Water Day, United Nations

Babu AG, Kim JD, Oh BT (2013) Enhancement of heavy metal phytoremediation by Alnus firmawith endophytic Bacillus thuringiensis GDB-1. J Hazard Mater 250:477–483

Bani A, Pavlova D, Echevarria G, Mullaj A, Reeves RD, Morel JL, Sulçe S (2010) Nickel hyperaccumulation by the species of Alyssum and Thlaspi (Brassicaceae) from the ultramafic soils of the Balkans. Bot Serb 34:3–14

Benazir JF, Suganthi R, Rajvel D, Pooja MP, Mathithumilan B (2010) Bioremediation of chromium in tannery effluent by microbial consortia. African J Biotech 9:3140–3143

Bereczky Z, Wang HY, Schubert V, Ganal M, Bauer P (2003) Differential regulation of Nramp and IRT metal transporter genes in wild type and iron uptake mutants of tomato. J Bio Chem 278:24697–24704

Bernal M, Testillano PS, Alfonso M, Del Carmen RM, Picorel R, Yruela I (2007) Identification and subcellular localization of the soybean copper P1B-ATPase GmHMA8 transporter. J Struct Biol 158:146–158

Bharagava RN, Mishra S (2018) Hexavalent chromium reduction potential of *Cellulosimicrobium* sp. isolated from common effluent treatment plant of tannery industries. Ecotoxicol Environ Safety 147:102–109

Bhardwaj RM (2005) Status of wastewater generation and treatment in India. IWG-Env, International work session on water statistics, Vienna, p 9. Available at http://unstats.un.org/unsd/environment/envpdf/pap_wasess3b6india.pdf

Bhargava A, Carmona FF, Bhargava M et al (2012) Approaches for enhanced phytoextraction of heavy metals. J Environ Manage 105:103–120

Bhattacharya A, Gupta A, Kaur A, Malik D (2019) Alleviation of hexavalent chromium by using microorganisms: insight into the strategies and complications. Water Sci Technol 79.3:411–424. https://doi.org/10.2166/wst.2019.060

Bhojiya AA, Joshi H, Upadhyay SK, Srivastava AK, Pathak VV, Pandey VC, Jain D (2021) Screening and optimization of zinc removal potential in Pseudomonas aeruginosa—HMR1 and its plant growth-promoting attributes. Bull Environ Contam Toxicol. https://doi.org/10.1007/s00128-021-03232-5

BIS (2012) Indian standard for drinking water, Bureau of Indian Standards (IS-10500), New Delhi

Black H (1995) Absorbing possibilities: phytoremediation. Environ Health Perspect 103:1106–1108

Bluskov S, Arocena JM (2005) Uptake, distribution, and speciation of chromium in Brassica juncea. Int J Phytoremediation 7:153–165

Bowen HJM (1979) Environmental chemistry of the elements. Academic press, London

Braud A, Jézéquel K, Bazot S and Lebeau T (2009) Enhanced phytoextraction of an agricultural Cr-, Hg- and Pb-contaminated soil by bioaugmentation with siderophore producing bacteria. Chemosphere 74: 280–286.

Bravin MN, Garnier C, Lenoble V, Gerard F, Dudal Y, Hinsinger P (2012) Root-induced changes in pH and dissolved organic matter binding capacity affect copper dynamic speciation in the rhizosphere. Geochim Cosmochim Ac 84:256–268

Buendía-González L, Orozco-Villafuerte J, Cruz-Sosa F, Barrera-Díaz C, Vernon-Carter E (2010) Prosopis laevigataa potential chromium (VI) and cadmium (II) hyperaccumulator desert plant. Bioresour Technol 101:5862–5867. https://doi.org/10.1016/j.biortech.2010.03.027

Cailliatte R, Schikora A, Briat JF, Mari S, Curie C (2010) High affinity manganese uptake by the metal transporter nramp1 is essential for Arabidopsis growth in low manganese conditions. Plant Cell 22:904–917

Cardoso P, Corticeiro S, Freitas R, Figueira E (2018) Different efficiencies of the same mechanisms result in distinct Cd tolerance T within Rhizobium. Ecotoxicol Environ Saf 150:260–269

Chaney RL, Angle JS, McIntosh MS, Reeves RD, Li YM, Brewer EP, Chen KY, Roseberg RJ, Perner H, Synkowski EC, Broadhurst CL, Wang S, Baker AJ (2005) Using hyperaccumulator plants to phytoextract soil Ni and Cd. Z Naturforsch C 60:190–198

Chaney RL, Broadhurst CL, Centofanti T (2010) Phytoremediation of soil trace elements. In: Hooda PS (ed) Trace elements in soils chichester. John Wiley & Sons, Inc. pp 311–352

Chaturvedi R, Favas P, Pratas J, Varun M, Paul MS (2018) Assessment of edibility and effect of arbuscular mycorrhizal fungi on T Solanum melongena L. grown under heavy metal(loid) contaminated soil. Ecotoxicol Environ Saf 148:318–326

Chehregani A, Malayeri BE (2007) Removal of heavy metals by native accumulator plants. Int J Agric Biol 9:462–465

Courbot M, Willems G, Motte P, Arvidsson S, Roosens N, Saumitou-Laprade P, Verbruggen N (2007) A major quantitative trait locus for Cd tolerance in Arabidopsis halleri colocalizes with HMA4, a gene encoding a heavy metal ATPase. Plant Physiol 144:1052–1065

CPCB (2009) Comprehensive environmental assessment of industrial clusters. Ecological impact assessment series: EIAS/5/2009–2010. Central pollution control board, ministry of environment and forests, Govt. of India, New Delhi

Cunningham SD, Ow DW (1996) Promises and prospects of phytoremediation. Plant Physiol 110:715. https://doi.org/10.1104/pp.110.3.715

Dakora FD, Phillips DA (2002) Root exudates as mediators of mineral acquisition in low-nutrient environments. Plant Soil 13:35–47

Dary M, Chamber-Pérez MA, Palomares AJ, Pajuelo E (2010) In situ phytostabilisation of heavy metal polluted soils using Lupinus luteus inoculated with metal resistant plant-growth promoting rhizobacteria. J Hazard Mater 177: 323–330

Deng Z, Cao L (2017) Fungal endophytes and their interactions with plants in phytoremediation: a review. Chemosphere 168:1100–1106

Deng Z, Cao L, Huang H, Jiang X, Wang W, Shi Y et al (2011) Characterization of Cd- and Pb-resistant fungal endophyte Mucor sp. CBRF59 isolated from rapes (Brassica chinensis) in a metal-contaminated soil. J Hazard Mater 185:717–724

Deng Z, Zhang R, Shi Y, Hu L, Tan H, Cao L (2013a) Characterization of Cd-, Pb-, Zn-resistant endophytic Lasiodiplodiasp. MXSF31 from metal accumulating Portulaca oleracea and its potential in promoting the growth of rape in metal-contaminated soils. Environ Sci Pollut Res 21:2346–2357

Deng Z, Zhang R, Shi Y, Hu L, Tan H, Cao L (2013b) Enhancement of phytoremediation of Cd- and Pb-contaminated soils by self-fusion of protoplasts from endophytic fungus Mucor sp. CBRF59. Chemosphere 91:41–47

Di Gregorio S, Barbafieri M, Lampis S, Sanangelantoni AM, Tassi E, Vallini G (2006) Combined application of Triton X-100 and Sinorhizobium sp. Pb002 inoculum for the improvement of lead phytoextraction by Brassica juncea in EDTA amended soil. Chemosphere 63:293–299

Dickinson NM, Pulford ID (2005) Cadmium phytoextraction using short-rotation coppice Salix: the evidence trail. Environ Int 31:609–613

DiDonato RJ, Roberts LA, Sanderson T, Eisley RB, Walker EL (2004) Arabidopsis yellow stripe-Like2 (YSL2): a metal-regulated gene encoding a plasma membrane transporter of nicotianamine metal complexes. Plant J 39:403–414

Dietz AC, SchnoorJL (2001) Advances in phytoremediation. Environmental health perspectives 109(suppl 1):163–168

Dixit R, Wasiullah MD, Pandiyan K, Singh UB, Sahu A, Shukla R, Singh BP, Rai JP, Sharma PK, Lake H, Paul D (2015) Bioremediation of heavy metals from soil and aquatic environment: an overview of principles and criteria of fundamental processes. Sustainability 7:2189–2212. https://doi.org/10.3390/su7022189

Dursun A, Uslu G, Cuci Y, Aksu Z (2003) Bioaccumulation of copper(II), lead(II) and chromium(VI) by growing Aspergillus niger. Process Biochem 38:1647–1651

Freeman JL, Michael WP, Nieman K, Salt DE (2005) Nickel and cobalt resistance engineered in Escherichia coli by overexpression of serine acetyltransferase from the nickel hyperaccumulator plant Thlaspigoes ingense. Appl Environ Microbiol 12:8627–8633

Fu YQ, Li S, Zhu HY, Jiang R, Yin LF (2012) Biosorption of copper(II) from aqueous solution by mycelial pellets of Rhizopus oryzae. Afr J Biotechnol 11:1403–1411

Garcia-Salgado S, Garcia-Casillas D, Quijano-Nieto MA, Bonilla-Simon MM (2012) Arsenic and heavy metal uptake and accumulation in native plant species from soils polluted by mining activities. Water Air Soil Pollut 223:559–572

Garg N, Aggarwal N (2011) Effects of interactions between cadmium and lead on growth, nitrogen fixation, phytochelatin, and glutathione production in mycorrhizal Cajanus cajan (L.) Millsp. J Plant Growth Regul 30:286–300

Gendre D, Czernic P, Conéjéro G, Pianelli K, Briat JF, Lebrun M, Mari S (2006) TcYSL3, a member of the YSL gene family from the hyperaccumulator Thlaspi caerulescens, encodes a nicotianamine- Ni/Fe transporter. Plant J 49:1–15

Ghosh M, Singh SP (2005) A review on phytoremediation of heavy metals and utilization of its byproducts. Appl Ecol Environ Res 3:1–18

Giasson P, Karam A, Jaouich A (2008) Arbuscular mycorrhizae and alleviation of soil stresses onplant growth. In: Siddiqui ZA, Akhtar MS, Futai K (eds) Mycorrhizae: sustainable agriculture and forestry. Springer, Dordrecht, pp 99–134

Gil-Cardeza ML, Müller DR, Amaya-Martin SM, Viassolo R, Gómez E (2018) Differential responses to high soil chromium of two arbuscular mycorrhizal fungi communities isolated from Cr-polluted and non-polluted rhizospheres of Ricinus communis. Sci Total Environ 625:1113–1121

Glick BR (2012) Plant growth-promoting bacteria: mechanisms and applications. Hindawi Publishing Corporation, Scientifica

Godfray HCJ, Beddington JR, Crute IR, Haddad L, Lawrence D, Muir JF, Pretty J, Robinson S, Thomas SM, Toulmin C (2010) Food security: the challenge of feeding 9 billion people. Science 327:812–818

Goher ME, El-Monem AMA, Abdel-Satar AM, Ali MH, Hussian AEM, Napiórkowska-Krzebietke A (2016) Biosorption of some toxic metals from aqueous solution using nonliving algal cells of Chlorella vulgaris. J Elementol 21(3):703–714

Gupta VK, Rastogi A (2008) Biosorption of lead from aqueous solutions by green algae Spirogyra species: kinetics and equilibrium studies. Jhazardous Mater 152(1):407–414

Gupta A, Meyer JM, Goel R (2002) Development of heavy metal resistant mutants of phosphate solubilizing Pseudomonas sp. NBRI4014 and their characterization. CurrMicrobiol 45:323–332

Gupta VK, Nayak A, Agarwal S (2015) Bioadsorbents for remediation of heavy metals: current status and their future prospects. Environ Eng Res 20:1–18

Hanikenne M, Kramer U, Demoulin V, Baurain D (2005) A comparative inventory of metal transporters in the green alga Chlamydomonas reinhardtiiand the red alga Cyanidioschizonmerolae. Plant Physiol 137:428–446

Hansen HK, Ribeiro A, Mateus E (2006) Biosorption of arsenic (V) with Lessonia nigrescens. Miner Eng 19(5):486–490

Hardoim PR, Overbeek LS, Elsas JD (2008) Properties of bacterial endophytes and their proposed role in plant growth. Trends Microbiol 16:63–471

Hemambika B, Rani MJ, Kannan VR (2011) Biosorption of heavy metals by immobilized and dead fungal cells: a comparative assessment. J Ecol Nat Environ 3:168–175

Hussain SA, Palmer DH, Moon S, Rea DW (2004) Endocrine therapy and other targeted therapies for metastatic breast cancer. Expert Rev Anticancer Ther 4:1179–1195

Ingle RA, Mugford ST, Rees JD, Campbell MM, Smith JAC (2005) Constitutively high expression of the histidine biosynthetic pathway contributes to nickel tolerance in hyperaccumulator plants. Plant Cell 17:2089–2106

Iqbal M, Edyvean R (2004) Biosorption of lead, copper and zinc ions on loofa sponge immobilized biomass of Phanero chaetechrysosporium. Miner Eng 17:217–223

Ishimaru Y, Suzuki M, Kobayashi T, Takahashi M, Nakanishi H, Mori S, Nishizawa NK (2005) OsZIP4, a novel zinc-regulated zinc transporter in rice. J Exp Bot 56:3207–3214

Jafari SA, Cheraghi S, Mirbakhsh M, Mirza R, Maryamabadi A (2015) Employing response surface methodology for optimization of mercury bioremediation by Vibrio parahaemolyticus PG02 in coastal sediments of Bushehr, Iran. CLEAN Soil, Air, Water 43(1):118–126

Järup L (2003) Hazards of heavy metal contamination. Br Med Bull 68:167–182. https://doi.org/10.1093/bmb/ldg032

Jing YD, Zhen Li HE, Yang XE (2007) Role of soil rhizobacteria in phytoremediation of heavy metal contaminated soils. J Zhejiang Univ Sci B 8:192–207

Kabata-Pendias A, Pendias H (1984) Trace elements in soils and plants. CPRC Press, Boca Raton, Fla

Kalve S, Sarangi BK, Pandey RA, Chakrabarti T (2011) Arsenic and chromium hyperaccumulation by an ecotype of Pteris vittata—prospective for phytoextraction from contaminated water and soil. Curr Sci India 100:888–894

Kang S, Lee J, Kim K (2005) Metal removal from wastewater by bacterial sorption: kinetics and competition studies. Environ Technol 26:615–624

Kawachi M, Kobae Y, Mimura T, Maeshima M (2008) Deletion of a histidine-rich loop of AtMTP1, a vacuolar Zn2þ/Hþ antiporter of Arabidopsis thaliana, stimulates the transport activity. J Biol Chem 283:8374–8383

Kerkeb L, Mukherjee I, Chatterjee I, Lahner B, Salt DE, Connolly EL (2008) Iron-induced turnover of the arabidopsis iron-regulated transporter1 metal transporter requires lysine residues. Plant Physiol 146:1964–1973

Khan FI, Husain T, Hejazi R (2004) An overview and analysis of site remediation technologies. J Environ Manag 71:95–112

Kim SY, Kim JH, Kim CJ, Oh DK (1996) Metal adsorption of the polysaccharide produced from Methylobacterium organophilum. Biotech Lett 18(10):1161–1164

Kim D, GustinJL LB, Persans MW, Baek D, Yun DJ, Salt DE (2004) The plant CDF family member TgMTP1 from the Ni/Zn hyperaccumulator Thlaspigoesingenseacts to enhance efflux of Zn at the plasma membrane when expressed in Saccharomyces cerevisiae. Plant J 39:237–251

Kim IH, Choi JH, Joo JO, Kim YK, Choi JW, Oh BK (2015) Development of a microbe-zeolite carrier for the effective elimination of heavy metals from seawater. J Microbiol Biotechnol 25:1542–1546

Koptsik G (2014) Problems and prospects concerning the phytoremediation of heavy metal polluted soils: a review. Eurasian Soil Sci 47:923–939. https://doi.org/10.1134/S1064229314090075

Krupa P, Kozdrój J (2007) Ectomycorrhizal fungi and associated bacteria provide protection against heavy metals in inoculated pine (Pinus sylvestris L.) seedlings. Water Air Soil Pollut 182:83–90

Kucharski R, Sas-Nowosielska A, Małkowski E, Japenga J, Kuperberg J, Pogrzeba M et al (2005) The use of indigenous plant species and calcium phosphate for the stabilization of highly metal-polluted sites in southern Poland. Plant Soil 273:291–305. https://doi.org/10.1007/s11104-004-8068-6

Kuiper I, Lagendijk EL, Bloemberg GV, Lugtenberg BJJ (2004) Rhizoremediation: a beneficial plant-microbe interaction. Mol Plant-Microbe Interact 17:6–15

Kuklinsky-Sobral J, Araujo WL, Mendes R, Geraldi IO, Pizzirani-Kleiner AA, Azevedo JL (2004) Isolation and characterization of soybean-associated bacteria and their potential for plant growth promotion. Environ Microbiol 6:1244–1251

Kumar KV, Srivastava S, Singh N, Behl HM (2009) Role of metal resistant plant growth promoting bacteria in ameliorating fly ash to the growth of Brassica juncea. J Hazard Mater 170:51–57

Kumar SS, Kadier A, Malyan SK, Ahmad A, Bishnoi NR (2017) Phytoremediation and rhizoremediation: uptake, mobilization and sequestration of heavy metals by plants. In: Singh DP (ed) Plant-Microbe interactions in agro-ecological perspectives Springer, pp 367–394. https://doi.org/10.1007/978-981-10-6593-4_15

Kumaran NS, Sundaramanicam A, Bragadeeswaran S (2011) Adsorption studies on heavy metals by isolated cyanobacterial strain (nostoc sp.) from uppanar estuarine water, southeast coast of India. J Appl Sci Res 7(11):1609–1615

Kupper H, Lombi E, Zhao FJ, Wieshammer G, McGrath SP (2001) Cellular compartmentation of Ni in the hyperaccumulators Alyssum lesbiacum A. Bertoloniiand Thlaspigoesingense. J Exp Bot 52:2291–2300

Lee S, Moon JS, Ko TS, Petros D, Goldsbrough PB, Korban SS (2003) Over expression of Arabidopsis phytochelatin synthase paradoxically leads to hypersensitivity to cadmium stress. Plant Physiol 131:656–663

Lee S, Kim YY, Lee Y, An G (2007) Rice P1B-type heavy-metal ATPase, OsHMA9, is a metal efflux protein. Plant Physiol 145:831–842

Li YM, Chaney R, Brewer E, Roseberg R, Angle JS, Baker A, Reeves R, Nelkin J (2003) Development of a technology for commercial phytoextraction of nickel: economic and technical considerations. Plant Soil 249:107–115

Li X, Zhang X, Yang Y, Li B, Wu Y, Sun H et al (2016) Cadmium accumulation characteristics in turnip landraces from China and assessment of their phytoremediation potential for contaminated soils. Front Plant Sci 7:1862. https://doi.org/10.3389/fpls.2016.01862

Liu LW, Li W, Song WP, Guo MX (2018) Remediation techniques for heavy metal–contaminated soils: principles and applicability. Sci Total Environ 633:206–219. https://doi.org/10.1016/j.scitotenv.2018.03.161

Lombi E, Zhao FJ, Dunham SJ, McGrath SP (2001) Phytoremediation of heavy metal-contaminated soils: natural hyperaccumulation versus chemical enhanced phytoextraction. J Environ Qual 30:1919–1926

Lone MI, He ZL, Stoffella PJ, Yang XE (2008) Phytoremediation of heavy metal polluted soils and water: progresses and perspectives. J Zhejiang Univ-Sci. B (biomed Biotechnol) 9:210–220. https://doi.org/10.1631/jzus.B0710633

Lopez-Millan AF, Ellis DR, Grusak MA (2004) Identification and characterization of several new members of the ZIP family of metal ion transporters in Medicago truncatula. Plant Mol Biol 54:583–596

Ma Y, Rajkumar M, Freitas H (2009) Inoculation of plant growth promoting bacterium Achromobacter xylosoxidans strain Ax10 for the improvement of copper phytoextraction by Brassica juncea. J Environ Manag 90:831–837

Ma Y, Prasad M, Rajkumar M, Freitas H (2011a) Plant growth promoting rhizobacteria and endophytes accelerate phytoremediation of metalliferous soils. Biotechnol Adv 29:248–258

Ma Y, Rajkumar M, Luo Y, Freitas H (2011b) Inoculation of endophytic bacteria on host and non-host plants-effects on plant growth and Ni uptake. J Hazard Mater 195:230–237

Malik G, Hood S, Majeed S, Pandey VC (2022). Understanding assisted phytoremediation: potential tools to enhance plant performance. In: Pandey VC (eds) Assisted phytoremediation. Elsevier, Amsterdam, pp 1–24. https://doi.org/10.1016/B978-0-12-822893-7.00015-X

Mane PC, Bhosle AB (2012) Bioremoval of some metals by living Algae Spirogyra sp. and Spirullina sp. from aqueous solution. Int J Environ Res 6(2):571–576

Mani D, Kumar C, Patel NK (2016) Integrated micro-biochemical approach for phytoremediation of cadmium and lead contaminated soils using Gladiolus grandiflorus L. cut flower. Ecotoxicol Environ Saf 124:435–446

Marmiroli M, Pietrini F, Maestri E, Zacchini M, Marmiroli N, Massacci A (2011) Growth, physiological and molecular traits in Salicaceae trees investigated for phytoremediation of heavy metals and organics. Tree Physiol 31:1319–1334

Marques AP, Rangel AO, Castro PM (2009) Remediation of heavy metal contaminated soils: phytoremediation as a potentially promising clean-up technology. Crit Rev Env Sci Technol 39:622–654. https://doi.org/10.1080/10643380701798272

Marschner H (1995) Mineral nutrition of higher plants. Oxford University Press, London

McGrath SP (1998) Phytoextraction for soil remediation. In: Brooks RR (ed) Plants that hyperaccumulate heavy metals. CABI Publishing, Wallingford, pp 261–287

Memon A, Aktopraklgil D, Ozdemir Z, Vertii A (2001) Heavy metal accumulation and detoxification mechanism in plants. Turk J Bot 25:111–121

Mesjasz-Przybyłowicz J, Nakonieczny M, Migula P, Augustyniak M, Tarnawska M, Reimold U et al (2004) Uptake of cadmium, lead nickel and zinc from soil and water solutions by the nickel hyperaccumulator Berkheyacoddii. Acta BiolCracoviensia Ser Bot 46:75–85

Milner MJ, Kochian LV (2008) Investigating heavy-metal hyperaccumulation using Thlaspi caerulescensas a model system. Ann Bot 102:3–13

Mitch ML (2002) Phytoextraction of toxic metals: a review of biological mechanism. J Environ Qual 31:109–120. https://doi.org/10.2134/jeq2002.1090

Mizuno T, Usui K, Horie K, Nosaka S, Mizuno N, Obata H (2005) Cloning of three ZIP/NRAMP transporter genes from a Ni hyperaccumulator plant Thlaspi japonicum and their Ni 2þ-transport abilities. Plant Physiol Biochem 43:793–801

Morel M, Crouzet J, Gravot A, Auroy P, Leonhardt N, Vavasseur A, Richaud P (2008) AtHMA3, a P1B-ATPase allowing Cd/Zn/Co/Pb vacuolar storage in Arabidopsis. Plant Physiol 149:894–904

Mosa KA, Saadoun I, Kumar K, Helmy M, Dhankher OP (2016) Potential biotechnological strategies for the cleanup of heavy metals and metalloids. Frontiers Plant Sci 7:303. https://doi.org/10.3389/fpls.2016.00303

Muradoglu F, Gundogdu M, Ercisli S, Encu T, Balta F, Jaafar HZE, Zia-Ul-Haq M (2015) Cadmium toxicity affects chlorophyll a and b content, antioxidant enzyme activities and mineral nutrient accumulation in strawberry. Biol Res 48:1–7. https://doi.org/10.1186/S40659-015-0001-3

Ortiz DF, Theresa R, McCue KF, Ow DW (1995) Transport of metal-binding peptides by HMT1, a fission yeast ABC-type vacuolar membrane protein. J Biol Chem 270(9):4721–4728

Pandey VC (2020) Phytomanagement of fly ash. Elsevier (Authored book), ISBN: 9780128185445. https://doi.org/10.1016/C2018-0-01318-3, pp 334

Pandey VC, Bajpai O (2019) Phytoremediation: from theory towards practice. In: Pandey VC, Bauddh K (eds) Phytomanagement of polluted sites. Elsevier, Amsterdam, pp 1–49. https://doi.org/10.1016/B978-0-12-813912-7.00001-6

Pandey VC, Singh V (2019) Exploring the potential and opportunities of recent tools for removal of hazardous materials from environments. In: Pandey VC, Bauddh K (eds) Phytomanagement of polluted sites. Elsevier, Amsterdam, pp 501–516. https://doi.org/10.1016/B978-0-12-813912-7.00020-X

Pandey VC, Singh N (2010) Impact of fly ash incorporation in soil systems. Agr Ecosyst Environ 136:16–27. https://doi.org/10.1016/j.agee.2009.11.013

Pathak S, Agarwal AV, Pandey VC (2020) Phytoremediation—a holistic approach for remediation of heavy metals and metalloids. In: Pandey VC, Singh V (eds) Bioremediation of pollutants. Elsevier, Amsterdam, pp 3–14. https://doi.org/10.1016/B978-0-12-819025-8.00001-6

Pierart A, Dumat C, Maes AQM, Sejalon-Delmas N (2018) Influence of arbuscular mycorrhizal fungi on antimony phyto-uptake and compartmentation in vegetables cultivated in urban gardens. Chemosphere 191:272–279

Plaza S, Tearall KL, Zhao FJ, Buchner P, McGrath SP, Hawkesford MJ (2007) Expression and functional analysis of metal transporter genes in two contrasting ecotypes of the hyperaccumulator Thlaspi caerulescens. J Exp Bot 58:1717–1728

Prasad MNV, Freitas HM (2003) Metal hyperaccumulation in plants—biodiversity prospecting for phytoremediation technology. Electron J Biotechnol 6:285–321

Puschenreiter M, Türktaş M, Sommer P, Wieshammer G, Laaha G, Wenzel WW, Hauser MT (2010) Differentiation of metallicolous and non-metallicolous Salix caprea populations based on phenotypic characteristics and nuclear microsatellite (SSR) markers. Plant, Cell Environ 33:1641–1655

Rai PK (2008) Phytoremediation of Hg and Cd from industrial effluents using an aquatic free floating macrophyte Azolla pinnata. Int J Phytorem 10:430–439

Rajkumar M, Ma Y, Freitas H (2008) Characterization of metalresistant plant-growth promoting Bacillus weihenstephanensis isolated from serpentine soil. Portugal J Basic Microbiol 48:500–508

Rajkumar M, Sandhya S, Prasad M, Freitas H (2012) Perspectives of plant-associated microbes in heavy metal phytoremediation. Biotechnol Adv 30(6):1562–1574

Rattan RK (2002) Heavy metals in environments-Indian scenario. Fertil News 47:21–40

Rattan RK (2005) Long-term impact of irrigation with sewage effluents on heavy metal content in soils, crops and groundwater—a case study. Agric Ecosys Environ 109:310–322

Rengel Z (2004) Heavy metals as essential nutrients. In: Prasad MNV (ed) Heavy metal stress in plants, 2nd edn. Springer, Berlin

Rensing C, Maier RM (2003) Issues underlying use of biosensors to reduce measure bioavailability. Ecotoxicol Environ Sat 56:140–147

Rodriguez L, Lopez-Bellido F, Carnicer A, Alcalde-Morano V (2003) Phytoremediation of mercury-polluted soils using crop plants. Fresen Environ Bull 9:328–332. https://doi.org/10.1007/s11356-019-06563-3

Román-Ponce B, Reza-Vazquez DM, Gutierrez-Paredes S, De Haro-Cruz MJ, Maldonado-Hernandez J, Bahena-Osorio Y et al (2017) Plant growth-promoting traits in rhizobacteria of heavy metal-resistant plants and their effects on Brassica nigra seed germination. Pedosphere 27:511–526

Romera E, González F, Ballester A, Blázquez M, Munoz J (2007) Comparative study of biosorption of heavy metals using different types of algae. BioresourTechnol 98:3344–3353

Roosens NH, Willems G, Saumitou-Laprade P (2008) Using Arabidopsis to explore zinc tolerance and hyperaccumulation. Trends Plant Sci 13:208–215

Rzymski P, Niedzielski P, Poniedziałek B, Klimaszyk P (2014) Bioaccumulation of selected metals in bivalves (Unionidae) and Phragmites australis inhabiting a municipal water reservoir. Environ Monitor Asses 186:3199–3212. https://doi.org/10.1007/s10661-013-3610-8

Saha JK, Panwar NR (2013) Environmental pollution in the country in relation to soil quality and human health. In: Manna MC, Biswas AK, Chaudhary RS, Lakaria BL, Rao AS (eds) Kundu S. IISS contribution in frontier areas of soil research, Indian Institute of Soil Science, Bhopal, India, pp 281–306

Saha JK, Selladurai R, Coumar MV, Dotaniya ML, Kundu S, Patra AK (2017) Soil pollution-an emerging threat to agriculture. Springer. pp 1–9

Sakakibara M, Ohmori Y, Ha NTH, Sano S, Sera K (2011) Phytoremediation of heavy metal contaminated water and sediment by Eleocharis acicularis. Clean Soil Air Water 39:735–741

Sancenon V, Puig S, Mateu-Andres I, Dorcey E, Thiele DJ, Penarrubia L (2004) The Arabidopsis copper transporter COPT1 functions in root elongation and pollen development. J Bio Chem 279:15348–15355

Sandarin TR, Hoffman DR (2007) Bioremediation of organic and metal contaminated environments: effects of metal toxicity, speciation and bioavailability or biodegradation. Environ Bioremd Technol 1–34

Sas-Nowosielska A, Galimska-Stypa R, Kucharski R, Zielonka U, Małkowski E, Gray L (2008) Remediation aspect of microbial changes of plant rhizosphere in mercury contaminated soil. Environ Monit Assess 137:101–109. https://doi.org/10.1007/s10661-007-9732-0

Schikora A, Thimm O, Linke B, Buckhout TJ, Müller M, Schmidt W (2006) Expression, localization, and regulation of the iron transporter LeIRT1 in tomato roots. Plant Soil 284:101–108

Sheoran V, Sheoran A, Poonia P (2009) Phytomining: a review. Miner Eng 22:1007–1019. https://doi.org/10.1016/j.mineng.2009.04.001

Shingu Y, Kudo T, Ohsato S, Kimura M, Ono Y, Yamaguchi I, Hamamoto H (2005) Characterization of genes encoding metal tolerance proteins isolated from Nicotiana glauca and Nicotiana tabacum. Biochem Biophys Res Commun 331:675–680

Silva P, Matos M (2016) Assessment of the impact of aluminum on germination, early growth and free proline content in Lactuca sativa L. Ecotoxicol Environ Saf 131:151–156

Simmons RW (2006) Impact of wastewater irrigation on Cd and Pb concentrations in rice straw and paragrass

Singh V, Chauhan PK, Kanta R, Dhewa T, Kumar V (2010) Isolation and characterization of Pseudomonas resistant to heavy metals contaminants. Int J Pharm Sci Rev Res 3:164–167

Srivastava M, Ma LQ, Santos JAG (2006) Three new arsenic hyperaccumulating ferns. Sci Total Environ 364:24–31

Srivastava S, Agrawal SB, Mondal MK (2015) A review on progress of heavy metal removal using adsorbents of microbial and plant origin. Environ Sci Pollu Res 22(20):15386–15415

Tastan BE, Ertûgrul S, Dönmez G (2010) Efectivebioremoval of reactive dye and heavy metals by Aspergillus versicolor. Bioresource Technol 10(3):870–876

Thambavani DS, Prathipa V (2012) Quantitative assessment of soil metal pollution with principal component analysis, geo accumulation index and enrichment index. Asian J Environ Sci 7(2):125–134

Tripathi M, Munot HP, Shouch Y, Meyer JM, Goel R (2005) Isolation and functional characterization of siderophore-producing lead- and cadmium-resistant Pseudomonas putida KNP9. Curr Microbiol 5:233–237

Ungureanu G, Santos S, Boaventura R, Botelho C (2015) Biosorption of antimony by brown algae S. muticum and A. nodosum. Environ Eng Manag J 14:455–463

USEPA (1999) Report on bioavailability of chemical wastes with respect to the potential for soil remediation. T28006: QT-DC-99-003260

van de Mortel JE, Villanueva LA, Schat H, Kwekkeboom J, Coughlan S, Moerland PD, Loren V, van Themaat E, Koornneef M, Aarts MGM (2006) Large expression differences in genes for iron and Zn homeostasis, stress response, and lignin biosynthesis distinguish roots of Arabidopsis thaliana and the related metal hyperaccumulator Thlaspicaerulescens. Plant Physiol 142:1127–1147

Wackernagel M, Schulz B, Deumling D et al (2002) Tracking the ecological overshoot of the human economy. PNAS, Proc Nat Acad Sci 99:9266–9271

Wang J, Feng X, Anderson CW, Xing Y, Shang L (2012) Remediation of mercury contaminated sites–a review. J Hazard Mater 221:1–18. https://doi.org/10.1016/j.jhazmat.2012.04.035

Wei S, Zhou Q, Saha UK (2008) Hyperaccumulative characteristics of weed species to heavy metals. Water Air Soil Pollut 192:173–181

Weyens N, Croes S, Dupae J, Newman L, van der Lelie D, Carleer R et al (2010) Endophytic bacteria improve phytoremediation of Ni and TCE co-contamination. Environ Pollut 158:2422–2427

Willems G, Drager DB, Courbot M, Gode C, Verbruggen N, Saumitou- Laprade P (2007) The genetic basis of Zn tolerance in the metallophyte Arabidopsis halleri ssp. Halleri (Brassicaceae): an analysis of quantitative trait loci. Genetics 176:659–674

Xiao H, Yin L, Xu X, Li T, Han Z (2008) The Iron-regulated transporter, MbNRAMP1, isolated from Malus baccatais involved in Fe, Mn and Cd trafficking. Ann Bot 102:881–889

Yang JX, Liu Y, Ye ZH (2012) Root-Induced changes of pH, Eh, Fe(II) and fractions of Pb and Zn in rhizosphere soils of four wetland plants with different radial oxygen losses. Pedosphere 22:518–527

Yang J, Li K, Zheng W, Zhang H, Cao X, Lan Y, Yang C, Li C (2015) Characterization of early transcriptional responses to cadmium in the root and leaf of Cd-resistant Salix matsudanaKoidz. BMC Genomics 16:705

Zhang X, Xia H, Li Z, Zhuang P, Gao B (2011) Identification of a new potential Cd-hyperaccumulator Solanum photeinocarpum by soil seed bank-metal concentration gradient method. J Hazard Mater 189:414–419

Zhang M, Xian S, Xu J (2020) Heavy metal pollution in the East China sea: a review. Mar Pollut Bull 159:111473

Zhao FJ, Ma JF, Meharg AA, McGrath SP (2009) Arsenic uptake and metabolism in plants. New Phytol 181:777–794

Zhu YG, Kneer R, Tong YP (2004) Vacuolar compartmentalization: a second-generation approach to engineering plants for phytoremediation. Trends Plant Sci 9:7–9

Zouboulis A, Loukidou M, Matis K (2004) Biosorption of toxic metals from aqueous solutions by bacteria strains isolated from metal-polluted soils. Process Biochem 39:909–916

Chapter 5
Cutting-Edge Tools to Assess Microbial Diversity and Their Function in Land Remediation

Indra Mani

Abstract Soil contamination caused by pollutants has been a great challenge for us. However, soil provides a vast shelter, which allows a co-occurrence of millions of microorganisms. These microbes play a critical role in the remediation of such contaminated land. There are various techniques available to evaluate microbial diversity (DNA and rRNA-based profiling) and their functions (functional genes). In addition, isolation of pure culture, 16S rDNA (for bacteria), and 18S rDNA (for fungi)-based identification and characterization have shifted to omics. For example, it has transformed from genomics to metagenomics, transcriptomics to metatranscriptomics, proteomics to metaproteomics, and metabolites to metabolomics to study microbial diversity and their function. These various omics methods are used to understand the microbial diversity, biomass, mineralization, detoxification, and nutrient cycling phenomenon. Currently, culture-independent-based molecular techniques prevailing tools to isolate and identify functional genes from the uncultured microbes. Continuous development of sequencing technology and in silico tools, which has accelerated the identification and characterization of complex microbial communities from various environmental samples. Therefore, the advancement of these technology would deliver meaningful insight to evaluate the microbial diversity and their function for land remediation. This chapter highlights various techniques from culture-dependent to culture-independent, which are to be used to assess the microbial diversity and their functions.

Keywords Microbial diversity · Metagenomics · Metatranscriptomics · Metaproteomics · Soil · Sequencing

5.1 Introduction

Soil is the major source of a variety of microorganisms such as viruses, archaea, bacteria, fungi, and other parasites. Approximately a gram of soil might comprise

---

I. Mani (✉)
Department of Microbiology, Gargi College, University of Delhi, Siri Fort Road, New Delhi 110049, India
e-mail: indra.mani@gargi.du.ac.in; indramanibhu@gmail.com

1000–10,000 species of unidentified prokaryotes (Torsvik et al. 1990). Microbes in soil play a vital role in soil fertility (O' Donnel et al. 2007), soil structure (Wright and Upadhyay 1998), plant health (Dodd et al. 2000), plant nutrition (Timonen et al. 1996), biogeochemical cycle (Wall and Virginia 1999), degradation of xenobiotic compounds (Barakat 2011), and land management (Nacke et al. 2011). Due to such great importance and so much complexity of microorganisms, it is very challenging to identify and characterize them. Interestingly advancement of genomics to metagenomics is very much helpful to characterize them (Mocali and Benedetti 2010; Huson et al. 2011; Mani 2020a; Gangotia et al. 2021; Gupta et al. 2021). Due to rapid progress in technology, that has enhanced the identification and characterization of microorganisms from any ecological samples. Soil contains a very important strain of microbes, which need to identify and use for the remediation of soil. Before, it was totally dependent on culture-based methods, which provides very trivial information about microorganisms. It might be due to a lack of numerous growth associated knowledge such as pH, temperatures, humidity, chemicals, and tracer molecules. However, culture-independent approaches (Metagenomics) are helpful to census the microbes in any environments (Schloss and Handelsman 2004; Schloss et al. 2016; Mani 2020b). Further, an advancement in the DNA sequencing technology and availability of international nucleotides sequence database collaboration (INSDC) provides an opportunity to assess the microbial diversity as well as the specific function of microbes. It provides established genome references, which are very important to analyze the microbial communities (microbiota) and their functions.

Metabolic networks such as the Kyoto Encyclopedia of Genes and Genomes (KEGG) and Clusters of Orthologous Group (COG) are available to understand the metabolic pathways involve in synthesis and degradation of particular molecules. In addition, various omics techniques like metatranscriptomics, metaproteomics, and metabolomics are very helpful to understand the microbial diversity and their function in soil environments. After exploring these omics, a particular stain can be identified, characterized, and utilized for the bioremediation of soil (Mani 2020c). To understand the diversity and functions of microorganisms in metals contaminated and non-contaminated soils, it would provide valuable information. Further, it can be utilized to analyze an abundance of the particular microorganisms and also helpful to discover potential pathways involve in the degradation of heavy metals.

5.2 Culture-Dependent Techniques

Cultivation of microorganisms for isolation, characterization, and identification is a gold standard approach for the detection of the pathogens (Rudkjøbing et al. 2016). There are various media used for isolation, characterization, and identification of microorganisms such as nutrient agar (NA), brain heart infusion (BHI) agar, Salmonella-Shigella (SS) agar, macConkey agar, mannitol salt agar, eosin methylene blue (EMB) agar, potato dextrose agar (PDA), trypticase soy agar, sabouraud dextrose agar, and many more selective, differential, enriched, and enrichment media. Several studies suggest that <0.1% of the microorganisms in soil are culturable using classical methods (Torsvik et al. 1990, 1994, 1996; Handelsman et al. 1998). However, molecular methods have the advantages to identify rapidly and cover more microbes, which may skip through a culture-based approach.

DNA markers are an appropriate tool in order to obtain information about gene flow, allele frequencies, and other parameters that are important in population biology (Neigel 1997). Ribosomal DNA (rDNA) is useful for phylogenetic analysis because different regions of the rDNA repeat unit evolve at very different rates. Therefore, regions of rDNA arrays that are particularly possible to generate informative data for almost any systematic question can be selected for analysis (Hillis and Dixon 1991). In addition, the islands of highly conserved sequences within most rRNA genes are very helpful for constructing "universal" primers, which can be used for sequencing either rRNA or rDNA from several species, for amplifying regions of interest by use of the polymerase chain reaction (PCR), or for use as probes in restriction enzyme analyses (Hillis and Moritz 1990). Remarkably, sequences of 16S rRNA gene uncover an information of microbial diversity "black box" that guide analysis of the previously unknown bacterial life and their function (Nelson et al. 2011; De Sundberg et al. 2013; De Vrieze et al. 2018). There are several molecular methods used for the analysis of identification and characterization of microorganisms from the soil.

5.2.1 Amplified Ribosomal DNA Restriction Analysis (ARDRA)

Amplified ribosomal DNA restriction analysis (ARDRA) is utilized to investigate the microbial diversity on the basis of DNA polymorphism (Deng et al. 2008). In this method, 16S rDNA is amplified by either genus specific primer or universal primer and processed with restriction endonucleases, followed by agarose gel electrophoresis or polyacrylamide gel electrophoresis (PAGE). DNA band profiles are used to genotyping the microbial community (Tiedje et al. 1999). ARDRA has been used to evaluate the microbial diversity in soil with changes in land use in Hawaii, USA (Nüsslein and Tiedje 1999), the Karst forest, China (Zhou et al. 2009), and arsenic affected Bangladesh soils (Sanyal et al. 2016). In addition, ARDRA-based study has isolated 358 isolates, which clustered into 35 groups from glacier foreland soils. These groups belong to 20 genera and six taxa such as *Betaproteobacteria, Actinobacteria, Alphaproteobacteria, Bacteroides, Deinococcus-Thermus*, and *Gammaproteobacteria* (Wu et al. 2018). The finding shows that ARDRA techniques could characterize the glacier foreland soils culturable microbial communities.

5.2.2 Ribosomal Intergenic Spacer Analysis (RISA)

Ribosomal intergenic spacer analysis (RISA) is another technique, which is based on ribosome DNA sequences. It is a culture-dependent technique, which used for the microbial community analyses (Sigler and Zeyer 2002). But in this technique, information coming from the spacer region of rDNA. In this technique, a pair of oligonucleotides primers (one from 16S and other from 23S rDNA) are required to amplify the internal transcribed spacer (ITS) (Borneman and Triplett 1997). The size of ITS ranges between 150 and 1500 bp, and it is a good candidate for an analysis of bacterial diversity (Sigler et al. 2002).

The ITS regions evolve rapidly and, hence, are useable as "high-resolution marker" in populations genetics (van Oppen et al. 2002). Although in the few cases, polymorphisms have been detected in these non-coding regions (Nichols and Barnes 2005). The ITS region has progressively been utilized for discrimination among bacterial species or strains, including *Mycobacterium* species (Roth et al. 1998), cyanobacteria (Boyer et al. 2001), acetic acid bacteria (Trcek 2005), and *Escherichia coli* strains (Gibreel and Taylor 2006), which cannot be easily distinguished by the 16S rRNA gene. Similarly, the identifications of closely related species based on only morphological characters are difficult in the case of the multi-species genus.

5.2.3 Random Amplified Polymorphic DNA (RAPD)

Random amplified polymorphic DNA (RAPD) is a molecular technique that used a decamer (10 nucleotides) primer for PCR amplification and followed by agarose gel electrophoresis. Comparative amplified fragments are used for the analysis of microbial diversity. These short primes randomly bind anywhere in genomic DNA at low melting temperature (Tm) (Franklin et al. 1999). RAPD has been used to analyze microbial diversity in the soil of arid zone plants (Sharma et al. 2013), viral diversity in soils (Srinivasiah et al. 2013), Panax ginseng rhizosphere, and non-rhizosphere soil (Li et al. 2012). Due to a limited resolving ability of RAPD and massive microorganisms, it needs to integrate with other advanced approaches.

5.3 Culture-Independent Techniques

There are numerous culture-independent methods, such as denaturing gradient gel electrophoresis (DGGE), terminal restriction fragment length polymorphism (TRFLP), and fluorescent in situ hybridization (FISH) have been utilized to investigate microbial diversity (Hwang et al. 2008; Rademacher et al. 2012; Klang et al. 2015).

5.3.1 Denaturing Gradient Gel Electrophoresis (DGGE)

Denaturing gradient gel electrophoresis (DGGE) is a method that has been exploited for species identification. There are various steps involved in this method, such as genomic DNA extraction, amplification of 16S rDNA sequences, and separation of amplified products by PAGE. The electrophoretic mobility of DNA fragments depends upon the melted double-stranded DNA in gel contains the linear gradient of DNA denaturant, formamide, and urea (Muyzer et al. 1993) or a linear temperature gradient (Muyzer and Smalla 1998). DGGE has been used to assess the microbial diversity for the sulfate-reducing bacteria (Kleikemper et al. 2002), *Gamma* and *Betaproteobactera* (Fahrenfeld et al. 2013), and for functional diversity in different contaminated sites (Ferris et al. 1996; Geets et al. 2006; Orlewska et al. 2018). It has been extensively used for the assessment of various microorganisms in different environmental samples.

5.3.2 Terminal Restriction Fragment Length Polymorphism (T-RFLP)

Terminal restriction fragment length polymorphism (T-RFLP) technique is based on PCR and restriction endonuclease digestion. After extraction of genomic DNA from any environmental sample, fluorescence labelled primers are utilized for amplification of 16S rDNA followed by restriction digestion. Analysis of separated fragments carried out by automated DNA sequencer, which provides the patterns of the peaks in the form of electropherogram (Thies 2007; Stenuit et al. 2008). The electropherogram peaks are identified through an available database for analysis of microbial diversity (Marsh et al. 2000). T-RFLP has been used to estimate the microbial diversity for the different groups such as eubacteria (Brunk et al. 1996), planctomycetes (Derakshani et al. 2001), methylotrophs and methanotrophs (Allen et al. 2007), aerobic and anaerobic hydrocarbon-degrading communities (Tipayno et al. 2012), and microbial diversity in anaerobic digestion (De Vrieze et al. 2018). This technique has been replaced with 16S rRNA gene sequencing because of its time-consuming and complex nature (De Vrieze et al. 2018). Another disadvantage of the method, it covers limited phylogenetic analysis due to short sequence reads (Marzorati et al. 2008). This technique facilitates the detection of different haplotypes from any environmental samples.

5.3.3 Fluorescence in situ Hybridization (FISH)

Fluorescence in situ Hybridization (FISH) is a molecular tool, which was developed by Langer-Safer et al. (1982). In this technique, a fluorescence dye labelled probes (DNA or cDNA) are used, which bind to the complementary region of the

DNA. The probes are prepared either by nick translation or PCR or tagged with biotin. After denaturation of DNA and probes, both allow for hybridization. After hybridization, followed by post-hybridization, samples examined under the fluorescence microscope (Amann et al. 1995; Mani et al. 2011). FISH, which can be used as a cultivation-independent approach for visualization, identification, and quantification of microorganisms in the medical and environmental sample. FISH has used to evaluation of microbial diversity in contaminated environments (Richardson et al. 2002), s-triazine herbicides treated soils (Caracciolo et al. 2010), methane-rich gas field in the Cook Inlet basin of Alaska (Dawson et al. 2012), and activated sludge from a nitrifying-denitrifying tank at the municipal wastewater treatment plant (WWTP) of Klosterneuburg, Austria (Lukumbuzya et al. 2019). Due to advancement in the FISH technique, multicolor FISH can be more suitable as compared to a classical FISH.

5.4 Cutting-Edge High-Throughput Culture-Independent Approach for Microbial Diversity

Presently, omics techniques like metagenomics, metatranscriptomics, metaproteomics, and metabolomics are very helpful to understand the microbial diversity and their function in soil (Fig. 5.1). For the bioremediation of soil, multi-omics approach can be utilized to screen potential microbial strain (Mani 2020c). These multi-omics are discussed in detail.

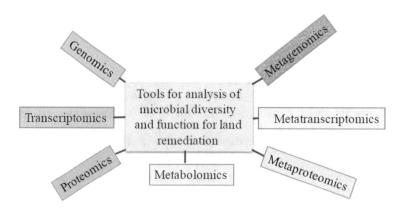

Fig. 5.1 A schematic diagram of different molecular tools that used for an assessment of the microbial diversity and their functions in the remediation of contaminated land

5.4.1 Metagenomics

Metagenomics is also known as environmental genomics or community genomics or population genomics. The term metagenomics was coined by Handelsman et al. (1998). A detail procedure of shotgun metagenomic sequencing (Regar et al. 2019) is given in Fig. 5.2. A shotgun metagenomic sequencing has been utilized to evaluate the microbial diversity from the pesticides contaminated and non-contaminated soil samples. Results have shown various abundance of microbes such as *Proteobacteria, Actinobacteria, Bacteroidetes, Firmicutes,* and *Acidobacteria* in both samples. Interestingly, substrate specific pathway degradative gene analysis has shown the presence of many genes for both upper and lower pathways. However, a smaller number of degradative genes have identified for the degradation of atrazine, styrene, naphthalene, and bisphenol (Regar et al. 2019). These xenobiotic degradative genes carrying microbes can be utilized for remediation of such pesticides contaminated lands.

Another shotgun metagenomics-based study has reported that bacterial and fungal microbes were associated with vineyards and forest land in Chile. In both habitats, the most abundant bacteria *Candidatus, Bradyrhizobium,* and *Solibacter*, and the fungus *Gibberella* were identified. Interestingly, metabolic diversity was different in the vineyards associated microbes while no difference was observed at the taxonomic level (Castañeda and Barbosa 2017). Swenson et al. (2018) analyzed the metagenomics to understand microbial diversity from the biological soil crust (biocrust). The study demonstrated that microbial diversity was directly linked with environmental chemistry in biocrust (Swenson et al. 2018). Soil microbial diversity greatly affected

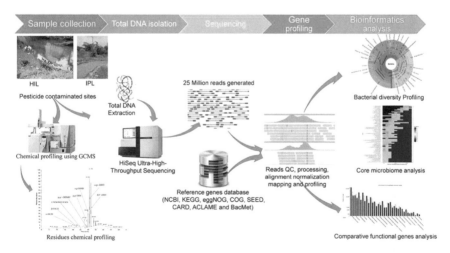

Fig. 5.2 A schematic presentation of a shotgun metagenomic approach for evaluation of the microbial diversity and their functions from the pesticides contaminated sites. Major steps such as sample collection, total DNA isolation, sequencing, gene profiling, and bioinformatics analysis have been shown (Regar et al. 2019. *Adapted with permission*)

at mining sites due to soil pollution. To understand the abundance of microbes, a study was conducted in zinc (Zn) and lead (Pb) contaminated soil using the 16S rDNA-based metagenomics approach. Results from the study, ten frequently detected bacteria, which included *Geobacter, Solirubrobacter, Edaphobacter, Gemmatiomonas, Pseudomonas, Xanthobacter, Sphingomonas, Ktedonobacter, Pedobacter*, and *Nitrosomonas* (Hemmat-Jou et al. 2018). This study demonstrates the bacterial profiling in Pb and Zn contaminated soils using a powerful tool like metagenomics.

To understand microbial diversity in the desert environment, a 16S rDNA sequence-based metagenomics approach has been utilized. For the analysis, soil samples were taken from the two quadrates desert environment (Thar Desert India) that face hot dry weather with fewer rain and intense temperatures. In this study, they utilized V3-V4 regions of 16S rDNA and Illumina next generation sequencing (NGS) to analyze bacterial diversity. They found the three phyla in abundance, *Actinobacteria, Proteobacteria,* and *Acidobacteria,* in both environments (Sivakala et al. 2018). Among these phyla, *Actinobacteria* is an important phylum based on their commercial value. A finding suggests that desert environments can be a good source to isolate an important microorganism to remediation of land.

A 16S rDNA-based high throughput sequencing method has been used to analyze alfalfa and barley rhizosphere microbial diversity in oil contaminated soil. A study reported that oil contaminated soil has higher abundance of oil-degrading microbes (*Alcanivorax* and *Aequorivita*) but reduced diversity as compared to oil non-contaminated samples. Moreover, two more phyla (*Thermi* and *Gemmatimonadetes*) were also present in the oil-contaminated soil (Kumar et al. 2018). These findings suggest that the presence of these oil-degrading microbes play a vital role in the degradation of hydrocarbon contamination in soil. The combination of metagenomics and in silico approaches have been used to identify novel genes, proteins, and enzymes from the diverse groups of microbes. With the help of the NGS and Sanger sequencing method, genome sequences are generated, and in silico method aids in the prediction of protein function. After all, it can be cloned and expressed in a particular host in in vitro. Such types of approaches can be used in any environmental samples (Calderon et al. 2019). Interestingly, this approach can be utilized to discover important enzymes from microbial diversity for the remediation of land.

A metagenomics method has been extended to understand the effect of altitude on microbial diversity in soil. Features of high altitude ecosystems are low temperature, decreased atmospheric pressure, variable precipitation, and soil nutrient stress (Morán-Tejeda et al. 2013). It has found the most abundant phyla of *Acidobacteria, Proteobacteria,* and *Actinobacteria* at high altitude land, whereas *Fermicutes* and *Bacteroidetes* at low altitude. The high throughput sequencing data analysis helped to identify a novel bacterial diversity at high altitude, which was missed by conventional methods (Kumar et al. 2019). Due to better survival of microbes at high altitudes under various variable conditions including soil nutrient stress, it would be beneficial to explore them further. Therefore, a study suggests that these groups of microbes can be used to remediation of hill agriculture land. A metagenomic method has been utilized to examine the microbial diversity in effluent contaminated constructed wetlands and in rhizosphere soil. Interestingly, the rhizosphere soils have shown the richness of microbial diversity as compared to wetlands. From functional analysis, it has been demonstrated that different xenobiotic degradation pathways are associated in the soils (Bai et al. 2014). The finding suggests that utilizing a recent tool to investigate the diversity of microbes on the sequence based as well as function based can be used to the remediation of effluents contaminated land.

A metagenomics-based study has analyzed the microbial diversity in Cadmium (Cd) contaminated soil. After comparison with non-contaminated soil, Feng et al. (2018) found that Cd-contamination significantly reduced the diversity of microorganism. Interestingly, they have found *Sulfuricella*, *Proteobacteria*, and *Thiobacillus* as major microbes which played an important role in the remediation of Cd-contaminated soil (Feng et al. 2018). These Cd resistant microbes can be further used in the remediation of Cd-contaminated land. Similarly, metagenomics study has performed in uranium contaminated soil to understand the functional and structural diversity of microbes. In uranium contaminated and non-contaminated soil, *Proteobacteria*, and *Actinobacteria* were common while *Alicyclobacillus*, *Robiginitalea*, and *Microlunatus* were present in the non-contaminated soil only. KEGG metabolic pathway database was used to analyze the metabolism of amino acids and signaling molecules (Yan et al. 2016). Common microbes such as *Proteobacteria* and *Actinobacteria* can be used in the remediation of uranium contaminated land.

A metagenomic study was reported from China utilizing mercury (Hg) contaminated soil. Analysis demonstrated the Hg affected microbial diversity, abundance, and functional aspects. In contaminated soil, *Firmicutes* and *Bacteroidetes* were abundance, and contamination of Hg also affected on different functional genes that involve in its transformation, such as methylation and reduction (Liu et al. 2018). Metagenomics methods have been used to evaluate the effects of natural groups of microbes and consortium microbes on the degradation of polycyclic aromatic hydrocarbon (PAH) in the contaminated soil. A study has demonstrated that the degradation of PAH was significantly higher by using microbial consortium as compared to other groups of microbes. At the gene level, variations in laccase, aromatic ring-hydroxylating dioxygenases (ARHD), salicylate, benzoate, and protocatechuate-degrading enzyme were found (Zafra et al. 2016). This study suggests that these potential gene producing microbes can be useful in remediation of PAH-contaminated land.

5.5 Cutting-Edge High-Throughput Culture-Independent Approach for Microbial Function

Assessments of functional characteristics of microorganisms are complex as compared to sequence-based study.

5.5.1 Metatranscriptomics

Metatranscriptomics is RNA-based methods used to analyze the taxonomic composition and profile of the microbial functions. There are various experimental steps involved in this approach, which need to be addressed while analyze through metatranscriptomics (Carvalhais and Schenk 2013; Jiang et al. 2016). The advancement in this technology is very promising to help to understand microbial function. A culture-independent method has been utilized to evaluate microbial diversity from halogen contaminated and non-contaminated German forest soils. Weigold et al. (2016) analyzed the genes encoding enzymes that are involved in halogenation and dehalogenation of the halogens. They determined that *Bradyrhizobium* and *Pseudomonas* genera were involved in these processes. Further, they found chloroperoxidases and haloalkane dehalogenases enzymes, which were responsible for the halogenation and dehalogenation of halogens in the contaminated forest soil (Weigold et al. 2016).

Metagenomics and metatranscriptomics methods have been used to examine microbial diversity and their functions in dissolved organic matter (DOM) from the soil samples. A study reported that there were great variations in microbial genera such as *Thermoleophilia, Syntrophobacterales, Spirochaeta, Geobacter,* and *Gaiella.* In this study, Li et al. (2018) found a correlation with the richness of microbial metabolic pathways lignolysis, methanogenesis, and fermentation in DOM of paddy soil samples. A metatranscriptomics-based study analyzed an environmental functional gene microarray (E-FGA) containing 13,056 mRNA microbial clones from different environmental samples. They have examined the E-FGA containing mRNA microbial clones by profiling the microbial activity of agricultural soils with a high or low flux of nitrous oxide (N_2O). Interestingly, 109 genes have been expressed and demonstrated significant variability with high and low N_2O emissions (McGrath et al. 2010). Such an approach may be useful to evaluate the functional activity of the microorganisms.

Shotgun metagenomic sequencing and metatranscriptomics studies have performed to analyze the rhizosphere microbial communities of *Archis hypogaea* (peanut plant), roots of plants grown in the soil of crop rotation, and peanut monocropping. Interestingly, in monocropping, an enrichment of different rare species occurred, but microbial diversity of rhizosphere had reduced. A further reduction occurred in the downregulation of genes in auxin and cytokinin and upregulation of genes related to other hormones (abscisic acid and salicylic acid) (Li et al. 2019). As the study suggested, plant rhizosphere microbiota and plant physiology were affected by land use history.

A metatrasncriptomics approach has been utilized to identify cadmium (Cd) tolerant genes from the contaminated sites. cDNA libraries of different sizes of yeast mRNA (from 0.1 kb to 4 kb) were developed. After screening of cadmium tolerant transcript through yeast complementation system, Thakur et al. (2018) have found that transformants ycf1ΔPLBe1 were capable to tolerate Cd in the range of 40–80 μM. Interestingly, a sequence of PLBe1 cDNA shown homology with AN1 type zinc finger protein of *Acanthameoba castellani*. In addition, it has also shown the tolerance against copper (Cu), cobalt (Co), and zinc (Zn) (Thakur et al. 2018). The finding suggests that PLBe1 can be a promising candidate for the multi-metal tolerant gene for remediation of the heavy metal contaminated lands. A metatranscriptomics study has identified an *Actinobacteria* as a most abundant family in hot desert soil samples. Interestingly, it found that chemoautotrophic carbon fixation genes were more expressed as compared to photosynthetic genes in these samples (León-Sobrino et al. 2019) indicating that chemoautotrophy could be alternative of photosynthesis in hot dessert soils.

Bragalini et al. (2014) developed a solution hybrid selection (SHS) technique, which is very effective for the recovery of eukaryotes cDNAs from soil extracted mRNA. The authors utilized this technique on endo-xylanases of Glycoside Hydrolase (GH) 11 gene family. Approximately 25% cloned cDNAs sequences were expressed in *Saccharomyces cerevisiae* (Bragalini et al. 2014). This technique can be utilized to explore eukaryotic microbial communities to the prospecting of land remediation related genes. A 16S rDNA and metatranscriptomics methods were used to evaluate microbial diversity and their functions in the sandy loam soil, which was treated with various concentrations (60–2000 mg/kg) of silver nanoparticles (AgNPs). Analysis has shown that it was very much upregulation in genes, which are involved in the heavy metal resistance (Meier et al. 2020). Finding suggests that multi-level concentration-based studies are important to assess microbial functions in a particular land site.

5.5.2 Metaproteomics

Another powerful tool of omics is metaproteomics that includes the study of all proteins which are directly recovered from any environmental samples. Metaproteomic approaches are undertaking microbial functional characteristics more directly as compared to metagenomics and metatranscriptomics. This method is used to understand the functional diversity of microorganisms in any particular site. A metaproteomics-based study analyzed the maize rhizosphere soil, where 696 proteins were discovered from 244 genus and 393 species (Renu et al. 2019). These important results can be helpful in designing experiments for other rhizosphere soil samples. Metaproteomics and phospholipid fatty-acids analysis has performed in petroleum polluted semiarid soil samples to understand the phylogenetic and physiological response of the microbiome. A 2016 study illustrated that petroleum contamination increases proteobacterial proteins while reducing the richness of *Rhizobiales* as compared to non-contaminated soil (Bastida et al. 2016). A metaproteomics method has been utilized to understand the effect of chlorophenoxy acid-degrading bacteria on the soil sample, which was treated with 2, 4-dichlorophenoxy acetic acid (2,4-D) for 22 days. They have identified the chlorocatechol dioxygenases enzymes from these samples (Benndorf et al. 2007). This enzyme can be further used for the treatment of 2,4-D contaminated land. Rotation of the plantation on a particular land may affect rhizosphere microbial diversity.

5.5.3 Metabolomics

Exometabolomics or metabolic footprinting is a sub-field of metabolomics, which is used to study extracellular metabolites (Allen 2003; Mapelli et al. 2008; Silva and Northen 2015). A detail procedure of metabolomics is given in Fig. 5.3. A study has analyzed exometabolome to understand the function of a microbial community of the biological soil crust (biocrust) (Swenson et al. 2018). Finding suggests that the microbial community is directly linked with environmental chemistry in biocrust. Gas chromatography-mass spectrometry (GC–MS) and liquid chromatography-mass spectrometry (LC–MS) were used to investigate the metabolites from the saprolite (chemically weathered rock) soil samples. In this study, 96 metabolites have been identified, including amino acids and their derivatives, nucleosides, sugar, alcohol, and carboxylic acids. After quantification of 25 metabolites, it has indicated an uneven quantitative distribution. There were two types of soil defined media (SDM 1 and SDM2) designed using these metabolites information. There were 30 different types of soil bacterial isolates grown on both media. However, a result has shown that SDM1 sustained growth of 13 isolates, and SDM2 supported the growth of 15 isolates (Jenkins et al. 2017). This information can be utilized to develop suitable media for the growth of promising microorganisms, which are potential candidates for land remediation.

5.6 Conclusion, Challenges, and Future Perspective

An increase of contamination in soil is a vast problem, and remediation of it a great challenge. Due to ubiquitous nature of the microbes in the environment, it plays an important role in the remediation of contaminated land. Further, microbes are an excellent source of enzymes that convert harmful metal into a neutral state. However, soil is a massive shelter of the diversity of culturable and unculturable microorganisms. Due to the complexity of microorganisms, it is very challenging to identify and characterize them. To understand the microbial diversity and their functions, various classical to advance techniques, including multi-omics are available. For the analysis of microbial diversity and their function, culture-dependent and culture-independent methods are being used. As metagenomics (culture-independent) molecular approach offers a powerful lens for viewing the microbial world and which is very promising to help to understand the questions like who are there? Or what are they doing? Therefore, the combined information of phylogenetic and functional aspects would provide thoughtful understandings about soil microorganisms. The

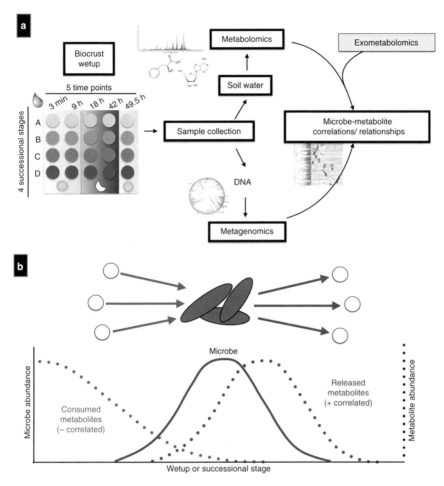

Fig. 5.3 Experimental workflow and biocrust microbe–metabolite relationship predictions. **a** Biocrust wetup metabolomics and metagenomics experimental setup and analysis. **b** Exometabolomics-based in situ microbe-metabolite relationship prediction (Swenson et al. 2018. *Adapted with permission*)

multi-omics methods such as metagenomics, metatranscriptomics, metaproteomics, and metabolomics are very helpful to screen potential microbes. Further, through the use of cutting-edge tools, a potential microbe can identify, characterize, modify, and construct microbial consortium to remediation of any contaminated land.

References

Allen J (2003) High-throughput classification of yeast mutants for functional genomics using metabolic footprinting. Nat Biotechnol 21(6):692–696

Allen JP, Atekwana EA, Atekwana EA, Duris JW, Werkema DD, Rossbach S (2007) The microbial community structure in petroleum-contaminated sediments corresponds to geophysical signatures. Appl Environ Microbiol 73(9):2860–2870

Amann RI, Ludwig W, Schleifer KH (1995) Phylogenetic identification and in situ detection of individual microbial cells without cultivation. Microbiol Rev 59:143–169

Bai Y, Liang J, Liu R, Hu C, Qu J (2014) Metagenomic analysis reveals microbial diversity and function in the rhizosphere soil of a constructed wetland. Environ Technol 35(17–20):2521–2527

Barakat MA (2011) New trends in removing heavy metals from industrial wastewater. Arab J Chem 4:361–377

Bastida F, Jehmlich N, Lima K, Morris BEL, Richnow HH et al (2016) The ecological and physiological responses of the microbial community from a semiarid soil to hydrocarbon contamination and its bioremediation using compost amendment. J Proteomics 135:162–169

Benndorf D, Balcke GU, Harms H, von Bergen M (2007) Functional metaproteome analysis of protein extracts from contaminated soil and groundwater. ISME J 1(3):224–234

Borneman J, Triplett EW (1997) Molecular microbial diversity in soils from eastern Amazonia: evidence for unusual microorganisms and microbial population shifts associated with deforestation. Appl Environ Microbiol 63(7):2647–2653

Boyer SL, Flechtner VR, Johansen JR (2001) Is the 16S–23S rRNA internal transcribed spacer region a good tool for use in molecular systematics and population genetics? A case study in Cyanobacteria. Mol Biol Evol 18:1057–1069

Bragalini C, Ribière C, Parisot N, Vallon L, Prudent E et al (2014) Solution hybrid selection capture for the recovery of functional full-length eukaryotic cDNAs from complex environmental samples. DNA Res 21(6):685–694

Brunk CF, Avaniss-Aghajani E, Brunk CA (1996) A computer analysis of primer and probe hybridization potential with bacterial small-subunit rRNA sequences. Appl Environ Microbiol 62(3):872-879

Calderon D, Peña L, Suarez A, Villamil C, Ramirez-Rojas A et al (2019) Recovery and functional validation of hidden soil enzymes in metagenomic libraries. Microbiologyopen 8(4):e00572

Caracciolo AB, Bottoni P, Grenni P (2010) Fluorescence in situ hybridization in soil and water ecosystems: a useful method for studying the effect of xenobiotics on bacterial community structure. Toxicol Environ Chem 92:567–579

Carvalhais LC, Schenk PM (2013) Sample processing and cDNA preparation for microbial metatranscriptomics in complex soil communities. Methods Enzymol 531:251–267

Castañeda LE, Barbosa O (2017) Metagenomic analysis exploring taxonomic and functional diversity of soil microbial communities in Chilean vineyards and surrounding native forests. Peer J 5:e3098

Dawson KS et al (2012) Quantitative fluorescence in situ hybridization analysis of microbial consortia from a biogenic gas field in Alaska's Cook Inlet basin. Appl Environ Microbiol 10:3599–3605

Derakshani M, Lukow T, Liesack W (2001) Novel bacterial lineages at the (sub) division level as detected by signature nucleotide-targeted recovery of 16S rRNA genes from bulk soil and rice roots of flooded rice microcosms. Appl Environ Microbiol 67(2):623–631

De Vrieze J, Ijaz UZ, Saunders AM et al (2018) Terminal restriction fragment length polymorphism is an "old school" reliable technique for swift microbial community screening in anaerobic digestion. Sci Rep 8:16818

Deng W, Xi D, Mao H, Wanapat M (2008) The use of molecular techniques based on ribosomal RNA and DNA for rumen microbial ecosystem studies: a review. Mol Biol Rep 35:265–274

Dodd JC, Boddington CL, Rodriguez A, Gonzalez-Chavez C, Mansur I (2000) Mycelium of arbuscular mycorrhizal fungi (AMF) from different genera: form, function and detection. Plant Soil 226(2):131–151

Fahrenfeld N, Zoeckler J, Widdowson MA et al (2013) Effect of biostimulants on 2, 4, 6-trinitrotoluene (TNT) degradation and bacterial community composition in contaminated aquifer sediment enrichments. Biodegradation 24:179–190

Feng G, Xie T, Wang X, Bai J, Tang L et al (2018) Metagenomic analysis of microbial community and function involved in cd-contaminated soil. BMC Microbiol 18(1):11

Ferris MJ, Muyzer G, Ward DM (1996) Denaturing gradient gel electrophoresis profiles of 16S rRNA-defined populations inhabiting a hot spring microbial mat community. Appl Environ Microbiol 62:340–346

Franklin RB, Taylor DR, Mills AL (1999) Characterization of microbial communities using randomly amplified polymorphic DNA (RAPD). J Microbiol Methods 35:225–235

Gangotia D, Gupta A, Mani I (2021) Role of bioinformatics in biological sciences. In: Advances in bioinformatics. Springer Nature, pp 37–57

Geets J et al (2006) DsrB gene-based DGGE for community and diversity surveys of sulfate-reducing bacteria. J Microbiol Methods 66:194–205

Gibreel A, Taylor DE (2006) Macrolide resistance in *Campylobacter jejuni* and Campylobacter coli. J Antimicrob Chemother 58:243–255

Gupta A, Gangotia D, Mani I (2021) Bioinformatics tools and softwares. In: Advances in bioinformatics. Springer Nature, pp 15–34

Handelsman J, Rondon MR, Brady SF, Clardy J, Goodman RM (1998) Molecular biological access to the chemistry of unknown soil microbes: a new frontier for natural products. Chem Biol 5(10):R245–R249

Hemmat-Jou MH, Safari-Sinegani AA, Mirzaie-Asl A, Tahmourespour A (2018) Analysis of microbial communities in heavy metals-contaminated soils using the metagenomic approach. Ecotoxicology 27(9):1281–1291

Hillis DM, Dixon MT (1991) Ribosomal DNA: molecular evolution and phylogenetic interference. Q Rev Biol 66(4):411–453

Hillis DM, Moritz C (1990) An overview of applications of molecular systematics. In: Hillis DM, Moritz C (eds) Molecular systematics. Sinauer Associates, Sunderland, pp 502–515

Huson DH, Mitra S, Ruscheweyh HJ, Weber N, Schuster SC (2011) Integrative analysis of environmental sequences using MEGAN4. Genome Res 21:1552–1560

Hwang K, Shin SG, Kim J, Hwang S (2008) Methanogenic profiles by denaturing gradient gel electrophoresis using order-specific primers in anaerobic sludge digestion. Appl Microbiol Biotechnol 80:269–276

Jenkins S, Swenson TL, Lau R, Rocha AR, Aaring A et al (2017) Construction of viable soil defined media using quantitative metabolomics analysis of soil metabolites. Front Microbiol 8:2618. https://doi.org/10.3389/fmicb.2017.02618

Jiang Y, Xiong X, Danska J, Parkinson J (2016) Metatranscriptomic analysis of diverse microbial communities reveals core metabolic pathways and microbiome-specific functionality. Microbiome 4:2

Klang J et al (2015) Dynamic variation of the microbial community structure during the long-time mono-fermentation of maize and sugar beet silage. Microb Biotechnol 8:764–775

Kleikemper J, Schroth MH, Sigler WV, Schmucki M, Bernasconi SM, Zeyer J (2002) Activity and diversity of sulfate-reducing bacteria in a petroleum hydrocarbon-contaminated aquifer. Appl Environ Microbiol 68(4):1516–1523

Kumar V, AlMomin S, Al-Aqeel H, Al-Salameen F, Nair S, Shajan A (2018) Metagenomic analysis of rhizosphere microflora of oil-contaminated soil planted with barley and alfalfa. PLoS ONE 13(8):e0202127

Kumar S, Suyal DC, Yadav A, Shouche Y, Goel R (2019) Microbial diversity and soil physiochemical characteristic of higher altitude. PLoS ONE 14(3):e0213844

Langer-Safer PR, Levine M, Ward DC (1982) Immunological method for mapping genes on Drosophila polytene chromosomes. Proc Natl Acad Sci USA 79(14):4381–4385

León-Sobrino C, Ramond JB, Maggs-Kölling G, Cowan DA (2019) Nutrient acquisition, rather than stress response over diel cycles, drives microbial transcription in a hyper-arid namib desert soil. Front Microbiol 10:1054

Li Y, Ying Y, Zhao D, Ding W (2012) Microbial community diversity analysis of Panax ginseng rhizosphere and non-rhizosphere soil using randomly amplified polymorphic DNA method. O J Gen 2:95–102

Li HY, Wang H, Wang HT, Xin PY, Xu XH et al (2018) The chemodiversity of paddy soil dissolved organic matter correlates with microbial community at continental scales. Microbiome 6(1):187

Li X, Jousset A, de Boer W, Carrión VJ, Zhang T, Wang X, Kuramae EE (2019) Legacy of land use history determines reprogramming of plant physiology by soil microbiome. ISME J 13(3):738–751

Liu YR, Delgado-Baquerizo M, Bi L, Zhu J, He JZ (2018) Consistent responses of soil microbial taxonomic and functional attributes to mercury pollution across China. Microbiome 6(1):183

Lukumbuzya M, Schmid M, Pjevac P, Daims HA (2019) Multicolor fluorescence in situ hybridization approach using an extended set of fluorophores to visualize microorganisms. Front Microbiol 10:1383

Mani I, Kumar R, Singh M, Nagpure NS, Kushwaha B, Srivastava PK, Rao DSK, Lakra WS (2011) Nucleotide variation and physical mapping of ribosomal genes using FISH in genus Tor (Pisces, Cyprinidae). Mol Biol Rep 38:2637–2647

Mani I (2020a) Metagenomic approach for bioremediation: challenges and perspectives. In: Bioremediation of pollutants. Elsevier, pp 275–285

Mani I (2020b) Current status and challenges of DNA sequencing. In: Advances in synthetic biology. Springer, Singapore, pp 71–80

Mani I (2020c) Biofilm in bioremediation. In: Bioremediation of pollutants. Elsevier, pp 375–385

Mapelli V, Olsson L, Nielsen J (2008) Metabolic footprinting in microbiology: methods and applications in functional genomics and biotechnology. Trends Biotechnol 26(9):490–497

Marsh TL, Saxman P, Cole J, Tiedje J (2000) Terminal restriction fragment length polymorphism analysis program, a web-based research tool for microbial community analysis. Appl Environ Microbiol 66:3616–3620

Marzorati M, Wittebolle L, Boon N, Daffonchio D, Verstraete W (2008) How to get more out of molecular fingerprints: practical tools for microbial ecology. Environ Microbiol 10:1571–1581

McGrath KC, Mondav R, Sintrajaya R, Slattery B, Schmidt S, Schenk PM (2010) Development of an environmental functional gene microarray for soil microbial communities. Appl Environ Microbiol 76(21):7161–7170

Meier MJ, Dodge AE, Samarajeewa AD, Beaudette LA (2020) Soil exposed to silver nanoparticles reveals significant changes in community structure and altered microbial transcriptional profiles. Environ Pollut 258:113816

Mocali S, Benedetti A (2010) Exploring research frontiers in microbiology: the challenge of metagenomics in soil microbiology. Res Microbiol 161:497–505

Morán-Tejeda EL, Moreno JI, Beniston M (2013) The changing roles of temperature and precipitation on snowpack variability in Switzerland as a function of altitude. Geophys Res Lett 40(10):2131–2136

Muyzer G, Smalla K (1998) Application of denaturing gradient gel electrophoresis (DGGE) and temperature gradient gel electrophoresis (TGGE) in microbial ecology. Antonie Van Leeuwenhoek 73:127–141

Muyzer G, Waal ECD, Uitterlinden AG (1993) Profiling of complex microbial populations by denaturing gradient gel electrophoresis analysis of polymerase chain reaction-amplified genes coding for 16S rRNA. Appl Environ Microbiol 59:695–700

Nacke H, Thürmer A, Wollherr A et al (2011) Pyrosequencing-based assessment of bacterial community structure along different management types in German forest and glassland soils. PLoS ONE 6:e17000

Neigel JE (1997) A comparison of alternative strategies for estimating gene flow from genetic markers. Annu Rev Ecol Syst 28:105–128

Nelson MC, Morrison M, Yu ZT (2011) A meta-analysis of the microbial diversity observed in anaerobic digesters. Bioresour Technol 102:3730–3739

Nichols S, Barnes PAG (2005) A molecular phylogeny and historical biogeography of the marine sponge genus Placospongia (Phylum Porifera) indicate low dispersal capabilities and widespread crypsis. J Exp Mar Biol Ecol 323:1–15

Nüsslein K, Tiedje JM (1999) Soil bacterial community shift correlated with change from forest to pasture vegetation in a tropical soil. Appl Environ Microbiol 65:3622–3626

O'Donnell GA et al (2007) Visualization, modelling and prediction in soil microbiology. Nat Rev Microbiol 5:9689–9699

Orlewska K, Piotrowska-Seget Z, Bratosiewicz-Wasik J, Cycon M (2018) Characterization of bacterial diversity in soil contaminated with the macrolide antibiotic erythromycin and/or inoculated with a multidrug-resistant Raoultella sp. strain using the PCR-DGGE approach. Appl Soil Ecol 126:57–64

Rademacher A, Nolte C, Schonberg M, Klocke M (2012) Temperature increases from 55 to 75 A degrees C in a two-phase biogas reactor result in fundamental alterations within the bacterial and archaeal community structure. Appl Microbiol Biotechnol 96:565–576

Regar RK, Gaur VK, Bajaj A, Tambat S, Manickam N (2019) Comparative microbiome analysis of two different long-term pesticide contaminated soils revealed the anthropogenic influence on functional potential of microbial communities. Sci Total Environ 681:413–423

Renu GSK, Rai AK, Sarim KM, Sharma A et al (2019) Metaproteomic data of maize rhizosphere for deciphering functional diversity. Data Brief 27:104574

Richardson RE, Bhupathi Raju VK, Song DL, Goulet TA, Alvarez-Cohen L (2002) Phylogenetic characterization of microbial communities that reductively dechlorinate TCE based upon a combination of molecular techniques. Environ Sci Technol 36:2652–2662

Roth A, Fischer M, Hamid ME et al (1998) Differentiation of phylogenetically related slowly growing mycobacteria based on 16S–23S rRNA gene internal transcribed spacer sequences. J Clin Microbiol 36:139–147

Rudkjøbing VB, Thomsen TR, Xu Y et al (2016) Comparing culture and molecular methods for the identification of microorganisms involved in necrotizing soft tissue infections. BMC Infect Dis 16:652

Sanyal SK, Mou TJ, Chakrabarty RP et al (2016) Diversity of arsenite oxidase gene and arsenotrophic bacteria in arsenic affected Bangladesh soils. AMB Expr 6:21

Schloss PD, Handelsman J (2004) Status of the microbial census. Microbiol Mol Biol Rev 68:686–691

Schloss PD, Girard RA, Martin T, Edwards J, Thrash JC (2016) Status of the archaeal and bacterial census: an update. Mbio 7(3):e00201–e00216

Sharma G, Verma HN, Sharma R (2013) RAPD analysis to Study metagenome diversity in soil microbial community of arid zone plants. Proc Natl Acad Sci India Sect B Biol Sci 83:135–139

Sigler WV, Zeyer J (2002) Microbial diversity and activity along the forefields of two receding glacier. Microb Ecol 43(4):397–407

Sigler WV, Crivii S, Zeyer J (2002) Bacterial succession in glacial forefield soils characterized by community structure, activity and opportunistic growth dynamics. Microb Ecol 44(4):306–316

Silva L, Northen T (2015) Exometabolomics and MSI: deconstructing how cells interact to transform their small molecule environment. Curr Opin Biotech 34:209–216

Sivakala KK, Jose PA, Anandham R, Thinesh T, Jebakumar SRD et al (2018) Spatial physiochemical and metagenomic analysis of desert environment. J Microbiol Biotechnol 28(9):1517–1526

Srinivasiah S et al (2013) Direct assessment of viral diversity in soils by random PCR amplification of polymorphic DNA. Appl Environ Microbiol 79(18):5450–5457

Stenuit B, Eyers L, Schuler L, Agathos SN, George I (2008) Emerging high throughput approaches to analyze bioremediation of sites contaminated with hazardous and/or recalcitrant wastes. Biotechnol Adv 26:561–575

Sundberg C et al (2013) 454 pyrosequencing analyses of bacterial and archaeal richness in 21 full-scale biogas digesters. FEMS Microbiol Ecol 85:612–626

Swenson TL, Karaoz U, Swenson JM et al (2018) Linking soil biology and chemistry in biological soil crust using isolate exometabolomics. Nat Commun 9:19

Thakur B, Yadav R, Fraissinet-Tachet L, Marmeisse R, Sudhakara Reddy M (2018) Isolation of multi-metal tolerant ubiquitin fusion protein from metal polluted soil by metatranscriptomic approach. J Microbiol Methods 152:119–125

Thies JE (2007) Soil microbial community analysis using terminal restriction fragment length polymorphisms. J Soil Sci Soc Am 71:579–591

Tiedje JM, Asuming-Brempong S, Nusslein K, Marsh TL, Flynn SJ (1999) Opening the black box of soil microbial diversity. Appl Soil Ecol 13:109–122

Timonen S, Finlay RD, Olsson S, Söderström B (1996) Dynamics of phosphorus translocation in intact ectomycorrhizal systems: non-destructive monitoring using a β-scanner. FEMS Microbiol Ecol 19(3):171–180

Tipayno S, Kim C-G, Sa T (2012) T-RFLP analysis of structural changes in soil bacterial communities in response to metal and metalloid contamination and initial phytoremediation. Appl Soil Ecol 61:137–146

Torsvik V, Gokseyr J, Daae FL (1990) High diversity in DNA of soil bacteria. Appl Environ Microbial 56:782–787

Torsvik V, Gokseyr J, Daae FL et al (1994) Use of DNA analysis to determine the diversity of microbial communitres. Beyond the biomass. John Wiley & Sons, Chichester, pp 39–48

Torsvik V, Sorheim R, Gokseyr J (1996) Total bacterial diversity in soil and sediment communities-a review. J Industr Microbial 17:170–l78

Trcek J (2005) Quick identification of acetic acid bacteria based on nucleotide sequences of the 16S–23S rDNA internal transcribed spacer region and of the PQQ-dependent alcohol dehydrogenase gene. Syst Appl Microbiol 28:735–745

van Oppen MJH, Wörheide G, Takahashi M (2002) Nuclear markers in evolutionary and population genetic studies of scleractinian corals and sponges. Proc 9th Int Coral Reef Symp. Bali 1:131–138

Wall DH, Virginia RA (1999) Controls on soil biodiversity: insights from extreme environments. Appl Soil Ecol 13(2):137–150

Weigold P, El-Hadidi M, Ruecker A, Huson DH, Scholten T et al (2016) A metagenomic-based survey of microbial (de)halogenation potential in a German forest soil. Sci Rep 6:28958

Wright SF, Upadhyaya A (1998) A survey of soils for aggregate stability and glomalin, a glycoprotein produced by hyphae of arbuscular mycorrhizal fungi. Plant Soil 198(1):97–107

Wu X, Zhang G, Zhang W et al (2018) Variations in culturable bacterial communities and biochemical properties in the foreland of the retreating Tianshan No. 1 glacier. Braz J Microbiol 49(3):443–451

Yan X, Luo X, Zhao M (2016) Metagenomic analysis of microbial community in uranium-contaminated soil. Appl Microbiol Biotechnol 100(1):299–310

Zafra G, Taylor TD, Absalón AE, Cortés-Espinosa DV (2016) Comparative metagenomic analysis of PAH degradation in soil by a mixed microbial consortium. J Hazard Mater 318:702–710

Zhou J, Huang Y, Mo M (2009) Phylogenetic analysis on the soil bacteria distributed in karst forest. Braz J Microbiol 40(4):827–837

Chapter 6
Endophytic Microbes and Their Role in Land Remediation

Satinath Das, Pralay Shankar Gorai, Linee Goswami, and Narayan Chandra Mandal

6.1 Introduction

Food security and "zero hunger challenge" are the most thriving topics of present era. It is listed second in the United Nations sustainable development goals. It aims to eradicate hunger, and malnutrition, that was estimated to be 12.5% of current global population i.e., 7.6 billion people in the year 2010–2012 (FAO 2012). According to UN report, 26.4% of global population affected by food insecurity in the year 2018; with an estimated demand of 9.7 billion by 2050 (DESA 2015). The pandemic apart from other factors like population rise, climate change, environmental stressors, land-use patterns, irrigation, post-harvest management techniques have been affecting the food production and supply chain throughout the world. Rapid industrialization and urbanization are also affecting agricultural productivity. The green revolution started in 1950–1960s targeted to increase agricultural productivity through adoption of new technologies, using high yielding crop varieties, increased use of inorganic fertilizers, agrochemicals and irrigated water supply. In 2014–15, with 250 million tons of food grain production, India is on the verge of becoming a food basket for the world. However, the challenge remains with over exploitation of agricultural lands resulting into loss of top-soil and reduced yield. At the same time, agricultural residues pose a grave danger to our fragile ecosystem. Prolonged application of inorganic fertilizer and agrochemicals exerts deleterious impact on soil health. Persistent pesticides, herbicides, fungicides etc. tend to cause loss of soil fertility over the years. These groups of contaminants are termed as emerging organic contaminants (EOCs). Apart from agricultural residues, emerging organic contaminants (EOCs) include industrial chemicals, surfactants, personal care products and pharmaceutical products etc. EOCs severely affect soil health (Hu et al. 2017; Usman et al.

S. Das · P. S. Gorai · L. Goswami · N. C. Mandal (✉)
Department of Botany, Mycology and Plant Pathology Laboratory, Visva Bharati, Santiniketan, West Bengal, India
e-mail: mandalnc@rediffmail.com

2017). Therefore, researchers have been extensively working on alternative bio-based options for sustainable pollution management without compromising the soil health and its fertility (Das et al. 2021).

Pollutants like toxic metals present in industrial wastes and effluents contaminate soil and surface water. These metals undergo phase distribution and speciation under variable environmental conditions. They enter the food chain, bioaccumulate, and magnify, thus create serious problems at trophic levels. For-example, in the year 1956, 1784 people died from consumption of organic mercury contaminated fish from Minamata bay of Japan, the episode infamously known as "Minamata disease", where the origin of methyl mercury was traced back to a chemical factory effluent (Nabi 2014). Similarly, other effluents with toxic metals like Cr, As, Pb, Se etc. also show tendencies toward bioaccumulation under different environmental condition (Gorai et al. 2020). However, there are few incidences where they entered into the food chain, causing harmful effect on living organisms. Therefore, safe removal techniques are the need of the hour.

Several researchers have been looking for sustainable remediation techniques. It involves both chemical and biological methods. Bioremediation is an emerging technique where biological methods are applied for a synergistic interaction between environmental contaminants and their cleaning process. Bioremediation techniques are of two types: microbe assisted remediation and phytoremediation. Microbe assisted remediation technique includes biostimulation, bioaugmentation, and intrinsic bioremediation. On the other hand, phytoremediation includes phytoextraction, phytostabilization, phytodegradation, phytostimulation, phytovolatilization, rhizofiltration, biological hydraulic containment, phytodesalinization. In the present chapter, we will discuss about few microbe-assisted remediation techniques, its present status and future scope.

Previous researchers have extensively worked with rhizospheric bacteria; but endophytic interaction and its implication in terms of bioremediation is a relatively new topic. Endophytes colonize easily, promote plant growth and enable to remediate the surrounding soil surface (Tong et al. 2017; Gorai et al. 2020). Therefore, they are more efficient and preferable over rhizospheric bacteria. At times, endophyte-plant interaction may lead to change in host plant metabolism and physiology (He et al. 2019). The altered metabolism can facilitate phytoextraction process and/or pollutant degradation in the substrate (Tripathi et al. 2017; Tong et al. 2017; Afzal et al. 2017). Among, all existing bioremediation techniques, endophyte assisted remediation techniques have tremendous scope and future use (Feng et al. 2017; Srivastava et al. 2020).

6.2 What Are Endophytic Microbes?

In the year 1886, Bary discovered the term "Endophyte". The term was originally derived from Greek words "endon" means "within", and "phyton" means "plants". Therefore, endophytes are those microbes that reside inside different parts of plant body. Later on, 1904, this was further re-discovered in Darnel, Germany (Tan and Zou 2001). Endophytes have been defined in various ways by several researchers depending upon their source of origin. Bacon and White 2000 defined endophytic microorganisms as "microbes that reside within living internal tissues of plants without causing any instant and overt negative effects". An alternative definition of endophytic fungi is "fungi that live for all or at least a significant part of their life cycle asymptomatically within plant tissues" (Wilson 1995). According to Carrol, endophytes are asymptomatic microbes that reside inside plants; while Petrini (1991) described that endophytic microbe are those microorganisms that living at least one part of their life cycle within the internal parts of plant tissues without imparting any harmful effects to the host plant. Wilson and Carrol (1997) depicted additional information regarding endophytes implied that a part or total life cycle of endophytic bacteria or fungi reside in the living tissues of host plants without causing any apparent or symptomatic infections. Different group of organisms are involved in endophytic association. These are bacteria, fungi, algae and oomycetes. Mostly bacteria and fungi are found to present as endophytic organisms in plants.

6.3 Effect of Endophytes in Soil Fertility Management

Change in land use pattern and simultaneous agricultural intensification exerts an unbearable pressure to the environment. Loss of agricultural land, results in extensive use of the remaining ones. However, with time, soil tends to lose its fertility. Continuous and rampant use of inorganic fertilizer including other agrochemicals contributes to the cause of fertility loss. Therefore, researchers around the Globe are looking for sustainable options to increase the crop yield without depleting any soil properties. It is quite obvious that consecutive cultivation without a fallow period or crop rotation leads to loss of top soil and eventually decreases fertility. Therefore, a gradual shift toward biological fertilizers like composts, organic manures, vermicompost, microbial consortiums, biofertilizers etc. is being explored. All biological techniques are found to be cost-effective, feasible and less harmful. Among biological techniques, endophytic microbes, in single inoculum or consortium, play vital role in soil quality improvement, when applied in a strategic manner. Therefore, the present study focuses on what role endophytic microorganisms play in soil fertility management, plant growth promotion and land remediation. Figure 6.1 represents how endophytic association regulates plant functioning and rhizo-spheric soil conditioning results in land remediation.

6.4 Nitrogen Fixation

Plants are unable to utilize atmospheric nitrogen directly. But nitrogen is an essential element for plant growth; therefore, atmospheric nitrogen needs to be fixed and

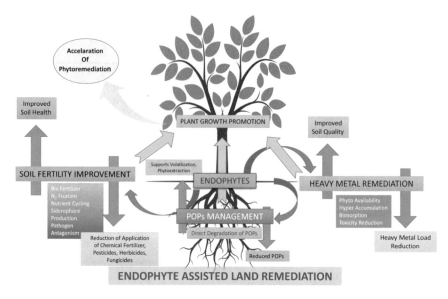

Fig. 6.1 Endophyte-assisted land remediation techniques

convert into bioavailable forms for plant uptake. Usually, to meet nitrogen requirement of crops, farmers apply inorganic nitrogen i.e., urea as N source following standard management practice. Rigorous application of nitrogen-based chemical fertilizer increases the risk of environmental pollution and decreases soil fertility. It also releases a great amount of greenhouse gases (NO_x) at the production site.

Soil microbes play an important role in N_2 fixation, assimilation and denitrification process. Exploration and strategic application of these microorganisms can assist in reducing soil nitrogen deficiency. Hurek and Reinhold-Hurek (2003) observed that under N_2-stressed condition, endophytic microorganisms are better promoter of plant growth than rhizospheric microbes.

Diazotrophs fix atmospheric nitrogen into ammonia. This ammonia gets oxidized to form nitrate that gets dissolved in the nutrient pool and become bioavailable. It undergoes further assimilation in plant body, forms amino acids that finally participates in protein synthesis, eventual plant growth. During endophytic nitrogen fixation, these microbes form nodule or oxygen free structure. These nodules are mostly infected with both Gram-positive and Gram-negative bacteria depending upon the host plant. All these microbe populations inside the root nodule, contribute to nitrogen fixation and are symbiotic in nature. According to Dobereiner et al. (1993) and Muthukumarasamy et al. (2007), *Gluconoacetobacter diazotrophicus* contribute approximately 150 kg N/H/year. During banana cultivation, an increase of bioavailable nitrogen at 79% and 11% was reported, when inoculated with *Agrobacteria* and *Azospirillum* respectively (Zuraida et al. 2000). Soybeans are extensively cultivated legumes around the world. The roots of soybean are found to form nodules with different strains of *Bradyrhizobium*, *Mesorhizobium*. Here also, different strains have different growth rate and different nitrogen fixing ability. Sainz et al. (2005) calculated a total of 142 kg N/H/year fixed nitrogen in Soybeans. These endophyte-plant symbiotic associations play impeccable role in N fixation and availability in soil. Table 6.2 listed N_2-fixing endophytic microorganisms with their respective host plants.

6.5 Biofertilizer

Biofertilizers are substances containing beneficial microbial inoculum. These microbes are efficient in enhancing N availability, P-solubilization and K-exchange in soil surface. They are environment friendly, and cost-effective (Kumar et al. 2017; Singh et al. 2011). The combined action of living microorganisms with soil or mineral substrate results in slow release of nutrients, and thus, enhance the rate of nutrient absorption by plants (Roychowdhury et al. 2017). It not only promotes plant growth but also increases soil fertility (Pal et al. 2015). Endophytic bacteria are capable of intensifying growth of non-leguminous crop improvement (Long et al. 2008; Sturz et al. 2000; Iniguez et al. 2004). Ngamau et al. (2014) described potential use of endophytic organisms as effective biofertilizer in the cultivation of banana. *Azospirillum brasiliens, Bacillus* sp.*, Barkholderia* sp.*, Citrobacter* sp.*, Enterobacter* sp. are some known endophytic bacteria isolated from banana plants. Shen et al. (2019) introduced the efficiency of *Rhizobium larrymoorei, Bacillus aryabhattai, Pseudomonas granadensis* and *Bacillus fortis* as potent biofertilizer in rice cultivation.

6.6 Pathogen Antagonism

A large number of endophytic organisms exhibit broad spectrum antimicrobial activities. Therefore, another beneficial trait of endophytes is pathogen antagonism i.e., reducing the pathogen load in soil and thus, improves soil fertility. They suppress plant pathogen growth via combined action of metabolite release and abiotic changes. They release metabolites like antibiotics, HCN, phenazines, pyoleutorin, pyrrolnitrin, 2, 4-diacetylphloroglucinol etc. (Lugtenberg and Kamilova 2009). Endophytes improve host plant resistance against the pathogens by delaying or defending the entry of pathogen into the plant systems (Walters et al. 2007). Endophytes present in the host plant tend to stimulate a group of elicitors to trigger plant's induced or innate defence mechanism. In due course, they also release a wide range of enzymes like phenylalanine ammonialyase, peroxidase, beta-glucanase, chitinase, ascorbate peroxidase, polyphenol oxidase and superoxide dismutase etc. Workers reported that *Pseudomonas fluroscence* are capable of inducing resistance in olive and tomato plant by activating defence enzymes (Siddiqui and Shaukat 2003; Gómez-Lama Cabanás et al. 2014). Endophytes use of plant secondary metabolites like alkaloids, steroids, terpenoids, flavonoids, phenols, phenolic acids and peptides against pathogens. Several studies confirmed the production of secondary metabolites could successfully reduce pathogen load in potato and turmeric cultivation (Sturz and Kimpinski 2004; Sessitsch et al. 2004; Vinayarani and Prakash 2018). Gorai et al. (2021) reported the control of early blight of potato caused by *Alternaria alternata* using endophytic bacteria *Bacillus velezensis* SEB1.

6.7 Siderophore Production

Siderophore is a low molecular weight iron chelating compound secreted by different microorganisms. It has very high and specific affinity to iron. It primarily forms complex with iron (Fe^{2+}) molecules and increases the availability and mobility of iron to the plants. Siderophores are produced by a number of plant growth promoting rhizobacteria as well as endophytic microbes. In *Cicer areatinum* and *Pisum sativum* endophytic bacterial strains are potent to produce more than 65 siderophore production units (Maheswari et al. 2019). Loaces et al. (2011) studied the diversity of siderophore producing endophytic strains in rice where *Pantoea* sp. was predominant over *Burkholderia sp., Pseudomonas sp., Enterobactor sp.* and *Sphignomonas sp.*

6.8 Nutrient Cycling

Nutrient cycling is of utmost important with regard to soil fertility management. It involves a continuous transfer of energy and mass among biotic and abiotic systems. Though energy transfer is unidirectional, mass transfer occurs in a continuous cycle. The process begins with degradation of dead biomass into smaller and simpler fractions. Such processes are managed by catalysts like different enzymes to facilitate faster break down of complex macromolecules and gradual microbial propagation. With eventual release of water-soluble fractions, nutrient fractions get dissolved in the soil nutrient pool and become readily available for the plants and other heterotrophs. Many saprophytic fungi and bacteria play important role in the degradation process (Carroll 1988). Promputtha et al. (2010) showed that endophytes can regulate nutrient cycling process. During litter degradation, Nair and Padmavathy (2014) observed that the endophytic organisms trigger the activities of saprophytic organisms to quicken the process. It has been observed that release of enzymes like cellulase, hemi-cellulase etc. accelerates the decomposition process and nutrient release. He et al. (2012) reported that the presence of endophytic microbes in the host body expedites the release of enzymes and their activity. Chen et al. (2020) reported that association of *Epichloe* endophytes promoting growth, metabolic activity and nutrient uptake in perennial ryegrass (*Lolium perenne*) in a low fertile soil environment. Presence of endophytes showed distinct positive impact on organic carbon content, major nutrient like N, P, K content, micronutrients like manganese (Mn) concentration in both root and shoot portions.

6.9 Plant–Endophytic Interaction and Their Role in Plant Growth Promotion

6.10 Plant–Endophytic Interactions

Figure 6.2 shows a detailed mechanism of plant–endophyte interaction and various mechanisms involved. Complex endophytic microbial communities colonize within

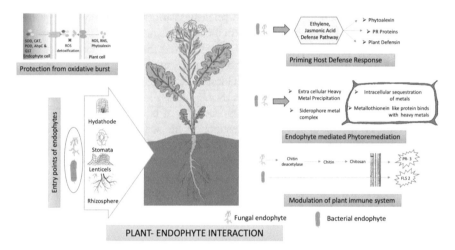

Fig. 6.2 Endophyte and plant interactions: mode of entry and mechanism of action. It is a pictorial representation showing multifaceted interaction of endophytes with host plants. (1) Endophytes prime the host plant's defensive responses against phytopathogens mediate intracellular responses and trigger ethylene/jasmonic acid transduction pathway. (2) Reactive oxygen species (ROS) and reactive nitrogen species (RNS), generated by the plant, are neutralized by the production of enzymes such as superoxide dismutase (SOD), catalase (CAT), peroxidase (POD), alkyl hydroperoxide reductase (AhpC) and glutathione-S-transferase (GSTs) in endophytes. (3) Fungal endophytes modulate the plant's immune system by the production of chitin deacetylases, which deacetylate chitosan oligomers and, hence, prevent themselves from being recognized by chitin-specific receptors (PR-3) of the plants that recognize chitin oligomers. Perception of flagellin (FLS 2) from endophytes also differs from phytopathogens. (7) Endophytic microbes alleviate metal phytotoxicity via extracellular precipitation, intracellular accumulation, sequestration, or biotransformation of toxic metal ions to less toxic or non-toxic forms. Where, ET, ethylene; JA, jasmonic acid; ROS, reactive oxygen species; SOD, superoxide dismutases; CatA, catalases; POD, peroxidases; AhpC, alkyl hyperoxide reductases; GSTs, glutathione-s-transferases; EF, effector protein; PR-3, chitin-specific receptors; FLS 2, flagellin; MT, metal transporters; IC, ion channels; CW, bacterial cell wall. (Adapted from source: Khare et al. 2018, Frontiers in Microbiology)

the plant tissues. They play major roles for the promotion of plant growth and development (Stone et al. 2000; Kobayashi and Palumbo 2000). Endophytes are ubiquitous in nature. Yet the mechanisms behind the endophytic microbe–plant interactions are hitherto unknown. They are in the primary stages of investigation and need more detailed works to understand these interactions (Strobel et al. 2004; Thomas and Soly 2009). On a simpler note, endophytic colonization means entry, growth and multiplication of endophytic organisms within host plants. Both the endophytic microbes and pathogen follow same mechanism during the entry within the host plant tissues (Gorai et al. 2020). But, one of the interesting points of endophytic microbial entry which markedly differs from pathogenic entry, host plant does not develop any resist power against the endophytes. Natural openings of plants like stomata, lenticels and hydathodes or any wounds caused by various pathogenic attack, soil particles or abiotic stresses generally use as routes for the entry of endophytes within host plant (Reinhold-Hurek and Hurek 1998). Behind these natural openings they are also eligible to take part in direct entry by releasing various plant cell wall degrading enzymes (Quadt-Hallmann et al. 1997; Reinhold-Hurek and Hurek 1998). In order to establish a successful endophytic colony, they need to cross few important steps like selection of the host, host-recognition and colonization on the targeted part and final entry into the host tissues respectively (Gorai et al. 2020). Plant secretes secondary metabolite in the form of root exudates. Some of these molecule act as signaling molecule which helps the chemotactic movements of endophytic microbes (Gorai et al. 2020). At first, they reach their destination site with the help of flagella and finally adhere with the surface using pilli (Zeidler et al. 2004). One of the excellent abilities of endophytic microbes is adaptive capabilities in highly diverse environment. Gorai et al. (2020) mentioned that with changes of different environmental factors like sudden changes in pH, carbon source, osmotic pressure, and oxygen availability of the surroundings, they can easily sustain and survive (Gorai et al. 2020). Endophytic microorganisms are very important to the plants, thus colonization of endophytes within the plant is very important for providing the benefits to the host plant. However, process of endophytic bacterial colonization within tissues of plant is quite complex and this includes several stages (Stępniewska and Kuźniar 2013).

Rhizosphere area around the plant root is inhabited by unique populations of microorganisms (Gorai et al. 2020). It was reported that plants release significant amounts of photosynthates or exudates like amino acids, organic acids, proteins etc. which act as signaling molecules to help the endophytic organisms in their chemotactic movements. Pattern and sites of colonization are specific for each endophytic strain (Zachow et al. 2015). When an endophytic strain attached to the host surface, it starts the penetration process for entering within the host tissues. Penetration process can occur through either active or passive ways. Penetration of endophytes occurs passively through the cracks of root tips or root regions caused by harmful organisms (Hardoim et al. 2008). On the other hand, active penetration occurs via attachment and proliferation of exogenous polysaccharides, lipopolysaccharides, structural components, quorum sensing that helps the endophytes to migrate and multiplication inside the tissues of plant (Böhm et al. 2007; Dörr et al. 1998; Duijff et al. 1997;

Suárez Moreno et al. 2010). After entering within the roots of host plant, endophytic bacteria can now migrate systematically to intercellular spaces of adjacent tissues by producing cell wall degrading enzymes like pectinase and cellulase (Compant et al. 2010) as well as to above ground tissues using flagella or through perforated plates of xylem tissues during transpiration (Compant et al. 2005; Sapers et al. 2005).

During the endophytic colonization process, microorganisms usually prefer the site of plant having thin surfaces such root hairs or apical part of root meristem. Reinhold-Hurek et al. (2006) described that endophyte *Azoarcus* sp. BH72 secretes lytic enzyme endoglucanase at entry site during colonization process. Suzuki et al. (2005) reported that endophyte *Streptomyces galbus* colonize in *Rhododendron* by using a non-specific wax degrading enzyme. Process of endophytic colonization depends upon several factors such as type of microbial strains, genotype of host plant, different biotic and abiotic factors, nutrients limitation etc. Till date, several researchers have indicated about the various routes of endophytic colonization inside the plants. For examples, endophyte *Ralstonia solanacearum* firstly attached to different parts of roots of host plant and enters by invasion of roots, then it migrates upwardly through xylem vessels (Alvarez et al. 2012). Another study stated that endophyte *Paraburkholderia phytofirmans* PsJN enters into the host cells through the layer exodermis of roots and crosses the cortical tissues, endodermal layers and finally moves upper part through xylem vessels (Compant et al. 2005, 2010). After the successful colonization within host tissues, endophytic microbes play multifaceted beneficial roles for the host plants. Endophytes can directly help the host plant by producing various plant growth promoting factors (Afzal et al. 2019) and by increasing nutrient uptake of the host plant (Vacheron et al. 2013). Indirectly, endophytic bacteria keep the host plant healthy by killing the pathogens and pests by nutrient restraint, by producing different kinds of antibiotics (Glick et al. 2007), siderophores (Lodewyckx et al. 2002), hydrolytic enzymes (Fan et al. 2002; Myo et al. 2019) and/or by inducing systemic resistance in plants (Kloepper and Ryu 2006).

6.11 Plant Growth Promotion

Diverse groups of beneficial microbial communities are found to inhabit in different locations of plant's body or its surface which are ranging from rhizosphere, phyllosphere to the endospheric regions (Feng et al. 2016). Most of these symbiotic organisms produce various substances which may promote plant growth and development. In endophyte–plant symbiotic relationships, both the partners are benefitted in which plants supply nutrients and provide shelter to the endophytes while indirectly endophytic organisms help the plants by increasing resistance against pathogen and herbivores (Bamisile et al. 2018). In addition, endophytes also increase the plant growth and development by increasing stress tolerance and nutrient uptake like nitrogen, iron and phosphorus by the plants especially in nutrient deficient conditions (Ji et al. 2014; Martinez-Klimova et al. 2017). It has been also reported earlier those different kinds of phytohormone such as auxin, gibberellin, cytokinine etc. are produced by some endophytes (Gohain et al. 2015; Pimentel et al. 2011). Beside the phytohormone production, some of them possess other plant growth properties like synthesis of 1-aminocyclopropane-1-carboxylate deaminase (ACCD), production of siderophores, solubilization of phosphates and production of antimicrobial metabolites etc. (Serepa-Dlamini 2020).

There are several endophytic bacteria play significant beneficial roles for the host plant growth promotion by various ways (Table 6.1) like production of phytohormones like Indole acetic acid (Gao and Tao 2012); ACC deaminase, (Karthikeyan et al. 2012; Ali et al. 2014; Glick, 2014), phosphate (P) solubilization, nitrogen fixation etc. For examples, endophytic bacteria *Bacillus subtilis* CNE 215 and *Bacillus lichenoformes* CRE1 isolated from chickpea were able to solubilize P and produce ammonia respectively (Saini et al. 2015). On the other hand, Egamberdieva et al. (2017) described that endophytic bacteria *Bacillus subtilis* NNU4 and *Archomobacter xylosoxidans* NNU2 isolated from chickpea show PGP properties like P solubilization, IAA production, siderophore production and HCN production. In addition, some of plant growth promoting endophytes showed excellent antagonistic activity against phytopathogens (Table 6.2).

6.12 Identification of Endophytes and Their Utilization Against Persistent Organic Pollutants

From last few decades of the twentieth century, impact of persistent organic pollutants became a matter of concern. Persistent organic pollutants (POPs) are toxic organic compounds present in the environment, used for anthropogenic purposes, transported by means of air or water. Transboundary movement of persistent organic pollutants makes them more dangerous than any other pollutant. During Stockholm convention, 1972, the "Dirty Dozen" term was coined to twelve POPs used extensively for industrial purpose. Their presence and magnification disrupt proper functioning of the ecosystem. Those synthetically produced toxic chemical substances are aldrin, endrin, dichlorodiphenyltrichloroethane (DDT), polychlorinated dibenzofurans (PCBs), hexachlorocyclohexane, mirex etc. They persist for a long time in the environment, hence termed as persistent organic pollutants (POPs) (Boudh et al. 2019). According to Oonnittan and Sillanpää (2020), POPs show salient features like acute toxicity, biomagnification and long-range transport. Most of the POPs are the outcome of different anthropogenic activities and the waste thus generated. POPs are resistant to any form of physical, chemical or photolytic degradation. Direct exposures of such POPs have drastic effect on living organism. In mammals, they behave as xenoestrogens, thus causes endocrinal malfunction, loss of body weight, ovarian cancer, congenital disease, low sperm count, damage of central nervous system etc. In other living organism, disruption in sexual reproduction, retarded growth, mutation

Table 6.1 List of various endophytic isolates and their plant growth promoting attributes

Host plants	Endophytes isolates	P solubilization	IAA production	Siderophore activity	N_2 fixation	Ammonia production	References
Oryza sativa	Paenibacillus kribbensis		+	+			Puri et al. (2018)
	Klebsiella pneumonia	+	+	+			
	Pseudomonas putida	+	+	+			
	Bacillus amyloliquefaciens	+	+	+	+		Verma et al. (2015)
Maize	Pseudomonas putida		+	+			Sandhya et al. (2017)
	Pseudomonas lini		+				
	Pseudomonas thivervalensis			+			
	Pseudomonas aeruginosa		+	+			
	Pseudomonas montelli		+	+			
Wheat	Acinetobacter guillouiae	+	+				Rana et al. (2020)
	Achromobacter xylosoxidans	+	+		+		Jha and Kumar (2009)
	Bacillus amyloliquefaciens	+	+	+	+		Verma et al. (2015)
	Acinetobacter lwoffii	+	+		+		Verma et al. (2015)
Chickpea	Archomobacter xylosoxidans NNU2	+	+	+			Egamberdieva et al. (2017)

(continued)

Table 6.1 (continued)

Host plants	Endophytes isolates	P solubilization	IAA production	Siderophore activity	N_2 fixation	Ammonia production	References
	Bacillus subtilis NNU4	+	+	+			Saini et al. (2015)
	Bacillus subtilis CNE215	+				+	
	Bacillus licheniformis CRE1	+				+	
	Bacillus licheniformis	+					
	Burkholderia cepacia		+				Shahid and Khan (2018)
Soybean	*Pseudomonas aeruginosa*	+	+	+			Kumawat et al. (2019)
	Bacillus subtilis			+			Singh et al. (2017)
	Bacillus megaterium	+	+	+	+		Subramanian et al. (2015)
	Methylobacterium oryzae	+	+	+	+		
Peanut	*Enterobacter asburiae*		+				Wang et al. (2013)
	Pantoea agglomerans		+				
	Sphingomonas azotifigens		+				
	Bacillus megaterium		+				

(continued)

Table 6.1 (continued)

Host plants	Endophytes isolates	P solubilization	IAA production	Siderophore activity	N$_2$ fixation	Ammonia production	References
	Bacillus arbutinivorans		+				
Faba bean	Pseudomonas yamanorum B12		+	+			Bahroun et al. (2018)
	Pseudomonas fluorescens B8P		+	+			
	Rahnella aquatilis B16C		+	+			

"+" indicates positive result and "−" indicates negative result

Table 6.2 List of N_2 fixing endophytic isolates and respective host plants

Endophytic organism	Host plant	Reference
Gluconacetobacter diazotrophicus	Sugarcane	James and Olivares (1998)
Azospirillum sp	Pineapple	Weber (1999)
Burkholderia sp	Pineapple	Weber (1999)
Herbaspirillum	Banana	Weber (1999)
Burkholderia sp	Rice	Baldani et al. (2000)
Microbacterium sp.	Sugarcane	Lin et al. (2012)
Paenibacillus sp	Poplar	Scherling et al. (2009)
Klebsiella oxytoca	Sugarcane	Govindarajan et al. (2007)
Klebsiella pneumoniae	Sugarcane	Govindarajan et al. (2007)
Rhizobium leguminosarum bv. trifolii	Rice	Yanni et al. (1997)
Frankia sp	*Alnus glutinosa*	Li et al. (1996)
Glomus fasciculatus	*Hippophaë rhamnoides*	Gardner et al. (1984)

etc. can directly be linked with the adverse effect of these persistent organic pollutants. Researchers have reported POPs multidirectional effect on basic agronomy like soil health, accumulation and contamination of food, genetic changes of soil microorganisms, disruption of normal soil biodiversity (Saha et al. 2017; Guo et al. 2012).

Because, POPs are resistant to other forms of physical, chemical or photolytic degradation methods, therefore, researchers have been concentrating on biological remediation of these pollutants. However, owing to their recalcitrant nature, POP bioavailability is almost negligible. Presence of excessive amount of POP in the environment, hinders plant growth and development, thus limiting phytoremediation process (Doty 2008). Therefore, even phytoremediation needs a co-metabolism assistant for this group of pollutants. In this part, endophytes and plant act synergistically. Here, metabolome i.e., mixed community of the host plant initiate oxidation of organic compounds present in the substrates and provide carbon and energy source for the microbes (Feng et al. 2017; Gerhardt et al. 2009; Pandey et al. 2009). Table 6.3 enlists groups of endophytic bacteria with their respective host plants that degrade persistent organic pollutants. The presence of endophytic microorganisms can be beneficial in two different ways. One is indirectly by supporting the plant growth and other is by direct degradation. Endophytes support plant growth and inhibit persistent organic pollutants through phytoremediation techniques like phytoextraction i.e., pollutants are absorbed and accumulate inside the plant tissue (Ali et al. 2013); phytovolatilization i.e., organic pollutants or contaminants are absorbed and released

Table 6.3 Endophytes and their host plant association for persistent organic pollutant degradation

Pollutants	Endophyte	Plant	References
Diesel	*Pseudomonas* sp. strain ITRI53 *Rhodococcus* sp. strain ITRH43	Ryegrass	Andria et al. (2009)
	Enterobacter ludwigii	Italian ryegrass, birds foot trefoil and alfalfa	Yousaf et al. (2011)
Hydrocarbon	*Bacillus* sp. *Pseudomonas* sp.	*Azadirachta indica*	Singh and Padmavathy (2015)
	Pseudomonas sp. strain ITRI53, *Pseudomonas* sp. strain MixRI75	Italian ryegrass (L. multiflorum var. Taurus)	Afzal et al. (2011), Afzal et al. (2012)
	Enterobacter ludwigii strains	*Lolium multiflorum, Lotus corniculatus,* and *Medicago sativa*	Yousaf et al. (2011)
	Pantoea sp. strain ITSI10, *Pseudomonas* sp. strain ITRI15	Italian rye grass (L. multiflorum var. Taurus) and birdsfoot trefoil (L. corniculatus var. Leo)	Yousaf et al. (2010a, b)
TCE	*Pseudomonas putida* W619-TCE	Poplar	Weyens et al. (2010a)
	Burkholderia cepacia VM1468 possessing (a) the pTOM-Bu61 plasmid	Yellow lupine	Weyens et al. (2010b)
	Enterobacter sp. strain 638	Poplar	Taghavi et al. (2011)
	Enterobacter sp. strain PDN3	Poplar	Kang et al. (2012)
Toluene	*Burkholderia cepacia*	*Zea mays Triticum aestivum*	Wang et al. (2010)
Chlorobenzoic acids	*Pseudomonas aeruginosa* R75; *Pseudomonas savastanoi* CB35	*Elymus dauricus*	Siciliano et al. (1998)
Pyrene	*Staphylococcus* sp. BJ106	*Alopecurus aequalis*	Sun et al. (2014)

(continued)

Table 6.3 (continued)

Pollutants	Endophyte	Plant	References
	Enterobacter sp. 12J1	Wheat (*Triticum* sp.) maize (*Z. mays*)	Sheng et al. (2008a, b)
2,4-Dichlorophenoxyacetic Acid	*Pseudomonas putida* VM1450	*Pisum sativum*	Germaine et al. (2006)
Catechol and phenol	*Achromobacter xylosoxidans*	*Ipomoea aquatica, Chrysopogon zizanioides, Phragmites australis*	Ho et al. (2009)
Naphthalene	*Pseudomonas putida*	*Pisum sativum*	Germaine et al. (2009)
2,4,6-Trinitrotoluene, hexahydro-1,3, 5-trinitro-1,3,5-triazine, octahydro-1,3,5,7-tetranitro-1,3,5-tetrazocine	*Methylobacterium populi* BJ001	*Populus alba*	Van Aken et al. (2004)
n-Hexadecane, PAH	*Pseudomonas* spp., *Brevundimonas* sp, *Pseudomonas rhodesiae*	*Medicago sativa, Puccinellia nuttaalliana, Festuca altaica, Lolium perenne, Thinopyrum ponticum*	Phillips et al. (2008)
Hexachlorocyclohexane	*Rhodococcus erythropolis* ET54b, *Sphingomonas* sp. D4	*Cytisusstriatus*	Becerra-Castro et al. (2013)
Fenpropathrin	*Klebsiella terrigena* E42; *Pseudomonas* sp. E46	*Spirodela polyrhiza*	Xu et al. (2015)

as volatile in atmosphere (Ferro et al. 2013), and/or transformation of complex toxic contaminants into simpler or non-toxic forms (Wiszniewska et al. 2016).

The intercellular spaces of plant tissue are enriched with sugars, nutrients, amino acids etc., therefore, it provides a safe environment for the endophytes to grow and populate (Bacon and Hinton, 2007). Studies have shown that endophytes readily use secondary metabolites like terpenes, flavonoids, salicylic acids and lignin derivatives synthesized inside host plant body and release a cluster of POP degrading enzymes (Feng et al. 2017; Jha et al. 2015). All these secondary metabolites serve either as an analogue of the POPs owing to their structural similarities or act as an intermediate, thus stimulate endophytic degradation (Jha et al. 2015). For example, metabolite like salicylate is involved in activating acquired systemic resistance in the host plant. It also tends to stimulate enzymes for naphthalene degradation (Singer et al. 2003). These metabolites provide carbon and energy source for microbial proliferation. And, these endophytic microbes receive pollutant degrading genes through horizontal gene transfer like pTOM-Bu61 plasmid (representing Toluene and TCE degradation) inside the host and can modulate the gene expression (Thijs et al. 2016; Taghavi et al. 2005). They release diverse array of catabolic enzymes like cytochrome P450 monooxygenase and co-enzymes like NAD/NADPH for metabolic degradation and detoxification of POPs (Liu et al. 2014; Zhu et al. 2016; Doty 2008). According to Siciliano et al. (2001), those endophytes, isolated from plants grown in hydrocarbon contaminated soil, are mostly capable of degrading hydrocarbons. It has been found that population of hydrocarbon degrading bacteria is much higher inside host plant tissue especially in root system, as compared to their rhizospheric soil. Researchers were able to isolate a number of potent crude oil degrading bacterial strains from the plants grown in crude oil contaminated soil (Yousaf et al. 2010a, b; Phillips et al. 2008). Germine et al. (2009) reported endophytic bacterial strains, isolated from poplar trees, were capable of degrading herbicide. Apart from this, several bacterial strains, isolated from the poplar trees, were capable of activating metabolic degradation of aromatic hydrocarbons like benzene, toluene, xylene etc. (Taghavi et al. 2011; Moore et al. 2006). Similarly, endophytic strains isolated from different wetland plants were capable of detoxifying a group of pesticides and organic hydrocarbons (Chen et al. 2012; Zhang et al. 2013).

The first in vitro study of POP degradation by endophyte was performed by Germaine et al. (2006). Here, the researcher inoculated *Pisum sativum* with endophytic bacterial strain *Pseudomonas putida* VM1450. It showed successful degradation of 2, 4-dichlorophenoxy acetic acid. Later on, Germaine et al. (2009) reported that another endophytic bacterial strain *Pseudomonas putida* VM1441 efficiently degraded naphthalene compounds from the soil surface. According to Andria et al. (2009) presence of organic pollutant in soil directly affects the colonization of endophyte in endo-sphere and POP degrading gene expression. Becerra-Castro et al. (2013) successfully exhibited cohort application of plant and endophytes to remediate hexachlorocyclohexane contaminated soil. They remarkably put an exemplary use of endophytes via consortium of *Rhodococcus erythropolis* ET54b and *Sphingomonas sp.* D4 inoculated inside the plant *Cytisuss triatus* grown in hexachlorocyclohexane contaminated soil. The consortium was successful in accelerating degradation of target pollutant.

Endophytes are getting attention for last few decades in the field of remediation. They show plant growth promoting activities, genetic diversity and stress tolerance. The synergistic action of host plant and endophytic bacteria for remediation of POPs is a very effective and sustainable approach. Selected endophytes can be genetically engineered for increased efficiency as a sole endophytes or cohort design for co-metabolism under different phytoremediation techniques. These models are now very promising and show an effective lineage, not only for the remediation of POPs but also for food safety.

6.13 Effect of Endophytes Against Heavy Metal Contaminated Soil

Soil pollution due to heavy metal contamination is a serious environmental hazard. Presence of heavy metals shows adverse effect on the trophic levels. It contaminates soil, surface water, agricultural crops, microbial ecosystem (Kidd et al. 2012). Presence of cadmium, lead, copper, chromium, and nickel above the permissible limit in the environment, exert harmful impact on living organism (Hemambika et al. 2011). Heavy metal toxicity is associated with long range contamination, non-degradation and bioaccumulation. Heavy metal toxicity retards plant growth by suppressing carbohydrate metabolism and photosynthesis (Becerril et al. 1988). It also affects the process of respiration (Keck 1978). The conventional methods to remediate heavy metal contamination are metal stabilization using soil amendments, soil washing with acid or chelators, reverse osmosis, evaporation, precipitation, electrochemical treatment, ion exchange and sorption (Kadirvelu et al. 2002; Luo et al. 2010). But these conventional methods are, chemical-dependent, exorbitant, also energy-expensive. They also contribute to generation of toxic sludge (Hemambika et al. 2011). Under such perspectives, bioremediation techniques are highly preferred over any other existing chemical technique. Endophyte mediated phytoremediation is an alternative approach for heavy metal removal from contaminated lands (Burges et al. 2016). Here, endophytes reduce the metal stress through reduced phytotoxicity and improved metabolic capabilities as growth promoter (GP) (Feng et al. 2017). Examples of potential endophytes are usually members of the genera *Pseudomonas* (Feng et al. 2017); *Rahnella* (He et al. 2019), *Bacillus* (Gorai et al. 2021) among all other microbes. These workers highlighted successful association of endophytes and plants for promising biological control methods. Table 6.4 elucidates endophytic association with their host plants actively involved in heavy metal remediation and potential mechanism involved.

Ma et al. (2016) described that endophyte plays an active role in metal detoxification via direct or indirect plant growth promotion and altered metal uptake mechanism. Govarthanan et al. (2016) reported a root endophytic bacteria *Paenibacillus* sp from *Tridax procumbens* were significantly able to remove Cu, Pb, As and Zn when incubated in vitro. Any change in temperature, pH and incubation period shows direct effect on the amount of heavy metal removal. The study recorded element removal percentage of up to 61.4% Cu, 37.3% As, 54.5% Zn and 37.5% Pb. Metal resistant endophytes promote plant growth via nitrogen fixation, production of siderophores, other phytohormones, solubilization of major nutrients viz. N, P, K, utilizing single N source in the form of 1-aminocyclopropane-1-carboxylic acid and biotransformation of N, P, K (Rajkumar et al. 2009).

Table 6.4 List of endophytes used for heavy metal remediation in soil

Host plant	Endophyte	Metal remediated	Mechanism	Reference
Brassica napus	*Pseudomonas fluorescens* G10, *Microbacterium* G16	Pb, Cd, Zn, Cu and Ni	Increased solubility, uptake of Pb	Sheng et al. (2008a, b)
Pteris vittata *Pteris multifida*	Proteobacteria and actinobacteria	As	As-V reduction, As-III oxidation	Zhu et al. (2014)
Alnus firma *Brassica napus*	*Bacillus* sp. MN3-4	Pb, Cd, Zn, Ni	Bio-removal, phytotoxicity reduction	Shin et al. (2012)
Alnus firma	*Bacillus thuringiensis* GDB-1	As, Cu, Cd, Ni, Zn, Pb	Bio-removal, increased bioaccumulation	Babu et al. (2011)
Solanum nigrum	*Serratia marcescens* LKR01, *Arthrobacter* sp. LKS02, *Flavobacterium* sp. LKS03, *Chryseobacterium* sp. LKS04	Pb, Zn, Cu, Cd	Decreased phytotoxicity, increased metal accumulation	Luo et al. (2011)
Solanum nigrum	*Pseudomonas* sp. Lk9	Cr, Cu, Cd, Zn	Improved heavy metal availability	Chen et al. (2014)
Lupinus luteus	*Burkholderia cepacia* L.S.2.4, *Herbaspirillum seropedicae* LMG2284	Pb, Cd, Co, Cu, Ni	Bio-removal, reduction of phytotoxicity	Lodewyckx et al. (2001)
Lycopersicon esculentum	*Methylobacterium oryzae* CBMB20, *Burkholderia* sp. CBMB40	Cd, Ni	Biosorption, removal of toxicity	Madhaiyan et al. (2007)
Miscanthus sinensis	*Pseudomonas koreensis* AGB-1	As, Cd, Pb, Zn	Increased metal uptake	Babu et al. (2015)
Sorghum bicolor	*Bacillus* sp. SLS18	Cd, Mn	Improved biomass production and total metal uptake	Luo et al. (2012)
Pelargonium graveolens	*Pseudomonas monteilii* PsF84, *Pseudomonas plecoglossicida* PsF610	Cr	Increased biomass, help Cr(IV) sequester	Dharni et al. (2014)

6.14 Phytoavailability

Transfer of heavy metals from soil to plant is dependent on bioavailability of the metals in the soil (Glick 2010). Other limiting factors such as redox potential, organic matter contents, soil particle size, nutrient dynamics, pH of soil etc. regulate metal availability in soil (Lebeau et al. 2008). Endophytes tend to reduce toxicity of pollutants inside host plant through several interwinding biochemical pathways. Studies showed that isolated heavy metal resistant endophytes promote plant growth and assist in phytoremediation of contaminated soil (Chen et al. 2014). Rajkumar et al. 2009 demonstrated that, by the secretion of low molecular weight organic acids and metal specific ligands, endophytic bacteria can increase metal and mineral solubilization. Production of organic acids by the root exudates alters soil pH. It plays a vital role in eventual nutrient solubilization and uptake. Endophytic bacteria can produce a wide range of chemicals like fatty acids, glycol lipids, mycolic acids, lipopeptides, polysaccharide protein complex, phospholipid etc. (Bannat et al. 2010) which can fasten the rate of phytoremediation as they increase the phytoavailability of the metals (Bacon and Hilton 2011). Rajkumar et al. 2009 stated that several biosurfactants are produced and released by groups of endophytic bacteria. These biosurfactants interact and form organo-metallic complex with insoluble metals. These metals are then gradually desorbed from the soil matrix. This process alters mobility and phytoavailability of metals. Hence, it accelerates the phytoremediation, especially phytoextraction of heavy metals. Application of such bacteria in soil can be beneficial from the aspect of heavy metal remediation. Babu et al. (2013) observed that endophytic *Bacillus thuringiensis* GDB-1 inoculation in *Alnus firma* removal up to 77% of Pb, 64% Zn, 34% As, 9% Cd, 8% Cu, and 8% Ni in metal amended mine tailing extract. The inoculum also facilitated P solubilization, ACC deaminase, Indole acetic acid production and activation, siderophore production. This resulted in 141% increase in root length, 144% increase in shoot height and 170% of dry biomass; thus, promoting overall crop health.

6.15 Hyper Accumulation and Biosorption

Other efficient bioremediation techniques are biosorption and hyper accumulation of pollutants. Hyperaccumulator plants effectively remove metals from contaminated surfaces. They are able to absorb selective metals even when their presence is below 1% in the substrate. Baker (2000) defined that if a plant is able to absorb 1% of Zn, 0.1% of nickel, cobalt, copper, lead and 0.01% of cadmium from the substrate, then that plant can be termed as a hyper accumulator. Hyperaccumulator plants reduce toxicity by reducing intracellular M-Cysteine and M-Methionine concentration (M representing metal), at times interfere with plant metabolism. For example, in selenium hyperaccumulator plant *Astagalus bisulcatus*, it has been observed that inoculation of genetically engineered *E. coli* increased its Se tolerance and decreased non-specific binding of Se to the proteins (Terry et al. 2000). Hyperaccumulators tend to uptake exceedingly high amount of one or more metals from the growing substrates and translocate, eventually accumulate in the shoot. Endophytes present in the hyperaccumulators tend to modulate the process of phytoextraction of heavy metals from the contaminated soil (Chen et al. 2014). In this study by Chen et al. (2014), bacterial endophyte *Pseudomonas* sp. Lk9 was found to increase the efficiency of *Solanum nigrum* L. for Cd accumulation up to 64% in the dry shoot. Similarly, another report suggested that *Serrartia* sp LRE07 is able to absorb more than 60% of cadmium and 35% of zinc in a mono metallic culture solution (Luo et al. 2011). In active biosorption, metal is slowly accumulated in intracellular space crossing the cell membrane. These metals are sequestered and accumulated inside the host body (Ma et al. 2011). On the contrary, in passive biosorption, entry of metal ions into a cell occurs without metabolite interactions (Vijayaraghavan and Yun 2008). Here, metals react with different functional groups on cell surface like hydroxyl, carbonyl, amine, phosphonate, sulfhydryl (Ma et al, 2011) and form complex structures, thus become unavailable.

According to Shin et al. (2012), inoculation of heavy metal resistant endophyte *Bacillus* sp. MN3-4 contributes to increase in the phytoremediation efficiency through intracellular Pb accumulation. Another report says *Bacillus thuringiensis* GDB-1 isolated from the root of *Pinus sylvestris* enhances the metal accumulation efficiency of *Alnus firma* (Babu et al. 2013). Sheng et al. (2008a, b) isolated and identified two endophytic bacterial strains capable of promoting Pb accumulation in *Brassica napus*. According to the report of Ma et al. (2015), the enhanced accumulation of cadmium (Cd), zinc (Zn) and lead (Pb) in plants was found to be controlled by the presence of heavy metal resistant endophytic bacteria *Bacillus* sp. It is evident from the past and recent studies that the endophytes play remarkable role in metal accumulation process supporting the phytoremediation methods and finally push to an improved and efficient heavy metal remediation technique.

6.16 Toxicity Reduction

Phytotoxicity is one of the critical factors for successful phytoremediation. Association of bacteria and plant plays a vital role in balancing the phytoremediation techniques and reducing metal toxicity. In this scenario, endophytic bacteria have some excellent host plant cohort backup that either leads to toxicity reduction or increased plant tolerance (Rajkumar et al. 2009). Recent studies revealed that mechanisms like extracellular precipitation (Babu et al. 2015), biotransformation of metal ions to non-toxic or less toxic forms (Zhu et al. 2014), intracellular accumulation (Shin et al. 2012) make those endophytes more relevant to metal remediation. Mindlin et al. (2002) said that, microorganisms develop heavy metal and antibiotic resistance, if they are synchronized with the ability to perform horizontal gene transfer (HGT). Recent studies revealed that endophytes tend to modulate activities of plant antioxidant enzymes like peroxidase (PO_x), catalase (CAT), super oxide dismutase (SOD), ascorbate peroxidase, as well as lipid peroxidation. These ROS activated enzymes play important role in plant defence mechanism. It has been reported that some endophytes promote DNA methylation in the form of metal resistance or detoxification process. Brown et al. (2003) and Cursino et al. (2000) stated that endophytic bacteria express different genes to convert mercury into non-toxic form. Studies showed genetically engineered endophyte-plant symbionts tend to improve phytoremediation efficiency of hyperaccumulator plants. Qiu et al. (2014) reported that introduction of gcsgs i.e., bifunctional glutathione-synthetase gene into *Enterobacter* sp. present as an endophyte symbiont in *Brassica juncea* increases plant's efficacy to remediate Cd and Pb from the soil.

6.17 Conclusion and Future Prospects

Microbial land remediation holds tremendous future potential. Endophytic microbes show traits that influence their exhibit for plant growth promoting activities. The plant–endophyte association also shows different mechanisms for pollutant removal and management. It is largely governed by the pollutant origin, concentration and fate. In the present scenario, besides the conventional methods of land remediation, application of endophytes is very promising because of its feasibility, non-harmful nature and cost-effectiveness. There is a vast field of endophytic population yet to be explored. Utilization of endophytic biofertilizer not only reduces the amount of chemical fertilizer, but also improves soil quality and agricultural productivity.

Phytoremediation techniques assisted by endophytes are environment friendly and sustainable in nature. Owing to its compatible nature, this symbiotic association is gaining popularity in the scientific community. Several researchers have been working on in order to understand the mechanism behind a successful endophyte–plant combination. At the same time, genetically engineered bacterial introduction can improve the efficacy of transgenic plants with regard to metal remediation and faster POP degradation. Recent development in the field of omics has enabled to maximize such understanding, harness beneficial traits and improve quality. However, challenges remain due to the diversity of endophytes. Screening for the most efficient and competent genera is cumbersome and tedious. Moreover, their population cannot be limited to *in-vitro* conditions. Main challenge remains with fact that how they respond in natural environment. Permission for introduction of transgenic plants and genetically engineered endophyte for field study is a matter of concern, however, it opens up the door to explore ideas and limitations of such studies. Researchers in the near future can work on developing field realistic variables for endophyte–plant partnership to execute and apply. The mechanisms also need an in-depth investigation, so that endophyte–plant potential can be realized, applied for an improved soil environment.

Acknowledgements LG would like to acknowledge the financial assistance received in the form of UGC Dr DS Kothari post-doctoral fellowship (BL/18-19/0215) for the year 2019–2022.

References

Afzal M, Yousaf S, Reichenauer TG, Kuffner M, Sessitsch A (2011) Soil type affects plant colonization, activity and catabolic gene expression of inoculated bacterial strains during phytoremediation of diesel. J Hazard Mater 186(2–3):1568–1575

Afzal M, Yousaf S, Reichenauer TG, Sessitsch A (2012) The inoculation method affects colonization and performance of bacterial inoculant strains in the phytoremediation of soil contaminated with diesel oil. Int J Phytorem 14(1):35–47

Afzal S, Begum N, Zhao H, Fang Z, Lou L, Cai Q (2017) Influence of endophytic root bacteria on the growth, cadmium tolerance and uptake of switchgrass (*Panicum virgatum* L.). J Appl Microbiol 123(2):498–510

Afzal I, Shinwari ZK, Sikandar S, Shahzad S (2019) Plant beneficial endophytic bacteria: Mechanisms, diversity, host range and genetic determinants. Microbiol Res 221:36–49

Ali H, Khan E, Sajad MA (2013) Phytoremediation of heavy metals—concepts and applications. Chemosphere 91(7):869–881

Ali S, Charles TC, Glick BR (2014) Amelioration of high salinity stress damage by plant growth-promoting bacterial endophytes that contain ACC deaminase. Plant Physiol Biochem 80:160–167

Alvarez F, Castro M, Principe A, Borioli G, Fischer S, Mori G, Jofre E (2012) The plant-associated *Bacillus amyloliquefaciens* strains MEP218 and ARP23 capable of producing the cyclic lipopeptides iturin or surfactin and fengycin are effective in biocontrol of sclerotinia stem rot disease. J Appl Microbiol 112(1):159–174

Andria V, Reichenauer TG, Sessitsch A (2009) Expression of alkane monooxygenase (alkB) genes by plant-associated bacteria in the rhizosphere and endosphere of Italian ryegrass (*Lolium multiflorum* L.) grown in diesel contaminated soil. Environ Pollut 157(12):3347–3350

Babu AG, Kim JD, Oh BT (2013) Enhancement of heavy metal phytoremediation by *Alnus firma* with endophytic *Bacillus thuringiensis* GDB-1. J Hazard Mater 250:477–483

Babu AG, Shea PJ, Sudhakar D, Jung IB, Oh BT (2015) Potential use of *Pseudomonas koreensis* AGB-1 in association with *Miscanthus sinensis* to remediate heavy metal (loid)-contaminated mining site soil. J Environ Manage 151:160–166

Bacon CW, Hinton DM (2007) Bacterial endophytes: the endophytic niche, its occupants, and its utility. Plant-associated bacteria. Springer, Dordrecht, pp 155–194

Bacon CW, Hinton DM (2011) In planta reduction of maize seedling stalk lesions by the bacterial endophyte *Bacillus mojavensis*. Can J Microbiol 57(6):485–492

Bahroun A, Jousset A, Mhamdi R, Mrabet M, Mhadhbi H (2018) Anti-fungal activity of bacterial endophytes associated with legumes against *Fusarium solani*: Assessment of fungi soil suppressiveness and plant protection induction. Appl Soil Ecol 124:131–140

Baker AJM (2000) Metal hyperaccumulator plants: a review of the ecology and physiology of a biological resource for phytoremediation of metal-polluted soils. In: Terry N, Bañuelos G (ed). Phytoremediation of contaminated soil and water, CRC Press, pp 85–107

Baldani VD, Baldani JI, Döbereiner J (2000) Inoculation of rice plants with the endophytic diazotrophs *Herbaspirillum seropedicae* and *Burkholderia* spp. Biol Fertil Soils 30(5):485–491

Bamisile BS, Dash CK, Akutse KS, Keppanan R, Wang L (2018) Fungal endophytes: beyond herbivore management. Fron Microbiol 9:544

Banat IM, Franzetti A, Gandolfi I, Bestetti G, Martinotti MG, Fracchia L, Smyth TJ, Marchant R (2010) Microbial biosurfactants production, applications and future potential. Appl Microbiol Biotechnol 87(2):427–444

Becerra-Castro C, Prieto-Fernández Á, Kidd PS, Weyens N, Rodríguez-Garrido B, Touceda-González M, Acea MJ, Vangronsveld J (2013) Improving performance of *Cytisus striatus* on substrates contaminated with hexachlorocyclohexane (HCH) isomers using bacterial inoculants: developing a phytoremediation strategy. Plant Soil 362(1):247–260

Becerril JM, Muñoz-Rueda A, Aparicio-Tejo P, Gonzalez-Murua C (1988) The effects of cadmium and lead on photosynthetic electron transport in. Plant Physiol Biochem 26(3):357–363

Böhm M, Hurek T, Reinhold-Hurek B (2007) Twitching motility is essential for endophytic rice colonization by the N_2-fixing endophyte *Azoarcus* sp. strain BH72. Mol Plant Microbe Interact 20(5):526–533

Boudh S, Singh JS, Chaturvedi P (2019) Microbial resources mediated bioremediation of persistent organic pollutants. In. Gupta V (ed) New and future developments in microbial biotechnology and bioengineering. Elsevier, pp 283–294

Brown NL, Stoyanov JV, Kidd SP, Hobman JL (2003) The MerR family of transcriptional regulators. FEMS Microbiol Rev 27(2–3):145–163

Burges A, Epelde L, Benito G, Artetxe U, Becerril JM, Garbisu C (2016) Enhancement of ecosystem services during endophyte-assisted aided phytostabilization of metal contaminated mine soil. Sci Total Environ 562:480–492

Carroll GC (1988) Fungal endophytes in stems and leaves: from latent pathogen to mutualistic symbiont. Ecol 69:2–9

Chen WM, Tang YQ, Mori K, Wu XL (2012) Distribution of culturable endophytic bacteria in aquatic plants and their potential for bioremediation in polluted waters. Aquat Biol 15(2):99–110

Chen L, Luo S, Li X, Wan Y, Chen J, Liu C (2014) Interaction of Cd-hyperaccumulator *Solanum nigrum* L. and functional endophyte *Pseudomonas* sp. Lk9 on soil heavy metals uptake. Soil Biol Biochem 68:300–308

Chen Z, Jin Y, Yao X, Chen T, Wei X, Li C, White JF, Nan Z (2020) Fungal endophyte improves survival of Lolium Perenne in low fertility soils by increasing root growth, metabolic activity and absorption of nutrients. Plant Soil 452:185–206

Chen L, Luo S, Li X, Wan Y, Chen J, Liu C (2014) Interaction of CD-hyperaccumulator *Solanum nigrum* L. and functional endophyte pseudomonas sp.. LK9 on soil heavy metals uptake. Soil Biol Biochem 68:300–308

Compant S, Reiter B, Sessitsch A, Nowak J, Clément C, Barka EA (2005) Endophytic colonization of *Vitis vinifera* L. by plant growth-promoting bacterium *Burkholderia* sp. strain PsJN. Appl Environ Microbiol 71(4):1685–1693

Compant S, Clément C, Sessitsch A (2010) Plant growth-promoting bacteria in the rhizo-and endosphere of plants: their role, colonization, mechanisms involved and prospects for utilization. Soil Biol Biochem 42(5):669–678

Cursino L, Mattos SV, Azevedo V, Galarza F, Bücker DH, Chartone-Souza E, Nascimento AM (2000) Capacity of mercury volatilization by mer (from *Escherichia coli*) and glutathione S-transferase (from *Schistosoma mansoni*) genes cloned in *Escherichia coli*. Sci Total Environ 261(1–3):109–113

Das, Satinath, Goswami, Linee, Bhattacharya, Satya Sundar, Mandal NC (2021) Chapter 12—biobased technologies to combat emerging environmental contaminants. In: Singh P, Hussain CM, Rajkhowa S (eds) Management of contaminants of emerging concern (CEC) in environment, Elsevier

DESA (2015) The world population prospects: 2015 Revision. Department of Economics and Social Affairs, United Nations, New York (https://www.un.org/en/development/desa/publications/2015.html)

Dharni S, Srivastava AK, Samad A, Patra DD (2014) Impact of plant growth promoting *Pseudomonas monteilii* PsF84 and *Pseudomonas plecoglossicida* PsF610 on metal uptake and production of secondary metabolite (monoterpenes) by rose-scented geranium (*Pelargonium graveolens* cv. *bourbon*) grown on tannery sludge amended soil. Chemosphere 117:433–439

Dobereiner J, Reis VM, Paula MA, Olivares F (1993) Endophytic diazotrophs in sugarcane cereals and tuber crops. In: Palacios R, Moor J, Newton WE (eds) New horizons in nitrogen fixation. Kluwer, Dordrecht, pp 671–674

Dörr J, Hurek T, Reinhold-Hurek B (1998) Type IV pili are involved in plant–microbe and fungus–microbe interactions. Mol Microbiol 30(1):7–17

Doty SL (2008) Enhancing phytoremediation through the use of transgenics and endophytes. New Phytol 179(2):318–333

Duijff BJ, Gianinazzi-Pearson VIVIENNE, Lemanceau P (1997) Involvement of the outer membrane lipopolysaccharides in the endophytic colonization of tomato roots by biocontrol *Pseudomonas fluorescens* strain WCS417r. New Phytol 135(2):325–334

Egamberdieva D, Wirth SJ, Shurigin VV, Hashem A, Abd_Allah EF (2017) Endophytic bacteria improve plant growth, symbiotic performance of chickpea (*Cicer arietinum* L.) and induce suppression of root rot caused by Fusarium solani under salt stress. Front Microbiol 8:1887

Fan Q, Tian S, Liu H, Xu Y (2002) Production of β-1, 3-glucanase and chitinase of two biocontrol agents and their possible modes of action. Chin Sci Bull 47(4):292–296

FAO (2012) The state of food insecurity in the world. Economic growth is necessary but not sufficient to accelerate reduction of hunger and malnutrition. Food and Agriculture Organization of the United Nations, Rome (https://www.hst.org.za/publications/NonHST%20Publications/i3027e.pdf)

Feng J, Xing W, Xie L (2016) Regulatory roles of microRNAs in diabetes. Int J Mol Sci 17(10):1729

Feng N-X, Yu J, Zhao H-M, Cheng Y-T, Mo C-H, Cai Q-Y, Li Y-W, Li H, Wong M-H (2017). Efficient phytoremediation of organic contaminants in soils using plant–endophyte partnerships. Sci Total Environ 583:352–368

Ferro AM, Kennedy J, LaRue JC (2013) Phytoremediation of 1, 4-dioxane-containing recovered groundwater. Int J Phytorem 15(10):911–923

Gao D, Tao Y (2012) Current molecular biologic techniques for characterizing environmental microbial community. Front Environ Sci Eng 6:82–97

Gardner IC, Clelland DM, Scott A (1984) Mycorrhizal improvement in non-leguminous nitrogen fixing associations with particular reference to Hippophaë rhamnoides L. In: Normand P, Pawlowski K, Dawson J (eds) Frankia symbioses. Springer, Dordrecht, pp 189–199

Gerhardt KE, Huang XD, Glick BR, Greenberg BM (2009) Phytoremediation and rhizoremediation of organic soil contaminants: potential and challenges. Plant Sci 176(1):20–30

Germaine KJ, Liu X, Cabellos GG, Hogan JP, Ryan D, Dowling DN (2006) Bacterial endophyte-enhanced phytoremediation of the organochlorine herbicide 2, 4-dichlorophenoxyacetic acid. FEMS Microbiol Ecol 57(2):302–310

Germaine KJ, Keogh E, Ryan D, Dowling DN (2009) Bacterial endophyte-mediated naphthalene phytoprotection and phytoremediation. FEMS Microbiol Lett 296(2):226–234

Glick BR (2010) Using soil bacteria to facilitate phytoremediation. Biotechnol Adv 28(3):367–374

Glick BR (2014) Bacteria with ACC deaminase can promote plant growth and help to feed the world. Microbiol Res 169(1):30–39

Glick BR, Stearns JC (2011) Making phytoremediation work better: maximizing a plant's growth potential in the midst of adversity. Int J Phytorem 13(sup1):4–16

Glick BR, Cheng Z, Czarny J, Duan J (2007) Promotion of plant growth by ACC deaminase-producing soil bacteria. In: Bakker PAHM, Raaijmakers JM, Bloemberg G, Höfte M, Lemanceau P, Cooke BM (eds) New perspectives and approaches in plant growth-promoting rhizobacteria research. Springer, Dordrecht, pp 329–339

Gohain A, Gogoi A, Debnath R, Yadav A, Singh BP, Gupta VK, Sharma R, Saikia R (2015) Antimicrobial biosynthetic potential and genetic diversity of endophytic actinomycetes associated with medicinal plants. FEMS Microbiol Lett 19:158

Gómez-Lama Cabanás C, Schilirò E, Valverde-Corredor A, Mercado-Blanco J (2014) The biocontrol endophytic bacterium *Pseudomonas fluorescens* PICF7 induces systemic defense responses in aerial tissues upon colonization of olive roots. Front Microbiol 5:427

Gorai PS, Ghosh R, Konra S, Mandal NC (2021) Biological control of early blight disease of potato caused by *Alternaria alternata* EBP3 by an endophytic bacterial strain *Bacillus velezensis* SEB1. Biol Control 156:104551

Gorai PS, Gond SK, Mandal NC (2020) Endophytic microbes and their role to overcome abiotic stress in crop plants. In: Jay Shankar S, Shobit Raj V (eds) Microbial services in restoration ecology. Elsevier, pp 109–122

Govarthanan M, Mythili R, Selvankumar T, Kamala-Kannan S, Rajasekar A, Chang YC (2016) Bioremediation of heavy metals using an endophytic bacterium *Paenibacillus* sp. RM isolated from the roots of *Tridax procumbens*. 3 Biotech 6(2):1–7

Govindarajan MA, Kwon SW, Weon HY (2007) Isolation, molecular characterization and growth-promoting activities of endophytic sugarcane diazotroph *Klebsiella* sp. GR9. World J Microbiol Biotechnol 23(7):997–1006

Guo H, Yao J, Cai M, Qian Y, Guo Y, Richnow HH, Blake RE, Doni S, Ceccanti B (2012) Effects of petroleum contamination on soil microbial numbers, metabolic activity and urease activity. Chemosphere 87(11):1273–1280

Hardoim PR, van Overbeek LS, van Elsas JD (2008) Properties of bacterial endophytes and their proposed role in plant growth. Trends Microbiol 16(10):463–471

He X, Han G, Lin Y, Tian X, Xiang C, Tian Q, Wang F, He Z (2012) Diversity and decomposition potential of endophytes in leaves of a *Cinnamomum camphora* plantation in China. Ecol Res 27(2):273–284

He W, Megharaj M, Wu CY, Suresh R, Bose SC, Dai C-C (2019) Endophyte-assisted phytoremediation: mechanisms and current application strategies for soil mixed pollutants. Crit Rev Biotechnol 40(1):31–45

Hemambika B, Rani MJ, Kannan VR (2011) Biosorption of heavy metals by immobilized and dead fungal cells: a comparative assessment. J Ecol Nat 3(5):168–175

Ho YN, Shih CH, Hsiao SC, Huang CC (2009) A novel endophytic bacterium, *Achromobacter xylosoxidans*, helps plants against pollutant stress and improves phytoremediation. J Biosci Bioeng 108

Hu Q, Zhao X, Yang XJ (2017) China's decadal pollution census. Nature 543(7646):491

Hurek T, Reinhold-Hurek B (2003) *Azoarcus* sp. strain BH72 as a model for nitrogen-fixing grass endophytes. J Biotechnol 106(2–3):169–178

Iniguez AL, Dong Y, Triplett EW (2004) Nitrogen fixation in wheat provided by *Klebsiella pneumoniae* 342. Mol Plant-Microbe Interact 17(10):1078–1085

James EK, Olivares FL (1998) Infection and colonization of sugar cane and other graminaceous plants by endophytic diazotrophs. Crit Rev Plant Sci 17(1):77–119

Jha P, Kumar A (2009) Characterization of novel plant growth promoting endophytic bacterium *Achromobacter xylosoxidans* from wheat plant. Microb Ecol 58(1):179–188

Jha P, Panwar J, Jha PN (2015) Secondary plant metabolites and root exudates: guiding tools for polychlorinated biphenyl biodegradation. Int J Environ Sci Technol 12:789–802

Ji SH, Gururani MA, Chun SC (2014) Isolation and characterization of plant growth promoting endophytic diazotrophic bacteria from Korean rice cultivars. Microbiol Res 169(1):83–98

Kadirvelu K, Senthilkumar P, Thamaraiselvi K, Subburam V (2002) Activated carbon prepared from biomass as adsorbent: elimination of Ni (II) from aqueous solution. Bioresour Technol 81(1):87–90

Kang JW, Khan Z, Doty SL (2012) Biodegradation of trichloroethylene by an endophyte of hybrid poplar. Appl Environ Microbiol 78(9):3504–3507

Karthikeyan B, Joe MM, Islam MR, Sa T (2012) ACC deaminase containing diazotrophic endophytic bacteria ameliorate salt stress in *Catharanthus roseus* through reduced ethylene levels and induction of antioxidative defense systems. Symbiosis 56(2):77–86

Keck RW (1978) Cadmium alteration of root physiology and potassium ion fluxes. Plant Physiol 62(1):94–96

Khare E, Mishra J, Arora NK (2018) Multifaceted interactions between endophytes and plant: developments and prospects. Front Microbiol 9:2732. https://doi.org/10.3389/fmicb.2018.02732

Kidd KA, Muir DC, Evans MS, Wang X, Whittle M, Swanson HK, Johnston T, Guildford S (2012) Biomagnification of mercury through lake trout (*Salvelinus namaycush*) food webs of lakes with different physical, chemical and biological characteristics. Sci Total Environ 438:135–143

Kloepper JW, Ryu CM (2006) Bacterial endophytes as elicitors of induced systemic resistance. In: Schulz BJE, Boyle CJC, Sieber TN (eds) Microbial root endophytes. Soil biology. Springer, Berlin, Heidelberg, pp 33–52

Kobayashi DY, Palumbo JD (2000) Bacterial endophytes and their effects on plants and uses in agriculture. In: White J, Bacon CW (eds) Microbial endophytes (1st ed). CRC Press, pp 199–233

Kumar M, Saxena R, Tomar RS (2017) Endophytic microorganisms: promising candidate as biofertilizer. In: Panpatte DG, Jhala YK, Shelat HN, Vyas RV (eds) Microorganisms for green revolution. Springer, Singapore, pp p77-85

Kumawat KC, Sharma P, Sirari A, Singh I, Gill BS, Singh U, Saharan K (2019) Synergism of *Pseudomonas aeruginosa* (LSE-2) nodule endophyte with *Bradyrhizobium* sp.(LSBR-3) for improving plant growth, nutrient acquisition and soil health in soybean. World J Microbiol Biotechnol 35(3):1–17

Lebeau T, Braud A, Jézéquel K (2008) Performance of bioaugmentation-assisted phytoextraction applied to metal contaminated soils: a review. Environ Pollut 153(3):497–522

Li, CY, Strzelczyk E, Pokojska A (1996). Nitrogen-fixing endophyte *Frankia* in polish *Alnus glutinosa* (L.) Gartn. Microbiol. Res. 151(4): 371–374

Lin L, Guo W, Xing Y, Zhang X, Li Z, Hu C, Li S, Li Y, An Q (2012) The actinobacterium *Microbacterium* sp. 16SH accepts pBBR1-based pPROBE vectors, forms biofilms, invades roots, and fixes N_2 associated with micropropagated sugarcane plants. Appl Microbiol Biotechnol 93(3):1185–1195

Liu J, Liu S, Sun K, Sheng Y, Gu Y, Gao Y (2014) Colonization on root surface by a phenanthrene-degrading endophytic bacterium and its application for reducing plant phenanthrene contamination. PLoS ONE 9:e108249

Loaces I, Ferrando L, Scavino AF (2011) Dynamics, diversity and function of endophytic siderophore-producing bacteria in rice. Microb Ecol 61(3):606–618

Lodewyckx C, Taghavi S, Mergeay M, Vangronsveld J, Clijsters H, Lelie DVD (2001) The effect of recombinant heavy metal-resistant endophytic bacteria on heavy metal uptake by their host plant. Int J Phytoremediation 3(2):173–187

Lodewyckx C, Vangronsveld J, Porteous F, Moore ER, Taghavi S, Mezgeay M, der Lelie DV (2002) Endophytic bacteria and their potential applications. Crit Rev Plant Sci 21(6):583–606

Long HH, Schmidt DD, Baldwin IT (2008) Native bacterial endophytes promote host growth in a species-specific manner; phytohormone manipulations do not result in common growth responses. PLoS ONE 3(7):e2702

Lugtenberg B, Kamilova F (2009) Plant-growth-promoting rhizobacteria. Annu Rev Microbiol 63:541–556

Luo JM, Xiao XIAO (2010) Biosorption of cadmium (II) from aqueous solutions by industrial fungus *Rhizopus cohnii*. T Nonferr Metals SOC. 20(6):1104–1111

Luo J, Tao Q, Jupa R, Liu Y, Wu K, Song Y, Li J, Huang Y, Zou L, Liang Y, Li T (2019) Role of vertical transmission of shoot endophytes in root-associated microbiome assembly and heavy metal hyperaccumulation in *Sedum alfredii*. Environ Sci Technol 53(12):6954–6963

Luo S, Wan Y, Xiao X, Guo H, Chen L, Xi Q, Zeng G, Liu C, Chen J (2011) Isolation and characterization of endophytic bacterium LRE07 from cadmium hyperaccumulator *Solanum nigrum* L. and its potential for remediation. Appl Microbiol Biotechnol 89(5):1637–1644

Ma Y, Prasad MNV, Rajkumar M, Freitas H (2011) Plant growth promoting rhizobacteria and endophytes accelerate phytoremediation of metalliferous soils. Biotechnol Adv 29(2):248–258

Ma Y, Oliveira RS, Nai F, Rajkumar M, Luo Y, Rocha I, Freitas H (2015) The hyperaccumulator *Sedum plumbizincicola* harbors metal-resistant endophytic bacteria that improve its phytoextraction capacity in multi-metal contaminated soil. J Environ Manage 156:62–69

Ma Y, Rajkumar M, Zhang C, Freitas H (2016) Beneficial role of bacterial endophytes in heavy metal phytoremediation. J Environ Manage 174:14–25

Madhaiyan M, Poonguzhali S, Sa T (2007) Metal tolerating methylotrophic bacteria reduces nickel and cadmium toxicity and promotes plant growth of tomato (*Lycopersicon esculentum* L.). Chemosphere 69(2):220–228

Maheshwari R, Bhutani N, Suneja P (2019) Screening and characterization of siderophore producing endophytic bacteria from *Cicer arietinum* and *Pisum sativum* plants. J Appl Biol Biotechnol 7:7–14

Martinez-Klimova E, Rodríguez-Peña K, Sánchez S (2017) Endophytes as sources of antibiotics. Biochem Pharmacol 15(134):1–17

Mindlin SZ, Bass IA, Bogdanova ES, Gorlenko ZM, Kalyaeva ES, Petrova MA, Nikiforov VG (2002) Horizontal transfer of mercury resistance genes in environmental bacterial populations. Mol Biol 36(2):160–170

Moore FP, Barac T, Borremans B, Oeyen L, Vangronsveld J, Van der Lelie D, Campbell CD, Moore ER (2006) Endophytic bacterial diversity in poplar trees growing on a BTEX-contaminated site: the characterisation of isolates with potential to enhance phytoremediation. Syst Appl Microbiol 29(7):539–556

Muthukumarasamy R, Kang UG, Park KD, Jeon WT, Park CY, Cho YS, Kwon SW, Song J, Roh DH, Revathi G (2007) Enumeration, isolation and identification of diazotrophs from Korean wetland rice varieties grown with long-term application of N and compost and their short-term inoculation effect on rice plants. J Appl Microbiol 102(4):981–991

Myo EM, Ge B, Ma J, Cui H, Liu B, Shi L, Jiang M, Zhang K (2019) Indole-3-acetic acid production by *Streptomyces fradiae* NKZ-259 and its formulation to enhance plant growth. BMC Microbiol 19:155

Nabi S (2014) Methylmercury and minamata disease. In: Toxic effects of mercury. Springer, New Delhi. https://doi.org/10.1007/978-81-322-1922-4_25

Nair DN, Padmavathy S (2014) Impact of endophytic microorganisms on plants, environment and humans. Sci World J 2014:1–11

Ngamau C, Matiru VN, Tani A, Muthuri, C (2014) Potential use of endophytic bacteria as biofertilizer for sustainable banana (Musa spp.) production. Afr J Hort Sci 8(1)

Oonnittan A, Sillanpää M (2020) Application of electrokinetic Fenton process for the remediation of soil contaminated with HCB. In: Sillanpää M (ed) Advanced water treatment. Elsevier, pp 57–93

Pal S, Singh HB, Farooqui A, Rakshit A (2015) Fungal biofertilizers in Indian agriculture: perception, demand and promotion. J Eco-Friendly Agri 10(2):101–113

Pandey J, Chauhan A, Jain RK (2009) Integrative approaches for assessing the ecological sustainability of in situ bioremediation. FEMS Microbiol Rev 33(2):324–375

Petrini O (1991) Fungal endophytes of tree leaves. In: Andrews J, Hirano SS (eds) Microbial ecology of leaves. Spring-Verlag, New York, pp 179–197

Phillips LA, Germida JJ, Farrell RE, Greer CW (2008) Hydrocarbon degradation potential and activity of endophytic bacteria associated with prairie plants. Soil Biol Biochem 40(12):3054–3064

Pimentel MR, Molina G, Dionisio AP, Maróstica MR, Pastore GM (2011) Use of endophytes to obtain bioactive compounds and their application in biotransformation process. Biotechnol Res Int 2011:576286. https://doi.org/10.4061/2011/576286

Promputtha I, Hyde KD, McKenzie EH, Peberdy JF, Lumyong S (2010) Can leaf degrading enzymes provide evidence that endophytic fungi becoming saprobes? Fungal Divers 41(1):89–99

Puri A, Padda KP, Chanway CP (2018) Nitrogen-fixation by endophytic bacteria in agricultural crops: recent advances. In: Amanullah FS (ed) Nitrogen in agriculture. Intech Open, London, GBR, pp 73–94

Qiu Z, Tan H, Zhou S, Cao L (2014) Enhanced phytoremediation of toxic metals by inoculating endophytic Enterobacter sp. CBSB1 expressing bifunctional glutathione synthase. J Hazardous Mater 267:17–20

Quadt-Hallmann A, Hallmann J, Kloepper JW (1997) Bacterial endophytes in cotton: location and interaction with other plant-associated bacteria. Can J Microbiol 43(3):254–259

Rajkumar M, Ae N, Freitas H (2009) Endophytic bacteria and their potential to enhance heavy metal phytoextraction. Chemosphere 77(2):153–160

Rana KL, Kour D, Kaur T, Sheikh I, Yadav AN, Kumar V, Suman A, Dhaliwal HS (2020) Endophytic microbes from diverse wheat genotypes and their potential biotechnological applications in plant growth promotion and nutrient uptake. In: Proceedings of the national academy of sciences, India section B: biological sciences, pp 1–11

Reinhold-Hurek B, Hurek T (1998) Life in grasses: diazotrophic endophytes. Trends Microbiol 6(4):139–144

Reinhold-Hurek B, Maes T, Gemmer S, Van Montagu M, Hurek T (2006) An endoglucanase is involved in infection of rice roots by the not-cellulose-metabolizing endophyte *Azoarcus* sp. strain BH72. Mol Plant Microbe Interact 19(2):181–188

Roychowdhury D, Mondal S, Banerjee SK (2017) The effect of biofertilizers and the effect of vermicompost on the cultivation and productivity of maize-a review. Adv Crop Sci Technol 5:1–4

Saha JK, Selladurai R, Coumar MV, Dotaniya ML, Kundu S, Patra AK (2017) Soil pollution-an emerging threat to agriculture (vol 10). Springer

Saini R, Kumar V, Dudeja SS, Pathak DV (2015) Beneficial effects of inoculation of endophytic bacterial isolates from roots and nodules in chickpea. Int J Curr Microbiol Appl Sci 4(10):207–221

Sainz JR, Zhou JC, Rodriguez-Navarro DN, Vinardell JM, Thomas-Oates JE (2005) Soybean cultivation and BBF in China. In: Werner D, Newton WE (eds) Nitrogen fixation in agriculture, forestry, ecology, and the environment. Springer, Dordrecht, pp 67–87

Sandhya V, Shrivastava M, Ali SZ, Prasad VSSK (2017) Endophytes from maize with plant growth promotion and biocontrol activity under drought stress. Russ Agric Sci 43(1):22–34

Sapers GM, Gorny JR, Yousef AE (2005) Microbiology of fruits and vegetables. CRC Press

Scherling C, Ulrich K, Ewald D, Weckwerth W (2009) A metabolic signature of the beneficial interaction of the endophyte *Paenibacillus* sp. isolate and in vitro–grown poplar plants revealed by metabolomics. Mol Plant-Microbe Interact 22(8):1032–1037

Serepa-Dlamini MH (2020) Culture-indepenadent characterization of endophytic bacterial communities associated with a South African medicinal plant, Dicoma anomala

Sessitsch A, Reiter B, Berg G (2004) Endophytic bacterial communities of field-grown potato plants and their plant-growth-promoting and antagonistic abilities. Can J Microbiol 50(4):239–249

Shahid M, Khan MS (2018) Glyphosate induced toxicity to chickpea plants and stress alleviation by herbicide tolerant phosphate solubilizing *Burkholderia cepacia* PSBB1 carrying multifarious plant growth promoting activities. 3 Biotech 8(2):1–17

Shen FT, Yen JH, Liao CS, Chen WC, Chao YT (2019) Screening of rice endophytic biofertilizers with fungicide tolerance and plant growth-promoting characteristics. Sustainability 11(4):1133

Sheng X, Chen X, He L (2008a) Characteristics of an endophytic pyrene-degrading bacterium of *Enterobacter* sp. 12J1 from *Allium macrostemon* Bunge. Int Biodeterior Biodegrad 62(2):88–95

Sheng XF, Xia JJ, Jiang CY, He LY, Qian M (2008b) Characterization of heavy metal-resistant endophytic bacteria from rape (*Brassica napus*) roots and their potential in promoting the growth and lead accumulation of rape. Environ Pollut 156(3):1164–1170

Shin MN, Shim J, You Y, Myung H, Bang KS, Cho M, Seralathan KK, Oh BT (2012) Characterization of lead resistant endophytic *Bacillus* sp. MN3-4 and its potential for promoting lead accumulation in metal hyperaccumulator *Alnus firma*. J Hazard Mater 199:314–320

Siciliano SD, Goldie H, Germida JJ (1998) Enzymatic activity in root exudates of Dahurian wild rye (*Elymus dauricus*) that degrades 2-chlorobenzoic acid. J Agric Food Chem 46(1):5–7

Siciliano SD, Fortin N, Mihoc A, Wisse G, Labelle S, Beaumier D, Ouellette D, Roy R, Whyte LG, Banks MK, Schwab P, Lee K, Greer CW (2001) Selection of specific endophytic bacterial genotypes by plants in response to soil contamination. Appl Environ Microbiol 67(6):2469–2475

Siddiqui IA, Shaukat SS (2003) Suppression of root-knot disease by *Pseudomonas fluorescens* CHA0 in tomato: importance of bacterial secondary metabolite, 2, 4-diacetylpholoroglucinol. Soil Biol Biochem 35(12):1615–1623

Singer AC, Crowley DE, Thompson IP (2003) Secondary plant metabolites in phytoremediation and biotransformation. Trends Biotechnol 21:123–130

Singh MJ, Padmavathy S (2015) Hydrocarbon Biodegradation by endophytic bacteria from neem leaves. LS Int J Life Sci 4(1):33–36

Singh JS, Pandey VC, Singh DP (2011) Efficient soil microorganisms: a new dimension for sustainable agriculture and environmental development. Agr Ecosyst Environ 140:339–353. https://doi.org/10.1016/j.agee.2011.01.017

Singh D, Rajawat MVS, Kaushik R, Prasanna R, Saxena AK (2017) Beneficial role of endophytes in biofortification of Zn in wheat genotypes varying in nutrient use efficiency grown in soils sufficient and deficient in Zn. Plant Soil 416(1):107–116

Srivastava S, Chaudhuri M, Pandey VC (2020) Endophytes—the hidden world for agriculture, ecosystem, and environmental sustainability. In: Pandey VC, Singh V (eds) Bioremediation of pollutants. Elsevier, Amsterdam. https://doi.org/10.1016/B978-0-12-819025-8.00006-5

Stępniewska Z, Kuźniar A (2013) Endophytic microorganisms—promising applications in bioremediation of greenhouse gases. Appl Microbiol Biot 97(22):9589–9596

Stone JK, Bacon CW, White JF, (2000) An overview of endophytic microbes: endophytism defined. In: White J, Bacon CW (eds) Microbial endophytes (1st ed). CRC Press, pp 29–33

Strobel G, Daisy B, Castillo U, Harper J (2004) Natural products from endophytic microorganisms. J Nat Prod 67(2):257–268

Sturz AV, Christie BR, Nowak J (2000) Bacterial endophytes: potential role in developing sustainable systems of crop production. Crit Rev Plant Sci 19(1):1–30

Sturz AV, Kimpinski J (2004) Endoroot bacteria derived from marigolds (*Tagetes* spp.) can decrease soil population densities of root-lesion nematodes in the potato root zone. Plant and Soil 262(1):241–249

Suárez-Moreno ZR, Devescovi G, Myers M, Hallack L, Mendonça-Previato L, Caballero-Mellado J, Venturi V (2010) Commonalities and differences in regulation of N-acyl homoserine lactone quorum sensing in the beneficial plant-associated *Burkholderia* species cluster. Appl Environ Microbiol 76(13):4302–4317

Subramanian P, Kim K, Krishnamoorthy R, Sundaram S, Sa T (2015) Endophytic bacteria improve nodule function and plant nitrogen in soybean on co-inoculation with *Bradyrhizobium japonicum* MN110. Plant Growth Regul 76(3):327–332

Sun K, Liu J, Jin L, Gao Y (2014) Utilizing pyrene-degrading endophytic bacteria to reduce the risk of plant pyrene contamination. Plant Soil 374(1):251–262

Suzuki T, Shimizu M, Meguro A, Hasegawa S, Nishimura T, Kunoh H (2005) Visualization of infection of an endophytic actinomycete *Streptomyces galbus* in leaves of tissue-cultured rhododendron. Actinomycetologica 19(1):7–12

Taghavi S, Barac T, Greenberg B, Borremans B, Vangronsveld J, van der Lelie D (2005) Horizontal gene transfer to endogenous endophytic bacteria from poplar improves. Appl Environ Microbiol 71(12):8500–8505

Taghavi S, Garafola C, Monchy S, Newman L, Hoffman A, Weyens N, Barac T, Vangrosveld J, van der Lelie D (2009) Genome survey and characterization of endophytic bacteria exhibiting a beneficial effect on growth and development of poplar trees. Appl Environ Microbiol 75(3):748–757

Taghavi S, Weyens N, Vangronsveld J, van der Lelie D (2011) Improved phytoremediation of organic contaminants through engineering of bacterial endophytes of trees. In: Pirttilä AM, Frank C (eds) Endophytes of forest trees. Springer, Dordrecht, pp 205–216

Tan RX, Zou WX (2001) Endophytes: a rich source of functional metabolites. Nat Prod Rep 18:448–459

Terry N, Zayed AM, de Souza MP, Tarun AS (2000) Selenium in higher plants. Annu Rev Plant Physiol Plant Mol Biol 51:401–432. https://doi.org/10.1146/annurev.arplant.51.1.401

Thijs S, Sillen W, Rineau F, Weyens N, Vangronsveld J (2016) Towards an enhanced understanding of plant–microbiome interactions to improve phytoremediation: engineering the metaorganism. Front Microbiol 7:341

Thomas P, Soly TA (2009) Endophytic bacteria associated with growing shoot tips of banana (*Musa* sp.) cv. Grand Naine and the affinity of endophytes to the host. Microb Ecol 58(4):952–964

Tong J, Miaowen C, Juhui J, Jinxian L, Baofeng C (2017) Endophytic fungi and soil microbial community characteristics over different years of phytoremediation in a copper tailings dam of Shanxi, China. Sci Total Environ 574:881–888

Tripathi V, Edrisi SA, Chen B, Gupta VK, Vilu R, Gathergood N, Abhilash PC (2017) Biotechnological advances for restoring degraded land for sustainable development. Trends Biotechnol 35(9):847–859

Usman M, Wakeel A, Farooq M (2017) India and Pakistan need to collaborate against pollution. Nature 552(7685):334

Vacheron J, Desbrosses G, Bouffaud ML, Touraine B, Moënne-Loccoz Y, Muller D, Wisniewski-Dyé F, Legendre L, Prigent-Combaret C (2013) Plant growth-promoting rhizobacteria and root system functioning. Front Plant Sci 4:356

Van Aken B, Yoon JM, Schnoor JL (2004) Biodegradation of nitro-substituted explosives 2, 4, 6-trinitrotoluene, hexahydro-1, 3, 5-trinitro-1, 3, 5-triazine, and octahydro-1, 3, 5, 7-tetranitro-1, 3, 5-tetrazocine by a phytosymbiotic *Methylobacterium* sp. associated with poplar tissues (*Populus deltoides*× *nigra* DN34). Appl Environ Microbiol 70(1):508–517

Verma P, Yadav AN, Khannam KS, Panjiar N, Kumar S, Saxena AK, Suman A (2015) Assessment of genetic diversity and plant growth promoting attributes of psychrotolerant bacteria allied with wheat (*Triticum aestivum*) from the northern hills zone of India. Ann Microbiol 65(4):1885–1899

Vijayaraghavan K, Yun YS (2008) Bacterial biosorbents and biosorption. Biotechnol Adv 26(3):266–291

Vinayarani G, Prakash HS (2018) Fungal endophytes of turmeric (*Curcuma longa* L.) and their biocontrol potential against pathogens *Pythium aphanidermatum* and *Rhizoctonia solani*. World J Microbiol Biotechnol 34(3):1–17

Walters D, Newton A, Lyon G (2007) Induced resistance for plant defence: a sustainable approach to crop protection. Blackwell, Wiley, p 272, ISBN: 978-1-405-13447-7

Wang Y, Li H, Zhao W, He X, Chen J, Geng X, Xiao M (2010) Induction of toluene degradation and growth promotion in corn and wheat by horizontal gene transfer within endophytic bacteria. Soil Biol Biochem 42(7):1051–1057

Wang S, Wang W, Jin Z, Du B, Ding Y, Ni T, Jiao F (2013) Screening and diversity of plant growth promoting endophytic bacteria from peanut. Afr J Microbiol Res 7(10):875–884

Weber OB, Baldani VLD, Teixeira KDS, Kirchhof G, Baldani JI, Dobereiner J (1999) Isolation and characterization of diazotrophic bacteria from banana and pineapple plants. Plant Soil 210(1):103–113

Weyens N, Croes S, Dupae J, Newman L, van der Lelie D, Carleer R, Vangronsveld J (2010a) Endophytic bacteria improve phytoremediation of Ni and TCE co-contamination. Environ Pollut 158(7):2422–2427

Weyens N, Truyen S, Dupae J, Newman L, Taghavi S, van der Lelie D, Carleer R, Vangronsveld J (2010b) Potential of the TCE-degrading endophyte *Pseudomonas putida* W619-TCE to improve plant growth and reduce TCE phytotoxicity and evapotranspiration in poplar cuttings. Environ Pollut 158(9):2915–2919

Wilson D (1995) Fungal endophytes which invade insect galls: Insect pathogens, benign saprophytes, or fungal inquilines? Oecologia 103:255–260

Wilson D, Carroll GC (1997) Avoidance of high-endophyte space by gall-forming insects. Ecol 78(7):2153–2163

Wiszniewska A, Hanus-Fajerska E, MUSZYŃSKA E, Ciarkowska K (2016) Natural organic amendments for improved phytoremediation of polluted soils: a review of recent progress. Pedosphere 26(1):1–12

Xu XJ, Sun JQ, Nie Y, Wu XL (2015) *Spirodela polyrhiza* stimulates the growth of its endophytes but differentially increases their fenpropathrin-degradation capabilities. Chemosphere 125:33–40

Yanni YG, Rizk RY, Corich V, Squartini A, Ninke K, Philip-Hollingsworth S, Orgambide G, De Bruijn FJ, Stoltzfus J, Buckley D, Schmidt TM, Mateos PF, Ladha JK, Dazzo FB (1997) Natural endophytic association between *Rhizobium leguminosarum* bv. *trifolii* and rice roots and assessment of its potential to promote rice growth. In: Ladha JK, De Bruijin FJ, Malik KA (eds) Opportunities for biological nitrogen fixation in rice and other non-legumes. Springer, Dordrecht, pp 99–114

Yousaf S, Andria V, Reichenauer TG, Smalla K, Sessitsch A (2010a) Phylogenetic and functional diversity of alkane degrading bacteria associated with Italian ryegrass (*Lolium multiflorum*) and Birdsfoot trefoil (*Lotus corniculatus*) in a petroleum oil-contaminated environment. J Hazard Mater 184(1–3):523–532

Yousaf S, Ripka K, Reichenauer TG, Andria V, Afzal M, Sessitsch A (2010b) Hydrocarbon degradation and plant colonization by selected bacterial strains isolated from Italian ryegrass and birdsfoot trefoil. J Appl Microbiol 109(4):1389–1401

Yousaf S, Afzal M, Reichenauer TG, Brady CL, Sessitsch A (2011) Hydrocarbon degradation, plant colonization and gene expression of alkane degradation genes by endophytic *Enterobacter ludwigii* strains. Environ Pollut 159(10):2675–2683

Zachow C, Jahanshah G, de Bruijn I, Song C, Ianni F, Pataj Z, Gerhardt H, Pianet I, Lämmerhofer M, Berg G, Gross H (2015) The novel lipopeptide poaeamide of the endophyte *Pseudomonas poae* RE* 1-1-14 is involved in pathogen suppression and root colonization. Mol Plant Microbe Interact 28(7):800–810

Zeidler D, Zähringer U, Gerber I, Dubery I, Hartung T, Bors W, Hutzler P, Durner J (2004) Innate immunity in *Arabidopsis thaliana*: lipopolysaccharides activate nitric oxide synthase (NOS) and induce defense genes. PNAS 101(44):15811–15816

Zhang X, Wang Z, Liu X, Hu X, Liang X, Hu Y (2013) Degradation of diesel pollutants in Huangpu-Yangtze River estuary wetland using plant-microbe systems. Int Biodeterior Biodegrad 76:71–75

Zhu LJ, Guan DX, Luo J, Rathinasabapathi B, Ma LQ (2014) Characterization of arsenic-resistant endophytic bacteria from hyperaccumulators *Pteris vittata* and *Pteris multifida*. Chemosphere 113:9–16

Zhu X, Jin L, Sun K, Li S, Ling W, Li X (2016) Potential of endophytic bacterium Paenibacillus sp. PHE-3 isolated from *Plantago asiatica* L. for reduction of PAH contam- ination in plant tissues. Int J Environ Res Public Health 13:633

Zuraida AR, Marziah M, Zulkifli Haji Shausuddin, Halimi S (2000) In: Wahad Z et al (eds) Proceedings of the first national banana seminar at Awana Gentig and Country resort. UPM, Serdang (MYS), p 343

Chapter 7
Fungal-Based Land Remediation

Soma Barman, Ratan Chowdhury, and Satya Sundar Bhattacharya

Abstract An organic and inorganic xenobiotic compound in agricultural land is a serious problem, mostly in agricultural countries like India. The land contaminated with toxic compound(s) like heavy metals (HM), polycyclic aromatic hydrocarbons (PAH) and polychlorinated biphenyls (PCBs) causes environmental hazards. Physical characteristics of soil (viz., pH, structural compositions, temperature and relative humidity) as well as intra- or extracellular fungal enzymes, other metabolic products play vital roles in bio-transformations of pollutants. Contaminants have negative impacts on both crop productivity and their internal quality. Mycoremediation is a novel technology for the reduction, biotransformation and eradication of PAHs, PCBs and HMs by the application of ecofriendly fungal organisms. Several macro-fungi like mushrooms, micro-fungi like *Trichoderma* spp., *Aspergillus* spp. helps to absorb HMs, thereby can be exploited as hyperaccumulator. Mushrooms secrete certain enzymes like laccase, manganese peroxidase, lignin peroxidase, cytochrome P450 monooxygenase, dehydrogenases, dioxygenase, epoxide hydrolases, FAD-dependent monooxygenases and glutathione transferase which are able to biodegrade the toxic agro-industrial wastes to products for plant growth and development. Moreover, mycorrhizal associations in higher plants help in biotransformation and biodegradation of harmful pollutants in the contaminated land.

Keywords Mycoremediation · Pollutants · Biodegradation · Biotransformation

---

S. Barman
Soil and Agro-Bioengineering Laboratory, Department of Environmental Science, Tezpur University, Assam 784028, India
e-mail: vb.somabarman@gmail.com

R. Chowdhury
Department of Botany, Rangapara College, Amaribari, Rangapara, Assam, India
e-mail: chowdhuryratan600@gmail.com

S. S. Bhattacharya (✉)
Department of Environmental Science, Tezpur University, Assam 784028, India
e-mail: satya72@tezu.ernet.in; evssatya@gmail.com

7.1 Introduction

Mycoremediation of land is an environmentally hospitable and useful approach to battle the ever-rising difficulty of land contamination. Bioremediation is considered as a green technology for the environmental cleanup of polluted land and water bodies (Perelo 2010). Bioremediation in polluted soils may result in a reduction below a safe threshold of pollutant levels (Alexander 1999). This technology can be employed for the onsite bioconversion of several agro-industrial contaminants viz., dyes, heavy metals (HMs), herbicidal and pharmaceutic effluents let out by several commercial sectors. The environmental pollution caused by these synthetic organic pollutants has become a foremost concern worldwide. Many of these xenobiotic compounds introduced to the nature are not easily degraded by the native microflora and fauna (Sullia 2004). Several classes of toxic chemicals viz., polycyclic aromatic hydrocarbons (PAH), polychlorinated biphenyls (PCB), pentachlorophenols, benzene, toluene, 1,1,1-trichloro–2,2-bis (4-chlorophenyl) ethane, ethylbenzene xylene, trinitrotoluene (TNT), that have been marked by United States Environmental Agency (USEPA) as priority pollutants due to their severe toxic effects on the environment and human health.

Fungi are a perfect group of microbial representative for the bioremediation of several toxic contaminants as they form hyphal network on the substratum, produce extracellular enzymes, resistance to heavy metals by the presence of metal-binding proteins, flexibility to changing temperature and pH (Khan et al. 2019; Kapahi and Sachdeva 2017; Singh et al. 2015; Bhattacharya et al. 2011a, b). Enzymatic activities of the mycelial fungi can break down organic pollutants into carbon dioxide and water. Sometimes, they can also be used in bioreactors for controlled fungal biomass and metabolites production, thereby employed to hasten the impairment of the toxicants (Aragão et al. 2020; Rodríguez Couto et al. 2006; Tekere 2019). These myco-bioreactors can also be used for ex-situ bioremediation of soil from PAH, agro-industrial wastes, tars, chlorinated compounds and explosive chemicals (Tekere 2019).

The key factors in fungal bioremediation of soil components are pH, temperature and metal speciation. This affects the transportation and take-up rate of pollutants (Liu et al. 2017; Rangel et al. 2018). The concentration of metal ions in soil is significantly reduced by organic substances and the crystalline form of clay minerals. The reduction in metal toxicity in the contaminants can be done by clay minerals that possess high cation exchange capacities (Sandrin and Maier 2003). Furthermore, the toxicity of the metals is reduced by the effect of organic contaminants to hamper the speciation, bio-accessibility (Ceci et al. 2019). The biodegradation of fossil fuel compounds and biotransformation of noxious metals is also determined by the pH. Community structure of microbes and the activities of the enzymes and metal contamination are also totally dependent on the changes in pH. Temperature indirectly affects the viscosity of the pollutants which indirectly influences the degradation process as the chemistry of pollutants and the overall diversity of fungi is totally affected by it (Rangel et al. 2018). There can be retardation of biodegradation in the affected soil;

this is due to the process as viscosity of petroleum escalates at lower temperatures and so instability is minimized. At around 30–40 °C temperature, the rate of degradation process for hydrocarbon pollutants is generally the highest in soil (Das and Chandran 2011). The dispersible efficiency of PAHs and mostly the pernicious metal ions sources at higher temperatures, which enhances their bioaccumulation. Whilst the microbial community structure is also affected at such high temperatures.

Generally, biotreatment process can be carried out based on the simultaneous effort of one organism or can be contributed by the combined action of different microorganisms and their metabolic pathway on the substrate. There are reports that many organisms either the eukaryotes or the prokaryotes have an innate ability to absorb lethal HMs ions. As eukaryotic organisms are more prone to toxic effects of metal compared to prokaryotes. As microorganisms may be advanced or eukaryotic or ancient cells that may be prokaryotic. So, the reciprocity with heavy metal ions by microorganisms is moderately dependent on nature of their cell such as eukaryotic organisms or may be prokaryotic. Among the probable approach of interaction is of intracellular eukaryotic chelation by numerous metal-attaching polypeptides and reducing the noxiousness by transformation into other chemical species or may be fungal steadily cast out of metal. Various fungi studied and strategically used for the purpose of bioremediation and treatments for heavy metals include *Penicillium canescens*, *Aspergillus versicolor*, *Aspergillus fumigatus*, *Saccharomyces cerevisiae* and *Candida utilis*, etc.

The present chapter illustrates different pollutants like PAH, PCBs; agrochemicals like pesticides, insecticides; pharmaceutical refuges viz., antibiotics, antifungal drugs; potentially toxic HMs, detergents, phthalates, dyes and their effects have been written brief. Their environmental toxicity, mechanism of action of myco-bioremediation was also described in detail.

7.2 Fungal Organisms and Bioremediation

(a) **Prospects and challenges**

The potential application of using the microbes either fungi or bacteria and their enzymes in different aspects such as industry, agriculture, pharmaceuticals and environments is considered as white biotechnology. Fungi have been widely used as potential candidates for bioremediation of contaminated environments. Fungi use their foremost enzymes essentially catalases, oxido reductase, peroxidases and laccasses which are naturally occurring to biodegrade different toxic compounds. These fungal enzymes react with xenobiotic contaminants like the synthetic ingredients and finally alter them from an intractable state to simple environment-safe conformation (Bollag 1992; Gianfreda and Rao 2004). Enzymes are much efficient to perform better functions compared to the toxic chemicals. The employment of enzymes in bioremediation is receiving much popularity among researchers as it does not produce any kind of perilous waste. The fungal organisms and their bio-active innate compounds are considered to play a significant role in environmental cleanup by displaying better performance, higher sustainability producing additionally industrially feasible products generated from conventional synthetic method.

The technology using fungi for bioremediation has gradually evolved with the advent of time. They are nowadays also used in different industrial parts, providing a key role in producing tremendously beneficial commercial products. The enzymes excreted by the cells permit additional efficient therapeutic practices for balancing the toxicity matter because they are able to accelerate the speed of biotransformation of these materials. Whilst using enzymes, it can reduce the cost as the requirements of heating the products is not needed. The method is inexpensive since the corresponding price of decontamination and the cost of derivation is reduced (Godfrey and Reichelt 1996; Gianfreda and Rao 2004). Fungi perform as a key role in biotreatment process as of their robust biology and diverse anabolic as well as catabolic ability. Although, the task using fungi as a device for biodegrading is a green and feasible way for restoration of polluted areas. The future prospects would depend on the research in this field which would considerably increase the chances in the expansion of modern novel expertise to reduce the impact of pollution.

(b) **Environmental consideration for optimum use**

It is no doubt that fungal remediation is one of the most versatile techniques for the removal of toxicity in sustainable way for different polluted areas. But for effective and optimum use of fungal inocula for mycoremediation environmental consideration cannot be left out. The effectiveness in bioremediation process depends on various properties such as the metabolic activity of microbes and ultimately the metabolism of microbes depends on pH, temperature, synthetic properties, substantial nature, dampness, type of soil, oxidation–reduction potential and appearance of macro and micronutrient. Also, on the factors such as bioavailability, concentration, potency and noxiousness of contaminants, it cannot be left out that the various substantial and synthetic properties of soil is also responsible for the biodegradation to function smoothly. In the soil, all of these which influence the metabolic action of microorganisms in the soil (Maloney 2001; Antizar-Ladislao et al. 2008; Lukic et al. 2017). Although a very optimum conditions are required for effective remediation. Degradation of pollutants rest on different factors like types of microorganisms used and what is their metabolic potential to act on the contaminants, the surface-active and chelating agents, genetic features is another factor and finally the extra- and intracellular enzymatic organizations of the fungus. As most of the enzymes are effective only at normal concentrations of contaminants, due to which when a fungus uses their enzymes to degrade, it becomes ineffective due to the low or much higher concentration of contaminants and thus the enzymes get deactivated. Fungus has a greater adaptability to different stress levels like low amount of pH, it can also tolerate variation in temperature, also if there is lower oxygen level in the environment and various other conditions such as sunlight variation can be easily adapted by fungus (Bamforth and Singleton 2005; D'Annibale et al. 2006; Bhattacharya et al. 2012). Comparatively, bacteria are less resistant to concentration of contaminants to that of fungus basically the filamentous fungi also they have greater potential to incorporate several enzymes responsible for neutralizing the toxicity (Harms et al. 2011).

7.3 Effectivity of Fungal Organism on Different Pollutants

Myco-bioremediation of different pollutants (Fig. 7.1) could be done by the following way.

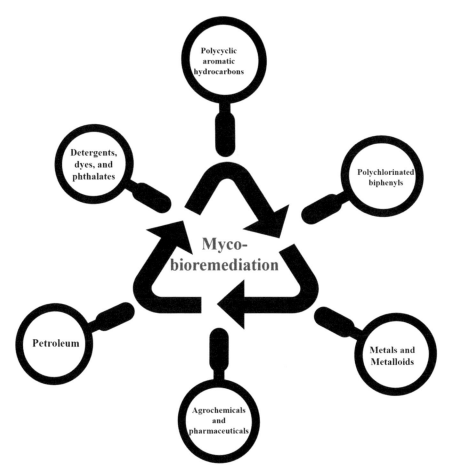

Fig. 7.1 Myco-bioremediation of environmental toxic pollutants

7.3.1 Polycyclic Aromatic Hydrocarbons (PAH) and Polychlorinated Biphenyls (PCBs)

PAHs are intractable environmental contaminants that are produced from the burning of fossil fuels and wood, coal mining, crude oil extracting, incomplete combustion of petroleum (Verdin et al. 2004). PAHs are different types of organic complexes with merged benzene rings (i.e. acenaphthene, anthracene, 2-methyl naphthalene, phenanthrene, chrysene, pyrene). These abundant toxicants can be divided into three main categories: petrogenic, biological and pyrogenic (Abdel-Shafy and Mansour 2016). Introduction to all of these pollutants can conquer the body's immunity and can result in cancers of different body parts like skin, lungs and stomach. In addition, PAHs can cause allergic reactions, inflammation of different organs, cataracts, lysis of RBC, breathing disorder and can harm kidneys. Upon successful entry to the environment, these may cause biomagnification which results in severe harm to our ecosystem. Henceforth, the safe exclusion of these tenacious, noxious toxicants from the environment is necessary (Abdel-Shafy and Mansour 2016). Lignin-degrading enzymes secreted by several fungal organisms are effective for the elimination of PAHs. The process is further consistent, cheap and biodegradable compared to other unadventurous methods for elimination of PAH from affected places. Some potent ligninolytic fungi were reported by Akhtar and Amin-ul Mannan (2020) to degrade both high as well as low molecular weight PAH by the action of several extracellular enzymes viz. laccase, lignin peroxidase and manganese peroxidase (Pozdnyakova 2012). Apart from these enzymes, cytochrome P450 monooxygenase also helps in PAH metabolism (Bhattacharya et al. 2013). *Dentipellis* sp. KUC8613 primarily employed cytochrome P450 monooxygenase followed by dehydrogenases, dioxygenase, glutathione transferase, epoxide hydrolases, FAD-dependent monooxygenases (Park et al. 2019). Some ligninolytic fungi were able to degrade PAH and release CO_2. But some fungi partially decompose PAH into diphenic acid, anthraquinone, phthalic acid which is then disintegrated by other soil-dwelling bacteria (Brodkorb and Legge 1992; Pozdnyakova 2012).

The phase-wise degradation of PAH by a white-rot fungus, *Phanerochaete chrysosporium* was well studied by (Bhattacharya et al. 2013). In the first phase, two genes viz., pah4 and pc2 under adequate nutrient conditions were upregulated. These two genes were coding for the enzyme cytochrome P-450 monooxygenase. In the second phase, under nutrient limiting conditions, ligninolytic enzymes were produced. The synergistic effect of these two enzymes leads to the effective degradation of PAHs. This theory effectively suggested a formulation for bioremediation of PAHs. Owing to the hydrophobicity of PAHs, some fungi overwhelmed the bioavailability by transporting it to other soil bacteria that sharing the same niche (Czaplicki et al. 2018; Schamfuß et al. 2013). *Punctularia strigosozonata* overexpressed small hydrophobin molecules during the deprivation of PAH in fuel (Young et al. 2015). The production of natural emulsifiers by some fungi in the presence of PAHs increased its bioavailability (Nikiforova et al. 2009). These emulsifying agents supported the solubility of PAH. Moreover, the biodegradation can be increased in slight addition of some carbohydrates like chitin and cellulose (Czaplicki et al. 2018). The addition of copper sulphate, coumaric acid, citric acid, glycerol, humic acid, ferulic acid, polyethylene glycol increased the degradation of PAH by supporting the expression of ligninolytic enzymes (Akhtara and Amin-ul Mannan 2020). Several biotic and abiotic environmental factors like salinity, cold temperature, presence of heavy metals in the soil restrict the degradation process. Under cold temperatures, the activity of some enzymes was suppressed and it increased the viscosity of pollutants like PAHs. *Trametes versicolor* is able to disintegrate PAHs but could not survive in cold temperatures. Contrarily, *Psathyrella* sp. can withstand in such environments (Robichaud et al. 2019). On the other hand, heavy metals slow down the biodegradation of PAH in ligninolytic fungi. It affects the structural rigidity and functional properties of the cell membranes, thereby hampers the metabolism, energy production, growth and development of fungi (Wu et al. 2016). However, some species of *Pleurotus* were reported to degrade PAH in heavy metal contaminated sites (Wu et al. 2016) due to the functioning of antioxidant enzymes (Vaseem et al. 2017). *Cochliobolus lunatus*, a marine fungus, was able to tolerate high concentrations of salt, varying temperatures and pH. It can be used for remediation of PAH in marine environments (Akhtara and Amin-ul Mannan 2020). A novel approach for mycoremediation of PAH was studied in existence of a plant, *Zea mays*. It facilitated in activation of the enzyme manganese peroxidase in fungi (Košnár et al. 2019).

The artificial compounds obtained through the chlorination reaction of biphenyls molecules are referred to PCBs. They are composed of a biphenyl molecule (i.e. two benzene rings connected by a C–C bond) that brings 1–10 chlorine atoms together. PCBs are one of the persistent organic pollutants (POPs) with high environmental toxicity (Lallas 2001). After get released into the environment they could enter inside the food chain and incorporated in the human's blood, breast milk and other tissues by ingesting fish, meat and dairy foodstuffs (Van den Berg et al. 2006). PCBs have several applications in industrial sectors, as fluids for heat transfer, dielectric, organic diluents, hydraulic, solvent extenders, flame-retardants, etc. Presently, PCBs are regarded as one of the most harmful pollutants in the world (Ross 2004). They accumulated in lipid and adipose tissues of animals including humans, soil organic matter (Danielovic et al. 2014). PCBs dysregulate the human immune system through immune suppression and/or stimulation and inflammation (Fisher and Fisher 2004). Bioremediation of PCBs by fungi particularly wood-decaying basidiomycetes were well studied (Stella et al. 2015). Ligninolytic strains of *P. chrysosporium* can mineralize PCB congeners and Aroclor 1254 (Chun et al. 2019). This fungus was also able to degrade higher concentrations (>10 ppm) of Aroclor 1242, 1254 and 1260 (Chun et al. 2019). The fungal mycelium can easily enter into the polluted substrate medium. Furthermore, the extracellular oxidative enzymes of fungi can scavenge even unusual bioavailable impurities by nonspecific reactions. Apart from *P. chrysosporium*, several other basidiomycetous fungi viz. *P. magnoliae, Lentinus edodes, T. versicolor, Irpex lacteus, Phlebia brevispora, Pleurotus ostreatus* could successfully removes PCBs from contaminated sites. *P. ostreatus* applied in the concentration of >2500 ppm in PCBs contaminated sites. All strains of *P. ostreatus* decomposed PCBs selectively at ortho > meta > para positions of the chlorine atoms.

Three potent strains of fungi causing white-rot, viz. *Bjerkandera adusta*, *T. versicolor*, *P. ostreatus* were reported to degrade more PCBs compared to *P. chrysosporium* (Chun et al. 2019). The spent mushroom substrate (SMS), of *P. ostreatus* produced lignocellulosic substances, which proved to be an efficient implement during bioremediation of PCBs (Moeder et al. 2005). The extracts from some white-rot fungi and their extracted enzyme laccases catalyze the effective degradation of hydroxylated PCBs as well as its congeners like Delors and Arochlors (Garcia-Delgado et al. 2015). Besides these, lignosulfonate, a persuader of lignolytic activity of *P. ostreatus* and *T. versicolor* was found to interrupt the degradation of PCBs (Gasecka et al. 2015).

7.3.2 Potentially Toxic Metals and Metalloids

Uptake of potentially toxic elements (PTE) like Cd, Cu, Cr, As and Pb causes immunity deficiency, psychological disorder, malnutrition, gastrointestinal cancer in humans. Furthermore, occurrence of cancer associated with undue intake of some heavy metals viz., Pb, Cu and Cd. Moreover, PTEs are found in different body parts like nails, hair, bone and then excreted through faces (Li et al. 2018). Mushrooms can accumulate high concentrations of HMs in their fruit bodies above the permitted range (Kalac and Svoboda 2000) thereby they can perform as an effective biosorption means (Das 2005). Due to their high metal accumulation capability and short duration of life, they are used as biosorbents. Different mushrooms belonging to the genera including *Agaricus*, *Armillaria*, *Boletus*, *Pleurotus*, *Polyporus*, *Russula*, *Termitomyces* have been studied by some researchers for the uptake of high concentrations of HMs (Raj et al. 2011). Some selected species of *Pleurotus* growing close vicinity polluted environments have greater efficiency to accumulate HMs in their fruit bodies. Barcan et al. (1998) reported that mushrooms can withstand and absorb more than 1540 times of nickel (Ni). *P. ostreatus* grown in metal scrap sites were able to accumulate Cu, Fe, Mn and Zn (Boamponsem et al. 2013). The metal accumulation potential of different species varies depending on the growth substrates of the ecosystems. Brunnert and Zadražil (1983) studied that the fruit bodies of *P. ostreatus* accumulate greater amount of Hg compared to Cd. On the other hand, *P. fabellatus* accumulate higher amount of Cd than Hg. *P. sajor-caju* can able to uptake highest amount of Cd and Cu in comparison to Co and Hg (Purkayastha et al. 1994). *P. pulmonarius* decreased the concentration of Cu, Ni and Mn in cement polluted soil whereas Pb in battery-contaminated soil (Adenipekun et al. 2011). The metals taken up by the mushrooms are distributed disproportionately throughout the fruiting body. The maximum amount of metals accumulates in the gills followed by pileus or cap and stipe. Different species of *Pleurotus* have been found to resist high concentrations of Cd (Kapahi and Sachdeva 2017). The metal uptake by the mushrooms leads to immobilization but its ingestion by animals results in biomagnification of higher trophic levels. So that, sometimes the market available fruiting bodies of *P. ostreatus* have been found to be risky for consumers (Quarcoo and Adotey 2013).

The biosorption efficacy of *P. florida* to absorb Cd was greater than Cr (Adhikari et al. 2004). Some mycelial fungi viz. *Aspergillus awamori*, *Penicillium* spp., *P. ostreatus* have amino, carboxylic, thiol, hydroxide, phosphate groups in their cell wall. These functional groups were involved in the biosorption of HMs contaminated sites (Javaid et al. 2011). The SMS of *Pleurotus* spp. has been exploited to remove Mn(II) from aqueous environment. The live mycelia of *Pleurotus* have been capable of removal of heavy metals from chemical-contaminated laboratory waste conditions (Arbanah et al. 2013). Pre-treatment of the fungal biomass with heat, acids or alkalies has a significant effect on heavy metal biosorption process. Das et al. (2007) reported that, enhancement of Cd biosorption after pre-treatment of living biomass of *P. florida* by physical and chemical methods.

7.3.3 Agrochemical and Pharmaceutical Refuges

Artificial and semi-synthetic agrochemicals, pharmaceutical drugs are notorious agents to contaminate the ecosystems. Sometimes they enter into via drinking water after dilution of cocktail of various drugs in unspecified concentrations. The perseverance of organic xenobiotic compounds in the atmosphere is a serious issue related to social and scientific community, because their probable mutagenic, carcinogenic and genotoxic potential. Thereby these refuges affect the ecosystem in a negative manner. The drugs with determined pharmacokinetic activities and extended tenancy in the environment for long time period. They are subjected to bioaccumulation/biomagnification in the food chain; thereby directly or indirectly affect the non-targeted organisms including animals and human beings.

Persistence of pesticides in the environment is caused by their complex structural properties or lack of biodegradation by microorganisms. Light and temperature could help in the physical degradation process whilst the microorganisms help in biodegradation process. Some fungal enzymes viz., dioxygenases, peroxidases and oxidases are capable of efficient biodegradation of pesticides compared to cytochrome P450. Some ligninolytic fungi viz., *Ganoderma australe*, *P. chrysosporium*, *P. ostreatus* and one saprobic fungus, *Fusarium ventricosum* produce laccase, Lignin peroxidase and dichlorohydroquinone dioxygenase. These enzymes have good biotransformation activity during pesticide degradation (Velázquez-Fernández et al. 2012). *P. chrysosporium* and *F. ventricosum* can degrade endosulfan by an intracellular peroxidase (Velázquez-Fernández et al. 2012). Fungal enzymes like dioxygenases and peroxidases are involved to degrade pentachlorophenol. A ligninolytic fungus, *G. australe*, isolated from gymnosperm, *Pinus pinea*, can degrade lindane (Rigas et al. 2007). Therefore, persistence of antimicrobial agents in the environmental sites causes the progression of multidrug resilient microbial strains. They can affect the animals and human being indirectly, resulting improved fatality. Some brown rot fungi, e.g. *Gloeophyllum striatum* can able to degrade fluoroquinolone enrofloxacin (Wetzstein et al. 1997). A white-rot fungus *Cyathus stercoreus* showed high lignocellulose degradation capability in vitro and has the ability to biodegrade enrofloxacin

(Wicklow et al. 1980). Pandey and Gundevia (2008) reported that *Periconiella* sp., a potent isolate from cow dung, was an excellent biodegrader of pharmaceutical and biomedical waste. It effectively degrades biomedical wastes within 50 days of incubation in vitro. *Aspergillus* spp., *Mucor* spp., *Rhizopus stolonifer, Rhizopus* sp. Can degrade hospital waste when mixed with cow dung slurry (Geetha and Fulekar 2008). Bioremediation of glyphosate, a foreign compound, is efficiently degraded by *Aspergillus flavus, Penicillium spiculisporus, P. verruculosum* which were previously isolated from herbicide-contaminated soil of agricultural farms (Eman et al. 2013).

7.3.4 Detergents, Dyes and Phthalates

The contamination of several types of detergents in our surrounding environment is a stern issue. However, some detergents are non-toxic and very easy to degrade, but they can damage the diversity of aquatic organisms. There are many physical procedures for treatment of detergents polluted wastewater, but in comparison, biological remediation strategies like mycoremediation can be a useful and lucrative process. Several ascomycetous mycelial fungi corresponding *Geotrichum candidum, Cladosporium cladosporioides and Penicillium verrucosum* were reported to destroy several commercially available detergents (Akhtarand Amin-ul Mannan 2020). Myco-bioremediation and differential degradation of different commercially available detergents by some limno fungi viz, *Acremonium strictum, Fusarium oxysporum, Mucor luteus, Cryptococcus neoformans, Aspergillus niger, Penicillium funiculum* was studied by Bharathkumari and Sivakami (2018). *F. oxysporum* can degrade detergents and its chemical constituents like sodium tripolyphosphate, ethoxylated oleyl-cetyl alcohol during its exponential growth phase in vitro (Jakovljević et al. 2014; Violeta et al. 2014).

Dye-yielding industries release several toxic dyes into the environment which pollute many natural water bodies. These dyes are physically, chemically and biologically resistant and persist in the ecosystem for long time. Many industries did not purify the effluents and release those into the nearby water bodies and contaminate them directly. Many of the effluents are carcinogenic to animal as well as humans (Ngieng et al. 2013). Fungal ligninolytic enzymes, laccase, peroxidase are acting on these dyes and degrade them (Yang et al. 2017). One potent soil isolate, *A. flavus* near a paper-making industry, was reported to degrade congo red dye (Bhattacharya et al. 2011a, b). *T. versicolor* can degrade several types of azo anthraquinone dyes (Yang et al. 2017). *Phlebia acerina* cleans up waste waterbodies containing harmful dyes and thereby decreased the toxicity (Kumar et al. 2018). Apart from that, an endophytic fungus *Marasmius cladophyllus* of *Melastoma malabathricum* detoxified various synthetic dyes (Ngieng et al. 2013). The high level and action of several lignin-degrading enzymes of fungi help in biodegradation of noxious dyes (Yang et al. 2017). So, it can be said that some endophytic and ligninolytic fungi can be applied as natural mode of treatment of industrial runoffs and lands contaminated with synthetic dyes.

Phthalates are employed in the manufacture of polyvinyl toys, tiles, films, capacitors and medical instruments. They provide elasticity to plastics. Though, their abandoned availability into the atmosphere provides opportunity to enter the food chain. Phthalates are potent carcinogens and sometimes diminish male fertility. They can lower down the level of different hormones like testosterone, thyroid hormone (Akhtar and Amin-ul Mannan 2020). There are several physical processes that can degrade phthalates, but these are lengthy methods (Kluwe et al. 1982). Mycoremediation help in the effective and quick degradation of phthalates. Some fungal derive enzymes like cutinase, esterase assists the mycoremediation process. Cutinase-mediated degradation are quick compared to estarses. Moreover, it does not produce any harmful intermediates. Cutinase of *F. oxysporum* can degrade various phthalates containing compounds viz, dipentyl-, di-hexyl-, dipropyl-, di-2-ethylhexyl-, butyl benzyl phthalate. *Purpureocillium lilacinum, Aspergillus parasiticus, A. japonicas, Penicillium brocae, Fusarium subglutinans* and *P. funiculosum* are some examples of phthalates degrading fungi (Pradeep and Benjamin 2012; Pradeep et al. 2013).

7.3.5 Petroleum

Several fungal species have been employed for bioremediation of lands contaminated with petroleum. These include mycelial fungi, yeast and mushrooms. *Penicillium* sp. and *Aspergillus* sp. are potent candidates for petroleum degradation (Dickson et al. 2019). Dilapidation of crude oil by a strain of *S. cerevisiae* isolated from Zobo, a fermented food was reported by Abioye et al. (2013). *Rhodotorula* sp., *Candida* sp. and *Torulopsis* sp. are some other examples of yeast involved in petroleum biodegradation (Dickson et al. 2019). Different fungal species and their associations isolated from cow dung manure viz. *Alternaria, Mucor, Aspergillus, Rhizopus, Cephalosporium, Thamnidum, Cladosporium, Geotrichum, Sporotrichum, Monilia, Penicillium* are used for mycoremediation of land contaminated with petroleum (Obire et al. 2008). Fungal peroxidases, laccases and lignin-degrading enzymes are responsible for the remediation of petroleum. Green tea residues, spent mushroom substrates can be used by the fungus for remediation of petrol polluted lands (Dickson et al. 2019).

7.4 Mechanism of Action: Hypotheses and Evidence

For effective bioremediation of the pollutants, fungus must enzymatically degrade it and transform them to non-toxic forms (Fig. 7.2). As there can be a variation in the kind of hazardous wastes that can be existing at a polluted area, so for effective remediation various types of microorganisms may be required. Careful selection needed to be done for employing organisms to the contaminated site for the process

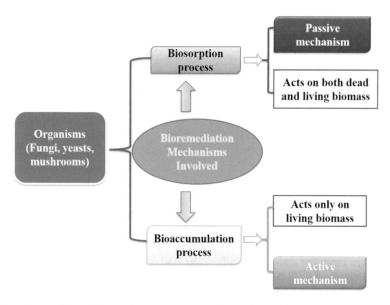

Fig. 7.2 A general flow chart on the mechanism involved in fungal bioremediation process

is totally reliant on the chemical composition of the contaminating substances and that varies accordingly the microbes involved. The efficiency is totally dependent on the mechanism process of how the degradation takes place with help of the particular microorganism. In a process, the microbes use molecular oxygen that reacts with the hydrocarbon and then it gives rise to midway products that eventually enter the overall energy resilient metabolic path of the particular cell. In some bacteria, chemotactic response is exhibited to look for the pollutant and advances towards it as they utilize the oil as their food. Some of the microbes have the ability to emulsify oil in water and thereby facilitate its successful removal by generating strong surface-active mixture.

Biotechnological approach for treatment of HMs can be an innovative approach where it can help to solve the problem of biodegradation associated with heavy metal contamination with maximum efficiency. In this regard, a more realistic address to regulate the activity of the microbes and the different metabolic pathways that obstruct the harmful action of the metals is considered. A process used sometimes during the bioremediation process where redesigned microbes can amend the hazardous contaminants to a very simple organic form by rising their solubility with the help of oxidation–reduction reaction. Apart from that microorganisms are been redesigned to escalate their resistance with other factors such as changes in pH and complexation reactions. Thus, this results in increased the solubility of the HMs and changes their inertness in the environment. The most benefit of fungi is that they are highly variable in terms of size and shape from mushrooms to microscopic moulds, thus they can accumulate metals on their cell surface. Fungi and yeast use the adsorption process where they can interact with the metal ions. The metal ions finally react with the proteins essential for the correct effectiveness of the cell.

7.4.1 Soil Systems

Fungi are ubiquitous eukaryotic organisms ranging from unicellular to multicellular bodies that freely advanced organisms (Gadd 2008, 2017; Stajich et al. 2010). Fungal colonization in soil results in the mix-up and assembles of soil particles and ultimately refinement of soil structure and thus also easing out the pollutant rate of utilization (Harms et al. 2011) Application of fungi is more advantageous in comparison to bacteria in the translocation of water, different nutrients and pollutants in soil systems (Boswell et al. 2003; Worrich et al. 2018). Fungal mycelia play a larger role in facilitating the transport of different substances responsible for degrading of pollutants, like some of the bacteria are transported over long distances in soil by this fungus which can facilitate the enhancement of bioremediation (Banitz et al. 2013; Kohlmeier et al. 2005; Wick et al. 2007), whereas some of the fungi can also remain and continue to hold out in the presence of various hazardous metals. This may be due to their genetic adaptability along with various other physiological and intrinsic biochemical properties. Environment also plays a major role in enhancing the bioremediation process by modifying the metal speciation also making the contaminants more bioavailable and less toxic (Gadd 1993, 2010; Glasauer et al. 2004; Sullivan and Gadd 2019). Most of the mycelial fungi like the *Penicilium* spp. and *Aspergillus* have been reported to have properties such as degradation of aliphatic hydrocarbons and various other PAH, phenols and chlorophenols by utilizing the carbon and energy sources from the toxic pollutants and remediating to less toxic products (Harms et al. 2011; Hofrichter et al. 1994; Pinedo-Rivilla et al. 2009). Reports on ureolytic fungi having the ability to precipitate the toxic metals as oxides or as metal carbonates cannot be ignored. They are found to immobilize metals basically the fungus *Neurospora crassa* has been utilized for this purpose. They are incubated in urea-supplemented media and this results in precipitation of the toxic metals (Li et al. 2014, 2015, 2016, 2019; Li and Gadd 2017a, b). In this process, the urea that is supplemented has heavy oil and Ca^{2+} in it. The mineral precipitation aggregates along the edges of heavy oil thus providing an additional energy source in the process of biomineralization (Haritash and Kaushik 2009; Vanholme et al. 2010). Due to which, many ligninolytic fungi, viz., *P. chrysosporium*, have been studied for deprivation of PAHs and other aromatic composite compounds because of the characteristic wide variety of elements that can be degraded by those organisms (Gadd 2001, 2004). A report published by Hong et al. (2010) is of a fungus *Fusarium solani* that was found in the soil of petroleum station. It was experimentally proved that; *Fusarium* spp. was able to accumulate more than 60% of metals (zinc and copper) in their body and able to degrade pyrene. Some of the fungi use non-detoxification process where high molecular mass PAHs are converted into less toxic water-soluble products viz. bezo α pyrene are converted to less carcinogenic and water-miscible products (Akhtar and Amin-ul Mannan 2020). The growth of certain soil fungi is highly limited in soil that contains both cadmium and phenanthrene compared to the soil with only one such as for example if only cadmium is present, it has a less drastic effect on growth of the fungus (Shen et al. 2005). The fungal cell membrane which are fat-soluble in nature link with the PAHs and this leads to permeability changes in the membrane which ultimately leads to penetration of noxious metals due to which the normal cellular functions are hampered. Fungal siderophores play a very crucial role in co-contaminated soil where there is limitation of Ferrous. They help in binding metals other than Fe(III) like zinc, copper, nickel, cadmium and lead and further satisfy the Fe requirement thus facilitating the biodegradation of different PAHs by the fungus (Ahmed and Holmström 2014).

7.4.2 Water Bodies

For the treatment of water bodies fungi usually the filamentous fungi have been widely used. For the treatment of waste streams, fungi use an enzyme-mediated process that delivers resolution to remediate toxic pollutants from the streams. Traditional biological treatment which uses bacteria for the treatment of wastewater generates a large amount of low-valued biomasses of bacteria. Thus, the cost incurred for the disposal and treatment together leads to extra amount nearly 40–60%. So, for a change in the cost incurred during wastewater treatment a different form of biomass instead of the bacterial biomass the fungal biomass can provide a greater value and this could suggestively transform the economics of the treatment process. Thus, fungi might be suggested for a much wider assistance with respect to that of bacteria in wastewater treatment processes. Genetic engineering is also used in cases where the normal fungus cannot be used for efficient degradation. It changes the gene function and thus the metabolic function of the fungus is also developed which allows them to accumulate more of the metals and thereby decontamination process of water bodies takes place more efficiently. Different fungal species are used in different kinds of treatment of wastewater systems. For example, *A. niger* is used in the treatment of industrial wastes such as the apple distillery wastes. From the report of National Collection of Industrial Microorganism, 2005, it was shown that fungus such as *Myrothecium verrucaria* and *Trametes hirsuta* is found to degrade cellulosic waste from different sources. *P. chrysoporium* are accountable for the dilapidation of lignin. Fungus also provides the dietary supplement for humans, as well as animals. They are known for their capability to derive valuable products that are biochemical in nature. Sometimes, fungi are cultured in industrial sector for production of variety of beneficial ingredients such as amino acids, enzymes, dyes, organic acids, organic alcohols and others (van Leeuwen et al. 2003). In this process, the fungus is grown in aseptic conditions using expensive chemical species for the commercial cultivation of various types of biochemicals used in pharmaceutical industry (Guest and Smith 2002; Stevens and Gregory 1987; Zheng et al. 2005). In addition to that during fungal treatment the fungus also helps in the process of separation of fungal biomass this helps in food supplements of different animals. Investigation carried out on the possibility of wastewater purification employing yeasts and moulds for the production of microbial biomass proteins (MBP). This led to the findings that yeasts are a perfect selection during the bioconversion process, as they have different matched qualities like they could be cultivated with ease and the rate of their outgrowth is quite faster compared to the other types of moulds and also their ability to grow in very low pH values (at pH 5 or less) (Jin et al. 1998, 1999; Zheng et al. 2005: Bergmann et al. 1988; Gonzalez et al. 1992). Furthermore, yeasts are quite less vulnerable to infection by some other microbes and yield healthy biomass (Satyawali and Balakrishnan 2007). Yeasts were considered more suitable than moulds for the production of MBP. However, the filament like nature of those fungi simplifies retrieval of the MBP from growth media. Thus, mycelial fungi could have stimulating effects for the treatment of industrial surplus water.

Fungi, in all phases of their life cycle, keep on producing the required enzymes for degradation and are exist even at little concentration of contaminant (Ryan et al. 2005). Biomass of fungi releases definite and non-definite extracellular enzymatic substances that have concerned the consideration by the scientists employed on deprivation of multifaceted high-molecular-weight organic products. For example, white-rot fungus secretes some highly oxidative extracellular enzymes which catalyze PAH deprivation through general oxidation reactions foremost to the development of variations of quinones and other aromatic products that are proficient of humiliating several natural and synthetic xenobiotic compounds (Giraud et al. 2001; Boyle et al. 1992; Elisa et al. 1991; Yesilada et al. 1999).

7.4.3 Contaminated Products

Contamination of products with different chemicals used in pesticides and dyes is one of the major concerns with respect to protection of the environment. They are responsible for the inhibition of electrons at the receptor site of PSII, thus this regulates the weeds in several crops. Photosystem II (PSII) is composed of multiple-subunit of pigment-protein composites. It is embedded in the chloroplast of thylakoid membranes of plants, cyanobacteria and algae. It catalyzes the water-splitting reaction, thereby produces molecular oxygen. Herbicides that inhibiting PSII are molecules that hinder photosynthetic electron transport by inhibiting D1 protein (Kyle 1985). Some examples of PSII inhibitors are triazines, pyridazinones, amides, phenyl-carbamates, uracils, benzothiadiazinones, nitriles, ureas, etc. (Forouzesh et al. 2015). These herbicides bind to D1 proteins of photosystem II and thereby inhibit photosynthesis by blocking electron transport system and promotes the formation of reactive molecules. That results in leakage of membrane allowing cell organelles to dehydration and rapid disintegration. They stop CO_2 fixation and energy generation required for plant growth (Battaglino et al. 2021). Susceptible plants exhibit interveinal or veinal chlorosis, necrosis depending on the type of herbicide application. Pesticides contain both inorganic and organic groups. They are a type of diverse group of chemicals (Verma et al. 2014). Pesticides start to accumulate in different products due to frequent and intense use and possibly transfer into the food chain. In fungi, different biochemical reactions like co-metabolism are responsible for the pesticides alteration. Ester hydrolysis, hydroxylation, alkylation or dealkylation, oxidation, dehalogenation, dehydrogenation, ether cleavage, reduction, ring cleavage condensation and conjugate origination are some biochemical responses in the fungal pesticides degradation (Bollag 1974). In the process of fungal degradation of pesticides, at first, the association of fungi with the pesticides involves immobilization and mobilization in the fungal environment, which leads to sorption to cell walls and assimilation of pollutants into the cell of fungi. Subsequently, alteration of chemicals along with deviation and soaking up takes place laterally with the reactions of pollutants with the fungal enzymes viz, extracellular oxidoreductases. These cell-bound enzymes permit fungi to take measures on several contaminated products (Harms et al. 2011).

Contaminated products can also be due to dyes or dyestuff used in various substances as colouring agents. They are basically used, to set forth colour permanently to different substances like cloth, paper, leather, etc. Worldwide, sometimes during the dyeing process due to some technical problems, it is evaluated that some amount of (5–10%) dyes are lost in the run-off during that course. Production of these dyes is also started with azulene synthesis which is very much harmful as they can contaminate various substances in the environment and that ultimately go into the food chain harming the overall health of the population. The scenario in India is much more complex as the dyeing facility here alone creates 32,000 m^3/day of wastewater. During textile processing wastewater generated is 450,000 m^3/day and along with them, the dyes are released which is more difficult for natural degradation as of their artificial derivation and composite aromatic structure (Liu and Tay 2004). Dyes are the most difficult component of wastewater contaminated from textile industries. Azo dyes, are among the most widely used and considered as non-biodegradable, harmful, persistent. Little is known about the consequence of textile dye on the PSII of plants. PSII is reported to play an vital role in higher plants to environmental stress (Baker 1991). The kinetics of the fluorescence of chlorophyll a is sensitive to stress and is hindered by the dyes. Contaminated dyes in aquatic environment decrease the permeability of light and negatively distress photosynthetic activity (Çiçek et al. 2012).

The degradation of various toxic compounds are solely dependent on the pH, moisture content of the compounds, also the fungal biomass, soil types and organic matter content. The data in case of fungal degradation of pesticides compared to that of bacteria are very less (Kullman and Matsumura 1996). However, there are some reports on some fungi like *Mucor alternans*, *F. oxysporum* and *Trichoderma viride* are able to degrade even DDT an organo-chlorine chemical insecticide (Anderson and Lichtenstein 1971; Engst and Kujawa 1968; Matsumura and Boush 1968).

7.5 Conclusion and Recommendations

Rejuvenation of our living atmosphere is a job of supreme status and utilizing the physicochemical methods solitarily convert the toxic contaminants from one form to another form then biotic methods convert them into non-toxic, environmentally safe byproducts. Such safety apprehensions are the prerequisite for the establishment as well as exploitation of practical and affordable practices for remedy for in vivo application. The bioremediation approach should be made on the type and toxicity of pollutants. Myco-bioremediation is one such technique that proposed the opportunity to abolish various toxicants into harmless residue using fungal associations and their metabolites. For successful myco-bioremediation, the advantages or disadvantages of the fungal organism is the only decisive factor on bioconversion proficiency. The adverse effects of the artificial environmental contaminants on the diversity of natural flora and fauna can be diminished by effective myco-bioremediation techniques for a dynamic, vigorous and innocuous future. Convincingly, myco-bioremediation has great potential to fight against the present pollution, abiotic stresses in terrestrial as well as aquatic ecosystems in an ecofriendly and sustainable way.

Acknowledgements Authors thankfully acknowledged Scientific and Engineering Research Board (SERB), National Post-Doctoral Fellowship, India (file no. PDF/2017/002639) for financial assistance.

References

Abdel-Shafy HI, Mansour M (2016) A review on polycyclic aromatic hydrocarbons: source, environmental impact, effect on human health and remediation. Egypt J Pet 25:107–123. https://doi.org/10.1016/j.ejpe.2015.03.011

Abioye OP, Akinsola RO, Aransiola SA, Damisa D, Auta SH (2013) Biodegradation of crude oil by *Saccharomyces cerevisiae* isolated from fermented Zobo (locally fermented beverage in Nigeria). Pak J Biol Sci 16:2058–2061

Adenipekun CO, Ogunjobi AA, Ogunseye AO (2011) Management of polluted soils by a white-rot fungus: *Pleurotus pulmonarius*. Assumption Univ Technol J 15(1):57–61

Adhikari T, Manna MC, Singh MV, Wanjari RH (2004) Bioremediation measure to minimize heavy metals accumulation in soils and crops irrigated with city effluent. J Food Agric Environ 2(1):266–270

Ahmed E, Holmström SJM (2014) Siderophores in environmental re-search: roles and applications. Microb Biotechnol 7(3):196–208. https://doi.org/10.1111/1751-7915.12117

Akhtar N, Amin-ul Mannan M (2020) Mycoremediation: expunging environmental pollutants. Biotechnol Rep 26:e00452. https://doi.org/10.1016/j.btre.2020.e00452

Alexander M (1999) Biodegradation and bioremediation, 2nd edn. Academic Press, Canada, p 453

Anderson JPE, Lichtenstein EP (1971) Effects of various soil fungi and insecticides on the capacity of *Mucor alternans* to degrade DDT. Can J Microbiol 18:553–560. https://doi.org/10.1139/m72-088

Antizar-Ladislao B, Spanova K, Beck AJ, Russel NJ (2008) Microbial community structure changes during bioremediation of PAHs in an aged coal tar contaminated soil by in vessel composting. Int J Biodeterior Biodegrad 61:357–364. https://doi.org/10.1016/j.ibiod.2007.10.002

Aragão MS, Menezes DB, Ramos LC (2020) Mycoremediation of vinasse by surface response methodology and preliminary studies in air-lift bioreactors. Chemosphere 244:125432. https://doi.org/10.1016/j.chemosphere.2019.125432

Arbanah M, Miradatul Najwa MR, Ku Halim KH (2013) Utilization of Pleurotusostreatus in the removal of Cr (VI) from chemical laboratory waste. Int Refreed J Eng Sci 2(4):29–39

Baker NR (1991) A possible role for photosystem II in environmental perturbations of photosynthesis. Physiol Plant 81:563–570

Bamforth SM, Singleton I (2005) Bioremediation of polycyclic aromatic hydrocarbons: current knowledge and future directions. J Chem Technol Biotechno: Int Res Process Environ Clean Technol 80(7):723736

Banitz T, Johst K, Wick LY, Schamfuß S, Harms H, Frank K (2013) Highways versus pipelines: contributions of two fungal transport mechanisms to efficient bioremediation. Env Microbiol Rep 5(2):211–218. https://doi.org/10.1111/1758-2229.12002

Barcan VS, Kovnatsky EF, Smetannikova MS (1998) Absorption of heavy metals in wild berries and edible mushrooms in an area affected by smelter emissions. Water Air Soil Pollut 103:173–195. https://doi.org/10.1023/A:1004972632578

Battaglino B, Grinzato A, Pagliano C (2021) Binding properties of photosynthetic herbicides with the QB site of the D1 protein in plant photosystem II: a combined functional and molecular docking study. Plants 10:1501

Bergmann FW, Abe J-I, Hizukuri S (1988) Selection of microorganisms which produce raw-starch degrading enzymes. Appl Microbiol Biotechnol 27:443–446. https://doi.org/10.1007/BF00451610

Bharathkumari K, Sivakami R (2018) Bioremediation of detergents using limnofungi. Int J Res 5(4):728–730

Bhattacharya S, Das A, Mangai G, Vignesh K, Sangeetha J (2011a) Mycoremediation of congo red dye by filamentous fungi. Braz J Microbiol 42:1526. https://doi.org/10.1590/S1517-838220110004000040

Bhattacharya S, Das A, Mangai G, Vignesh K, Sangeetha J (2011b) Mycoremediation of congo red dye by filamentous fungi. Braz J Microbiol 42(4):1526–1536. https://doi.org/10.1590/S1517-838220110004000040

Bhattacharya S, Prashanthi K, Palaniswamy M, Angayarkanni J, Das A (2012) Mycoremediation of Benzo[a]Pyrene by *Pleurotusostreatus* isolated from Wayanad district in Kerala. India. Int J Pharm Biol Sci 2(2):8493. https://doi.org/10.1007/s13205-013-0148-y

Bhattacharya SS, Syed K, Shann J, Yadav JS (2013) A novel P450-initiated biphasic process for sustainable biodegradation of benzo[a]pyrene in soil under nutrient-sufficient conditions by the white rot fungus *Phanerochaete chrysosporium*. J Hazard Mater 261:675–683. https://doi.org/10.1016/j.jhazmat.2013.07.055

Boamponsem GA, Obeng AK, Osei-Kwateng M, Badu AO (2013) Accumulation of heavy metals by *Pleurotus ostreatus* from soils of metal scrap sites. Int J Curr Res Rev 5(4):01–09

Bollag JM (1974) Microbial transformation of pesticides. Adv Appl Microbiol 18:75–130. https://doi.org/10.1016/s0065-2164(08)70570-7

Bollag JM (1992) Enzymes catalysing oxidative coupling reactions of pollutants. In: Sigel H, Sigel A (eds) Metal ions in biological systems. Marcel Dekker Inc., NY, pp 206–217

Boswell GP, Jacobs H, Davidson FA, Gadd GM, Ritz K (2003) Growth and function of fungal mycelia in heterogeneous environments. Bull Math Biol 65(3):447–477. https://doi.org/10.1016/S0092-8240(03)00003-X

Boyle CD, Kropp BR, Reid ID (1992) Solubilization and mineralization of lignin by white-rot fungi. Appl Environ Microbiol 58:3217–3224

Brodkorb TS, Legge RL (1992) Enhanced biodegradation of phenanthrene in oil tar-contaminated soils supplemented with *Phanerochaete chrysosporium*. Appl Environ Microbiol 58:3117–3121

Brunnert H, Zadražil F (1983) The translocation of mercury and cadmium into the fruiting bodies of six higher fungi. A comparative study on species specificity in five lignocellulolytic fungi and the cultivated mushroom *Agaricus bisporus*. Eur J Appl Micorbiol Biotechnol 17:358–364. https://doi.org/10.1007/BF00499504

Ceci A, Pinzari F, Russo F, Persiani MA, Gadd GM (2019) Roles of saprotrophic fungi in biodegradation or transformation of organic and inorganic pollutants in co-contaminated sites. Appl Microbiol Biotechnol 103(1):53–68. https://doi.org/10.1007/s00253-018-9451-1

Chun SC, Muthu M, Hasan N, Tasneen S, Gopal J (2019) Mycoremediation of PCBs by *Pleurotus ostreatus*: possibilities and prospects. Appl Sci 9:4185. https://doi.org/10.3390/app9194185

Çiçek N, Efeoğlu B, Tanyolaç F, Ekmekçi Y, Strasser RJ (2012) Growth and photochemical responses of three crop species treated with textile azo dyes. Turk J Bot 36:529–537

Czaplicki LM, Dharia M, Cooper EM, Fergusson PL, Gunsch C (2018) Evaluating the mycostimulation potential of select carbon amendments for the degradation of a model PAH by an ascomycete strain enriched from a superfund site. Biodegradation 29:463–471. https://doi.org/10.1007/s10532-018-9843-z

D'Annibale A, Rosetto F, Leonardi V, Federici F, Petruccioli M (2006) Role of autochthonous filamentous fungi in bioremediation of a soil historically contaminated with aromatic hydrocarbons. Appl Environ Microbiol 72(1):2836. https://doi.org/10.1128/AEM.72.1.28-36.2006

Danielovic I, Hecl J, Danilovic M (2014) Soil contamination by PCBs on a regional scale: the case of Strazske, Slovakia. Pol J Environ Stud 23:1547–1554

Das N (2005) Heavy metals biosorption by mushrooms. Indian J Natl Prod Resour 4:454–459

Das N, Chandran P (2011) Microbial degradation of petroleum hydrocarbon contaminants: an overview. Biotechnol Res Int 2011:1–13. https://doi.org/10.4061/2011/941810

Das N, Charumathi D, Vimala R (2007) Effect of pretreatment on Cd^{2+} biosorption by mycelia biomass of *Pleurotus florida*. Afr J Biotechnol 6:2555–2558. https://doi.org/10.5897/AJB2007.000-2407

Dickson UJ, Michael C, Mortimer RJG, Marcello DB, Nicholas R (2019) Mycoremediation of petroleum contaminated soils: progress, prospects and perspectives. Environl Sci Proc-Imp 21:1446–1458. https://doi.org/10.1039/c9em00101h

Elisa E, Vanderlei PC, Nelson D (1991) Screening of lignin degrading fungi for removal of color from kraft mill wastewater with no additional extra carbon source. Biotechnol Lett 13:571–576. https://doi.org/10.1007/bf01033412

Eman A, Abdel-Megeed A, Suliman A-MA, Sadik MW, Sholkamy EN (2013) Biodegradation of glyphosate by fungal strains isolated from herbicides polluted-soils in Riyadh area. Accessed 19 Nov 2019. http://www.ijcmas.com

Engst R, Kujawa M (1968) EnzymatischerAbbau des DDT durchSchimmelpilze. 3. Mitt. Darstellung des 2,2-Bis (p-chlorophenyl) acetaldehydes (DDHO) und seine BedentungimAbbaucyclus. Nahrung Chem Physiol Technol 12:783–785. https://doi.org/10.1002/food.19680120807

Fisher SG, Fisher RI (2004) The epidemiology of non-Hodgkin's lymphoma. Oncogene 23:6524–6534. https://doi.org/10.1038/sj.onc.1207843

Forouzesh A, Zand E, Soufizadeh S, Samadi Foroushani S (2015) Classification of herbicides according to chemical family for weed resistance management strategies-an update. Weed Res 55:334–358

Gadd GM (1993) Interactions of fungi with toxic metals. New Phytol 124:25–60

Gadd GM (2001) Fungi in bioremediation. Cambridge University Press, Cambridge. https://doi.org/10.1017/CBO9780511541780

Gadd GM (2004) Mycotransformation of organic and inorganic substrates. Mycologist 18(2):60–70. https://doi.org/10.1017/S0269915X04002022

Gadd GM (2008) Fungi and their role in the biosphere. In: Jorgensen SE, Fath B (eds) Encyclopedia of ecology. Elsevier, Amsterdam, pp 1709–1717

Gadd GM (2010) Metals, minerals and microbes: geomicrobiology and bioremediation. Microbiol 156(3):609–643. https://doi.org/10.1099/mic.0.037143-0

Gadd GM (2017) Geomicrobiology of the built environment. Nat Microbiol 2(4):1–9. https://doi.org/10.1038/nmicrobiol.2016.275

Garcia-Delgado C, Yunta F, Eymar E (2015) Bioremediation of multi-polluted soil by spent mushroom (*Agaricus bisporus*) substrate: polycyclic aromatic hydrocarbons degradation and Pb availability. J Hazard Mater 300:281–288. https://doi.org/10.1016/j.jhazmat.2015.07.008

Gasecka M, Drzewiecka K, Siwulski M, Sobieralski K (2015) Evaluation of polychlorinated biphenyl degradation through refuse from *Pleurotus ostreatus*, *Lentinula edodes* and *Agaricus bisporus* production. Folia Hort 27:135–144. https://doi.org/10.1515/fhort-2015-0023

Geetha M, Fulekar MH (2008) Bioremediation of pesticides in surface soil treatment unit using microbial consortia. Afr J Environ Sci Technol 2(2):036–045

Gianfreda L, Rao MA (2004) Potential of extra cellular enzymes in remediation of polluted soils: a review. Enzyme Microb Tech 353:39–354. https://doi.org/10.1016/j.enzmictec.2004.05.006

Giraud F, Guiraud P, Kadri M, Blake G, Steiman R (2001) Biodegradation of anthracene and fluoranthene by fungi isolated from an experimental constructed wetland for wastewater treatment. Water Res 35:4126–4136. https://doi.org/10.1016/s0043-1354(01)00137-3

Glasauer S, Beveridge TJ, Burford EP, Harper FA, Gadd GM (2004) Metals and metalloids, transformations by microorganisms. In: Hillel D, Rosenzweig C, Powlson DS, Scow KM, Singer MJ, Sparks DL, Hatfield J (eds) Encyclopedia of soils in the environment. Elsevier, Amsterdam. pp 438–447. https://doi.org/10.1016/B978-0-12-409548-9.05217-9

Godfrey T, Reichelt J (1996) Introduction to industrial enzymology. Godfrey T, Reichelt J (eds) Industrial enzymology: the application of enzymes in industry, 2nd edn. Nature, New York

Gonzalez MP, Siso MIG, Murado MA, Pastrana L, Montemayor ML, Mirón J (1992) Depuration and valuation of mussel-processing wastes: characterization of amylolytic post incubates from different species grown on an effluent. Bioresour Technol 42:133–140. https://doi.org/10.1016/0960-8524(92)90072-6

Guest RK, Smith DW (2002) A potential new role for fungi in a wastewater MBR biological nitrogen reduction system. J Environ Eng Sci 1:433–437. https://doi.org/10.1139/s02-037

Haritash AK, Kaushik CP (2009) Biodegradation aspects of polycyclic aromatic hydrocarbons (PAHs): a review. J Hazard Mater 169(1):1–15. https://doi.org/10.1016/j.jhazmat.2009.03.137

Harms H, Schlosser D, Wick LY (2011) Untapped potential: exploiting fungi in bioremediation of hazardous chemicals. Nat Rev Microbiol 9(3):177–192. https://doi.org/10.1038/nrmicro2519

Hofrichter M, Bublitz F, Fritsche W (1994) Unspecific degradation of halogenated phenols by the soil fungus *Penicillium frequentans* Bi 7/2. J Basic Microbiol 34(3):163–172. https://doi.org/10.1002/jobm.3620340306

Hong JW, Park JY, Gadd GM (2010) Pyrene degradation and copper and zinc uptake by *Fusarium solani* and *Hypocrea lixii* isolated from petrol station soil. J Appl Microbiol 108(6):2030–2040. https://doi.org/10.1111/j.1365-2672.2009.04613.x

Jakovljević V, Milićević JM, Stojanović JD, Solujić SR, Miroslav V (2014) The influence of detergent and its components on metabolism of *Fusarium oxysporum* in submerged fermentation. Hemijska Industrija 68:465–473. https://doi.org/10.2298/HEMIND130620071J

Javaid A, Bajwa R, Shafque U, Anwar J (2011) Removal of heavy metals by adsorption on *Pleurotus ostreatus*. Biomass Bioenergy 35:1675–1682. https://doi.org/10.1016/j.biombioe.2010.12.035

Jin B, van Leeuwen J, Patel B, Yu Q (1998) Utilization of starch processing wastewater for production of microbial biomass protein and fungal α-amylase by *Aspergillus oryzae*. Bioresour Technol 66:201–206. https://doi.org/10.1016/S0960-8524(98)00060-1

Jin B, van Leeuwen J, Patel B, Doelle HW, Yu Q (1999) Production of fungal protein and glucoamylase by *Rhizopus oligosporus* from starch processing wastewater. Process Biochem 34:59–65. https://doi.org/10.1016/S0032-9592(98)00069-7

Kalac P, Svoboda L (2000) A review of trace element concentrations in edible mushrooms. Food Chem 69:273–281. https://doi.org/10.1016/S0308-8146(99)00264-2

Kapahi M, Sachdeva S (2017) Mycoremediation potential of *Pleurotus* species for heavy metals: a review. Bioresour Bioprocess 4:32. https://doi.org/10.1186/s40643-017-0162

Khan I, Aftab M, Shakir S, Ali M, Qayyum S, Rehman MU, Haleem K, Tauseef I (2019) Mycoremediation of heavy metal (Cd and Cr) polluted soil through indigenous metallotolerant fungal isolates. Environ Monit Assess 191. https://doi.org/10.1007/s10661-019-7769-5

Kluwe WM, McConnell EE, Huff JE, Haseman JK, Douglas JF, Hartwell WV (1982) Carcinogenicity testing of phthalate esters and related compounds by the National Toxicology Program and the National Cancer Institute. Environ Health Perspect 45:129–133. https://doi.org/10.1289/ehp.8245129

Kohlmeier S, Smits TH, Ford RM, Keel C, Harms H, Wick LY (2005) Taking the fungal highway: mobilization of pollutant-degrading bacteria by fungi. Environ Sci Technol 39(12):4640–4646. https://doi.org/10.1021/es047979z

Košnár Z, Cástková T, Wiesnerová L, Praus L, Jablonský I, Koudela M, Tlustoš P (2019) Comparing the removal of polycyclic aromatic hydrocarbons in soil after different bioremediation approaches in relation to the extracellular enzyme activities. J Environ Sci 76:249–258. https://doi.org/10.1016/j.jes.2018.05.007

Kullman SW, Matsumura F (1996) Metabolic pathway utilized by *Phanerochete chrysosporium* for degradation of the cyclodiene pesticide endosulfan. Appl Environ Microbiol 62:593–600. https://doi.org/10.1128/AEM.62.2.593-600.1996

Kumar R, Negi S, Sharma P, Prasher IB, Chaudhary S, Dhau JS, Umar A (2018) Wastewater cleanup using *Phlebia acerina* fungi: an insight into mycoremediation. J Environ Manage 228:130–139. https://doi.org/10.1016/j.jenvman.2018.07.091

Kyle DJ (1985) The 32000 Dalton Qb protein of photosystem II. Photochem Photobiol 41:107–116

Lallas PL (2001) The Stockholm convention on persistent organic pollutants. Am J Int Law 95:692–708. https://doi.org/10.2307/2668517

Li QW, Csetenyi L, Gadd GM (2014) Biomineralization of metal carbonates by *Neurospora crassa*. Environ Sci Technol 48(24):14409–14416. https://doi.org/10.1021/es5042546

Li Q, Csetenyi L, Paton GI, Gadd GM (2015) $CaCO_3$ and $SrCO_3$ bioprecipitation by fungi isolated from calcareous soil. Environ Microbiol 17(8):3082–3097. https://doi.org/10.1111/1462-2920.12954

Li Q, Liu D, Jia Z, Csetenyi L, Gadd GM (2016) Fungal biomineralization of manganese as a novel source of electrochemical materials. Curr Biol 26(7):950–955. https://doi.org/10.1016/j.cub.2016.01.068

Li Q, Gadd GM (2017a) Biosynthesis of copper carbonate nanoparticles by ureolytic fungi. Appl Microbiol Biotechnol 101(19):7397–7407. https://doi.org/10.1007/s00253-017-8451-x

Li Q, Gadd GM (2017b) Fungal nanoscale metal carbonates and production of electrochemical materials. Microb Biotechnol 10(5):1131–1136. https://doi.org/10.1111/1751-7915.12765

Li Y, Wang Z, Qin F, Fang Z, Li X, Li G (2018) Potentially toxic elements and health risk assessment in farmland systems around high-concentrated arsenic coal mining in Xingren, China. J Chem 2018:10. https://doi.org/10.1155/2018/2198176

Li Q, Liu D, Chen C, Shao Z, Wang H, Liu J, Zhang Q, Gadd GM (2019) Experimental and geochemical simulation of nickel carbonate mineral precipitation by carbonate-laden ureolytic fungal culture supernatants. Environ Sci Nano 6(6):1866–1875. https://doi.org/10.1039/C9EN00385A

Liu Y, Tay JH (2004) State of the art biogranulation technology for wastewater treatment. Biotechnol Adv 22:533–563. https://doi.org/10.1016/j.biotechadv.2004.05.001

Liu S-H, Zeng G-M, Niu Q-Y, Liu Y, Zhou L, Jiang L-H, Tan X-F, Xu P, Zhang C, Cheng M (2017) Bioremediation mechanisms of combined pollution of PAHs and heavy metals by bacteria and fungi: a mini review. Bioresour Technol 224:25–33. https://doi.org/10.1016/j.biortech.2016.11.095

Lukic B, Panico A, Huguenot D, Fabbricino M, Van Hullebusch ED, Esposito G (2017) A review on the efficiency of land farming integrated with composting as a soil remediation treatment. Environ Technol Rev 6(1):94116. https://doi.org/10.1080/21622515.2017.1310310

Maloney S (2001) Pesticide degradation. In: Gadd G (ed) Fungi in bioremediation. Cambridge University Press, Cambridge

Matsumura F, Boush GM (1968) Degradation of insecticides by a soil fungus, *Trichoderma viride*. J Econ Entomol 61(3):610–612. https://doi.org/10.1093/jee/61.3.610

Moeder M, Cajthaml T, Koeller G, Erbanová P, Sasek V (2005) Structure selectivity in degradation and translocation of polychlorinated biphenyls (Delor 103) with a *Pleurotus ostreatus* (oyster mushroom) culture. Chemosphere 61:1370–1378. https://doi.org/10.1016/j.chemosphere.2005.02.098

Ngieng NS, Zulkharnain A, Roslan HA, Husaini A (2013) Decolourisation of synthetic dyes by endophytic fungal flora isolated from Senduduk Plant (Melastoma malabathricum). Biotechnology 1–7. https://doi.org/10.5402/2013/260730

Nikiforova SV, Pozdnyakova NN, Turkovskaya OV (2009) Emulsifying agent production during PAHs degradation by the white rot fungus *Pleurotus ostreatus* D1. Curr Microbiol 58:554–558. https://doi.org/10.1007/s00284-009-9367-1

Obire O, Anyanwu E, Okigbo R (2008) Saprophytic and crude oil degradation fungi from caw dung dropping as bioremediation agents. Int J Agric Technol 4(2):81–89

Pandey A, Gundevia HS (2008) Role of the fungus—*Periconiella* sp. in destruction of biomedical waste. J Environ Sci Eng 50(3):239–240

Park H, Min B, Jang Y, Kim J, Lipzen A, Sharma A, Andreopoulos B, Johnson J, Riley R, Spatafora JW, Henrissat B, Kim KH, Grigoriev IV, Kim JJ, Choi IG (2019) Comprehensive genomic and transcriptomic analysis of polycyclic aromatic hydrocarbon degradation by a mycoremediation fungus, *Dentipellis* sp. KUC8613. Appl Microbiol Biotechnol 103:8145–8155. https://doi.org/10.1007/s00253-019-10089-6

Perelo LW (2010) Review: in situ and bioremediation of organic pollutants in aquatic sediments. J Hazard Mat 177:81–89

Pinedo-Rivilla C, Aleu J, Collado I (2009) Pollutants biodegradation by fungi. Curr Org Chem 13(12):1194–1214. https://doi.org/10.2174/138527209788921774

Pozdnyakova NN (2012) Involvement of the ligninolytic system of white-rot and litter-decomposing fungi in the degradation of polycyclic aromatic hydrocarbons. Biotechnol Res Int 1–20. https://doi.org/10.1155/2012/243217

Pradeep S, Benjamin S (2012) Mycelial fungi completely remediate di(2-ethylhexyl)phthalate, the hazardous plasticizer in PVC blood storage bag. J Hazard Mater 235–236:69–77. https://doi.org/10.1016/j.jhazmat.2012.06.064

Pradeep S, Faseela P, Josh MKS, Balachandran S, Devi RS, Benjamin S (2013) Fungal biodegradation of phthalate plasticizer in situ. Biodegradation 24:257–267. https://doi.org/10.1007/s10532-012-9584-3

Purkayastha RP, Mitra AK, Bhattacharyya B (1994) Uptake and toxicological effects of some heavy metals on *Pleurotus sajor-caju* (Fr.) Singer. Ecotoxicol Environ Safe 27:7–13. https://doi.org/10.1006/eesa.1994.1002

Quarcoo A, Adotey G (2013) Determination of heavy metals in *Pleurotus ostreatus* (Oyster mushroom) and *Termitomyces clypeatus* (Termite mushroom) sold on selected markets in Accra, Ghana. Mycosphere 4(5):960–967

Raj DD, Mohan B, Vidya Shetty BM (2011) Mushrooms in the remediation of heavy metals from soil. Int J Environ Pollut Control Manag 3(1):89–101

Rangel DEN, Finlay RD, Hallsworth JE, Dadachova E, Gadd GM (2018) Fungal strategies for dealing with environment and agriculture induced stresses. Fungal Biol 122(6):602–612. https://doi.org/10.1016/j.funbio.2018.02.002

Rigas F, Papadopoulou K, Dritsa V, Doulia D (2007) Bioremediation of a soil contaminated by lindane utilizing the fungus *Ganoderma australe* via response surface methodology. J Hazard Mater 140(1–2):325–332. Epub 24 Oct 2006. https://doi.org/10.1016/j.jhazmat.2006.09.035

Robichaud K, Stewart K, Labrecque M, Hijri M, Cherewyk J, Amyot M (2019) An ecological microsystem to treat waste oil contaminated soil: using phytoremediation assisted by fungi and local compost, on a mixed contaminant site, in a cold climate. Sci Total Environ 672:732–742. https://doi.org/10.1016/j.scitotenv.2019.03.447

Rodríguez Couto S, Rodríguez A, Paterson RRM (2006) Laccase activity from the fungus *Trametes hirsuta* using an air-lift bioreactor. Lett Appl Microbiol 42:612–616. https://doi.org/10.1111/j.1472-765X.2006.01879.x

Ross G (2004) The public health implications of polychlorinated biphenyls (PCBs) in the environment. Ecotoxicol Environ Saf 59:275–291. https://doi.org/10.1016/j.ecoenv.2004.06.003

Ryan DR, Leukes WD, Burton SG (2005) Fungal bioremediation of phenolic wastewaters in an airlift reactor. Biotechnol Prog 21:1068–1074. https://doi.org/10.1021/bp049558r

Sandrin TR, Maier RM (2003) Impact of metals on the biodegradation of organic pollutants. Environ Health Perspect 111(8):1093–1101. https://doi.org/10.1289/ehp.5840

Satyawali Y, Balakrishnan M (2007) Wastewater treatment in molasses-based alcohol distilleries for COD and color removal: a review. J Environ Manage 86:481–497. https://doi.org/10.1016/j.jenvman.2006.12.024

Schamfuß S, Neu TR, Van Der Meer JR, Tecon R, Harms H, Wick LY (2013) Impact of mycelia on the accessibility of fluorene to PAH-degrading bacteria. Environ Sci Technol 47:6908–6915. https://doi.org/10.1021/es304378d

Shen G, Cao L, Lu Y, Hong J (2005) Influence of phenanthrene on cadmium toxicity to soil enzymes and microbial growth. Environ Sci Pollut Res 12(5):259–263. https://doi.org/10.1065/espr2005.06.266

Singh M, Srivastava PK, Verma PC, Kharwar RN, Singh N, Tripathi RD (2015) Soil fungi for mycoremediation of arsenic pollution in agriculture soils. J Appl Microbiol 119:1278–1290. https://doi.org/10.1111/jam.12948

Stajich JE, Wilke SK, Ahrén D et al (2010) Insights into evolution of multicellular fungi from the assembled chromosomes of the mushroom *Coprinopsis cinerea* (*Coprinus cinereus*). Proc Natl Acad Sci USA 107(26):11889–11894. https://doi.org/10.1073/pnas.1003391107

Stella T, Covinoa S, Burianová E, Filipová A, Křesinová Z, Voříšková J, Větrovský T, Baldrian P, Cajthaml T (2015) Chemical and microbiological characterization of an aged PCB-contaminated soil. Sci Total Environ 533:177–186. https://doi.org/10.1016/j.scitotenv.2015.06.019

Stevens CA, Gregory KF (1987) Production of microbial biomass protein from potato processing wastes by *Cephalosporium eichhorniae*. Appl Environ Microbiol 53:284–291

Sullia SB (2004) Environmental applications of biotechnology. Asian J Microbiol Biotechnol Environ Sci 4:65–68

Sullivan TS, Gadd GM (2019) Metal bioavailability and the soil microbiome. Adv Agronomy 155:79–120. https://doi.org/10.1016/bs.agron.2019.01.004

Tekere M (2019) Microbial bioremediation and different bioreactors designs applied. Biotechnol Bioeng Intech Open. https://doi.org/10.5772/intechopen.83661

Van den Berg M, Birnbaum LS, Denison M, De Vito M, Farland W, Feeley M, Fiedler H, Hakansson H, Hanberg A, Haws L, Rose M, Safe S, Schrenk D, Tohyama C, Tritscher A, Tuomisto J, Tysklind M, Walker N, Peterson RE (2006) The 2005 World Health Organization reevaluation of human and mammalian toxic equivalency factors for dioxins and dioxin-like compounds. Toxicol Sci 93:223–241. https://doi.org/10.1093/toxsci/kfl055

Van Leeuwen J, Hu Z, Yi TW, Pometto AL III, Jin B (2003) Kinetic model for selective cultivation of microfungi in a microscreen process for food processing wastewater treatment and biomass production. Acta Biotechnol 23:289–300. https://doi.org/10.1002/abio.200390036

Vanholme R, Demedts B, Morreel K, Ralph J, Boerjan W (2010) Lignin biosynthesis and structure. Plant Physiol 153(3):895–905. https://doi.org/10.1104/pp.110.155119

Vaseem H, Singh VK, Singh MP (2017) Heavy metal pollution due to coal washery effluent and its decontamination using a macrofungus, *Pleurotus ostreatus*. Ecotoxicol Environ Saf 145:42–49. https://doi.org/10.1016/j.ecoenv.2017.07.001

Velázquez-Fernández JB, Martínez-Rizo AB, Ramírez-Sandoval M, Domínguez-Ojeda D (2012) Biodegradation and bioremediation of organic pesticides, Chap 12, pp 253–272. https://doi.org/10.5772/46845

Verdin A, Sahraoui ALH, Durand R (2004) Degradation of benzo[a]pyrene by mitosporic fungi and extracellular oxidative enzymes. Int Biodeterior Biodegradation 53:65–70. https://doi.org/10.1016/j.ibiod.2003.12.001

Verma JP, Jaiswal DK, Sagar R (2014) Pesticide relevance and their microbial degradation: a-state-of-art. Rev Environ Sci Biotechnol. https://doi.org/10.1007/s11157-014-9341-7

Violeta DJ, Jasmina MM, Jelica DS, Slavica R, Miroslav MV (2014) The influence of detergent and its components on metabolism of Fusarium oxysporum in submerged fermentation. Hem Ind 68(4):465–473. https://doi.org/10.2298/HEMIND130620071J

Wetzstein HG, Schmeer N, Karl W (1997) Degradation of the fluoroquinolone enrofloxacin by the brown rot fungus *Gloeophyllum striatum*: identification of metabolites. Appl Environ Microbiol 63(11):4272–4281

Wick LY, Remer R, Würz B, Reichenbach J, Braun S, Schafer F, Harms H (2007) Effect of fungal hyphae on the access of bacteria to phenanthrene in soil. Environ Sci Technol 41(2):500–505. https://doi.org/10.1021/es061407s

Wicklow DT, Detroy RW, Jessee BA (1980) Decomposition of lignocellulose by *Cyathusstercoreus* (Schw.) de Toni NRRL 6473, a "white rot" fungus from cattle dung. Appl Environ Microbiol 40:169–170

Worrich A, Wick LY, Banitz T (2018) Ecology of contaminant biotransformation in the mycosphere: role of transport processes. Adv Appl Microbiol 104:93–133. https://doi.org/10.1016/bs.aambs.2018.05.005

Wu M, Xu Y, Ding W, Li Y, Xu H (2016) Mycoremediation of manganese and phenanthrene by *Pleurotus eryngii* mycelium enhanced by Tween 80 and saponin. Appl Microbiol Biotechnol 100:7249–7261. https://doi.org/10.1007/s00253-016-7551-3

Yang SO, Sodaneath H, Lee JI, Jung H, Choi JH, Ryu HW, Cho KS (2017) Decolorization of acid, disperse and reactive dyes by *Trametes versicolor* CBR43. J Environ Sci Health 52:862–872. https://doi.org/10.1080/10934529.2017.1316164

Yesilada O, Sik S, Sam M (1999) Treatment of olive oil mill wastewater with fungi. Turk J Biol 23:231–240

Young D, Rice J, Martin R, Lindquist E, Lipzen A, Grigoriev I, Hibbett D (2015) Degradation of bunker C fuel oil by white-rot fungi in sawdust cultures suggests potential applications in bioremediation. PLoS ONE 10:e0130381. https://doi.org/10.1371/journal.pone.0130381

Zheng S, Yang M, Yang Z (2005) Biomass production of yeast isolated from salad oil manufacturing wastewater. Bioresour Technol 96:1183–1187. https://doi.org/10.1016/j.biortech.2004.09.022

Chapter 8
Microbial Detoxification of Contaminated Land

Nazneen Hussain, Linee Goswami, and Satya Sundar Bhattacharya

8.1 Introduction

Land pollution is a widely considered subject of global concern. Since the onset of industrialization, the problem of land pollution has been a matter of great concern and the issue has only been increasing every passing year. Land pollutants are the potent source of toxic and persistant organic pollutants (POPs) (Paul et al. 2019). Persistant organic pollutants are toxic chemical organic compounds that are highly resistant to environmental degradation. It adversely affects human health and the surrounding environment (Kevin 2021). These land pollutants are a storehouse of numerous contaminants which are taken up by the plants and animals. These organisms are later consumed by human beings which thereby affects and disturbs the equilibrium of the natural ecosystem. The toxic components of polluted land broadly involve heavy metals, petroleum hydrocarbons, polyaromatic hydrocarbons (PAH), pesticides, insecticides, halogenated and non-halogenated compounds (Saravanan et al. 2021; Gaine et al. 2021; Meagher 2000; Allen 2002). With rapid development of industry and agriculture, production and extensive use of artificial products have led to massive buildup of xenobiotics. Anthropogenic activities such as fossil fuel combustion, mining, corrosion, smelting, and waste disposal, greatly contribute to land pollution (Gaine et al. 2021; Thassitou and Arvanitoyannis 2001). According to the reports, approximately 2.5 million tons of pesticides are consumed throughout the world (Danila et al. 2020; FAO 2002). Based on the statistical data on total consumption of pesticides, Europe occupies the first position utilizing 45% of the

---

N. Hussain
Deparment of Biosciences, Assam Don Bosco University, Guwahati, Assam, India

L. Goswami
Department of Biology, School of Science and Technology, Örebro University, Örebro, Sweden

S. S. Bhattacharya (✉)
Department of Environmental Science, Tezpur University, Tezpur, Assam, India
e-mail: evssatya@gmail.com

pesticides, tailed by USA (24%) and others. The use of pesticides in Asia, particularly in India, has been increasing at an alarming rate. Currently, India as a pesticide producer holds twelfth position in the world and is the largest in Asia. Recent surveys have also confirmed that the presence of pesticides and groundwater aquafiers are potent pollutants in water bodies and river beds (Tang et al. 2021; Shafi et al. 2020; Leong et al. 2007; Zakaria et al. 2003). These pollutants reach the water bodies in the form of toxic effluents and landfill leachate composed of persistent chemical components. Reports available in World Bank studies discussed the occurrence of dieldrin, polychlorinated biphenyls (PCBs) in rivers. The accumulation of these harmful contaminants by the aquatic and marine bodies has posed serious threats to their lives. The primary source of pesticides arises from the activities performed in the agricultural land (Zakaria et al. 2003). The presence of POPs in aquatic and terrestrial species has also been confirmed by various studies. According to Zakaira et al. (2003), dolphins are highly exposed to large amounts of DDT, chlordane, aldrin, and dieldrin in India. Based on reports, pesticide poisoning is one of the primary causes of increased death rates in developing countries (Zhong and Zhang 2020; WHO 1990). The linear rise in cancer cases and increasing death rate on exposure to high percentage of pesticides has been a matter of great concern (Hites 2021; Gong et al. 2021; WHO 1990; UNEP 1993). Major sources of xenobiotics are found in the production and manufacturing units of pesticide industries. Production farm workers, sprayers, and loaders are at high risk to pesticide consumption. This is because during formulation and application of the product the risk associated with the process is much more hazardous. Additionally, scientific reviews have also reported the contamination of heavy metals in agricultural land. These heavy metalloids found in the soil surface when taken up by the crops impose serious risks to crop growth and global food security (Rodríguez Eugenio et al. 2018). The heavy metalloids that are most widely available in the soil ecosystem are Pb, Cd, Cr, As, Cu, and Hg. Among these heavy metals, Cd is extensively encountered in agricultural lands where rice crops are grown. India, being an agrarian country and rice being the staple food; the concentration of cadmium continues to rise exponentially despite the implementation of various Environmental Regulatory Acts. Soil is the largest reservoir of carbon pool which holds almost five times more than the total mass of atmospheric carbon. The aggravating degradation of agricultural land because of the overload of contaminants might deteriorate the role and the mechanism of natural carbon cycle (Kopittke et al. 2019). Marrugo-Negrete et al. (2017), have also reported that the occurrence of heavy metals could also be derived from geogenic sources. The presence of geogenic heavy metals largely regulates the soil property and thus influences the soil quality (Clemens 2006). Thus, explosive increases in the level of contaminants over a short period of time have deteriorated the in-built self-remediating capacity of the environment, and hence the accumulation of pollutants. The use of living organisms as a model to undergo the process of biotransformation of toxic contaminants is a very effective bioremediation technique. Among all the bioremediation strategies, bioremediation assisted via microbes is a popular technique to transform highly toxic components into less toxic components (Garbisu and Alkorta 2003; Adhikari et al. 2004). These diverse ranges of microbial species could largely facilitate the

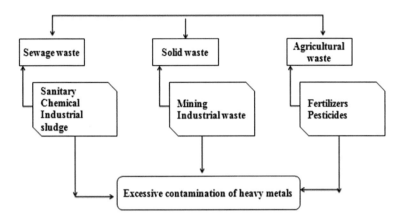

Fig. 8.1 An overview of all kinds of land waste

remediation of toxic contaminants both via active and passive processes (Bhatt et al. 2021; Scott and Karanjkar 1992). The microbial remediation process is conducted by various methods. Monitored natural attenuation and engineered microbial attenuation are the two popular techniques of microbial remediation. Usage of genetically modified microorganisms is another advanced biotechnological approach to degrade hazardous wastes (Dilek et al. 1998). Availability of carbon sources is the primary requirement of all the approaches to initiate the degradation process (Vidali 2001). Both indigenous and extraneous microorganisms are used for the remediation technique (Prescott et al. 2002). The mechanism of microbial remediation either participates in the oxidation–reduction reaction or is involved in the catalytic reaction that includes synthesis/degradation of organic compounds (Gadd 2000; Rajendran et al. 2003). The overview of all kinds of land wastes that undergoes microbial remediation is described in Fig. 8.1. *Pseudomonas putida* is the first patented microorganism registered in the year 1974 known for the degradation of petroleum (Glazer and Nikaido 2007). Presently, various reports are available on the usage of both aerobic and anaerobic microorganisms for the degradation of wide range of toxic components. Few potent examples of microorganisms that belong to the broadly classified groups mentioned above are: *Mycobacterium, Alcaligens, Methylotrophs, Phanaerochaete chrysosporium, Rhodococcus* (EPA 2003; Zeyaullah et al. 2009). Thus, this study is an attempt to highlight the effectiveness of microbial bioremediation and their associated mechanism of degradation.

8.2 Environmental Impact of Pesticides

Pesticides are globally used chemicals that are extensively used in agricultural land to safeguard the harvested crops from pests. Pesticides are classified into different types

Table 8.1 List of pesticides and their examples

Pesticides	Examples
Organophosphorus	Diazinon, dichlorvos, dimethoate, malathion, parathion
Carbamate	Carbaryl, propoxur
Organochlorine	DDT, methoxychlor, toxaphene, mirex, kepone
Cyclodienes	Aldrin, chlordane, dieldrin, endrin, endosulfan, heptachlor
Herbicides	Chlorophenoxy acids, hexachlorobenzene (HCB)
Nitrogen-based	Picloram, Atrazine, diquat, paraquat
Organophosphates	Glyphosate (Roundup)
Nitrogen-containing	Triazines, dicarboximides, phthalimide
Wood preservatives	Creosote, hexachlorobenzene
Botanicals	Perethrin, permethrin
Antimicrobial	Chlorine, quaternary alcohols

of chemicals based on their structural composition. It includes herbicides, insecticides, fungicides, and rodenticides (Table 8.1). The structural classification varies based on the presence of organochlorine, organophosphorus, pyrethroids, carbamates, and nitrogen-based compounds (Gilden et al. 2010). During 2001 amendment, Stockholm Convention has listed 12 persistent organic pollutants (POP) to cause adverse effects on the human health and ecosystem (Table 8.2). These POPs are sectioned under three different categories which include: pesticides, industrial chemicals, and by-products. The application, storage, and disposal approaches of POPs have posed a serious toxicity risk to living organisms and the environmental land (Fantke et al. 2012; Pieterse et al. 2015; Torres et al. 2013). The consumption of these pesticides has become such an inevitable practice that with each passing year the usage of xenobiotics has been increasing at an alarming rate (Abhilash and Singh 2009). As per reports, pesticide consumption was found to be highest in China followed by Korea, Japan, and India. Interestingly, among all the types of pesticides used in India, insecticides are mostly used which account to 76% as against 44% of global usage (Mathur 1999). These toxic recalcitrant compounds have become an issue of serious global concern because of their persistency, long-range transportability via air and water, and lipophilic; thus accumulate in the fatty tissues of living organisms (Torres et al. 2013; Pieterse et al. 2015). The impact of pesticides in these three components of ecosystems is described in detail. Even after banning few pesticides, significant amount of these toxic chemicals is observed in the major components of soil interaction system which includes: atmosphere, pedosphere, and hydrosphere.

Table 8.2 Persistant organic pollutants listed in Stockholm Convention amendment

Sl. No.	Item	Type of chemicals
	2001 amendment	
1	Aldrin	Pesticide
2	Dieldrin	Pesticide
3	Endrin	Pesticide
4	Chlordane	Pesticide
5	Heptachlor	Pesticide
6	HCB	Pesticide
7	Mirex	Pesticide
8	Toxaphene	Pesticide
9	DDT	Pesticide
10	PCBs	Industrial and by-product
11	Polychlorinated dibenzo-p-dioxins (PCDD)	By-product
12	Polychlorinated dibenzofurans (PCDF)	By-product
	2009 amendment	
1	Chlordecone (Kepone)	Pesticide
2	Lindane	Pesticide
3	α- HCH	Pesticide and by-products
4	B- HCH	Pesticide and by-products
5	Hexabromobiphenyl	Pesticide, industrial, and by-product
6	Tetra-BDE and penta-BDE	Pesticide, industrial, and by-product
7	Hexa-BDE and hepta-BDE	Pesticide, industrial, and by-product
8	PFOs and its salts	Pesticide, industrial, and by-product
9	PFOSF	Industrial
10	Pentachlorobenzene	Pesticide, industrial and by-product
	2011 amendment	
1	Endosulfan	Pesticide

8.2.1 Atmospheric Contamination

Pesticides once applied to the soil, it volatilizes from the soil and contaminates the atmospheric environment adversely (USGS 1995). The emission and evapo-transmission of pesticide particulates largely affect the air quality and human health (Gavrilescu 2005; Usman and Usman 2013). Usman and Usman (2013) monitored the percentage of DDT and HCH in air and it was observed to be ~5.930 ng m^{-3} and ~11.45, respectively. Similarly, researchers recorded the concentration of DDT, chlordane, HCH, and endosulfan ranging from 250 pg m^{-3} to 6110 pg m^{-3}, 240 to 4650 pg m^{-3}, 890 to 17,000 pg m^{-3}, and 40 to 4650 pg m^{-3} respectively, in India (Gavrilescu 2005).

8.2.2 Surface and Ground Water Contamination

The primary cause of water body contamination is mostly due to agricultural and industrial runoff, soil erosion from contaminated sites, percolation through rainwater, and leaching (Barbash and Resek 1996). Widely reported insecticide and herbicide in the water bodies across the US are 2,4-D, diuron, chlorpyrifos, and diazinon followed by Trifluralin and 2,4-D (U.S. Geological Survey 1998; Bevans et al. 1998; Fenelon et al. 1998). Extensive use of these toxic chemical pollutants was found to be detrimental to the aquatic environment. According to USGS reports, 143 pesticides and 21 transformed products were recorded in the ground water of 43 states across the nation. In Bhopal, Madhyapradesh, India, 58% of groundwater was contaminated with organochlorine pesticide that exceeded the EPA standards (Kole and Bagchi 1995). Similarly, another study monitored and reported the concentration of endosulfan and DDT metabolites to be excessively high across all the states in India (Bakore et al. 2004). The main problem of ground water contamination with toxic chemicals is because it takes years dissipate the contaminants (O'Neil et al. 1998; USEPA 2001).

8.2.3 Land Surface Contamination

Soil contamination is documented by many workers because of the occurrence of pesticides and transformed products derived from pesticides (Roberts 1998; Roberts and Hutson 1999). These hydrophobic compounds are highly persistent and tightly bound to the soil. The binding efficiency of the pesticides with soil largely depends on the soil-pesticide interaction. Generally, the higher the organic matter content in soil, the higher will be adsorption properties of the pesticides and transformed products. The adsorption property of the ionizable pesticides is again inversely proportional to the soil pH (Andreu and Pico' 2004). The changes in the conformation of the pesticides include various metabolic pathways like hydrolysis, methylation, etc. These pathways generate toxic phenolic compounds as end products (Barcelo and Hennion 1997). The concentration of pesticides and their transformed products are characterized by determining water solubility, soil sorption constant, and partition coefficient of octanol/water. Indiscriminate use of pesticides in the soil has detrimental effects on the soil biota. The inherent population of bacterial and fungal communities is largely affected due to high dosage of pesticide application (Pell et al. 1988). These contaminants present in the soil inhibit the natural carbon and nitrogen cycle and reduces the growth and activity of the microorganisms involved in the process (Santos and Flores 1995). The organic compounds present in the pesticides target the tissues of the living organisms and disrupt the endocrine receptors and cause severe hormonal imbalance (Hurley et al. 1998). The other side effects associated with the accumulation of these harmful chemicals are immune-suppressive disorders, reproductive abnormalities, decreased memory, and cancer (Crisp et al. 1998; Hurley et al. 1998).

8.3 Sources of POP and Their Current Status

The complex status of persistent organic pollutants in the world needs scientific planning to harmonize the soil and revive its health. The chief components of POPs are polychlorinated dibenzodioxins (PCDD), polychlorinated biphenyls (PCBs), and polychlorinated dibenzofurans (PCDF). The Stockholm Convention identified 22 POPs as listed in Table 8.2. Among the toxic POPs, twelve POPs namely aldrin, dieldrin, endrin, chlordane, heptachlor, HCB, mirex, toxaphene, DDT, PCBs, PCDDs, and PCDFs are considered as "dirty dozens". Additionally, twelve more compounds were added to the existing list, which included chlordecone, lindane, α-HCH, β-HCH, hexabromobiphenyl, tetra-BDE, penta-BDE, hexa-BDE, hepta-BDE, PFOs, and its salts, PFOSF, and pentachlorobenzene. The overdose of these compounds to combat disease and augment plant growth has caused serious threat to the soil ecosystem.

8.3.1 Pesticide Toxicity and Its Degradation

The persistence nature of pesticides is because of their compact physicochemical properties. The chlorinated pesticide is highly lethal. It imparts its toxicity at a significant level to the soil ecosystem disturbing the equilibrium of every living organism. The major routes of exposure to these chemicals are through food chain series, dermal contact, respiratory tract, etc. Microorganisms ubiquitously play an important role in the degradation wide range of POPs. The bioremediation ability of microorganisms depends on the structure and presence of functional group of the toxic group. The recalcitrant nature of the pesticides is also because of the presence of anionic species in the compounds. Microorganisms, with the help of their electron-donating and electron-accepting capability, can remediate the toxic contaminant. The degradation of these recalcitrant compounds by the microorganisms leads to the breakdown of parent compounds by producing carbon dioxide and water (Ref). The efficiency of this oxidation reaction depends on various environmental factors. This technique of decaying harmful pollutants using microbes can happen by two different processes. One being, the natural microorganisms stimulate the degradation of pollutants by utilizing the available nutrients in the contaminated land and the other being, isolating useful degrading microorganisms from one site and transporting into the polluted land. The latter process is termed as "bio-augmentation". Apart from bioremediation, there are different terminologies used based on the target organism involved in the remediation strategy. Few microbes involved in the process of bioremediation is mentioned under Table 8.3. Among all the bacterial classes; *Sphingomonas* of alphaproteobacteria; *Burkholderia* of betaproteobacteria; *Pseudomonas, Acenetobacter* of gamma proteobacteria and *Flavobacterium* are considered to be efficient microdegraders (Mamta and Khursheed 2015; Kafilzadeh et al. 2015).

Table 8.3 List of microorganisms involved in the degradation of hazardous chemicals

Microorganisms	Toxic compounds
Flavobacterium spp.	Organophosphate
Cunniughamela elegans and *Candida tropicalis*	PCBs (Polychlorinated Biphenyls) and PAHs (Polycyclic Aromatic Hydrocarbons)
Pseudomonas, Arthrobacter, Citrobacter, Vibrio	Phenylmercuric acetate
Alcaligenes spp. and *Pseudomonas* spp.	PCBs, halogenated hydrocarbons and alkylbenzene, sulphonates, PCBs, organophosphates, benzene, anthracene, phenolic compounds, 2,4 D, DDT, and 2,4,5-trichlorophenoxyacetic acid, etc.
Actinomycetes	Raw rubber
Nocardia, Pseudomonas	Detergents
Trichoderma, Pseudomonas	Malathion
Arthrobacter, Bacillus	Endrin
Closteridium	Lindane
Escherichia, Hydrogenomonas, Saccharomyces	DDT
Mucor	Dieldrin
Phanerochaete chrysoporium	Halocarbons such as lindane, pentachlorophenol,
P. sordida and *Trametes hirsute*	DDT, DDE, PCBs, 4,5,6-trichlorophenol, 2,4,6-trichlorophenol, dichlorphenol, and chlordane
Closteridium	Lindane
Arthrobacter and *Bacillus*	Endrin
Trichoderma and *Pseudomonas*	Malathion
Zylerion xylestrix	Pesticides/Herbicides (Aldrin, dieldrin, parathion, and malathion)
Mucor	Dieldrin
Yeast (*Saccharomyces*)	DDT
Phanerochaete chrysoporium	Halocarbons such as lindane, pentachlorophenol
P. sordida and *Trametes hirsute*	DDT, DDE, PCBs, 4,5,6-trichlorophenol, 2,4,6-trichlorophenol, dichlorphenol, and chlordane
Pseudomonas spp.	Benzene, anthracene, hydrocarbons, PCBs
Alcaligenes spp.	Halogenated hydrocarbons, linear alkylbenzene sulfonates, polycyclic aromatics, PCBs
Arthrobacter spp.	Benzene, hydrocarbons, pentachlorophenol, phenoxyacetate, polycyclic aromatic
Bacillus spp.	Aromatics, long-chain alkanes, phenol, cresol
Corynebacterium spp.	Halogenated hydrocarbons, phenoxyacetates

(continued)

Table 8.3 (continued)

Microorganisms	Toxic compounds
Flavobacterium spp., *Methosinus* sp., *Methanogens, Azotobacter* spp.	Aromatics
Rhodococcus spp.	Naphthalene, biphenyl
Mycobacterium spp.	Aromatics, branched hydrocarbons benzene, cycloparaffins
Nocardia spp.	Hydrocarbons
Xanthomonas spp.	Hydrocarbons, polycyclic hydrocarbons
Streptomyces spp.	Phenoxyacetate, halogenated hydrocarbon, diazinon
Candida tropicalis	PCBs, formaldehyde
Cunniughamela elegans	PCBs, polycyclic aromatics, biphenyls
Pseudomonas spp.	Benzene, anthracene, hydrocarbons, PCBs
Alcaligenes spp.	Halogenated hydrocarbons, linear alkylbenzene sulfonates, polycyclic aromatics, PCBs
Arthrobacter spp.	Benzene, hydrocarbons, pentachlorop henol, phenoxyacetate, polycyclic aromatic Aromatics, long-chain alkanes, phenol, cresol
Bacillus spp.	Halogenated hydrocarbons, phenoxyacetates
Corynebacterium spp.	Aromatics
Flavobacterium spp.	Aromatics, Naphthalene, biphenyl
Azotobacter spp.	Aromatics, branched hydrocarbons benzene, cycloparaffins
Rhodococcus spp.	Hydrocarbons
Mycobacterium spp.	Hydrocarbons, polycyclic hydrocarbons
Nocardia spp.	Phenoxyacetate, halogenated hydrocarbon diazinon
Methosinus sp.	PCBs, formaldehyde, biphenyl, polycyclic aromatics

8.4 Factors Responsible for Microbial Degradation

All micro-organisms have different metabolic rates and hence different growth patterns. The presence of different microorganisms contributes to adequate biodegradation since the process involves various microdegraders belonging to different bacterial classes. These variant microdegraders involve reactions like cleavage, oxidation, reduction, biotransformation, volatilization, biosorption, and bioleaching. Physicochemical parameters of the environmental matrix-like pH, oxygen, temperature, substrate availability, moisture content, types of carbon sources, largely regulate the metabolic features of the microbes which later influence the rate of degradation. Sites contaminated with pesticides often undergo leaching of particles. The leachate has sufficiently high pH, resulting in the lowering of the degradation rate. Thus, the choice of pH depends on the selection of microorganisms chosen for the degradation process. Soil texture, permeability, and its bulk density are major soil properties that affect the rate of degradation. These soil properties are again determined by the moisture content, oxygen concentration, and nutrient availability in soil. Scientists have reported various mathematical models that describe the how the role of moisture content determines the degradation rate in polluted land (Raymond et al. 2001). It is reported that soil with low permeability hinders the bioremediation process. Anoxic sites with low oxygen concentration play a significant role in the process of bioremediation. The bioremediation process that takes place in sites contaminated with polyaromatic hydrocarbon is generally hindered because of low oxygen concentration since aerobic microorganisms are chosen for PAH degradation (Liu and Cui 2001). Limited oxygen concentration and low moisture content agglomerates the soil, thereby decreasing the bioremediation process (Cho et al. 2000). Availability of substrates/nutrients and their concentration largely affect the rate of degradation. For example: if the sites contaminated with PAH is supplemented with inorganic nutrients like nitrogen could enhance the degradation efficiency of most of the microorganisms (Zhou and Hua 2004).

8.5 Microbial Enzymes Used in Pesticide Bioremediation

Microbial degradation largely depends on enzymatic degradation. Microbial enzymes act as the biological catalysts that facilitate the conversion of persistent environmental pollutants to innocuous products (Table 8.4). These catalysts can be applied to a wide range of substances because of its chemical, regional, and stereo selectivity. It has great potential to effectively transform the pollutants in the biota by extensive transformations of the toxicological-based structural properties of the lethal compounds. According to Langerhoff et al. (2001), both aerobic and anaerobic degradations are essential to attain proper mineralization. Based on reports, aerobic degradation targets the cleavage of aliphatic and aromatic metabolites attached with

Table 8.4 Microbial enzymes involved in the degradation of various land contaminants

Enzyme	Used in industry
Dioxygenases	Synthetic industry, pharmaceutical industry
Laccases	Food industry, paper and pulp industry, textile industry, nanotechnology, synthetic industry, bioremediation, cosmetics
Peroxidases	Food industry, paper and pulp industry, textile industry, pharmaceutical industry
Oxidoreductases	Synthetic industry
Oxygenases	Synthetic industry
Monooxygenases	Synthetic industry

the compounds while anaerobic degradation results in dechlorination (Singh et al. 1999; Baczynski et al. 2010).

8.5.1 Oxidoreductase

Oxidases are the class of enzymes that utilizes molecular oxygen (O_2) as an electron acceptor to catalyze the oxidation–reduction reaction. The oxygen utilized in the process is either reduced to water (H_2O) or hydrogen peroxide (H_2O_2). Microorganisms such as bacteria and fungi undergo the detoxification process of toxic components via oxidative coupling (Bollag and Dec 1998). The biochemical reactions catalyzed by the enzymes help the microorganisms to derive energy and cleave the bonds to assist the transfer of electrons from one substrate to another. Thus, via oxidation–reduction mechanism, microorganisms contribute in converting the contaminants to harmless oxidized compounds (Gianfreda et al. 1999). Furthermore, oxidoreductases also participate in the detoxification of pesticides wherein phenolic/anilinic groups are attached to the moieties.

8.5.2 Oxygenase

Oxygenases belong to the group of oxydoreductase enzymes that involves oxidation of reduced substrates using FAD/NADH/NADPH as co-substrates. These enzymes actively participate in the cleavage of aromatic rings by introduction of molecular oxygen and regulating the metabolism of organic compounds. These classes of enzymes are also active against chlorinated aliphatics and halogenated methanes, ethanes, and ethyles by undergoing dehalogenation reactions (Fetzner and Lingens 1994). These enzymes are active against wide range of pollutants such a; herbicides, fungicide, insecticide (Fetzner and Lingens 1994; Fetzner 2003; Arora et al. 2009). Oxygenases are divided into two types based on the number of oxygen atoms involved in the process of oxygenation; namely monooxygenases and the other dioxygenases.

Monooxygenases include wide range of superfamily that catalyze substrates containing alkanes, fatty acids, and also steroids utilizing molecular oxygen. Due to wide range of stereo and region selectivity monooxygenases are often classified as biocatalysts (Cirino and Arnold 2002; Arora et al. 2009). Moreover, monooxygenases effectively degrade the aliphatic groups attached to the target compounds via hydroxylation, desulphurization, and denitrification (Arora et al. 2009). It comprises of two groups depending on the presence of cofactor; namely flavin-dependent monooxygenases and P450 monooxygenases. As the name suggests, flavin-dependent monooxygenases involve flavin as the prosthetic group whereas for P450 monooxygenases heme acts as the coenzyme in all living organisms. The reaction for flavin-dependent monooxygenases is catalyzed by NADP/NADPH. An enzyme called methane monooxygenase enzyme is a promising degrader of hydrocarbons which includes methanes, alkanes, and alkenes (Fox et al. 1990; Grosse et al. 1999). Under aerobic conditions, monooxygenase undergoes dehalogenation; whereas for anaerobic conditions, it catalyzes reductive dechlorination (Jones et al. 2001).

Dioxygenases are multicomponent enzyme that participates enantiospecifically on wide range of substrates via oxygenation. Dioxygenases that act on aromatic hydrocarbon belong to the family of Rieske non-heme iron oxygenases. The mononuclear iron attached to the Rieske cluster is present in the alpha subunit. These microbial enzymes are present mostly in the soil that converts aromatic substrates to aliphatic products (Que and Ho 1996).

8.5.3 Cytochrome P450

The cytochrome P450 belongs to the family of hememonooxygease which utilizes molecular oxygen to oxidize substrates and participate in the degradation process (Morant et al. 2003; Urlacher et al. 2004). It has a broad substrate range that includes 200 families for both prokaryotes and eukaryotes. Cytochrome P450 requires NAD(P)H, as a cofactor, which remains non-covalently bound to catalyze the reaction. One potential cytochrome P459 extracted from *P. putida* is known to have significant contribution in detoxifying chlorinated pollutants, namely pentachlorobenzene and hexachlorobenzene (Chen et al. 2002). Another variant of cytochrome P450 extracted from *Sphingobium chlorophenolicum* efficiently degrades hexachlorobenzene (Yan et al. 2006).

8.5.4 Peroxidase

Peroxidases are ubiquitously distributed in nature catalyzed by oxidation of organic and inorganic substrates and reduction of peroxides. The peroxidase enzyme can either be heam or non-heam proteins. These enzymes are produced from varied number of microorganisms such as *Bacillus subtilis*, *Citobacter* sp., *Bacillus sphaericus*, *Cyanobacteria*, *Strptomyces* sp., *Candida krusei*, *Anabaena* sp., and also yeasts (Hiner et al. 2002; Koua et al. 2009). Peroxidases extracted from fungal species work best on organophosphorus pesticides (Piontek et al. 2001). According to reports, *Caldariomyces fumago* is an efficient bio-degrader of organophosphorus-based substrates. These enzymes also help in transforming PAHs to less toxic products via oxidation (Hinter et al. 2002). The heme-based peroxidases have been classified into three groups on the basis of sequence comparison. Class I group of heme peroxidases is an intracellular enzyme that includes ascorbate peroxidase, cytochrome c, and also catalase-peroxidase. Class II group includes secretory fungal enzymes that is primarily involved in lignin degradation and Class III is a group of peroxidases extracted from plants such as horse radish peroxidase, soyabean (Hiner et al. 2002; Koua et al. 2009). Non-hemeperoxidases comprise five different families namely thiol peroxidase, NADH peroxidase, alkylhydroperoxidase, and non-haem halo peroxidase. Among all the five different types of non-based peroxidase, thiol peroxidase is the largest to have two subfamilies in the group.

Microbial extracted peroxidase enzymes are further classified into lignin peroxidase enzyme, manganese peroxidase enzyme, and versatile peroxidase enzyme. Lignin peroxidase is extracted efficiently by white-rot fungus via secondary metabolism (Yoshida 1998). It is majorly involved in degradation of plant cell walls. The reaction converts hydrogen peroxide to water by accepting electrons from the oxidized form of LiP. This facilitates LiP to oxidize phenolic compounds into less toxic forms (Piontek et al. 2001). Manganese-based peroxidase is an extracellular protein secreted by basidiomycetes. This group of heme enzymes oxidizes Mn^{2+} which promotes the production of MnP and thus acts as substrates whereas MnP formed from Mn^{3+} helps in the chelation of oxalate compounds and helps in the biodegradation of xenobiotic compounds (Tsukihara et al. 2006). Microbial versatile peroxidases oxidizes Mn^{2+}, methoxybenzenes, and phenolic compounds similar to that of MnP, LiP. Unlike, lignin and manganese-based peroxidase this class of versatile peroxidase also participates in degrading non-phenolic dimers (Ruiz-Duenas et al. 2007).

8.5.5 Laccases

Laccases belong to the group of ubiquitous enzymes catalyzed by oxidation of reduced phenolic and aromatic substrates. It comprises multicopper oxidases extracted from wide range of prokaryotes as well as eukaryotes (Mai et al. 2000). The enzymes extracted from microorganisms is capable of producing both extracellular and intracellular laccases and thus proficient in oxidizing wide range of compounds such as aminophenols, polyphenols, methoxyphenols, diamines, polyamines, (Ullah et al. 2000; Rodriguez Couto and Toca Herrera 2006). The performance of the enzymes widely depends in pH and presence of chemicals such as halides, hydroxide, and cyanide (Xu 1996; Kim et al. 2002) in the environment.

8.5.6 Hydrolases

Hydrolases are broad group of enzymes used in pesticide bioremediation majorly involved in the catalysis of the hydroxylation of thioesters, C–H bonds, peptide bonds, esters, carbon–halide bonds, etc. Interestingly, this reaction is undergone in the absence of any redox cofactors. The hydrolytic enzyme is quite popular in the bioremediation process because of its availability, resistance against water-miscible solvents, economic viability, and eco-friendly properties.

8.5.7 Haloalkane Dehalogenases

Lindane, a highly persistant insecticide (an isomer of hexachlorocyclohexane) is widely used against pests. The bacterial genes involved in the degradation of hexachlorocyclohexane (HCH) have been studied extensively. Lin A and Lin B are the two key enzymes responsible for the biodegradation process. Lin A helps in the detoxification of γ-HCH whereas Lin B also known as haloalkane dehalogenase (Dhl A) is isolated from *Xanthobacter autotrophicus* (Nagata et al. 1993a, b). The enzyme also participates in the degradation of β-HCH and δ-HCH. The mechanism of β-HCH degradation is catalyzed by *Sphingomonas paucimobilis* whereas for δ-HCH *Sphingobium indicum* B90A and *Sphingobium japonicum* UT26 (Nagata et al. 2005; Sharma et al. 2006). The enzyme mediates the conversion of 2,3,5,6-tetrachloro-1,4-cyclohexadiene to 3,6-dichloro-2,5-dihydoxy-1,4-cyclohexadiene (Negri et al. 2007) (fig). The remediation of β-HCH via the enzyme LinB is initially converted to pentachlorocyclohexanol, which is further transformed to tetrachlorocyclohexanol. LinA enzyme, encoded by the Lin operon, mediates the very first step of dehydrochlorination of γ-HCH (Nagata et al. 1993a, b). This enzyme belongs to the family of scytalone dehydratase and naphthalene dioxygenase (Nagata et al. 2001). However, complete detoxification of γ-HCH involves both LinA and LinB. Raina et al. (2007) has observed another potent bacteria named *Sphingobium indicum* involved in significant reduction of HCH from the soil surface. Lin operon-mediated genes activate the mechanism of the remediation process.

8.6 Impact of Genetic Engineering in Pesticide Degradation

Several reports have described the importance on the genetic basis and the role of catabolic genes in pesticide biodegradation (Table 8.5). The genetically engineered microbial genes responsible for bioremediation are either extrachromosomal or genomic that work via recombinant DNA technology. In in-situ bioremediation technique, the most vital point is to understand how the mechanism of how the microbial genes interact with the environment to biodegrade the toxic pollutants (Hussain et al. 2018). The microbial enzymes isolated from plasmids, encoding opd gene have

Table 8.5 List of genetically modified organisms used in bioremediation

Microorganisms	Compound degraded
P. putida	Camphor degradation
P. oleovarans	Alkane degradation
P. cepacea	2,4,5-Trichlorophenoxyacetic acid degradation
P. mendocina	Trichloroethylene degradation
P. diminuata	Parathion (pesticide) degradation

varied genetic diversity. The opd gene is present in a highly conserved region in a bacterial plasmid which is widely responsible for the degradation of organophosphorus hydrolase (OPH) (Yan et al. 2006). The gene that encodes organophosphorus hydrolase has a promoter sequence that includes 996 nucleotides. Another important enzyme named, methyl parathion hydrolase (MPH) encoded by mpd gene is isolated from Acrobacter and Brucella widely participates in the biodegradation process. These microorganisms have interesting properties to respond to various kinds of stresses and adapt itself to the polluted environment. The molecular adaption of these genes to achieve fitness in toxic environment draws attention to the development of strategies for the optimization of metabolic pathways and minutely characterizes the genes involved in the bioremediation process (Cho et al. 2004). The modified organisms developed by the insertion of gene of interest can perform the best only if it is introduced within its regulatory network (Cho et al. 2000). The degradation of organophosphate-based pesticides can be enhanced by site-directed mutagenesis. The modified organisms developed on the basis of variants of organophosphate can hydrolyze methyl parathion, an insecticide to 25 fold higher than the wild type (Cho et al. 2004). Cho et al. (2004) has also described in his paper the improvement in the performance of chlorpyrifos hydrolysis by the variants of OPH compared to the wild type. *Variovorax* sp. is well known for the degradation of Linuron, a phenylurea-based herbicide. The libA gene of *Variovorax* sp. strain SRS16 participates in the bioremediation process by undergoing hydrolysis. Another gene named Hyl A that also encodes for hydrolase enzyme is found in *Variovorax* sp. Strain WLD1 (Van Der et al. 1992). A herbicide named aryloxyphenoxypropanoate (AOPP) is degraded by Rhodococcusruber, strain JPL-2. The feh gene that encodes carboxyesterase initiates the bioremediation (Zhang et al. 2002). Another xenobiotic metabolite, petachlorophenol and chlorpyrifos can be efficiently degraded by *Streptomyces* sp. A5 and *Streptomyces* sp. M7, respectively (Zhang et al. 2002).

8.7 Impact of Heavy Metal on Land Quality

With the development of the global economy, heavy metal contamination has been exponentially increasing leading to the deterioration of the ecosystem (Sayyed and Sayadi 2011; Prajapati and Meravi 2014; Zojaji et al. 2014). Excessive anthropogenic activities, improper management of chemical and industrial waste, mining activities, atmospheric deposition are various sources of heavy metal contamination (Zhang et al. 2011). In addition, the fly ash deposits by coal thermal power stations are a major anthropogenic source of heavy metal pollution across the globe (Pandey and Singh 2010; Pandey 2020). These toxic chemicals once accumulated in the body of living organisms disturb the entire food chain reaction. It risks the human life by attacking the cellular components and inducing stress even if it is accumulated at a very trace amount in the body (Ayangbenro and Babalola 2017; Wang et al. 2017). These harmful chemicals largely affect the liver and bones, damage the nervous system, blocks the functional property of essential enzymes (Moore 1990; Ewan and Pamphlett 1996). Few metals which are listed as carcinogens are also noted to cause reproductive disorders. Reports are available on various physical and chemical methods to remove heavy metal (Azimi et al. 2017; Zamri et al. 2017; Fang et al. 2017). However, these technologies have many limitations which as a result generate toxic sludge as the by-product. Thus, an eco-friendly approach to bioremediate heavy metals is a worldwide accepted technology. Microbial-mediated bioremediation of heavy metals is in high demand because of their high-efficiency and cost-effective properties (Leal-Gutierrez et al. 2021; Pandey and Singh 2019). The microorganisms' work on the remediation process by transforming the active form of metals to inactive form. These organisms manage to build up resistance against heavy metals by changing their metabolic activity when the active form of metal act on the microorganisms. Heavy metals come in contact with the soil ecosystem through different pathways described in Fig. 8.1. Industrial activities like mining builds up elevated level of metal concentration in land and also in the wetlands (Bourrin et al. 2021; Leal-Gutierrez et al. 2021; DeVolder et al. 2003). Lead and zinc are the two excessively found metals during ore mining. These metals are later assimilated by the plants and thereby pose serious risk to the secondary level of organisms which includes human being and animals (Basta and Gradwohl 1998). Toxic metals such as chromium, arsenic, nickel, and lead are massively found in tanning, textile, and petrochemical industries. Extensive availability of these hazardous metals can cause serious threat to ecological equilibrium. Solid waste or sewage sludge such as animal waste and municipal solid waste leads to the accumulation of large amounts of heavy metals such as copper, arsenic, cadmium, lead, nickel, mercury, chromium, selerium, and molybdenum in soil (Basta et al. 2005; Sumner 2000). Generally, animal waste are considered to be organic manures for crop growth; however, induction of growth promoters in the diet could participate in the contamination of heavy metals such as arsenic, zinc, and copper in soil (Sumner 2000; Chaney and Oliver 1996). Thus, practice on application of animal waste as organic manure has become a matter of great concern. Additionally, toxic metals available in the land surface could also contaminate groundwater by leaching (McLaren et al. 2004). Fertilizers and pesticides are two sources of agricultural waste that generate large amount of metal deposition in soil. Large quantities of fertilizers (N, P, and K) are added in the crop field to enhance crop growth. However, along with crop growth these fertilizers adversely affect the soil quality and texture. Phosphate-based fertilizers largely build up the concentration of toxic metals in soil such as cadmium, mercury, and lead (Raven et al. 1998). Moreover, insecticides and fungicides also have heavy metals like Cu, Hg, Mn, Pb, or Zn in it. Insecticides used in fruit orchards contain high concentration of lead arsenate. Reports suggest that the pesticides used in banana crops in New Zealand and Australia are mostly arsenic-based (McLaren et al. 2004).

8.7.1 Microorganisms Mediated Heavy Metal Bioremediation Approaches

Microorganisms as bioremediators have gained enough popularity because of their outstanding performance in heavy metal detoxifying strategies. However, choosing an efficient approach is a major challenge (Fig. 8.2). Each of these approaches is discussed underneath.

(a) *Biosparging*–Biosparging is an in-situ-based bioremediation technology where indigenous microorganisms are used to decontaminate polluted land. It is considered to be an advanced form of air sparging since nutrients are also pumped along with air in a saturated zone (USEPA 1995; Muehlberger et al. 1997). It takes about six months to two years to complete the process. The technique works efficiently in permeable soil.

(b) *Bioventing*–The process of bioventing includes injection of air in a contaminated land. This technique degrades volatile organic contaminants and the effectiveness of this technique is reported under in-situ conditions (USEPA 1997). Bioventing is widely used for degrading petroleum products like diesel. The treatment process varies from six months to two years. However, limitation of this technology that this technique cannot be applied to sites with low permeability and high clay content (USEPA 1995).

(c) *Biostimulation*–Biostimulation induces the activity of the microorganisms by addition of rate-limiting nutrients. These artificially added nutrients enhance the performance of the microorganisms by modifying the in-situ environmental

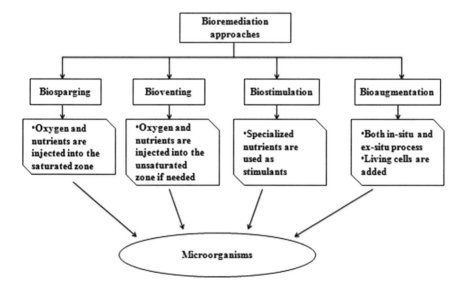

Fig. 8.2 An overview of bioremediation approaches catalyzed by microorganisms

condition which thereby augments the bioremediation potential of the inhabited microbial population (Nikolopoulou and Kalogerakis 2008; Prince 1997). Land contaminated with petroleum hydrocarbon has sufficient amount of C-source for the microorganisms, however, these lands are deficit of nitrogen and phosphorous. The richness of these environments could be stimulated by the addition of nutrients (Sarkar et al. 2005). Besides nutrients, temperature is another parameter that has a considerable effect on biodegradation. The temperature of an environment is correlated with other parameters such as viscosity and solubility properties in water (Atlas 1981). Additionally, biosurfactants are also used to increase the biodegradation rate (Bordoloi and Konwar 2009; Ron and Rosenberg 2002). Addition of nitrogen and phosphorous along with biosurfactants could stimulate the performance of the indigenous microorganisms (Nikolopoulou and Kalogerakis 2008). This could be considered one of the promising approaches to accelerate the process of microbial bioremediation (Baek et al. 2007).

(d) *Bioaugmentation*–Bioaugmentation is one of the successful bioremediation strategies implemented to clean polluted land. The various approaches to perform bioremediation are: addition of pre-adapted a single bacterial strain or in consortium, addition of genetically engineered microorganisms (El Fantroussi and Agathos 2005). Thus, choosing an appropriate technique is quite a challenging task. For eg. if a land is contaminated with wide range of pollutants then a multi-component system could be a better representation of detoxifying land (Ledin 2000). According to study, a consortium-based approach is more advantageous than pure culture-based approaches since it has high metabolic activity (Nyer et al. 2002; Hussain et al. 2018). Alisi et al. (2009) has reported the reduction of isoprenoid and total hydrocarbon concentration to 60% and 75%, respectively in 42 days using microbial consortium. Additionally, Li et al. (2009) have documented that use of *Bacillus* sp., *Zoogloea* sp., and *Flavobacterium* as a microbial consortium in PAH degradation could enhance the degradation rate to 41.3%. Insufficient amount of substrates, competition among species could be few troubleshoot of the process. To overcome these drawbacks, application of carrier material could act as a physical support for the biomass as well as it could increase the survival rate of the organisms by providing better moisture and nutrients (Mishra et al. 2001).

Recent studies have shown that heavy metal removal by living/dead microorganisms are the most widely used technique in bioremediation (Table 8.6). The most primary mechanism of detoxifying heavy metals is by the adsorption of metal components through the slimy layers of bacteria with the help of functional groups like carboxyl, amino, phosphate, and sulfate. The microorganisms involved in the bioremediation process are known to utilize the metal adsorption techniques. For example, *Pseudomonas aeruginosa*, participates widely in the detoxification of mercury by adsorbing mercury ions. These ions once adsorbed by the microorganisms are accumulated by sulfhydral groups present in cysteine-rich protein (Yin et al. 2016). *P. putida* is also one of the efficient mercury detoxifying organisms and with the help of

Table 8.6 List of microorganisms that utilizes heavy metals

Metals	Microorganism
Cu, Hg, Pb	*Ganodermaapplantus*
Cd, Pb	*Phormidiumvalderium*
Cd, Co, Cu, Ni	*Stereumhirsutum*
Ag, Hg, P, Cd, Pb, Ca	*Rhizopusarrhizus*
Cd, Cu, Zn	*Pleurotusostreatus*
Cd, Zn Zn, Ag, Th, U	*Aspergilusniger*
Cd, U, Pb, Co, Ni, Cd	*Citrobacter* spp.
U, Cu, Ni	*Zooglea* spp.
Cu, Zn	*Bacillus* spp.
U, Cu, Ni	*Pseudomonas aeruginosa*
Co, Ni, Cd	*Zooglea* spp.
Cd, U, Pb	*Citrobacter* spp.
Au, Cu, Ni, U, Pb, Hg, Zn	*Chlorella vulgaris*
Cd, Zn Zn, Ag, Th, U	*Aspergillus niger*
Cd, Cu, Zn	*Pleurotus ostreatus*
Ag, Hg, P	*Rhizopus arrhizus*
Cd, Pb, Ca	*Stereum hirsutum*
Cd, Co, Cu, Ni	*Phormidium valderium*
Cd, Pb	*Ganoderma applantus*
Cu, Hg, Pb	*Volvariella volvacea*
Zn, Pb, Cu	*Daedalea quercina*
Co, Ni	*Zooglea* spp.
Cd, U	*Citrobacter* spp.
Cu, Zn	*Bacillus* spp.
Au	*Chlorella vulgaris*
Ag, Hg	*Rhizopus arrhizus*
Th	*Aspergillus niger*
As	*E. coli* strain
Cd	*Alcaligenes eutrophus*; *Stenotrophomonas* sp.
Cr	*Bacillus coagulans, B. Megaterium*
Cr, Cd, Ni	*Desulfovibrio desulfuricans*
Cu	*B. subtilis, Micrococcus luteus, Pseudomonas stutzeri*; *Enterobacter* sp.
Hg	*Pseudomonas K-62*
Pb	*Bacillus* sp.; *P. Aeruginosa*
Zn	*Thiobacillus ferrooxidans*; *Acinetobacter* sp.

(continued)

Table 8.6 (continued)

Metals	Microorganism
Cr, Cd	*Aspergillus* sp., *Rhizopus* sp.
Cr, Pb	*A. lentulus*
Cd, Cu, Pb	*Penicillium chrysogenum*
Cu(II), Pb(II)	*A. niger*
Cd, Hg	*Pleurotus sapidus*
Ni, Cd, Zn, Pb, Cu	*Rhizopus arrhizus*
Zn	*Trametes versicolor*

reductase, the microorganism participates in the reduction of Hg(II) to Hg0 (Wang et al. 2014). The use of mining sites is a global challenge due to the presence of heavy metals and needs urgent attention for their restoration (Ahirwal and Pandey 2021). Therefore, researchers revealed that the Pseudomonas aeruginosa-HMR1 is helpful to remove heavy metals and exhibits plant growth-promoting attributes, and can be used for mine land restoration for agro-forestry purposes (Bhojiya et al. 2021). Interestingly, bacteria named *Arthrobacter viscosus* can detoxify Cr(VI) by reducing it to Cr(III) from its surface in both living and dead forms. Quiton et al. (2018) have reported the use of *Staphylococcus epidermidis* biofilm for the elimination of Cr(VI) from solution. The removal of Zn(II) and Cd(II) is processed by *Rhodobacter capsulatus* and *Bacillus cereus,* respectively (Magnin et al. 2014). *Bacterial firmus* participates in the oxidation and reduction of As(III) and Cr(VI), respectively to decontaminate waste water from heavy metal pollutants (Bachate et al. 2012). The modification in the structural composition of heavy metals upon the action of reductase enzyme helps in the detoxification of heavy metals and thus enhances the bioremediation process via bacterial species. Additionally, extracellular polymeric substances (EPS) found on the microbial cell surface plays an important role in the adsorption of heavy metals. These EPS is made up of carbohydrates, lipids, proteins because of which ionic interaction is generated by the functional groups which thereby helps in the remediation process (Sheng et al. 2013; Wang et al. 2014). Heavy metal adsorption by exopolysaccharides catalyzed by the ionic interaction of the functional groups (carboxyl, phosphonate, amine, hydroxyl) is an energy-independent process that depends on adsorption isotherm models (Kim et al. 1996). Thus, the efficient soil microorganism is a new dimension and eco-friendly approach for sustainable agriculture and environmental development (Singh et al. 2011).

Fungi and algae are also widely known for their capability to accumulate heavy metals from the environment. The primary constituents of fungal cells are composed of chitin, glucuronic acid, phosphate, and polysaccharide. Fungi widely use the technique of metal speciation to implement bioremediation approaches (Lovley and Coates 1997; Eccles 1999). The mechanism involves the management of the transport of metal species via mobilization-immobilization method largely depending on the soluble and insoluble (White et al. 1998; Vachon et al. 1994). The carboxyl, phosphoryl, amine, hydroxyl groups attached to the cellular components actively perform the ion exchange mechanism (Purchase et al. 2009). *Aspergilus niger* is a prolific Pb(II) removing fungus with the help of their biosorption capacities. Another fungal strain named, *Aspergilus fumigates* accumulates Cr(VI) following Freundlich isotherm (Ramrakhiani et al. 2011; Dhal and Pandey 2018). Trichoderma has been employed for bioremediating Cd(II) by Langmuir and Freundlich isotherm model (Amirnia et al. 2015; Bazrafshan et al. 2016). Similarly, *Saccharomyces cerevisiae* is reported to accumulate copper, zinc, and cadmium from a high alkaline environment. The process of accumulation by the microorganisms can be enhanced on addition of sodium chloride (Li et al. 2013). Algae are another group of microorganisms that has a very good property of adsorbing heavy metals. *Fucus vesiculosus* and *Cladophora fascicularis* help in the bioremediation of Pb(II) (Bilal et al. 2018; Poo et al. 2018). The peptide molecules present in the algal environments prevent the organism from heavy metal stress once accumulated (Bilal et al. 2018). *Saccharina japonica* and *Sargassum fusiforme* is well known for the removal of zinc, copper and cadmium (Poo et al. 2018; Rugnini et al. 2018).

8.7.2 Bioremediation of Toxic Metals Through Microbial Biomass and Their Enzymes

Bioremediation is an efficient technique for detoxifying harmful metals from soil system. This technique has offered various potential for its significant advancement. Further research on the mechanism of microbial remediation by various microbial communities could enhance the degradation of pollutants. The inoculation of microbial enzymes is another most favored approach used in the bioremediation processes. The defense pathway of microbial enzymes by changing the redox state of the ions present in the toxic components via oxidation–reduction reaction could efficiently decrease their toxicity (Prokop et al. 2003; Cho et al. 2004). The specialized genes present in the microorganisms regulate the activity of the microbial enzymes and prevent it from environmental stress. For example, *Bacillus* sp. and *Micrococcus* sp. have resistance against Hg and As ions, respectively (Mc Loughlin et al. 2005; Behrens et al. 2007).

Microorganisms have the capacity to produce high-yield biomass. This microbes-based biomass can be utilized as an effective adsorbent that could adsorb heavy metals from the pollutants (Yang et al. 2003). The process is initiated by the binding sites present in the cellular structures of the microorganisms. The two pathways associated with this process are: the active uptake mechanism and the passive uptake mechanism. When heavy metals get entrapped into the cellular membrane with the involvement of biological metabolic cycle it is referred as "active uptake" mechanism. Additionally, when the uptake of the heavy metal into the cellular structure is independent of the metabolic cycle it is termed as "active uptake" mechanism. Scientific reports have suggested that microbial remediation via cells follows biphasic uptake of metals. This means, the biosorption of heavy metals takes place rapidly at the initial phase and gradually the process slows down as the reaction proceeds. It is noteworthy to mention that the process of biosorption is quite sensitive to pH, ionic strength, and presence or absence of organic/inorganic ligands. These physiological parameters are practical limitations associated with this technique. However, these challenges can be met by proper strain selection with enhanced accumulation property. The efficiency of the process could be augmented by inoculation of consortium of strains with high metal biosorption properties (Pumpel et al. 2001). Generally, the naturally formed biomass does not have high adsorption capacity for heavy metal remediation. Thus, acid or alkali base treatments are used to enhance the adsorption capacity of the biomass (Cho et al. 2004). Under acid treatment, the biomass becomes positively charged which thereby generates a strong electrostatic attraction between biomass and heavy metals by opening up additional adsorption sites (Jackson et al. 2006, 2005). Similarly, under alkali treatment, the negative charge of the biomass is increased and thus can generate attraction of the positively charged heavy metals (Jackson et al. 2005; Afriat et al. 2006). Sodium hydroxide, sodium bicarbonate, disodium carbonate is widely used in alkali treatment (de Souza et al. 1996). Thus, optimization of the process in regard to biomass growth, cell development, and constant metal removal property would make the process technically advanced and economically viable.

8.8 Future Prospects

Bioremediation is no doubt a high-efficiency and low-cost technique. Compared to all other remediation technologies, microbe-associated bioremediation is the most feasible approach to pollutant degradation. The usage of appropriate consortium species based on the nature of contaminants has proved the technique to be far more profitable in degradation of complex substances. Considering the environmental factors such as pH, temperature, and ionic strength the rate of bioremediation process is altered and makes the technique quite a challenging one. Despite its wide application, there are still few bottlenecks that need further elucidation to enhance the bioremediation process. Treatment technologies that can deal with the toxicity of pollutants need a special attention in genetic engineering. The overexpression of genes involved in the remediation process can augment the rate of removal efficiency and microbial resistance against pollutants. Research on enzymatic function and their genomic advancement could be another prospect of potential for microbial remediation. This machinery of this technique follows either of the two paths among which one catalyzes the reaction in presence of catalyst and the other describes the quantitative efficiencies under controlled environment. One very important criterion of enzymatic degradation is the requirement of high catalytic efficiency under low substrate concentration. However, implementation of bioremediation technology independent of any cofactors makes the technique more viable and cost-effective. Cofactors free enzymatic method eliminates the regulatory issues involved in genetically modified technologies. This accelerates the solubility and hence the expression of the enzymes. Moreover, use of consortium microorganisms could be another efficient mode of microbial remediation that needs further research. Thus, this process of remediation is a promising and ideal way to greener pastures. Regardless of any mode of microbial bioremediation that is used, the technology offers an efficient and cost-effective way to treat contaminated land.

References

Abhilash PC and Singh N (2009) Pesticide use and application: an Indian scenario. J Hazard Mater 165(1–3):1–12

Adhikari T, Manna MC, Singh MV, Wanjari RH (2004) Bioremediation measure to minimize heavy metals accumulation in soils and crops irrigated with city effluent. Food, Agriculture and Environment 2:266–270

Afriat L, Roodveldt C, Manco G, Tawfik DS (2006) The latent promiscuity of newly identified microbial lactonases is linked to a recently diverged phosphotriesterase. Biochem 45:13677–13686

Ahirwal J, Pandey VC (2021) Restoration of mine-degraded land for sustainable environmental development. Restor Ecol 29(4):e13268. https://doi.org/10.1111/rec.13268

Alisi C, Musella R, Tasso F, Ubaldi C, Manzo S, Cremisini C, Sprocati AR (2009) Bioremediation of diesel oil in a cocontaminated soil by bioaugmentation with a microbial formula tailored with native strains selected for heavy metals resistance. Sci Total Environ 407:3024–3032. https://doi.org/10.1016/j.scitotenv.2009.01.011

Allen HE (2002) Bioavailability of metals in terrestrial ecosystems: importance of partitioning for bioavailability to invertebrates, microbes, and plants. Society of Environmental Toxicology and Chemistry SETAC, Pensacola, Fla

Amirnia S, Ray MB, Margaritis A (2015) Heavy metals removal from aqueous solutions using *Saccharomyces cerevisiaein* a novel continuous bioreactor—biosorption system. Chem Eng J 264:863–872

Andreu V, Pico Y (2004) Determination of pesticides and their degradation products in soil: critical review and comparison of methods. Trends Anal Chemistry 23(10–11):772–789

Arora PK, Kumar M, Chauhan A, Raghava GP, Jain RK (2009) OxDBase: a database of oxygenases involved in biodegradation. BMC Res Notes 2:67

Atlas RM (1981) Microbial-degradation of petroleum-hydrocarbons—an environmental perspective. Microbiol Rev 45:180–209

Ayangbenro AS, Babalola OO (2017) A new strategy for heavy metal polluted environments: a review of microbial biosorbents. Int J Environ Res Public Health 14:94

Azimi A, Azari A, Rezakazemi M, Ansarpour M (2017) Removal of heavy metals from industrial waste waters: a review. Chem Bio Eng Rev. 4:37–59

Bachate SP, Nandre VS, Ghatpande NS, Kodam KM (2012) Simultaneous reduction of Cr(VI) and oxidation of As(III) by *Bacillus firmus* TE7 isolated from tannery effluent. Chemosphere 90:2273–2278

Baczynski TP, Pleissner D, Grotenhuis T (2010) Anaerobic biodegradation of organochlorine pesticides in contaminated soil—significance of temperature and availability. Chemosphere 78(1):22–28

Baek KH, Yoon BD, Kim BH, Cho DH, Lee IS, Oh HM, Kim HS (2007) Monitoring of microbial diversity and activity during bioremediation of crude OH-contaminated soil with different treatments. J Microbiol Biotechnol 17:67–73

Barbash JE, Resek EA (1996) Pesticides in ground water. Ann Arbor Press, Chelsea, MI

Barcelo D, Hennion MC (1997) Trace determination of pesticides and their degradation products in water. Elsevier, Amsterdam

Basta NT, Gradwohl R (1998) Remediation of heavy metal-contaminated soil using rock phosphate. Better Crops 82(4):29–31

Basta NT, Ryan JA, Chaney RL (2005) Trace element chemistry in residual-treated soil: key concepts and metal bioavailability. J Environ Qual 34(1):49–63

Bazrafshan E, Zarei AA, Mostafapour FK (2016) Biosorption of cadmium from aqueous solutions by Trichoderma fungus: kinetic, thermodynamic, and equilibrium study. Desalin Water Treat 57:14598–14608

Behrens MA, Mutlu N, Chakraborty S, Dumitru R, Jiang WZ, La Vallee BJ, Herman PL, Clemente TE, Weeks DP (2007) Dicamba resistance: enlarging and preserving biotechnology-based weed management strategies. Science 316:1185–1188

Bevans HE, Lico MS, Lawrence SJ (1998) Water quality in the Las Vegas Valley area and the Carson and Truckee Riverbasins, Nevada and California, 1992–96. Reston, VA: USGS.U.S. Geological Survey Circular, pp 1170

Bhatt P, Verma A, Gangola S et al (2021) Microbial glycoconjugates in organic pollutant bioremediation: recent advances and applications. Microb Cell Fact 20:72

Bhojiya AA, Joshi H, Upadhyay SK, Srivastava AK, Pathak VV, Pandey VC, Jain D (2021) Screening and optimization of zinc removal potential in Pseudomonas aeruginosa-HMR1 and its plant growth-promoting attributes. Bull Environ Contam Toxicol. https://doi.org/10.1007/s00 128-021-03232-5

Bilal M, Rasheed T, Sosa-Hernández JE, Raza A, Nabeel F, Iqbal H (2018) Biosorption:an interplay between marine algae and potentially toxic elements—a review. Mar Drugs16:65

Bollag JM, Dec J (1998) Use of plant material for the removal of pollutants by polymerization and binding to humic substances. Technical report R-82092, Center for Bioremediation and Detoxification Environmental Resources Research Institute, The Pennsylvania State University, University Park, Pa, USA

Bordoloi NK, Konwar BK (2009) Bacterial biosurfactant in enhancing solubility and metabolism of petroleum hydrocarbons. J Hazard Mater 170:495–505. https://doi.org/10.1016/j.jhazmat.2009.04.136

Bourrin F, Uusoue M, Artigas MC et al (2021) Release of particles and metals into seawater following sediment resuspension of a coastal mine tailings disposal off Portman Bay. Southern Spain. Environ Sci Pollut Res 28:47973–47990

Chaney RL, Oliver DP (1996) Sources, potential adverse effects and remediation of agricultural soil contaminants. In: Naidu R (ed) Contaminants and the soil environments in the Australia-Pacific region. Kluwer Academic Publishers, Dordrecht, pp 323–359

Chen X, Christopher A, Jones JP, Bell SG, Guo Q, Xu F, Roa Z, Wong LL (2002) Crystal structure of the F87W/ Y96F/V247L mutant of cytochrome P-450 cam with 1,3,5-trichlorobenzene bound and further protein engineering for the oxidation of pentachlorobenzene and hexachlorobenezene. J Biol Chem 277:37519–37526

Cho YG, Rhee SK, Lee ST (2000) Effect of soil moisture on bioremediation of chlorophenol-contaminated soil. Biotech Lett 22(11):915–919

Cho CM, Mulchadnani A, Chen W (2004) Altering the substrate specificity of organophosphorus hydrolase for enhanced hydrolysis of chlorpyrifos. Appl Environ Microbiol 70:4681–4685

Cirino PC, Arnold FH (2002) Protein engineering of oxygenases for biocatalysis. Curr Opin Chem Biol 6(2):130–135

Clemens S (2006) Toxic metal accumulation, responses to exposure and mechanisms of tolerance in plants. Biochimie 88:1707–1719

Crisp TM, Clegg ED, Cooper RL, Wood WP, Anderson DG, Baeteke KP, Hoffmann JL, Morrow MS, Rodier DJ, Schaeffer JE, Touart LW, Zeeman MG, Patel YM (1998) Environmental endocrine disruption: an effects assessment and analysis. Environ Health Perspect 106:11

Danila V, Kumpiene J, Kasiuliene A, Vasarevicius S (2020) Immobilisation of metal(loid)s in two contaminated soils using micro and nano zerovalent iron particles: evaluating the long-term stability. Chemosphere 248:126054 de Souza ML, Sadowsky MJ, Wackett LP (1996) Atrazine chlorohydrolase from *Pseudomonas sp.* strain ADP: gene sequence, enzyme purification, and protein characterization. J Bacteriol 178:4894–4900

DeVolder PS, Brown SL, Hesterberg D, Pandya K (2003) Metal bioavailability and speciation in a wetland tailings repository amended with biosolids compost, wood ash, and sulfate. J Environ Qual 32(3):851–864

Dhal B, Pandey BD (2018) Mechanism elucidation and adsorbent characterization for removal of Cr(VI) by native fungal adsorbent. Sustainable Environ Res 28:289–297

Dilek FB, Gokcay CF, Yetis U (1998) Combined effects of Ni(II) and Cr(VI) on activated sludge. Water Res 32:303–312

Donmez G, Aksu Z (2001) Bioaccumulation of copper(II) and nickel(II) by the non-adapted and adapted growing Candida spp. Water Res 35(6):1425–1434

Eccles H (1999) Treatment of metal-contaminated wastes: why select a biological process? Trends Biotechnol 17:462–465

El Fantroussi S, Agathos SN (2005) Is bioaugmentation a feasible strategy for pollutant removal and site remediation? Curr Opin Microbiol 8:268–275. https://doi.org/10.1016/j.mib.2005.04.011

Ewan KB, Pamphlett R (1996) Increased inorganic mercury in spinal motor neurons following chelating agents. Neurotoxicology 17:343–349

Fang J, Li X, Li S, Pan X, Zhang X, Sun J, Shen W, Han L (2017) Internal pore decoration with polydopaminen a no-particle on polymeric ultrafiltration membrane for enhanced heavy metal removal. Chem Eng J 314:38–49

Fantke P, Friedrich R, Jolliet O (2012) Health impact and damage cost assessment of pesticides in Europe. Environ Int 49:9–17

Fenelon JM (1998) Water quality in the White River Basin, Indiana, 1992–96. Reston, VA: USGS.U.S. Geological Survey Circular 1150

Fetzner S (2003) Oxygenases without requirement for cofactors or metal ions. Appl Microbiol Biotechnol 60(3):243–257

Fetzner S, Lingens F (1994) Bacterial dehalogenases: biochemistry, genetics, and biotechnological applications. Microbiol Rev 58(4):641–685

Food and Agricultural Organization (FAO) (2002) FAO/WHO global forum of food safety regulators. Marrakech, Morocco, 28–30 Jan 2002

Fox BG, Borneman JG, Wackett LP, Lipscomb JD (1990) Haloalkene oxidation by the soluble methane monooxygenase from Methylosinus trichosporium OB3b: mechanistic and environmental implications. Biochemistry 29(27):6419–6427

Gadd GM (2000) Bioremedial potential of microbial mechanisms of metal mobilization and immobilization. Curr Opin Biotechnol 11:271–279

Ganie AS, Bano S, Khan N, Sultana S, Rehman Z, Rahman M, Sabir S, Coulon F, Khan MZ (2021) Nanoremediation technologies for sustainable remediation of contaminated environments: recent advances and challenges. Chemosphere 275:130065

Garbisu C, Alkorta I (2003) Review basic concepts on heavy metal soil bioremediation. The European Journal of Mineral Processing and Environmental Protection 3:58–66

Gavrilescu M (2005) Fate of pesticides in the environment and its bioremediation. Eng Life Sci 5:497–526

Gianfreda L, Xu F, Bollag JM (1999) Laccases: a useful group of oxidoreductive enzymes. Bioremediat J 3(1):1–25

Gilden RC, Huffling K, Sattler B (2010) Pesticides and health risks. J Obstet Gynecol Neonatal Nurs 39(1):103–110

Glazer AN, Nikaido H (2007) Microbial biotechnology: fundamentals of applied microbiology, 2nd edn. Cambridge University Press, Cambridge, pp 510–528

Gong P, Xu H, Wang YC, Guo L, Wang XP (2021) Role of forests in the cycling of persistent organic pollutants: absorption and release. Nat Rev Earth Environ 2:182–197

Grosse S, Laramee L, Wendlandt KD, Mc Donald IR, Miguez CB, Kleber HP (1999) Purification and characterization of the soluble methane monooxygenase of the type II methanotrophic bacterium *Methylocystis sp.* strain WI 14. Appl Environ Microbiol 65(9):3929–3935

Hiner ANP, Ruiz JH, Rodri JN et al (2002) Reactions of the class II peroxidases, lignin peroxidase and Arthromycesramosus peroxidase, with hydrogen peroxide: catalase-like activity, compound III formation, and enzyme inactivation. J Biol Chem 277(30):26879–26885

Hites RA (2021) Polycyclic aromatic hydrocarbons in the atmosphere near the Great Lakes: why do their concentrations vary? Environ Sci Technol (this special issue)

Hurley PM, Hill RN, Whiting RJ (1998) Mode of carcinogenic action of pesticides inducing thyroid follicular cell tumors in rodents. Environ Health Perspect 106:437

Hussain N, Das S, Goswami L, Das P, Sahariah B, Bhattacharya SS (2018) Intensification of vermitechnology for kitchen vegetable waste and paddy straw employing earthworm consortium: assessment of maturity time, microbial community structure, and economic benefit. J Clean Prod 182:414–426

Jackson CJ, Carr PD, Kim HK, Liu JW, Herrald P, Mitic N, Schenk G, Smith CA, Ollis DL (2006) Anomalous scattering analysis of *Agrobacterium radiobacter* phosphotriesterase: the prominent role of iron in the heterobinuclear active site. Biochem J 397:501–508

Jackson C, Kim HK, Carr PD, Liu JW, Ollis DL (2005) The structure of an enzyme-product complex reveals the critical role of a terminal hydroxide nucleophile in the bacterial phosphotriesterase mechanism. Biochem Biophys Acta 1752:56–64

Jones JP, O'Hare EJ, Wong LL (2001) Oxidation of polychlorinated benzenes by genetically engineered CYP101 (cytochrome P450cam). Eur J Biochem 268(5):460–1467

Kafilzadeh F, Ebrahimnezhad M, Tahery Y (2015) Isolation and identification of endosulfan-degrading bacteria and evaluation of their bioremediation in Kor River, Iran. Osong Public Health Res Perspect 6(1):39–46

Kevin CJ (2021) Persistent organic pollutants (POPs) and related chemicals in the global environment: some personal reflections. Environ Sci Technol 55(14):9400–9412

Kim JS, Park JW, Lee SE, Kim JE (2002) Formation of bound residues of 8-hydroxybentazon by oxidoreductive catalysts in soil. J Agric Food Chem 50(12):3507–3511

Kim SY, Kim JH, Kim CJ, Oh DK (1996) Metal adsorption of the polysaccharide produced from *Methylobacterium organophilum*. Biotechnol Lett 18:1161–1164

Kole RK, Bagchi MM (1995) Pesticide residues in the aquatic environment and their possible ecological hazards. J Inland Fish Soc India 27(2):79–89

Kopittke PM, Menzies NW, Wang P, McKenna BA, Lombi E (2019) Soil and the intensification of agriculture for global food security. Environ Int 132:105078

Koua D, Cerutti L, Falquet L et al (2009) Peroxi base: a database with new tools for peroxidase family classification. Nucleic Acids Res 37(1):261–266

Kour D, Kaur T, Devi R et al (2021) Beneficial microbiomes for bioremediation of diverse contaminated environments for environmental sustainability: present status and future challenges. Environ Sci Pollut Res 28:24917–24939

Langerhoff A, Charles P, Alphenaar A, Zwiep G, Rijnaarts H (2001) Intrinsic and stimulated in situ biodegradation of Hexachlorocyclohexane (HCH). 6th international HCl and pesticides forum, Poland, pp 132–185

Leal-Gutierrez MJ, Cuellar-Briseño R, Castillo-Garduno AM et al (2021) Precipitation of heavy metal ions (Cu, Fe, Zn, and Pb) from mining flotation effluents using a laboratory-scale upflow anaerobic sludge blanket reactor. Water Air Soil Pollut 232:197

Ledin M (2000) Accumulation of metals by microorganisms—processes and importance for soil systems. Earth Sci Rev 51:1–31. https://doi.org/10.1016/S0012-8252(00)00008-8

Leong KH, Benjamin Tan LL, Mustafa AM (2007) Contamination levels of selected organochlorine and organophosphate pesticides in the Selangor River, Malaysia between 2002 and 2003. Chemosphere 66(6):1153–1159

Li XJ, Lin X, Li PJ, Liu W, Wang L, Ma F, Chukwuka KS (2009) Biodegradation of the low concentration of polycyclic aromatic hydrocarbons in soil by microbial consortium during incubation. J Hazard Mater 172:601–605. https://doi.org/10.1016/j.jhazmat.2009.07.044

Li C, Xu Y, Wei J, Dong X, Wang D, Liu B (2013) Effect of NaCl on the heavy metal tolerance and bioaccumulation of *Zygosaccharomyces rouxii* and *Saccharomyces cerevisiae*. Bioresour Technol 143:46–52

Liu L, Cui GB (2001) Biodegradation of PAHs in soil-water-microbes system. Acta Pedologica Sinica (in Chinese) 38(4):558–568

Lovley DR, Coates JD (1997) Bioremediation of metal contamination. Curr Opin Biotechnol 8:285–289

Magnin JP, Gondrexon N, Willison JC (2014) Zinc biosorption by the purple non sulfur bacterium *Rhodobacter capsulatus*. Can J Microbiol 60:829–837

Mai C, Schormann W, Milstein O, Huttermann A (2000) Enhanced stability of laccase in the presence of phenolic compounds. Appl Microbiol Biotechnol 54(4):510–514

Mamta RJR, Khursheed AW (2015) Bioremediation of pesticides under the influence of bacteria and fungi, Chap. 3. In: Handbook of research on uncovering new methods for ecosystem management through bioremediation, pp 51–72

Marrugo-Negrete J, Pinedo-Hernández J, Díez S (2017) Assessment of heavy metal pollution, spatial distribution and origin in agricultural soils along the Sinu River Basin, Colombia. Environ Res 154:380–388

Mathur SC (1999) Future of Indian pesticide industry in next millennium. Pesticide Information 24(4):9–23

Mc Loughlin SY, Jackson C, Liu JW, Ollis DL (2005) Increased expression of a bacterial phosphotriesterase in *Escherichia coli* through directed evolution. Protein Expr Purif 41:433–440

McLaren RG, Clucas LM, Taylor MD, Hendry T (2004) Leaching of macronutrients and metals from undisturbed soils treated with metal-spiked sewage sludge. 2. Leaching of metals. Australian Journal of Soil Research 42(4):459–471

Meagher RB (2000) Phytoremediation of toxic elemental and organic pollutants. Curr Opin Plant Biol 3:153–162

Mishra S, Jyot J, Kuhad RC, Lal B (2001) In situ bioremediation potential of an oily sludge-degrading bacterial consortium. Curr Microbiol 43:328–335

Moore JW (1990) Inorganic contaminants of surface water residuals and monitoring priorities. Springer, New York, pp 178–210

Morant M, Bak S, Moller BL, Werck-Reichhart D (2003) Plant cytochromes P450: tools for pharmacology, plant protection and phytoremediation. Curr Opin Biotechnol 14:151–162

Muehlberger EW, Harris K, Hicks P (1997) In situ biosparging of a large scale dissolved petroleum hydrocarbon plume at a southwest lumber mill. In: TAPPI proceedings—environmental conference and exhibition, vol 1

Nagata Y, Imai R, Sakai A, Fukuda M, Yano K, Takagi M (1993a) Isolation and characterisation of Tn5-induced mutants of *Pseudomonas paucimobilis* UT26 defective in γ-hexachlorocyclohexane dehydrochlorinase (LinA). Biosci Biotechnol Biochem 57:703–709

Nagata Y, Nariya T, Ohtomo R, Fukuda M, Yano K, Takagi M (1993b) Cloning and sequencing of a dehalogenase gene encoding an enzyme with hydrolase activity involved in the degradation of γ-hexachlorohexane in *Pseudomonas paucimobilis*. J Bacteriol 175:6403–6410

Nagata Y, Mori K, Takagi M, Murzin AG, Damborsky J (2001) Identification of protein fold and catalytic residues of γ-hexachlorocyclohexane dehydrochlorinaseLinA. Proteins 45:471–477

Nagata Y, Prokop Z, Sato Y, Jerabek P, Kumar A, Ohtsubo Y, Tsuda M, Damborsky J (2005) Degradation of β-hexachlorcyclohexane by haloalkane dehalogenase Lin B from *Sphingomonas paucimobilis* UT26. Appl Environ Microbiol 71:2183–2185

Negri A, Marco E, Damborsky J, Gago F (2007) Stepwise dissection and visualization of the catalytic mechanism of haloalkane dehalogenase LinB using molecular dynamics simulations and computer graphics. J Mol Graph Model 26:643–651

Nikolopoulou M, Kalogerakis N (2008) Enhanced bioremediation of crude oil utilizing lipophilic fertilizers combined with biosurfactants and molasses. Mar Pollut Bull 56:1855–1861. https://doi.org/10.1016/j.marpolbul.2008.07.021

Nyer EK, Payne F, Suthersan S (2002) Environment vs. bacteria or let's play 'name that bacteria'. Ground Water Monit Remediat 23:36–45

O'Neil W, Raucher R, Wayzata MN (1998) Groundwater policy education project; groundwater public policy leaflet series no. 4: the costs of groundwater contamination

Pandey VC (2020) Phytomanagement of fly ash. Elsevier (Authored book), p 334. ISBN: 9780128185445. https://doi.org/10.1016/C2018-0-01318-3

Pandey VC, Singh N (2010) Impact of fly ash incorporation in soil systems. Agr Ecosyst Environ 136:16–27. https://doi.org/10.1016/j.agee.2009.11.013

Pandey VC, Singh V (2019) Exploring the potential and opportunities of recent tools for removal of hazardous materials from environments. In: Pandey VC, Bauddh K (eds) Phytomanagement of polluted sites. Elsevier, Amsterdam, pp 501–516. https://doi.org/10.1016/B978-0-12-813912-7.00020-X

Paul K, Chattopadhyay S, Dutta A, Krishna AP, Ray S (2019) A comprehensive optimization model for integrated solid waste management system: a case study. Environ Eng Res 24:220–237. https://doi.org/10.4491/eer.2018.132

Pell M, Stenberg B, Torstensson L (1988) Potential denitrification and nitrification tests for evaluation of pesticide effects in soil. Ambio 27:24–28

Pieterse B, Rijk IJC, Simon E, van Vugt-Lussenburg BMA, Fokke BFH, van der Wijk M, Besselink H, Weber R, van der Burg B (2015) Effect-based assessment of persistent organic pollutant- and pesticide dumpsite using mammalian CALUX report

Piontek K, Smith AT, Blodig W (2001) Lignin peroxidase structure and function. Biochem Soc Trans 29(2):111–116

Poo KM, Son EB, Chang JS, Ren X, Choi YJ, Chae KJ (2018) Biochars derived from wasted marine macro-algae (*Saccharina japonica* and *Sargassum fusiforme*) and their potential for heavy metal removal in aqueous solution. J Environ Manage 206:364–372

Prajapati SK, Meravi N (2014) Heavy metal speciation of soil and *Calotropis procera* from thermal power plant area. Proceedings of the International Academy of Ecology and Environmental Sciences 4(2):68–71

Prescott LM, Harley JP, Klein DA (2002) Microbiology, 5th edn. McGraw Hill, New York, pp 10–14

Prince RC (1997) Bioremediation of marine oil spills. Trends Biotechnol 15:158–160

Prokop Z, Moninvoca M, Chaloupkova R, Klvana M, Nagata Y, Janssen D, Damborsky B (2003) Catalytic mechanism of the haloalkane dehalogenase LinB from *Sphingomona spaucimobilis* UT26. J Biol Chem 278:45094–45100

Pumpel T, Ebner C, Pernfu BB, Schinner F, Diels L, Keszthelyi Z et al (2001) Treatment of rinsing water from electroless nickel plating with a biologically active moving-bed sand filter. Hydrometallurgy 59:383–393

Purchase D, Scholes LN, Revitt DM, Shutes RBE (2009) Effects of temperature on metal tolerance and the accumulation of Zn and Pb by metal-tolerant fungi isolated from urban run off treatment wetlands. J Appl Microbiol 106:1163–1174

Que L, Ho RYN (1996) Dioxygen activation by enzymes with mononuclear non-heme iron active sites. Chem Rev 96(7):2607–2624

Quiton KG, Doma Jr B, Futalan CM, Wan MW (2018) Removal of chromium(VI) and zinc(II) from aqueous solution using kaolin-supported bacterial biofilms of Gram-negative *E. coli* and Gram-positive *Staphylococcus epidermidis*. Sustainable Environ Res 5:206–213

Raina V, Suar M, Singh A, Prakash O, Dadhwal M, Gupta SK, Dogra C, Lawlor K, Lal S, van der Meer JR, Holliger C, Lal R (2007) Enhanced biodegradation of hexachlorocyclohexane (HCH) in contaminated soils via inoculation with *Sphingobium indicum* B90A. Biodegradation 19:27–40

Rajendran P, Muthukrishnan J, Gunasekaran P (2003) Microbes in heavy metal remediation. Indian J Exp Biol 41:935–944

Ramrakhiani L, Majumder R, Khowala S (2011) Removal of hexavalent chromium by heat inactivated fungal biomass of *Termitomyces clypeatus*: surface characterization and mechanism of biosorption. Chem Eng J 171:1060–1068

Raven PH, Berg LR, Johnson GB (1998) Environment, 2nd edn. Saunders College Publishing, New York

Raymond JW, Rogers TN, Shonnard DR, Kline AA (2001) A review of structure-based biodegradation estimation methods. J Hazard Mater 84:189–215

Roberts T R (1998) Metabolic pathway of agrochemicals. Part I. In: Herbicides and plant growth regulators. The Royal Society of Chemistry, Cambridge, UK

Roberts TR, Hutson DH (1999) Metabolic pathway of agrochemicals Part II. In: Insecticides and fungicides. The Royal Society of Chemistry, Cambridge, UK

Rodrıguez Couto S, Toca Herrera JL (2006) Industrial and biotechnological applications of laccases: a review. Biotechnology Advances 24(5):500–513

Rodríguez Eugenio N, McLaughlin M, Pennock D (2018) Soil pollution: a hidden reality, FAO

Ron EZ, Rosenberg E (2002) Biosurfactants and oil bioremediation. Curr Opin Biotechnol 13:249–252

Rugnini L, Costa G, Congestri R, Antonaroli S, diToppi LS, Bruno L (2018) Phosphorus and metal removal combined with lipid production by the green microalga Desmodesmus sp.: an integrated approach. Plant Physiol Biochem 125:45–51

Ruiz-Duenas FJ, Morales M, Perez-Boada M et al (2007) Manganese oxidation site in Pleurotuseryngii versatile peroxidase: a site-directed mutagenesis, kinetic, and crystallographic study. Biochemistry 46(1):66–77

Santos A, Flores M (1995) Effects of glyphosate on nitrogen fixation of free-living heterotrophic bacteria. Lett Appl Microbiol 20:349–352

Saravanan A, Kumar PS, Jeevanantham S, Karishma S, Tajsabreen B, Yaashikaa PR, Reshma B (2021) Effective water/wastewater treatment methodologies for toxic pollutants removal: processes and applications towards sustainable development. Chemosphere 280:130595

Shafi A, Bano S, Sabir S, Khan MZ, Rahman MM (2020) Carbon-based material for environmental protection and remediation. Intechopen

Sarkar D, Ferguson M, Datta R, Birnbaum S (2005) Bioremediation of petroleum hydrocarbons in contaminated soils: comparison of biosolids addition, carbon supplementation, and monitored natural attenuation. Environ Pollut 136:187–195. https://doi.org/10.1016/j.envpol.2004.09.025

Sayyed MRG, Sayadi MH (2011) Variations in the heavy metal accumulations within the surface soils from the Chitgar industrial area of Tehran. Proceedings of the International Academy of Ecology and Environmental Sciences 1(1):36–46

Scott JA, Karanjkar AM (1992) Repeated cadmium biosorption by regenerated *Enterobacter aerogenes* biofilm attached to activated carbon. Biotechnol Lett 14:737–740

Sharma P, Raina V, Kumari R, Shweta M, Dogra C, Kumari H, Kohler HPE, Holliger C, Lal R (2006) Haloalkane dehalogenase Lin B is responsible for β- and δ-hexachlorocyclohexane transformation in *Sphingobium indicum* B90A. Appl Environ Microbiol 72:5720–5727

Sheng GP, Xu J, Luo HW, Li WW, Li WH, Yu HQ, Xie Z, Wei SQ, Hu FC (2013) Thermodynamic analysis on the binding of heavy metals onto extracellular polymeric substances (EPS) of activated sludge. Water Res 47:607–614

Singh BK, Kuhad RC, Singh A, Lal R, Tripathi KK (1999) Biochemical and molecular basis of pesticide degradation by microorganisms. Crit Rev Biotechnol 19(3):197–225

Singh JS, Pandey VC, Singh DP (2011) Efficient soil microorganisms: a new dimension for sustainable agriculture and environmental development. Agr Ecosyst Environ 140:339–353. https://doi.org/10.1016/j.agee.2011.01.017

Sumner ME (2000) Beneficial use of effluents, wastes, and biosolids. Commun Soil Sci Plant Anal 31(11–14):1701–1715

Tang FHM, Lenzen M, Bratney A, Maggi F (2021) Risk of pesticide pollution at the global scale. Nat Geosci 14:206–210

Thassitou PK, Arvanitoyannis IS (2001) Bioremediation: a novel approach to food waste management. Trends in Food Sci Technol 12:185–196

Torres JPM, Froes-Asmus CIR, Weber R, Vijgen JMH (2013) Status of HCH contamination from former pesticide production and formulation in Brazil—a task for Stockholm Convention implementation. Environ Sci Pollut Res 20:1951–1957

Tsukihara T, Honda Y, Sakai R, Watanabe T, Watanabe T (2006) Exclusive overproduction of recombinant versatile peroxidase MnP 2 by genetically modified white rot fungus. Pleurotusostreatus. Journal of Biotechnology 126(4):431–439

U.S. Geological Survey (1995) Pesticides in ground water. U.S. Geological Survey Fact Sheet FS, pp 244–295

U.S. Geological Survey (1998) National water-quality assessment. Pesticide National Synthesis Project. Pesticides in surface and ground water of the United States. Summary of results of the National Water Quality Assessment Program

Ullah MA, Bedford CT, Evans CS (2000) Reactions of pentachlorophenol with laccase from Coriolusversicolor. Appl Microbiol Biotechnol 53(2):230–234

UNEP (1993) The Aral Sea: diagnostic study for the development of an action plan for the conservation of the Aral Sea. Nairobi, Kenya: United Nations Environment Programme (UNEP)

Urlacher VB, Lutz-Wahl S, Schmid RD (2004) Microbial P450 enzymes in biotechnology. Appl Microbiol Biotechnol 64:317–325

USEPA (1995) How to evaluate alternative cleanup technologies for underground storage tank sites. Office of Solid Waste and Emergency Response, US Environmental Protection Agency. Publication # EPA 510-B-95-007, Washington, DC

USEPA (2001) Source water protection practices bulletin: Managing small-scale application of pesticides to prevent contamination of drinking water. DC: Office of Water (July), Washington

Usman S, Usman B (2013) New method of soil classification in defining the dynamic condition of agricultural surface soils. J Environ Sci Toxicol Food Technol 2:32–42

Vachon PR, Tyagi RD, Auclair JC, Wilkinson KJ (1994) Chemical and biological leaching of aluminium from red mud. Environ Sci Technol 28:26–30

Van der Meer JR, De-Vos WM, Harayama S, Zehnder AJB (1992) Molecular mechanisms of genetics adaptation to xenobiotic compounds. Microbiol Rev 56:677–694

Vidali M (2001) Bioremediation: an overview. Pure Applied Chemistry 73:1163–1172

Wang J, Li Q, Li MM, Chen TH, Zhou YF, Yue ZB (2014) Competitive adsorption of heavy metal by extracellular polymeric substances (EPS) extracted from sulfate reducing bacteria. Bioresour Technol 163:374–376

Wang X, Zhang C, Qiu B, Ashraf U, Azad R, Wu J, Ali S (2017) Biotransfer of Cd along a soil-plant-mealybug-ladybird food chain: a comparison with host plants. Chemosphere 168:699–706

White C, Sharman AK, Gadd GM (1998) An integrated microbial process for the bioremediation of soil contaminated with toxic metals. Nat Biotechnol 16:572–575

WHO/UNEP Working Group (1990) Public health impact of pesticides used in agriculture. World Health Organization, Geneva

Xu F (1996) Catalysis of novel enzymatic iodide oxidation by fungal laccase. Appl Biochem Biotechnol 59(3):221–230

Yan DZ, Lui H, Zhou NY (2006) Conversion of *Sphingobium chlorophenolicum* ATCC 39723 to a hexachlorobenzene degrader by metabolic engineering. Appl Environ Microbiol 72:2283–3228

Yang H, Carr PD, Mc Loughlin SY, Liu JW, Horne I, Qiu X, Jeffries CM, Russell RJ, Oakeshott JG, Ollis DL (2003) Evolution of an organophosphate-degrading enzyme: a comparison of natural and directed evolution. Protein Eng 16:135–145

Yin K, Lv M, Wang Q, Wu Y, Liao C, Zhang W, Chen L (2016) Simultaneous bioremediation and biodetection of mercury ion through surface display of carboxyl esterase E2 from *Pseudomonas aeruginosa* PA1.Water Res 103:383–390

Yoshida S (1998) Reaction of manganese peroxidase of *Bjerkandera adusta* with synthetic lignin in acetone solution. J Wood Sci 44(6):486–490

Zakaria Z, Heng LY, Abdullah P, Osman R, Din L (2003) The environmental contamination by organochlorine insecticides of some agricultural areas in Malaysia. Mal J Chem 5:078–085

Zamri MFMA, Kamaruddin MA, Yusoff MS, Aziz HA, Foo KY (2017) Semi-aerobic stabilized land fill leachate treatment by ion exchange resin: isotherm and kinetic study. Appl Water Sci 7:581–590

Zeyaullah M, Atif M, Islam B, Azza S, Abdelkafel SP, ElSaady MA, Ali A (2009) Bioremediation: a tool for environmental cleaning. Afr J Microbio Res 3:310–314

Zhang WJ, Jiang FB, Ou JF (2011) Global pesticide consumption and pollution: with China as a focus. Proceedings of the International Academy of Ecology and Environmental Sciences 1(2):125–214

Zhang YX, Perry K, Vinci VA, Powell K, Stemmer WPC, del Cardayree SB (2002) Genome shuffling leads to rapid phenotypic improvement in bacteria. Nature 415:644–646

Zhong AK, Zhang G (2020) Evidence for major contributions of unintentionally-produced PCBs in the air of China: implications for the national source inventory. Environ Sci Technol 54:2163–2171

Zhou QX, Hua T (2004) Bioremediation: a review of applications and problems to be resolved. Prog Nat Sci 14(11):937–944

Zojaji F, Hassani AH, Sayadi MH (2014) Bioaccumulation of chromium by Zea mays in wastewater-irrigated soil: an experimental study. Proceedings of the International Academy of Ecology and Environmental Sciences 4(2):62–67

Chapter 9
Vermi-Remediation of Metal(loid)s Contaminated Surfaces

Linee Goswami, Subhasish Das, Nazneen Hussain, and Satya Sundar Bhattacharya

Abstract Remediation of metal and metal(oid) contaminated surfaces are very critical for the environment. Extensive use of chemicals, mining, and industrial operations including developmental activities deplete soil quality over time. Several workers developed a number of remedial techniques. However, bio-based remediation was regarded as the safest option so far. Among all available bio-remediation techniques, vermi-remediation was found to be the best suitable, not only in terms of time and cost-effectiveness, but also long-term impact on the soil ecosystem. Here, in this review, we have summarized works done over the years on vermi-remediation. We tried to find out the knowledge gap in the existing works in order to establish a full-proof system in the near future.

Keywords Vermi-remediation · Soil health · Nutrient cycling · Earthworm database · Waste management

9.1 Introduction

Vermi-remediation is a combination of two Latin words "Vermi" meaning "worms" and "remedium" meaning "removal of unwanted substances"; the term coined by Edwards and Arancon (2006). Later, several researchers used this term to define the "use of the detoxifying potential of earthworms" to reduce pollutant load from environmental compartments. Das et al. (2015) used the term "vermiremediation"

L. Goswami (✉)
Department of Biology, School of Science and Technology, Örebro University, Örebro, Sweden
e-mail: linee5.evs@gmail.com

S. S. Bhattacharya
Department of Environmental Science, Tezpur University, Tezpur, Assam, India

S. Das
Department of Environmental Science, Mizoram University (PUC), Aizwal, Mizoram, India

N. Hussain
Department of Biosciences, Assam Don Bosco University, Guwahati, Assam, India

to describe the feasibility of *Metaphire posthuma* in bioremediation of water treatment sewage sludge. Rodriguez-Campos et al. (2014) used the term to describe the role of earthworms in soil remediation and their help in faster degradation of slow-recyclable contaminants. It can easily be explained as "earthworm-based" bioremediation techniques. The process involves earthworm life cycle, feeding behavior, metabolism, and excretion/secretion. Shi et al. (2019) described the life events involved in the vermi-remediation process as: vermi-extraction, vermi-accumulation, vermi-transformation, and drilodegradation. Vermi-accumulation and vermiextraction refer to processes where earthworms ingest and store considerable amounts of contaminants in their body. All these processes are broadly compiled under a single umbrella known as "vermicomposting".

Global census for earthworm population recorded around 3200 species around the world. Out of this, about 500 species are found only in India (Goswami and Mondal 2015). Popularly known as, "Friends of the Farmers", earthworms improve soil quality via improved aeration and balanced nutrition. Earthworms have been classified into detrivores and geophagous based on their food preferences (Lee 1995). Bouche (1977) extended another classification where earthworms are differentiated based on their habitat preference viz. epigeic, anecics, and endogeics. The epigeic earthworm dwells on the upper soil surface, thriving on plant litter (detritivores). Anecics reside below O-horizon and dwell on the soil organic matter (geophytophagous); while endogeic earthworms live in the remote horizons surviving mostly on soil (geophagous) (Tisdale and Oades 1982; Ismail 1997). Epigeic and anecics earthworms are found to be most suitable for vermicomposting (Ismail 1997). However, Lavelle and Martin (1992) studied the influence of endogeic earthworms on soil organic matter in order to analyze their feasibility as vermi-agent.

In the context of vermicomposting technology, several different earthworm species have been reared commercially. Among them, epigeic earthworms (for e.g., *Eisenia fetida, Eudrillus eugeniae*, and *Perionyx excavatus*) have been used widely in recycling of industrial and domestic refuses. Epigeics show proficient waste degradation capacity. They break down complex waste material into fine and granular forms. Yet, works related to their roles in improvement of soil structural stability and physical attributes are scanty. Over the years, benefits of epigeic earthworms on soil fertility management and crop production were validated by several workers (Haque 2006; Wang et al. 2010; Ansari and Ismail 2012; Singh et al. 2012; Anton et al. 2014; Liu et al. 2014). Many authors reported that *E. fetida* vermicompost amendments in soil led to considerable enhancement in soil humification, metabolic quotient, enzyme activity, and soil plasticity (Masciandaro et al. 2000; Garcia Massinyi et al. 2003; Bhattacharya et al. 2012; Singh et al. 2013). High occurrence of nutrient solubilizing microorganisms reported under different treatment combinations of Eisenia vermicompost (Goswami et al. 2013). These microbes play pivotal roles in sustenance of the soil biodiversity and boost crop production (Aira et al. 2002; Arancon et al. 2008; Das et al. 2016). However, despite of varying nature of work done on different aspects of vermicomposting, very limited works have been done to cover long-term impact on soil quality. Prior to that, few limitations of the vermicomposting process need to be properly addressed for its successful large-scale implementation. They are:

1. High feedstock (especially cow dung) demand and cost of continuous supply of the animal manures at the batch reactors.
2. Higher time required for monoculture vermicompost preparations.
3. Lack of pointers to assess the solid retention times for the batch reactors.

In this context, this review intends to discuss the possibilities of commercial vermicomposting techniques for conversion of metal-contaminated substrates and their probable inclusion in the circular economy.

9.2 Factors Governing the Process of Vermicomposting

Vermicomposting is largely governed by species abundance and selection, substrate composition, and atmospheric conditions like moisture content, humidity, etc. Earthworms are rapacious feeders. They can potentially comminute multifarious organic materials within a short time frame. Bhawalkar (1993) observed that earthworms with 1000 kg of biomass require 500 kg day^{-1} amount of food. But they utilize only a minute fraction of such huge consumption. They excrete 95% of the ingested food as vermicast (Kale 1993). While ingestion, the food material gets grounded into finer and soluble forms (Edwards and Lofty 1972). Because earthworm gut harbor diverse microflora (Senapati 1992; Wallwork 1984); naturally the vermicast gets loaded with microbes that aid in further rapid decomposition of the substrate (Altavinyte and Vanagas 1982). Hanc et al. (2019) reported that the maturity and stability of vermicompost can be analyzed through indicators like C/N ratio, N/NH^{4+}, and N/NO^{3-} ions, dissolved organic carbon, and substrate ion exchange capacity.

According to Shinde et al. (1992) vermicompost expedites nutrient (N, P, K) mineralization in soil as compared to farmyard manure. Rasal et al. (1988) reported that introduction of earthworms into sugarcane bagasse showed increase in N solubilization and simultaneously decreased C:N ratio. Similarly, Senapati et al. (1980) showed how earthworm incubation led to rapid reduction in C/N ratio in the substrate. However, a few contradictory reports by workers like Satchell (1983) and Graff (1971) showed that there can be an increase in C:N ratio in the vermicast post-vermicomposting. Graff (1971) stated that vermicasts with high cation exchange capacity show higher concentrations of exchangeable cations like calcium, magnesium, and potassium and available phosphorus. Bano and Devi (1996) observed considerable amounts of nitrogen and potassium in vermicast. Such changes in nutrient availability also varied from substrate to substrate. Kale and Bano (1986) reported 3% Nitrogen (N) in vermicompost. On the other hand, Senappa and Kale (1995) during their studies on aromatic herbal wastes vermicomposting, reported only 0.8–1.0% N in the vermicast. Scott (1988) while studying the nature and properties of cattle manures under composting and vermicomposting, observed that nitrogen availability increases significantly under vermicomposting; whereas occurrence of P and

K showed different trends. Landgraf et al. (1998) established an inverse relationship between humic acid composition and N content in vermicompost. Lavelle and Barois (1988) have stated several potential benefits of earthworm casts in tropical soils. They said that it contributes to increased porosity, water retention capacity, nutrient homeostasis, and improved plant growth. Hanc et al. (2019) reported that Eisenia andrei-mediated horse manure vermicompost transforms aliphatic humic components into aromatic ones with oxygenated functional groups. They observed that tryptophane-like fluorophores were rapidly converting into humic-like fluorophores during the period of vermicomposting. Therefore, study of these indicators is of utmost importance to analyze the vermicomposting status.

9.3 Role of Earthworms in a Vermi-Reactor: Benefits and Limitations

Earthworms can survive within a broad range of temperature 5–29 °C (Sinha et al. 2010); prefers dark and moist habitat with temperature varying from 20 to 25 °C and function within 60–75% moisture content. Studies showed that under optimum temperature, moisture, and sufficient feeding material, earthworms can reproduce at a rate of 28 times in every six months. An average life expectancy of an earthworm is 220 days. They can produce 300–400 hatchlings within that period (Hand 1988; Sinha et al. 2010). Earthworms can shred the ingested food materials to finer forms with the help of the gizzard. The amount of food, they swallow every day is quite large. Earthworms' body walls can absorb dissolved nutrients from the substrate (Sinha et al. 2010). Afterward, intestinal fluids get mixed with the swallowed and absorbed food materials. The intestine serves as a storehouse of microbes and digestive enzymes. Homeostasis and temperature are the limiting factors for proper functioning of earthworm gut microbial activity. Their unique temperature regulatory mechanism prevents enzyme deactivation in the gut, that in turn assists in faster breakdown of complex organic compounds (Prabha et al. 2007). Earthworms and its gut microflora collectively produce wide range of digestive enzymes. For example, clusters of digestive enzymes responsible for carbohydrate metabolism, nitrogen assimilation, and phosphate solubilization (Prabha et al. 2007; Shweta 2012; Das et al. 2021). These include for example, amylase, cellulase, and xylanase are responsible for complex carbohydrate metabolism. Nitrate reductase assists in N assimilation; whereas acid and alkaline phosphatase in P solubilization (Prabha et al. 2007). All these enzymes together with gut-microbes attribute toward rendering a vital service to the terrestrial ecosystems. How the different ecosystem functions are facilitated by soil-dwelling earthworms are shown in Fig. 9.1.

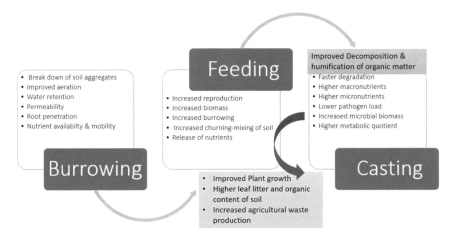

Fig. 9.1 Flow chart of earthworm activities in soil surfaces (*Source* Bhatnagar and Palta 1996)

9.4 Vermi-Remediation and Solid Waste Management

Solid waste generation due to anthropogenic activities is an ever-increasing phenomenon. It contributes to greenhouse gas (GHG) emissions. In India, as per a report published by NIRD & PR (2016), annual rural solid waste generation is about 0.3–0.4 million tons. This has a significant and expensive global footprint. Hoornweg and Bhada-Tata (2012) reported an average global expenditure of 205.4 USD per ton of solid waste for its proper management and disposal. This expenditure is happened to raise further up to five folds by 2025–2030 (Das et al. 2019). Open burning and transportation of MSWs is the primary source of black carbon and CO_2 to act as immediate and long-term pollution hazards in urban areas (Gupta et al. 2015). The GHGs CO_2 and CH_4 are fundamental sources of global rise in temperature and climate change (Khatib 2011). Despite the significance of sustainable waste management, the technological standards of these practices are poor in most countries except a few (Mallawarachch and Karunasena 2012; Hasan 2011).

Waste generation is intricately related to environmental pollution (Brunner 2013). Municipal solid waste runoffs are responsible for leaching of metals and toxic compounds. These compounds contaminate soil and water (both surface and ground) sources (Das et al. 2021). A study carried out in the city of Riyadh, researchers showed leachates collected from municipal landfills with chemical oxygen demand range up to 22.3 g L^{-1}, presence of heavy metals like cadmium (10 ppm), chromium (242 ppm), copper (234 ppm) and nickel (166 ppm) (Al-Wabel et al. 2011). Fan et al. (2006) showed variations among landfill leachate composition in Taiwan. In the study, leachate generated from closed landfills showed different physicochemical characteristics as compared to those generated from mixture of MSW and bottom ash and direct MSW dump. Direct dumps showed higher load of toxic elements like lead (Pb), chromium (Cr), and cadmium (Cd). Landfills, incineration, and direct dumps have been the primary methods of solid waste disposal (Hasan 2011). However, these methods have dire impacts on the environmental compartments. In this context, indigenous technologies like vermicomposting and aerobic composting look promising to recycle MSW (Arebey et al. 2011).

As stated earlier, vermitechnology is a process carried out by the joint action of earthworms and the gut-microbiota to decompose miscellaneous solid wastes (Bhattacharya and Kim 2016). Vermicomposting of waste accelerates organic matter mineralization rate as well as humification process. Thus, a detox product, rich in nutrients and phytohormones, is formed. Such products have higher beneficial microbe load and stable humic components (Singh and Sharma 2002). Vermiremediation of industrial solid wastes was reported by several researchers across the globe (Goswami et al. 2014; Das et al. 2016, 2015; Hickman and Reid 2008a, b). Table 9.2 summarizes a gist of some case studies where vermicomposting is successfully adopted for transformation of several industrial solid wastes (Bhattacharya and Kim 2016).

9.5 Vermicompost and Soil Health Management

9.5.1 Impact on Nutrient Cycling

Advantages of vermicomposting on specific soil properties are well documented (Brady 1984; Tisdale and Nelson 1975). Due to high nutrient status, vermicompost application is likely to increase soil fertility as compared to other organic amendments (Satchell 1983). Edwards and Burrows (1988a, b) reported significantly high concentration of major nutrients in vermicompost applied crop fields. Kale (1993) stated that vermicompost may be used as conditioner in the field to prevent organic carbon deficiency and soil erosion. Soil vermicompost application adds large numbers of beneficial bacteria (Atlavinyte et al. 1971). Kale (1993) stated that since existing microbial population fails to remain active in tropical soils due to lack of active energy demand, regular applications of vermicast likely to improve soil's physico-chemical and biological attributes. Buchanan and Gliessman (1990) reported positive effects of vermicompost application in a P-fatigued soil. On the other hand, Mitchell and Alter (1993) used vermicompost in suppressing Al toxicity in acidic soils. In a study, on nutrient status of rice soils under treatment with vermicompost and inorganic fertilizers, Vasanthi and Kumaraswamy (1996) reported vermicompost treated soil to exhibit higher amount of macro and micronutrients than the soil receiving only inorganic NPK. Higher nutrient status of vermicompost tends to contribute to the profound growth of other agricultural cash crops like potatoes, tomatoes, cabbage, chilli, etc. (Bhattacharya et al. 2012; Goswami et al. 2017). Goswami et al. (2017) reported positive influence of vermicompost on the yield of winter vegetables like tomato and cabbage. Desai (1993) reported good yields of capsicum followed by tomato from a vermicompost treated plot. Similar results were also evident post-vermicompost application in cultivation of Coccinia cordifolia, onion, okra, lettuce, etc. (Desai 1993; Khamkar 1993; Seno et al. 1995; Jambhekar 1966). Galli et al. (1990) reported a considerable increase in protein synthesis in lettuce following the use of vermicast. Szczech and Brzeski (1994) observed the potential of vermicompost as biopesticide against some plant diseases. Edwards and Burrows (1988a, b) reported considerably higher rate of seedling emergence under vermicompost application on performance of ornamental plants. They reported that some of the plants like salvia, petunia, etc. exhibit early flowering under vermi-treatments. This behavioral attribute correlated with the release of phytohormones under the influence of vermicompost. Reddy (1988) reported similar benefits of vermicompost on other ornamental plants. Further analysis by Senapati (1993) showed the impact of vermicompost on crop production summarized with narrowing fertilization gap. It improved moisture content, enhanced essential element availability, and restricted leaching of nutrients. In addition, presence of various plant growth-promoting substances in vermicompost has been attributed to improve crop growth and yield, as reported by different workers (Tomati et al. 1983; Springett and Syres 1978). Sahariah et al. (2020) reported that vermi-amendments help in soil sustenance under a rice-based agro-ecosystem.

9.5.2 Mechanism of Heavy Metal Detoxification: Role Metallothionein Isoforms in Vermi-Remediation

Soil is a heterogeneous substrate where earthworms reside. Elemental composition of soil depends upon the parent rock composition. Therefore, it invariably controls the presence of absence of certain elements in the surface. Evolution of defense mechanism and physiological functioning of an earthworm favors them to survive under diverse environmental conditions (Li et al. 2011). For example, low metal concentration in the substrate often gets compensated by overexpression of molecular signatures at cellular level in earthworms. Whereas exceeding the tolerance limit for metal stress leads to activation of stress proteins or upregulation of distinct transport channels to excrete the toxic elements from their body (Dallinger and Hockner 2013). Fredericq (1878) made a significant discovery of presence of proteolytic enzymes in earthworms. Hussain et al. (2016) reported that earthworm alimentary tract harbors several other digestive enzymes. Afterward, detailed investigation showed their potential participation in earthworm stress physiology.

Among all metal inducible proteins, metallothioneins are widely studied ones owing to their excellent cation sequestering capacity. These proteins maintain non-essential metal homeostasis and carry out detoxification process. They are ubiquitous in nature (Struzenbaum et al. 1998). Struzenbaum et al. (1998) isolated and sequenced earthworm metallothionein (MT) using advance techniques like q-PCR and directed differential display. During the study, it was ascertained that MTs have low molecular weight (6–13 KDa) and has two isoforms (α and β). Localization of metallothionein in earthworm intestines showed their presence, particularly in the gut epithelium, coelomocytes, nephridia, and typhlosole (Dallinger et al. 1997; Sturzenbaum et al. 2001, 2013; Brulle et al. 2006). They bear high sulfur content in the form of cysteine residues; whereas aromatic and histidine residues are absent (Demuynck et al. 2006). They have been found to have high affinities toward the divalent and trivalent metals (Kägi 1991). MTs detoxify toxic metals like Cd and Hg by chelating them in their thiolate bonds forming organometallic complexes (Masters et al. 1994; Dabrio et al. 2002).

Apart from MT, Liebeke et al. (2013) reported a phytochelatin-mediated $As^{3+/5+}$ (arsenic) detoxification pathway in earthworm gut. Goswami et al. (2016) reported presence of a high molecular weight non-metallothionein protein, in *E. fetida*, capable of sequestering Cd^{2+} from the substrate. Later on, Hussain et al. (2021) isolated and identified N-terminal sequences of 15 amino acids rich in glutamic acid. It was hypothesized that the identified protein sequesters Cd^{2+} ions through glutamic acid-based charge-dependent channels. All of these metal inducible proteins can efficiently bind toxic metals and help in converting them into non-reactive and non-toxic forms inside the cell (Liebeke et al. 2013; Chiaverini and Ley 2009; Das et al. 2020). Chelated metal compounds get stored in the free-floating chloragogenous tissues of the coelomic fluid (Vijver et al. 2003; Sturzenbaum et al. 2013) and remain as immobilized entities for a long period of time.

MTs in *E. fetida* are reported as sensitive biomarkers of Cd exposure (Demuynck et al. 2006). Morgan et al. (2004) detected MT-induced chloragogen tissues in the whole body of *Lumbricus rubellus* under metal exposure. Goswami et al. (2014) during the study of vermi-remediation of coal ashes, reported induction of MT proteins in directly exposed *E. fetida*. The radioactive biomarker study used to quantify induced metallothionein. It vindicated that level of production of MT in *E. fetida* increases with duration of exposure to metal-rich CA. The study concluded that application of vermi-technology is a feasible option to augment environmental compatibility of metal-rich coal ash as substrate. Figure 9.2 shows how earthworm gut-microbiota affects the metabolism that in turn enables the worm to sequester toxic metal and sanitize the substrate (Wang et al. 2021).

A study, reported by Bhattacharya and Chattopadhyay (2006) showed significant reduction in solubility of non-essential metals like Pb, Cr, and Cd in Eisenia vermicompost. Maity et al. (2009) observed bioaccumulation of Cd and Zn by *Lampito mauritii*, whereas Nannoni et al. (2011) reported *Allolobophora rosea* and *Nicodrilus caliginosus* capable of accumulating heavy metals significantly. Interestingly, Goswami et al. (2013) have observed differential bioaccumulation tendency for specific metals under *E. fetida* vermicompost. Factors like interspecific dietary preferences, biochemical demands, physiological functions, and morphological attributes greatly influence the bioaccumulation behavior of earthworms (Goswami et al. 2016). Nannoni et al. (2011) opined that earthworms tend to accumulate non-essential elements in high quantity as compared to that of essential elements (Fe, Mn, and Zn). Maity et al. (2009) reported metals like As, Hg, Al, Fe tend to have higher detoxification rates as they form organometallic complexes with metallothionein isoforms.

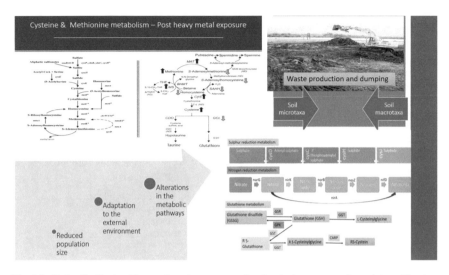

Fig. 9.2 Role of soil microbiota and earthworm gut microbes in the process of metal detoxification: activation of metabolic pathways in earthworm body when exposed to Vanadium contaminated surfaces (*Source* Wang et al. 2021)

9.6 Detoxification of Soil Environment

The unique methods of sequestration and immobilization of toxic metals by earthworms have been studied extensively in the recent past (Goswami et al. 2014; Dai et al. 2004; Maity et al. 2009; Nannoni et al. 2011; Sturzenbaum et al. 2013; Chachina et al. 2015). When earthworms ingest a metal-rich substrate and excrete a detox product, studies confirmed that the surface becomes rich in nutrients, PGPR microbes, and humified carbons. Now, applications of such substrates showed positive influence on the soil properties. Tables 9.1 and 9.2 summarizes the probable mechanisms of earthworm-pollutant interactions in soil and vermibeds.

Earthworm behavior immensely contributes toward building soil's physical properties. Due to their constant burrowing behavior, both porosity and moisture retention capacity increase in a surface. Therefore, when soil surface is exposed to any sort of xenobiotic substance, it gets dissolved easily and gradually starts decomposing. Earthworm accelerates this process either directly or indirectly. Soil formation is largely governed by the weathering process (Brady 1984). Therefore, parental substrate composition determines the elemental composition of soil. Plants contribute significantly to the natural polyphenols in the soil system (Das et al. 2020). Anthropogenic activities contribute to the presence of petroleum, nanoparticles, persistent organic compounds, etc. in the soil systems (Chachina et al. 2016; Goswami et al. 2017; Boyle et al. 1997).

Sizmur and Hodson (2009) suggested that earthworm-mediated soil physicochemical changes tend to influence metal mobility in the surface. Ma et al. (2002) reported *Pheretima sp.* significantly reduced both exchangeable and bound forms of Zn from

Table 9.1 Role of earthworms in pollutant detoxification and their fate post vermi-remediation

Sl No.	Earthworm species	Pollutant	Source of pollutants	Fate	References
1	*Eisenia fetida*	Pb, Zn	Smelting industry	Selective accumulation of Pb in the gut	Vasanthi and Kumaraswamy (1996)
	Lumbricus rubellus				
2	*Lumbricus rubellus*	Cu, Zn, Pb, Cd, Mn	–	Selective accumulation of Pb, Mn, Cd in gut	Ghosh et al. (1999)
	Dendrobaena veneta			(*L. rubellus* > *D. veneta* > *E. tetraedra*)	
	Eiseniella tetraedra				
3	*Eisenia Andrei*	Pb, As, Cd, Ni, Cr, Cu, Zn	Heavy engineering zone	Significant accumulation of Pb, As, Cd, Ni, Cr	Desai (1993)
	Eisenia hortensis	Pb, Cd, Cu, Zn, As, Sb	Lead recycling industry	Significant accumulation of Pb, Cd, Cu, As, Sb	Khamkar (1993)
	Lumbricus terrestris				
4	*Aporrectodea caliginosa*	Zn, Cd, Pb, Cu	Metallurgy unit	Significant accumulation of Zn, Cd, Pb, Cu	Senapati (1992)
	Lumbricus rubellus				
5	*Dendrobaena rubida*	Pb, Zn	Mine spoilage	Higher gut accumulation of Pb than Zn	Matsumoto (1982)
6	*Lumbricus terrestris*	Pb, Zn	Lead mine	Comparatively similar accumulation of Pb and Zn	Brady (1984)
7	*Allolobophora sp.*	Cd, Pb, Zn	Industrial sludge	Significantly higher Cd and Pb accumulation	Seno et al. (1995)
	Lumbricus terrestris				
8	*Allolobophora caliginosa*	DDT, dieldrin	Pesticides	Approximately 66% degradation of dieldrin	Jambhekar (1966)
9	*Eisenia Andrei*	Dimethoate	Pesticides	Degradation of dimethoate	Galli et al. (1990)

(continued)

Table 9.1 (continued)

Sl No.	Earthworm species	Pollutant	Source of pollutants	Fate	References
10	*Eisenia Andrei*	Chlorpyrifos	Pesticides	Break down of Chlorpyrifos	Szczech and Brzeski (1994)

Table 9.2 Industrial wastes and vermi-remediation processes

Earthworm species	Industrial waste	Mode of action	Reference
Eisenia fetida	Olive mill sludge	Significant degradation of alpechin	Mallawarachchi and Karunasena (2012)
Eisenia fetida	Wastewater sludge	Reduction in Cu and Cr contents	Das et al. (2015)
Eisenia fetida	Thermal power plant ash	Stabilization of Cu, Fe, Mn, Zn bioavailability	Bhattacharya and Chattopadhyay (2006)
Eisenia fetida	Leather processing waste	Degradation of amide complexes and chromate content	Nunes et al. (2016)
Eisenia fetida	Distillery waste	Stabilizes nutrient content	Mahaly et al. (2018)
Eisenia andrei	Grape-marc	Develop polyphenol-free fertiliser	Domínguez et al. (2014)
Eisenia fetida	Sago waste	Stabilize N, Ca, and S content	Subramanian et al. (2010)
Eisenia fetida	Paper mill waste	Reduce Cd, Cr, Ni, Pb concentration	Goswami et al. (2013)
Metaphire posthuma	Jute mill waste	Significant accumulation of Pb and Cr	Das et al. (2015)
Eudrilus eugeniae	Textile sludge	Prominent bioaccumulation of Pb, Zn, Cd, and Cr	Paul et al. (2018)
Eudrilus eugeniae	Tannery sludge	Reduction in Cr (89%), Cd (88%) and Zn (79%) contents	Goswami et al. (2018)
Eudrilus eugeniae			
Eisenia fetida	Tea-industry coal ash	Reduction in Cr and increase in Fe, Mn, Zn contents	Goswami et al. (2014)

contaminated surface. Ireland (1979) reported significant bioaccumulation of Pb, Mn, and Cd by *Lumbricus rubellus, Dendrobaena veneta,* and *Eiseniella tetraedra,* when collected from polluted soil near a smelting industry. In a similar work, Roberts and Johnson (1978) showed a relatively higher accumulation of Pb than Zn by earthworm species Dendrobaena rubida exposed to mine spoilage polluted soils (Table 9.1). Such selectivity varied greatly among different earthworm species and soil types (Udovic and Lestan 2007).

Earthworms can survive and proliferate under petroleum-based hydrocarbons [e.g., poly aromatic hydrocarbons (PAH) and polychlorinated biphenyls (PCBs)] contaminated soils (Rodriguez-Campos et al. 2014; Singer et al. 2001; Hickman and Reid 2008a, b). According to Rodriguez-Campos et al. (2014), *Lumbricus terrestris* can immobilize PAH and PCB (Table 9.1). Shan et al. (2011) reported that earthworms efficiently sorb chlorophenols in soil; thereby reduced bioavailability. On the contrary, Verma and Pillai (1991) exhibited earthworm facilitated solubilization of DDT and HCH (pesticides) in soil. Pesticides, like chlorpyrifos, glyphosate, atrazine degradation efficiencies of earthworms were studied extensively by several workers in recent past (Santos et al. 2011; Tejada et al. 2011; Buch et al. 2013). Chachina et al. (2016) showed earthworm species like (e.g., *E. fetida, Eisenia andrei, Dendrobena veneta*) could be useful proposition for bioremediation of petroleum, diesel, and engine oil-contaminated soils. They observed a 99% decline in petroleum-based hydrocarbon contents in soil after 22 weeks of incubation with *E. fetida*. On the other hand, engine lubricant oil concentrations were reduced by a sharp 60–90%. It was interesting as the study was conducted in the temperate soils of Russia, place of their origin. These findings clearly indicate that earthworms are capable enough to detoxify surfaces. However, despite of the highly fluctuating response toward external factors, these exotic earthworm species can be widely used for reclamation of polluted soils.

Liebeke et al. (2015) reported presence of a hitherto unknown metabolite, drilodefensin, in earthworm gut that readily degrades polyphenols in soil environment. It was opined that because of this metabolite, earthworms are highly tolerant to the plant-exuded toxins. Biswas et al. (2018) reported induction of multiple heavy metal resistant genes in worm gut, when they are exposed to metal-contaminated soils. Several researchers have studied expression of genes like sod (superoxide dismutase gene), CYP450 (cytochrome P450), gst (Glutathione S-transferase) associated with reactive oxygen species and oxidative stress in earthworms exposed to certain abiotic factors (Shi et al. 2018; Sun et al. 2020). They are responsible for maintaining homeostasis in earthworm gut. But very limited reports are available to explain the mechanisms involved in these pathways. Therefore, intensive and inclusive research is needed to understand how these functional genes regulate earthworm behavior and their adaptive abilities under different environmental conditions.

9.7 Future Prospects

Even though, earthworm database has recently been compiled. Yet, till date, these databases are limited to only Lumbricidae and Megascolecidae. Under the heading of Global Biodiversity information facility, it holds information of about 105 worm-relevant datasets out of the total of 47,088 datasets. These data can provide information about taxonomic distribution and importance of worms found in America and European countries only (Sun et al. 2020). Among all of them, only three datasets namely Lumbribase (earthworms.org), Earthworm species (earthworm.uw.hu), and E-growth (http://www.jerome-mathieu.com/) are available, where sequences isolated from unknown earthworms can be uploaded and compared. Even though, E-growth used to collect datasets from 1900 to 2016 published in all earthworm-related findings; yet, it has a record of only 1073 growth curves of 51 species of earthworms residing under different habitats (Sun et al. 2020). Lack of information hinders compilation of heterogeneity of the global earthworm population data. Therefore, to prepare a comprehensive report, more detailed studies are needed. Molecular signatures of earthworms around the world need to be analyzed. These will keep a stock of the present status quo of these sensitive organisms.

Acknowledgements First author LG would like to acknowledge the financial assistance received in the form of UGC Dr DS Kothari post-doctoral fellowship (BL/18-19/0215) for the year 2019–2022.

References

Aira M, Monroy F, Dominguez J, Mato S (2002) How earthworm density affects microbial biomas and activity in pig manure. Eur J Soil Biol 38:7–10. https://doi.org/10.1016/S1164-5563(01)01116-5

Al-Wabel MI, Al Yehya WS, AL-Farraj AS, El-Maghraby SE (2011) Characteristics of landfill leachates and bio-solids of municipal solid waste (MSW) in Riyadh City, Saudi Arabia. J Saudi Soc Agricult Sci 10:65–70

Ansari AA, Ismail SA (2012) Role of earthworms in vermitechnology. J Agric Technol 8(2):403–415

Anton D, Matt D, Pedastsaar P, Bender I, Kazimierczak R, Roasto M, Kaart T, Luik A, Pussa, T (2014) Three-year comparative study of polyphenol contents and antioxidant capacities in fruits of tomato (Lycopersicon esculentum Mill.) Cultivars grown under organic and conventional conditions. J Agric Food Chem 62:5173–5180

Arancon NQ, Edwards CA, Babenko A, Cannon J, Galvis P, Metzger JD (2008) Influences of vermicomposts, produced by earthworms and microorganisms from cattle manure, food waste and paper waste, on the germination, growth and flowering of petunias in the greenhouse. Appl Soil Ecol 39(1):91–99

Arebey M, Hannan MA, Basri H, Begum RA, Abdullah H (2011) Integrated technologies for solid waste bin monitoring system. Environ Monit Assess 177(1–4):399–408. https://doi.org/10.1007/s10661-010-1642-x

Atlavinyte O, Vanagas J (1982) The effect of earthworms on the quality of barley and rye and grain. Pedobiologia 23(3):256–262

Atlavinyte O, Daciulyte J, Lugauskas A (1971) Correlation between the number of earthworms, microorganisms and vitamin B_{12} in soil fertilized with straw. Liet TSR Mokslu Akad Darb Ser B 3:43–56

Bano K, Devi L (1996) Vermicompost and its fertility aspects. In: Proceeding of national seminar on organic farming and sustainable agriculture, Bangalore, p 37

Bhatnagar RK, Palta R (1996) Earthworm vermiculture and vermicomposting. Kalyani Publishers, Ludhiana

Bhattacharya SS, Chattopadhyay GN (2006) Effect of vermicomposting on the transformation of some trace elements in fly ash. Nutr Cycl Agroecosyst 75:223–231. https://doi.org/10.1007/s10 705-006-9029-7

Bhattacharya SS, Iftikar W, Sahariah B, Chattopadhyay GN (2012) Vermicomposting converts fly ash to enrich soil fertility and sustain crop growth in red and lateritic soils. Resour Conserv Recycl 65:100–106

Bhattacharya SS, Kim K-H (2016) Utilization of coal ash: is vermitechnology a sustainable avenue? Renew Sustain Energy Rev 58:1376–1386. https://doi.org/10.1016/j.rser.2015.12.345

Bhattacharya SS, Kim KH, Ullah MA, Goswami L, Sahariah B, Bhattacharyya P, Cho SB, Hwang OH (2016) The effects of composting approaches on the emissions of anthropogenic volatile organic compounds: a comparison between vermicomposting and general aerobic composting. Environ Pollut 208:600–607. https://doi.org/10.1016/j.envpol.2015.10.034

Bhawalkar VU, Bhawalkar US (1993) Vermiculture: the bionutrition system; National seminar on indigenous technology for sustainable agriculture, I.A.R.I, New Delhi, 23–24 March, pp 1–8

Biswas JK, Banerjee A, Rai M, Naidu R, Biswas B, Vithanage M, Dash MC, Sarkar SK, Meers E (2018) Potential application of selected metal resistant phosphate solubilizing bacteria isolated from the gut of earthworm (*Metaphire posthuma*) in plant growth promotion. Geoderma 330:117–124

Bouche MB (1977) Strategies *lombriciennes*. In: Lohm U, Persson T (eds) Soil organisms as components of ecosystems. Ecological bulletin, vol 25. Stockholm, Sweden, pp 122–132

Boyle KE, Curry JP, Farrell EP (1997) Influence of earthworms on soil properties and grass production in reclaimed cutover peat. Biol Fertil Soils 25(1):20–26

Brady NC (1984) The nature and properties of soils. Macmillan Publishing Company, NY, USA

Brulle F, Mitta G, Cocquerelle C, Vieau D, Lemière S, Leprêtre A, Vandenbulcke F (2006) Cloning and real-time PCR testing of 14 potential biomarkers in Eisenia fetida following cadmium exposure. Environ Sci Technol 40(8):2844–2850

Brunner PH (2013) Cycles, spirals and linear flows. In: waste management & research: the journal of the international solid wastes and public cleansing association, ISWA. https://doi.org/10.1177/0734242X13501152

Buch AC, Brown GG, Niva CC, Sautter KD, Sousa JP (2013) Toxicity of three pesticides commonly used in Brazil to *Pontoscolex corethrurus* and *Eisenia andrei*. Appl Soil Ecol 69:32–38. https://doi.org/10.1016/j.apsoil.2012.12.011

Buchanan MA, Gliessman SR (1990) The influence of conventional and compost fertilization on phosphorus use efficiency by broccoli in a phosphorus deficient soil. Am J Altern Agric 5(1):38–46

Chachina SB, Voronkova NA, Baklanova ON (2015) Biological remediation of the engine lubricant oil-contaminated soil with three kinds of earthworms, *Eisenia fetida, Eisenia andrei, Dendrobena veneta*, and a mixture of microorganisms. Procedia Engineering 113:113–123. https://doi.org/10.1016/j.proeng.2015.07.302

Chachina SB, Voronkova NA, Baklanova ON (2016) Biological remediation of the petroleum and diesel contaminated soil with earthworms *Eisenia Fetida*. Procedia Engineering 152:122–133. https://doi.org/10.1016/j.proeng.2016.07.642

Chiaverini N, Ley MD (2009) Protective effect of metallothionein on oxidative stress-induced DNA damage. Free Radical Res. https://doi.org/10.3109/10715761003692511

Dabrio M, Rodriguez AR, Bordin G, Bebianno MJ, de Ley M, Sestakova I, Vasak M, Nordberg M (2002) Recent developments in quantification methods for metallothionein. J Inorg Biochem 88:123–134

Dai J, Becquerb T, Rouiller HJ, Reversata G, Reversata FB, Lavelle P (2004) Influence of heavy metals on C and N mineralisation and microbial biomass in Zn-, Pb-, Cu-, and Cd-contaminated soils. Appl Soil Ecol 25(2):99–109

Dallinger R, Berger B, Hunziker P, Kagi JH (1997) Metallothionein in snail Cd and Cu metabolism. Nature 388:237–238

Dallinger R, Hockner M (2013) Evolutionary concepts innecotoxicology: tracing the genetic background of differential cadmium sensitivities in invertebrate lineages. Ecotoxicology. https://doi.org/10.1007/S10646-013-1071-Z

Das S, Bora J, Goswami L, Bhattacharyya P, Raul P, Kumar M, Bhattacharya SS (2015) Vermiremediation of water treatment plant sludge employing *Metaphire posthuma*: a soil quality and metal solubility prediction approach. Ecol Eng 81:200–206. https://doi.org/10.1016/j.ecoleng.2015.04.069

Das S, Deka P, Goswami L, Sahariah L, Hussain N, Bhattacharya SS (2016) Vermiremediation of toxic jute mill waste employing *Metaphire posthuma*. Environmental Science Pollution. https://doi.org/10.1007/s11356-016-6718-x

Das, S Lee S-H, Kumar, P Kim, K-H, Lee, S-S, Bhattacharya SS (2019) Solid waste management: scope and the challenge of sustainability. J Cleaner Prod 228:658–678. ISSN 0959-6526. https://doi.org/10.1016/j.jclepro.2019.04.323

Das S, Teron R, Duary B, Bhattacharya SS, Kim K-H (2019) Assessing C–N balance and soil rejuvenation capacity of vermicompost application in a degraded landscape: a study in an alluvial river basin with Cajanus cajan. Environ Res 177:108591. ISSN 0013-9351. https://doi.org/10.1016/j.envres.2019.108591

Das S, Goswami L, Bhattacharya SS (2020) Vermicomposting: earthworms as potent bioresources for biomass conversion, Chap. 3. In: Kataki R, Pandey A, Khanal SK, Pant D (eds) Current developments in biotechnology and bioengineering, Elsevier, pp 79–102. ISBN 9780444643094

Das S, Sarkar S, Das M, Banik P, Bhattacharya SS (2021) Influence of soil quality factors on capsaicin biosynthesis, pungency, yield, and produce quality of chili: an insight on Csy1, Pun1, and Pun12 signaling responses. Plant Physiol Biochem 166:427–436. ISSN 0981-9428. https://doi.org/10.1016/j.plaphy.2021.06.012

De Fratis JR, Banerjee MR, Germida JJ (1997) Phosphate-solubilizing rhizobacteria enhance the growth and yield but not phosphorus uptake of canola (*Brassica napus L.*). Biology and Fertility of Soils 24(4):358–364

Demuynck S, Grumiaux F, Mottier V, Schikorski D, Lemière S, Leprêtre A (2006) Metallothionein response following cadmium exposure in the oligochaete *Eisenia fetida*. Comparative Biochemistry and Physiology, Part C 144:34–46

Desai A (1993) Vermiculture application in horticulture—the experience of farmers of Navasari, Gujarat. Congo Traditional Science and Technology of India, IIT, Bombay

Domínguez J, Martínez-Cordeiro H, Álvarez-Casas M, Lores M (2014) Vermicomposting grape marc yields high quality organic biofertiliser and bioactive polyphenols. Waste Manag Res32(12):1235–1240. https://doi.org/10.1177/0734242X14555805

Edwards CA, Lofty JR (1972) Biology of earthworms. Chapman & Hall, London, p 283

Edwards CA, Burrows I (1988a) The potential of earthworm composts as plant growth media. In: Edwards CA, Neuhauser E (eds) Earthworms in waste and environmental management. SPB Academic Press, The Hague, pp 21–32

Edwards CA, Burrows I (1988b) The potential of earthworm composts as plant growth media. In: Edwards CA, Neuhauser EF (eds) Earthworms in waste and environmental management. SPB Academic Publ. Co., The Hague, Netherlands, pp 211–219

Edwards CA, Arancon NQ (2006) The science of vermiculture: the use of earthworms in organic waste manangement. In: Guerrero RDIII, Guerrero-del Castillo MRA (eds), vermi technologies for developing countries. Proceedings of the international symposium-workshop on vermi technologies for developing countries. Nov. 16-18, 2005, Los Baños, Laguna, Philippines, Philippine Fisheries Association, Inc. (2006), pp 1–30

Fan H, Shu H-Y, Yang H.-S, Chen W-C (2006) Characteristics of landfill leachates in central Taiwan. Sci Total Environ 361:25e37. https://doi.org/10.1016/j.scitotenv.2005.09.033

Fredericq L (1878) La digestion des matieresalbuminoides chez quelquesinvertebres. Arch Zool Exp Genet 7:391

Galli E, Tomati U, Grappelli E, Di Lena G (1990) Effect of earthworm casts on protein synthesis in *Agaricus bisporus*. Biol Fertil Soils 9(4):290–291

Garcia Massinyi P, Tataruch F, Slameka J, Toman R, Jurik R (2003) Accurilulation of lead, cadmium, anal mercury in liver and kidney of the brown hare (Lepuseuropaeus) in relation to the season, age, and sex in the West Slovakian Lowland. J Environ Sci Health Part A-Toxic/Hazard Subst Environ Eng A38(7):1299–1309

Ghosh M, Chattopadhyay GN, Baral K (1999) Transformation of phosphorus during vermicomposting. Biores Technology 69(2):149–154

Goswami L, Patel AK, Dutta G, Bhattacharyya P, Gogoi N, Bhattacharya SS (2013) Hazard remediation and recycling of tea industry and paper mill bottom ash through vermiconversion. Chemosphere 92:708–713

Goswami L, Sarkar S, Mukherjee S, Das S, Barman S, Raul P, Bhattacharyya P, Mandal NC, Bhattacharya S, Bhattacharya SS (2014) Vermicomposting of tea factory coal ash: metal accumulation and metallothionein response in *Eisenia fetida* (Savigny) and *Lampito mauritii* (Kinberg). Biores Technol 166:96–102. https://doi.org/10.1016/j.biortech.2014.05.032

Goswami R, Mondal CK (2015) A study on earthworm population and diversity with special reference to physicochemical parameters in different habitats of south 24 parganas district in west. Bengal Records of the Zoological Survey of India 115(1):31–38

Goswami L, Pratihar S, Dasgupta S, Bhattacharyya P, Mudoi P, Bora J, Bhattacharya SS, Kim KH (2016) Exploring metal detoxification and accumulation potential during vermicomposting of Tea factory coal ash: sequential extraction and fluorescence probe analysis. Nat Sci Rep. https://doi.org/10.1038/srep30402

Goswami L, Kim K-H, Deep A, Das P, Bhattachary SS, Kumar S, Adelodun AA (2017) Engineered nano particles: nature, behavior, and effect on the environment. J Environ Manage 196:297–315. https://doi.org/10.1016/j.jenvman.2017.01.011

Goswami L, Mukhopadhyay R, Bhattacharya SS, Das P, Goswami R (2018) Detoxification of chromium-rich tannery industry sludge by *Eudrillus eugeniae*: insight on compost quality fortification and microbial enrichment. Biores Technol 266:472–481. https://doi.org/10.1016/j.biortech.2018.07.001

Graff O (1971) Do earthworms influence plant nutrition? Landbauforschung Volkenrode 21:103–108

Gupta N, Yadav KK, Kumar V (2015) A review on current status of municipal solid waste management in India. J Environ Sci 37(1):206–217

Hanc A, Enev V, Hrebeckova T, Klucakova M, Pekar M (2019) Characterization of humic acids in a continuous-feeding vermicomposting system with horse manure. Waste Manage 99:1–11

Hand P (1988) Earthworm biotechnology. In: Greenshields R (ed) Resources and application of biotechnology: the new wave. MacMillan Press Ltd., US

Hoornweg D, Bhada-Tata P (2012) What a waste: a global review of solid waste management

Hasan SE (2011) Public awareness is key to successful waste management. Journal of Environmental Science and Health, Part A 39(2):483–492

Haque KMF (2006) Yield and nutritional quality of cabbage as affected by nitrogen and phosphorus fertilization. Bangladesh J SciInd Res 41:41–46

Hickman JA, Reid BJ (2008a) The co-application of earthworms (*Dendrobaena veneta*) and compost to increase hydrocarbon losses from diesel contaminated soils. Environ Int 34(7):1016–1022

Hickman ZA, Reid BJ (2008b) Earthworm assisted bioremediation of organic contaminants. Environ Int 34(7):1072–1081

Hussain N, Singh A, Saha S, Kumar MVS, Bhattacharya P, Bhattacharya SS (2016) Excellent N-fixing and P-solubilizing traits in earthworm gut-isolated bacteria: a vermicompost based assessment with vegetable market waste and rice straw feed mixtures. Biores Technol 222:165–174. https://doi.org/10.1016/j.biortech.2016.09.115

Hussain N, Chatterjee SK, Maiti TK, Goswami L, Das S, Deb U, Bhattacharya SS (2021) Metal induced non-metallothionein protein in earthworm: a new pathway for cadmium detoxification in chloragogenous tissue. J Hazard Mater 401:123357

Ireland MP (1979) Metal accumulation by the earthworms Lumbricus rubellus, Dendrobaena veneta and Eiseniella tetraedra living in heavy metal polluted sites. Environ Pollut 19:201–206

Ismail SA (1997) Vermicology: the biology of earthworms. Orient longman Press, Hyderabad, p 92

Jambhekar H (1966) Effect of vermicompost on short duration crops. In: Veeresh GK, Shivashanka K (eds) National seminar on organic farming and sustainable agriculture, Association for Promotion of Organic farming, Bangalore, pp 9–11

Kägi JHR (1991) Overview of metallothionein. Meths Enzymol 205:613–626

Kale RD (1993) Earthworms: the biological tools for the healthy and productive soils. In: Proceedings of congress on traditional sciences and technologies of India. Indian Institute of Technology. Bombay. Publ. PPST Foundation, Madras and IIT, Bombay, pp 10.8–10.13

Kale RD, Bano K (1986) Proceedings of national seminar on organic waste utilization, pp 151–160

Khamkar MG (1993) Vegetable farming using venniculture. Gong Traditional Science and Technology India, IIT, Bombay

Khatib IA (2011) Municipal solid waste management in developing countries: future challenges and possible opportunities. INTECH Open Access Publisher

Landgraf MD, da Silva SC, Rezende MOO (1998) Mechanism of metribuzin herbicide sorption by humic acid samples from peat and vermicompost. Analytica Chimica Acta 368(1,2):155–164

Lavelle P, Barois I (1988) Potential use of earthworms in tropical soils. In: Edwards CA, Neuhauser EF (eds) Earthworms in waste and environmental management. SPB, The Hague, pp 273–279

Lavelle P, Martin A (1992) Small-scale and large-scale effects of endogeic earthworms on soil organic matter dynamics in soils of the humid tropics. Soil Biol Biochem 24(12):1491–1498

Lee KE (1995) Earthworms and sustainable land use. In: Hendrix PF (ed) Earthworm ecology and biogeography in North America. Lewis Publishers, Boca Raton, pp 215–234

Li W, Wang C, Sun Z (2011) Vermipharmaceuticals and active proteins isolated from earthworms. Pedobiologia 54:S49–S56

Liebeke M, Garcia-Perez I, Anderson CJ, Lawlor AJ, Bennett MH, Morris CA, Kille P, Svendsen C, Spurgeon DJ, Bundy JG (2013) Earthworms produce phytochelatins in response to arsenic. PLoS ONE. https://doi.org/10.1371/journal.pone.0081271

Liebeke M, Strittmatter N, Fearn S, Morgan A-J, Kille P, Fuchser J, Wallis D, Palchykov V, Robertson J, Lahive E, Spurgeon DJ, McPhail D, Takats Z, Bundy JG (2015) Unique metabolites protect earthworms against plant polyphenols. Nature Communication. https://doi.org/10.1038/ncomms8869

Liu C-W, Sung Y, Chen B-C, Lai H-Y (2014) Effects of nitrogen fertilizers on the growth and nitrate content of lettuce (Lactuca sativa L.). Int J Environ Res Public Health 11:4427–4439

Ma Y, Dickinson N, Wong M (2002) Toxicity of Pb/Zn mine tailings to the earthworm Pheretima and the effects of burrowing on metal availability. Biol Fertil Soils 36(1):79–86

Mahaly M, Senthilkumar AK, Arumugam S, Kaliyaperumal C, Karupannan N (2018) Vermicomposting of distillery sludge waste with tea leaf residues. Sustain Environ Res 28(5):223–227. ISSN 2468-2039. https://doi.org/10.1016/j.serj.2018.02.002

Maity S, Bhattachary S, Chaudhury S (2009) Metallothionein response in earthworms *Lampito mauritii* (Kinberg) exposed to fly ash. Chemosphere 77(3):319–324

Mallawarachchi H, Karunasena G (2012) Electronic and electrical waste management in Sri Lanka: suggestions for national policy enhancements. Resource, Conservation, and Recycling 68:44–53. https://doi.org/10.1016/j.resconrec.2012.08.003

Masters BA, Kelly EJ, Quaife CJ, Brinster RL, Palmiter RD (1994) Targeted disruption of metallothionein I and II genes increases sensitivity to cadmium. Proc Natl Acad Sci USA 91:584

Masciandaro G, Ceccanti B, Garcia C (2000) In situ vermicomposting of biological sludges and impacts on soil quality. Soil Biol Biochem 32:1015–1024

Matsumoto G (1982) Comparative study on organic constituents in polluted and unpolluted inland aquatic environments-IV Indicators of hydrocarbon pollution for waters. Water Res 16(11):1521–1527

Mitchell A, Alter D (1993) Suppression of labile aluminium in acidic soils by the use of vermicompost extract. Commun Soil Sci Plant Anal 24(11–12):1171–1181

Morgan AJ, Sturzenbaum SR, Winters C, Grime GW, Abd Aziz NA, Kille P (2004) Differential metallothionein expression in earthworm (Lumbricusrubellus) tissues. Ecotoxicol Environ Safety 57:11–19

Nannoni F, Protano G, Riccobono F (2011) Uptake and bioaccumulation of heavy elements by two earthworm species from a smelter contaminated area in northern Kosovo. Soil Biol Biochem 43(12):2359–2367

NIRD and PR (2016) Solid waste management in rural areas a step-by-step guide for gram panchayats a companion to the facilitators of swachh bharat mission (gramin), centre for rural infrastructure national institute of rural development & Panchayati raj Rajendranagar, Hyderabad - 500 030 www.nird.org.in

Nunes RR, Bontempi RM, Mendonça G, Galetti G, Rezende MO (2016) Vermicomposting as an advanced biological treatment for industrial waste from the leather industry. J Environ Sci Health B 51(5):271–277. https://doi.org/10.1080/03601234.2015.1128737

Paul S, Das S, Raul P, Bhattacharya SS (2018) Vermi-sanitization of toxic silk industry waste employing *Eisenia fetida* and *Eudrilus eugeniae*: substrate compatibility, nutrient enrichment and metal accumulation dynamics. Biores Technol 266:267–274. https://doi.org/10.1016/j.biortech.2018.06.092

Prabha ML, Jayaraaj IA, Jeyaraaj R, Rao S (2007) Comparative studies on the digestive enzymes in the gut of earthworms, *Eudrilluseugeniae* and *Eisenia fetida*. Indian J Biotechnol 6:567–569

Rasal PH, Kalbhor HB, Shingte VV, Patil PL (1988) Development of technology for rapid composting and enrichment. In: Sen SP, Patil P (eds) Biofertilizers—potentialities and problems. Plant Physiology Forum and Naya Prakash, Kolkata, India, pp 254–258

Reddy MV (1988) The effect of casts of Pheretima alexandri (Beddard) on the growth of *Vinca rosea*, and *Oryza sativa L*. In: Edwards CA, Neuhauzer EF (eds) Earthworms in waste and environmental management. SPB Academic Publishing, The Hague, pp 241–248

Roberts RD, Johnson MS (1978) Dispersal of heavy metals from abandoned mine workings and their transference through terrestrial food chains. Environ Pollut 16(4):293–310

Rodriguez-Campos J, Dendooven L, Alvarez-Bernal D, Contreras-Ramos SM (2014) Potential of earthworms to accelerate removal of organic contaminants from soil: a review. Appl Soil Ecol 79:10–25. https://doi.org/10.1016/j.apsoil.2014.02.010

Sahariah B, Das S, Goswami L, Paul S, Bhattacharyya P, Bhattacharya SS (2020) An avenue for replacement of chemical fertilization under rice-rice cropping pattern: Sustaining soil health and organic C pool via MSW-based vermicomposts. Arch Agron Soil Sci 66:10:1449–1465. https://doi.org/10.1080/03650340.2019.1679782

Santos MJ, Morgado R, Ferreira NG, Soares AM, Loureiro S (2011) Evaluation of the joint effect of glyphosate and dimethoate using a small-scale terrestrial ecosystem. Ecotoxicol Environ Saf 74(7):1994–2001

Satchell JE (1983) Earthworm ecology-from Darwin to vermiculture. Chapman and Hall Ltd., London, pp 1–5

Scott MA (1988) The use of worm-digested animal waste as a supplement to peat in loamless composts for hardy nursery stock. In: Edwards CA, Neuhauser EF (eds) Earthworms in environmental and waste management. SPB Academic Publisher, The Netherlands, pp 221–229

Seenappa SN, Kale RD (1995) Efficiency of earthworm *Eudrilus eugeniae* in converting the solid waste from aromatic oil extraction unit into vermicompost. Paper presented and accepted in the 3rd international conference in appropriate waste technologies for developing countries, Nagpur, India

Senapati BK (1992) Vermibiotechnology: an option for recycling of cellulosic waste in India. In: Subba Rao MS, Balgopalan C, Ramakrishnan SV (eds) New trends in biotechnology. Oxford and IBH publishing Co. Pvt. Ltd., New Delhi, pp 347–358

Senapati BK (1993) Earthworm gut contents and its significance. In: Ghosh AK (ed) Earthworm resources and vermiculture. Zoological Survey of India, Kolkata, India, pp 97–99

Senapati BK, Dash MC, Rane AK, Panda BK (1980) Observation on the effect of earthworms in the decomposition process in soil under laboratory conditions. Comp Physiol Ecol 5:140–142

Seno S, Saliba GG, de Paula FJ, Koga PS (1995) Utilization of phosphorus and farmyard manure in garlic cultivation. Hortic Bras 13(1):196–199

Shan J, Xu J, Zhou W, Ji L, Cui Y, Guo H, Ji R (2011) Enhancement of chlorophenol sorption on soil by geophagous earthworms (*Metaphire guillelmi*). Chemosphere 82(2):156–162

Shi Y, Xu X, Chen J, Liang R, Zheng X, Shi Y, Wang Y (2018) Antioxidant gene expression and metabolic responses of earthworms (*Eisenia fetida*) after exposure to various concentrations of hexabromocyclododecane. Environ Pollut 232:245–251

Shi Z, Liu J, Tang Z, Zhao Y, Wang C (2019) Vermiremediation of organically contaminated soils: concepts, current status, and future perspectives. Appl Soil Ecol. https://doi.org/10.1016/j.apsoil.2019.103377

Shinde PH, Nazirker RB, Kadam SK, Khaire VM (1992) Evaluation of vermicompost. In: Patil PS (ed) Proceedings of the national seminar on organic farming, Pune, India, pp 54–55

Shweta M (2012) Cellulolysis: a transient property of earthworm or symbitotic/ingested microorganisms? Int J Sci Res Publ 2:1–8

Singer AC, Jury W, Luepromchai E, Yahng CS, Crowley DE (2001) Contribution of earthworms to PCB bioremediation. Soil Biol Biochem 33(6):765–776

Singh A, Sharma S (2002) Composting of a crop residue through a treatment with microorganisms and subsequent vermicomposting. Biores Technol 85(2):107–111

Singh R, Divya S, Awasthi A, Kalra A (2012) Technology for efficient and successful delivery of vermicompost colonized bioinoculants in Pogostemoncablin (patchouli) Benth. World J Microbiol Biotechnol 28:323–333

Singh R, Singh R, Soni SK, Singh SP, Chauhan UK, Kalra A (2013) Vermicompost from biodegraded distillation waste improves soil properties and essential oil yield of Pogostemoncablin(patchouli) Benth. Appl Soil Ecol 70:48–56

Sinha RK, Chauhan K, Valani D, Chandran V, Soni BK, Patel V (2010) Earthworms: Charles Darwin's 'unheralded soldiers of mankind': protective & productive for man & environment. J Environ Protection 1:251–260

Sizmur T, Hodson ME (2009) Do earthworms impact metal mobility and availability in soil?—a review. Environ Pollut 157(7):1981–1989

Springett JA, Syres DK (1978) Effect of earthworm cast on ryegrass seedlings. In: Proceedings of second Australian conference on grassland invertebrate ecology, pp 44–46

Sturzenbaum SR, Winters C, Galay M, Morgan AJ, Kille P (2001) Metal ion trafficking in earthworms—identification of a cadmium specific metallothionein. J Biol Chem 276(36):34013–34018

Sturzenbaum SR, Hockner M, Panneerselvam A, Levitt J, Bouillard J-S, Taniguchi S, Dailey L-A, Khanbeigi RA, Rosca EV, Thanou M, Suhling K, Zayats AV, Green M (2013) Biosynthesis of luminescent quantum dots in an earthworm. Nature 8(1):57–60

Stürzenbaum SR, Kille P, Morgan AJ (1998) The identification, cloning and characterization of earthworm metallothionein. FEBS Lett 431:437–442

Subramanian S, Sivarajan M, Saravanapriya S (2010) Chemical changes during vermicomposting of sago industry solid wastes. J Hazard Mater 179(1–3):318–322. ISSN 0304-3894. https://doi.org/10.1016/j.jhazmat.2010.03.007

Sun M, Chao H, Zheng X, Deng S, Ye M, Hu F (2020) Ecological role of earthworm intestinal bacteria in terrestrial environments: a review. Sci Total Environ 740:140008

Szczech M, Brzeski MW (1994) Vermicompost—fertilizer or biopesticide? Zesz. Nauk. AR Krakowie 292(41):77–83

Tejada M, Gomez I, Toro M (2011) Use of organic amendments as a bioremediation strategy to reduce the bioavailability of chlorpyrifos insecticide in soils. Effects on soil biology. Ecotoxicology and Environmental Safety 74(7):2075–2081

Tisdale JM, Oades JM (1982) Organic matter and water stable aggregates in soil. J Soil Sci 33(2):141–163

Tisdale SL, Nelson WL (1975) Soil fertility and fertilizers, 2nd edn. Macmillan Publishing Co., New York

Tomati V, Grapelli A, Gelli E (1983) Fertility factors in earthworm humus. In: Tomati V, Grapelli A (eds) Proceedings of international symposium on agricultural environment. Prospects in earthworm farming, Rome. Tinolitografia Euromadera, Modena

Udovic M, Lestan D (2007) The effect of earthworms on the fractionation and bioavailability of heavy metals before and after soil remediation. Environ Pollut 148:663–668

van Gansen P (1963) Structure and function of the digestive canal of the earthworm *Eisenia foetida* Savigny. Annales De La Société Royale Zoologique De Belgique 93:1–120

Vasanthi D, Kumaraswamy K (1996) Organic farming and sustainable agriculture. National seminar. G.B.P.UAT., Pantnagar, p 40

Verma A, Pillai M (1991) Bioavailability of soil-bound residues of DDT and HCH to earthworms. Curr Sci 61(12):840–843

Vijver MG, Vink JP, Miermans CJ, van Geste CA (2003) Oral sealing using glue; a new method to distinguish between intestinal and dermal uptake of metals in earthworms. Soil Biol Biochem 35(1):125–132

Wallwork JA (1984) Earthworm biology (1st Indian edn), Gulab Vazirani for Arnold-Heinemann, New Delhi, Zoological Survey of India, Calcutta, pp 27–31

Wang D, Shi Q, Wang X, Wei M, Hu J, Liu J, Yang F (2010) Influence of cow manure vermicompost on the growth metabolite contents, and antioxidant activities of Chinese cabbage (Brassica campestrisssp. chinensis). Biol Fertil Soils 46:689–696

Wang X, Xia R, Sun M, Hu F (2021) Metagenomic sequencing reveals detoxifying and tolerant functional genes in predominant bacteria assist *Metaphire guillelmi* adapt to soil vanadium exposure. J Hazard Mater 415:125666. https://doi.org/10.1016/j.jhazmat.2021.125666

Chapter 10
Fly Ash Management Through Vermiremediation

Sanat Kumar Dwibedi and Vimal Chandra Pandey

Abstract Fly ash (FA) is an inevitable byproduct from the coal-fired thermal power plants that need timely, effective and safe disposal in many developing countries. It is an amorphous ferro-alumino silicate material similar to soil having practically all the elements except organic carbon, nitrogen and phosphorous. Although in many developed countries its use has reached saturation but technologically-starved poor countries are still lagging far behind in its resourceful use. Its use in cement-concrete, and land and mine filling have been widely accepted but in agriculture, this chemically heterogeneous material deserves cautious consideration. At low concentration, FA alters soil physicochemical properties and thus, acts as soil ameliorant or conditioner. However, its use at higher rate is restricted due to presence of heavy metals that affect soil biosphere and limits plant growth. Hence, remediation of toxic metal ions for sustainable agricultural intervention is a prerequisite in FA-contaminated soils or dumpsites. Like phytoremediation, earthworms with unique accumulation, extraction, transformation, conversion, degradation and stimulation properties could also be engaged in remediation of FA. In this chapter, attempts have been made to elucidate various mechanisms and processes involved in vermiremediation, and the advantages, disadvantages and future prospects of this innovative technology.

Keywords Amendment · Bioaccumulation · Earthworm · Fly ash · Heavy metal · Vermiremediation

10.1 Introduction

With the burgeoning global population, the demands for food have increased tremendously over last few decades beyond the yielding ability of many crops. Increase in

---

S. K. Dwibedi
College of Agriculture, Odisha University of Agriculture and Technology, Bhubaneswar, India

V. C. Pandey (✉)
Department of Environmental Science, Babasaheb Bhimrao Ambedkar University, Lucknow, Uttar Pradesh, India
e-mail: vimalcpandey@gmail.com

the current global food production for feeding the teaming millions is the greatest challenge before us (Dwibedi 2018). The pressure on land for higher productivity per unit area and time is increasing day by day, resulting in more dependence on chemical fertilizers, synthetic pesticides, hormones and probiotics at the cost of environmental health and sustainability. The land is degrading and becoming less productive which needs bio-physical amelioration for bringing back to its pristine conditions. Furthermore, the greed for energy, under the veil of pseudo civilization, prosperity and economic development, has been driving us towards peril (Dwibedi and Sahoo 2017).

Although the global primary energy consumption in 2018 recorded sharp decline in coal share (27%), it still ranks next to petroleum oil (34%) (International Energy Agency 2020). However, other alternative energy sources such as nuclear and hydrothermal power require sophisticated technologies and huge initial investments that are beyond the reach of many developing countries. Therefore, production of ash (bottom and fly ash), is an inevitable byproduct from the coal-fired thermal power plants that need safe, timely and effective disposal. Combustion of pulverized sub-bituminous coal (lignite) in thermal power plants results in generation of this end residue (Basu et al. 2009). Fly ash (FA) is an amorphous ferro-alumino silicate material similar to soil with all the elements except organic carbon, P and N (Tripathy and Sahu 1997; Pandey and Singh 2010; Pandey 2020a, b, c, d). It has been categorized 'under high volume low effect waste under Hazardous Waste (Management and Handling and Trans-boundary Movement) Rules, 2008' (Parab et al. 2012). Its production along with power generation in thermal power plants over decades of economic developments, both by developed and developing countries has been a necessary evil. This problematic 'solid waste' across the globe has now acquired the status of 'resource material' due to innovative uses in cement-concrete, land and mine filling, agriculture, etc. Its utilization in European countries is almost 100% while in developing countries like India lower percentage is being utilized in spite of its higher production (Dwibedi and Sahoo 2017).

10.2 Properties of FA

The physical, chemical and mineralogical properties of FA (Fisher et al. 1978; Page et al. 1979; Adriano et al. 1980; Carlson and Adriano 1993; Pandey 2020a) depend on the quality of coal, extent of thermal combustion and storage-handling methods. Therefore, ash compositions vary with burning of anthracite, bituminous and lignite coals. Elements present in coal are intense in FA. Physically, FA is very fine with mean diameter of <10 μm, light in texture. It has low to moderate bulk density (BD) and more surface area. Its water holding capacity is of 49–66% on the weight basis (Sharma and Kalra 2006). Its pH ranges from 4.5 to 12 largely depending on the S content in the coal. FA is chemically heterogeneous in nature as it contains variable proportions of different trace and heavy metals such as Be, B, Cd, Cr, Co, Hg, Mo, Mn, Pb and oxides Al, Ca, Fe and Si.

Incorporation of FA alters physicochemical properties of soil and works as soil conditioner or modifier (Kalra et al. 1998; Pandey and Singh 2010; Pandey 2020b). It alters the texture of soil (Kalra et al. 2000), reduces BD and increases porosity, water holding capacity and aeration due to its silty nature. Kuchawar et al. (1997) and Bhaisare et al. (1999) have shown an increase in cation-exchange capacity (CEC) as a result of FA amendment in soil. It also improves soil bacteria count and enzyme activity viz. dehydrogenase, urease and alkaline phosphatase that promote plant growth (Yeledhalli et al. 2007). Comparative physicochemical properties of soil and FA, and also FA in combination with press mud (PM) have been depicted under Table 10.1 (Singh and Pandey 2013).

According to the Intergovernmental Panel on Climate Change (IPCC), lime application for soil amelioration releases carbon dioxide (CO_2) gas which ultimately adds to global warming. In United States of America, the Environment Protection Authorities (EPA) has estimated emission of 9 Tg (teragram $= 1.012$ g $= 106$ t) of CO_2 from an approximate 20 Tg of agricultural lime applied in 2001. FA could be the befitting substitute for it minimizing global warming process (West and McBride 2005). It has also been estimated that 1 tonne of FA has the ability to sequester up to 26 kg of CO_2 (i.e. 38.46 tonnes of FA per tonne of CO_2 sequestered).

10.3 Verms as Bioreactor

Earthworms, regarded as the intestine of earth (Aristotle), are the terrestrial invertebrates, belonging to the phylum Annelida, and class Oligochaeta and they have more than 3000 species across the globe (Berridge 2020). They act as bioreactors in recycling the organic wastes to reusable plant nutrients at a very low or marginal cost of production and because of that, they act as 'farmers' friends'. Wastes from the agricultural field after harvest, and urban and rural solid organic wastes can very well be used in vermicomposting. Vermicomposting of agricultural residues and its effects on plant growth, microbial population and nutrient transformation at different concentrations in soil rhizosphere have been studied with much attention and interest.

10.4 Research Status on FA Use and Vermiremediation

The research on FA use began in late 1970s to evaluate its suitability for improving soil environment and increasing crop productivity (Dwibedi and Sahoo 2017). In developed countries, its utilization is more than 70% but in developing countries; it is still less than 5%. FA may be applied as soil amendment along with organic substrates such as farmyard manure, compost and microbial culture. A lot of research on use of FA in agricultural crops such as rice, maize, grams, beans, vegetables, etc. in pot culture and field trials has already been conducted. Its far-reaching consequences on soil bio-physicochemical properties have also been evaluated in long-term experiments.

Table 10.1 Comparative ash/soil properties with different levels of FA treatment

Characteristics	FA	Soil			
(A) *Physicochemical properties of Indian fly ash and soil* (*source* Kumar et al. 2000; Goyal et al. 2002)					
Bulk density (g cc^{-1})	<1.0	1.33			
Water holding capacity (%)	35–40	<20			
Porosity (%)	50–60	<25			
K (%)	0.19–3.0	0.04–3.0			
P (%)	0.004–0.8	0.005–0.2			
S (%)	0.1–1.5	0.01–0.2			
Metals (mg kg^{-1})					
Zn	14–1000	2–100			
Mn	100–3000	100–4000			
Fe	36–1333	10–300			
Cu	1–26	0.7–40			
B	46–618	0.1–40			
(B) *Soil properties and metal composition as influenced by combined application of FA and press mud (PM)* (*source* Singh and Pandey 2013)					
Treatments parameters	Control	PM + FA (10 t ha^{-1})	PM + FA (50 t ha^{-1})	PM + FA (100 t ha^{-1})	P value
Soil properties					
pH	6.9 ± 1.3	7.1 ± 1.6	8 ± 1.8	8.3 ± 1.5	<0.01
EC (ds m^{-1})	2.4 ± 0.7	3.5 ± 0.2	6.3 ± 0.6	6.9 ± 0.8	<0.01
Soil moisture (%)	17.2 ± 1.2	25.7 ± 2.2	28.4 ± 2.3	28.5 ± 2.1	<0.01
Inorganic-N (NH_4-N and NO_3-N)	32 ± 1.2	22.2 ± 1.3	26.2 ± 1.3	26.6 ± 1.3	<0.01
Metal (μg^{-1})					
Cr	3.68 ± 0.33	4.37 ± 0.23	5.64 ± 0.48	7.6 ± 0.63	<0.01
Cd	1.8 ± 0.06	2.43 ± 0.19	3.6 ± 0.63	4.12 ± 0.45	<0.01
Cu	4.34 ± 0.58	5.23 ± 0.33	6.23 ± 0.48	7.89 ± 0.23	<0.01
Ni	5.52 ± 0.46	7.2 ± 0.33	9.06 ± 0.35	12.21 ± 0.42	<0.01
Methanotrophs number (× 10^4 g^{-1} of soil)	23.4 ± 6.1	53 ± 11.5	29.4 ± 6.1	25.2 ± 6.1	<0.05

Modified from Source Bhattacharya and Kim (2016)

The role of FA in reclamation of acidic and sodic soils has been well acclaimed. Its utilization in agriculture has been a proven support as it improves physicochemical properties of soil resulting in better fertility and increased crop yield (Rautaray et al. 2003). However, heavy metal accumulation with FA amendment is a great concern. Researchers are in view of its application in lower concentrations as soil microbial population and availability of plant nutrients are affected at higher concentrations.

Earthworms are the ecological engineers having profound role in amelioration of soil physical, chemical and biological properties (Shi et al. 2017). The significant role played by earthworms in soil fertility enhancement, biodiversity restoration and detoxification of contaminated soil was studied since early 1800s (Edwards 2004) while much stress on soil remediation was given during 1980s (Sinha et al. 2010). In the recent past, 'vermiremediation', a new approach has been invoked (Gupta and Garg 2009). Attempts have also been made to study the composting behaviour of earthworms at varying levels of FA substrates to prepare vermi-ash.

10.4.1 FA Impact on Soil Characteristics

FA has tremendous potential as a valuable resource in agriculture, building, road and bridge construction and other related areas. Its soil amending and nutrient-enriching properties contribute to agricultural production (Pandey 2020c). It contains considerable quantities of both macro and micronutrients (Singh et al. 1997) which when applied to soil sustain crop growth and development, even in poor soils. As mentioned above, FA is deficient in N, P and organic matter and hence, its amendments with organic materials or microbial inoculants help in plant growth. Its possible agricultural applications such as liming material, fertilizer and physical amendment have been illustrated by many researchers. For effective and efficient vermiremediation of FA, it is imperative to understand the effects of FA on soil properties and agricultural crops as remediated land may simultaneously or subsequently be brought under cultivation. A brief review of the earlier studies on FA use in agriculture is hereunder for general reference.

FA is helpful in increasing the physical properties of soil that ultimately improve soil fertility and enhance crop yield (Rautaray et al. 2003). FA amendment in sunflower fields decreases BD of the soil (Pani et al. 2015). Wong and Wong (1990) noticed alteration in soil texture, bulk density and porosity. FA addition in sandy soil alters soil texture and increases micro-porosity (Ghodrati et al. 1995). Increase in porosity and decrease in bulk density in soil was also reported by Zibilski et al. (1995). Water holding capacity of soil increases with FA amendment in sunflower fields (Pani et al. 2015; Parab et al. 2012). FA amendment in clay soil improves infiltration whereas in the coarse soil it reduces infiltration as reported by Dhindsa et al. (2016).

The pH of soil (pH 6.65) increases with the addition of FA (pH 7.56) due to acid-neutralizing capacity of the latter one in presence of oxides of Ca and Mg in it. The soil becomes more alkaline with FA amendment in sunflower fields (Pani et al. 2015). Such increase in pH was also reported by Lee et al. (2006) and Sarkar et al. (2012). However, Sikka and Kansal (1995) reported no significant increase in pH with FA amendment. Electrical conductivity (EC) of the soil (281 dS cm^{-1}) increases with the addition of FA (600 dS cm^{-1}) in radish field, possibly due to precipitation of soluble cations (Singh et al. 2011a, b) and binding of metal ions to soil separates that facilitates ready availability of plant nutrients (Pani et al. 2015) in FA amended soils. However, elevated EC may suppress normal plant growth (Singh et al. 2011a, b). Organic carbon (OC) decreases with increase in FA concentration in radish (Sarkar et al. 2012) whereas in brinjal, the value of OC increases with FA (Singh et al. 2011a, b). FA improves nutrient levels in soil (Rautaray et al. 2003). Singh et al. (2011a, b) have observed increase in availability of N, P, K, Co, Ni, Cu, Zn, Mo, Al, V, Se, etc. as well as toxic metals such as Cr, Pb and As with addition of FA at different grades. Sarkar et al. (2012) have reported increase in availability of Na, K, Ca, Mg and Fe with significant reduction in total N, available P and OC under FA soil amendment. FA is also used to rectify B and S deficiencies in soil (Chang et al. 1977). P availability increases with the addition of FA (Lee et al. 2006). Reddy et al. (2010) have reported 'the highest available N (224.6 kg ha^{-1}), P (24.6 kg ha^{-1}), K (366.7 kg ha^{-1}), S (8.80 mg kg^{-1}), Fe (10.62 mg kg^{-1}) and Zn (0.95 mg kg^{-1}) content after harvest of rice crop with application of FA at 15 t ha^{-1} + FYM at 10 t ha^{-1} (FA$_{15}$ + FYM$_{10}$), which were at par with FA$_{10}$ + FYM$_{10}$'. However, Sikka and Kansal (1995) reported no significant increase in available N and P in soil with the addition of FA whereas the available K increased.

The nematode population as observed in Chandrapura Thermal Power Station, reduced significantly (Singh et al. 2011a, b) with 40% FA amendment (Ahmad and Alam 1997; Khan et al. 1997) due to inhibitory effect (Khan et al. 1997; Tarannum et al. 2001) of FA. The carbon dioxide efflux from the soil as an indirect method of knowing soil biotic activities increased with 0–100 t ha^{-1} addition of FA than 400–700 t ha^{-1} amendments. Several metals present at potentially toxic levels in FA might have suppressed soil heterotrophic microbial activities at higher levels (Arthur et al. 1984).

10.4.2 FA in Agriculture

Direct use of FA in crop fields is not so promising due to poor bioavailability of plant nutrients such as C, N and P that inhibit mineralization through reduced microbial activities (Lazcano 2009/66). When applied to soil directly, it severely inhibits microbial process, N cycle and enzyme activity (Lazcano 2009). Pandey et al. (2009a) observed accumulation of Fe, Zn, Cu, Cd and Cr in *Cajanas cajan* when the soil was mixed with FA. FA amendment affects rice germination count in initial stage but after 115 h, it picks up again equalizing with the untreated soil. Such delay in germination could be due to increase in soil impedance/ strength (Kalra et al. 1997). However, no such inhibitory effect is noticed in green gram, golden gram and black gram at 0, 10, 20, 30, 40 and 50% FA amendment, except at 100%; possibly due to balance between growth promoters and inhibitors (Singh et al. 2011a, b). The highest rice seed germination is at 20 and 30% FA amendment (Adriano and Weber 2001) while the lowest is at 100% (Panda and Tikadar 2014). Germination of rice and maize in wet season is less sensitive to moderate FA than dry season (Kalra et al. 1998) whereas germination decreases with further increase in ash concentration (Panda and Tikadar 2014).

Shoot and root length of green gram, golden gram and black gram increase with application of FA and the maximum length occurs at 30–40% while in radish, FA shortens plant height (Singh et al. 2011a, b). Shoot length of *Luffa cylindrica* increases up to 180 t ha^{-1} FA but at higher dose, the plant shortens (Singh et al. 2011a, b). At 25% FA, taller rice plants with longer roots are observed compared to no or higher levels (Panda and Tikadar 2014). Tiller count in rice goes on increasing with the addition of FA up to 75 t ha^{-1} (Priatmadi et al. 2015) but on further addition, it declines (Sarkar et al. 2012). Chlorophyll a and b and carotenoid pigment concentration in chickpea, golden gram and black gram improves significantly at moderate levels of FA (120–180 t ha^{-1}) but at 240 t ha^{-1}, the pigmentation decreases (Singh et al. 2011a, b). Dry matter accumulation in rice seedlings reduces with increase in concentration of FA from 25 to 100% in rice nursery (Panda and Tikadar 2014). FA and FYM amendments enhance the rates of N transformation processes, plant available-N and paddy productivity (Singh and Pandey 2011) and can be used to enrich nutrient-poor soils for crop productivity and yields. The mixture of FA and press mud shows positive effect on crop growth, physicochemical, microbial and enzymatic activities of sodic soil (Singh et al. 2016a). The mixture of 40% soil + 20% FA + 40% vermicompost is proved as most promising blend for wet rice nursery raising and for remediating the coal FA in agricultural production system (Dwibedi et al. 2021). Recently, it is proved that phytoremediated FA can be used as a fertilizer up to 100% for peas farming as metal concentrations was reported either below detection limit or below the WHO permissible limit (Bhattacharya et al. 2021). The application of FA for agriculture production is explored in great depth using the facts of plants, amendments, FA doses range and remark (Pandey et al. 2009b).

10.5 What is Vermiremediation?

The term 'vermiremediation' has come from two Latin words: 'vermis' means 'worm' and 'remedium' means 'correct' or 'remove an evil' (Shi et al. 2020). The term was coined by Edward and Arancon (2006) while Rodriguez-Campos et al. (2014) first attempted to define it as 'the use of earthworms for removing contaminants (Sinha et al. 2008) or not recyclable compounds (Gupta and Garg 2009) from the soil'. However, a better definition by Shi et al. (2020) has come up later which expresses 'vermiremediation as an earthworm-based bioremediation technology that makes use of the earthworm's life cycle (i.e. feeding, burrowing, metabolism and secretion) or their interaction with other abiotic and biotic factors to accumulate and extract, transform, or degrade contaminants in the soil environment'. As per this definition, few synonymous terms viz. vermiaccumulation and vermiextraction, vermitransformation, vermiconversion and drilodegradation or drilostimulation could be used to understand the mechanisms and processes of vermiremediation (Shi et al. 2020).

Vermiaccumulation and vermiextraction, similar to term phytoaccumulation, refer to the process of ingestion of contaminants (organic and inorganic) from the soil by earthworms and accumulation of pollutants in their body parts (Shi et al. 2020). Accumulation of contaminants occurs through preferential dermal or intestinal sequestration involving sub-organismic (preclitellum, clitellum, post-clitellum), tissue (body wall, gut, body fluids) and sub-cellular (intra and extracellular fractions) body parts of the earthworm (Shi et al. 2020). The process of biotransformation of contaminants by earthworms into harmless products by enzymes (such as peroxidases) and microbes (bacteria and fungi) in the alimentary canal and ultimately egested out as compost is known as vermitransformation or vermiconversion (Panda and Tikadar 2014). Drilodegradation or drilostimulation refers to the microbial decomposition, degradation or elimination of toxic materials by microbes present in the drilosphere, the 2 mm thick zone of earthworm burrow wall (Bouché 1975; Brown et al. 2000). Drilospheric soil is rich in earthworm mucus and casts that stimulate microbial growth which subsequently promotes the growth of protozoa and nematodes (Stromberger et al. 2012). Drilosphere, a habitat rich in energy and nutrients, mostly C and N, acts as hotspot for soil microbial communities (Kuzyakov and Blagodatskaya 2015). The nutrients are mixtures of low-molecular organic acids such as amino acids, nucleic acid derivatives, carbohydrates, phenolics and enzymes (Zhang et al. 2009). The labile organic carbon supply in drilosphere can sustain microbial communities that supplement utilizable sources of energy (Tiunov and Scheu 1999). And hence, drilospheric microorganisms have tremendous ability to remediate the potential pollutants (Shi et al. 2020).

10.5.1 Advantages of Vermiremediation

Vermiremediation is an emerging concept that needs rigorous investigation and exploration for gaining ecological milestones over conventional physicochemical methods. Primarily, it is one of the cheapest, easiest, efficient and in some cases, the fastest way of remediating the contaminated land without disturbing the topsoil. Furthermore, it is not substrate-specific, rather a useful technology for treating a wider range of hazardous pollutants. Synthetic insecticides, herbicides, polycyclic aromatic hydrocarbons (PAHs), polychlorinated biphenyls (PCBs), crude oil and FA in soil can be removed by engaging earthworms (Rodriguez-Campos et al. 2014). It is environmentally sustainable self-regenerating *in-situ* approach to remediate polluted land. Furthermore, vermiremediation enhances soil quality through addition of organic matter, supplementation of plant nutrients and proliferation of biodiversity (Sinha et al. 2008).

10.5.2 Limitations of Vermiremediation

Vermiremediation technology has its own limitations as it can only be applicable in moderately or slightly contaminated soils that allow survival of the earthworms. In severely contaminated soil, earthworms may not survive due to toxic effects of the pollutants (Rodriguez-Campos et al. 2014; Shi et al. 2019). Vermiremediation is also restricted to the earthworm habitats depending on the species used and ambient environmental conditions-beyond which its efficacy is limited. Earthworms are categorized into epigeic, anecic and endogeic groups (Fig. 10.1) depending on the species used, body size, mobility, fecundity, habitat, feeding and burrowing behaviour, casting activity, etc. (Lazcano et al. 2009) of the earthworm. *Dendrobaena octaedra, Dendrobaena attemsi, Dendrodrilus rubidus, Eiseniella tetraedra, Heliodrilus oculatus, Lumbricus rubellus, Lumbricus castaneus, Lumbricus festivus, Lumbricus friendi, Lumbricus rubellus, Satchellius mammalis, Eisenia fetida* and *Eudrilus euginae* live on the upper layer of the soil profile and feed mainly on organic debris and thus are classified as detritivores under epigeic group. Endogeic (means within the earth) earthworms such as *Allolobophora chlorotica, Apporectodea caliginosa, Apporectodea icterica, Apporectodea rosea, Drawida grandis, Murchieona muldali, Octolasion cyaneum, Octolasion lacteum, Anecies longa, Anecies nocturna* and *Octochaectona thurstoni* remain deep inside the soil and are geophagus in nature. Whereas anecics or anegeic (out of the earth) earthworms, e.g. *Aporrectodea longa, Aporrectodea nocturna, Lumbricus friend, Lumbricus terrestris* and *Letmpito mauritii*

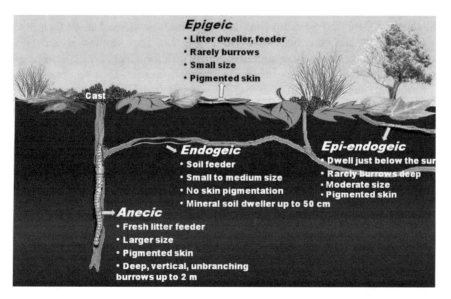

Fig. 10.1 Three major ecological groups of earthworms identified basing on feeding and burrowing behavior. *Source* Adapted and modified from Brown and Sherlock (2021)

are sub-surface dwellers and are phyto-geophagus in nature (Brown and Sherlock 2021; Bhattacharya and Kim 2016). Vermiremediation potential is dependent on food abundance and feeding preference of earthworm species (Curry and Schmidt 2007). Earthworms are sensitive to temperature, moisture and other climatic and seasonal conditions that may inhibit their survivability thereby affecting the vermiremediation process (Butt and Lowe 2011). Additionally, accumulated contaminants in earthworms can become a potential threat if get transferred into food chain under mismanagement in disposal schedule (Shi et al. 2014).

10.6 Biology of Earthworm and Its Functional Significance in Waste Degradation

Before getting into the process of vermiremediation in FA-contaminated soils, it is imperative to know the biology of earthworm and the mechanism of waste degradation with relation to soil health. They prefer moist and dark habitats with optimum moisture of 60–75% and their skin is permeable for which they need moist environment to prevent from drying out (Shi et al. 2020). Although they can survive temperature range of 5–35 °C, but the optimum is 20–25 °C. Most of them prefer neutral pH and C/N ratio of 2–8 (Sharma and Garg 2018). Within a life span of 220 days, they produce 300–400 offspring (Shi et al. 2020). They are bisexual and under ideal soil temperature, moisture, pH and food availability they can multiply 2^8 times in every six months (Shi et al. 2020). They mostly feed on detritus materials, living bacteria, fungi, protozoa, nematodes and many other microorganisms (Sharma and Garg 2018). Earthworms have digestive tubes housed inside their thick cylindrical muscular outer body tube (Berridge 2020). They swallow considerable amount of food materials along with soil through their mouth present at 1st segment and shred down by gizzard present at 8th or 8th to 9th segment. The elementary canal of earthworm includes mouth (1st), buccal cavity (2nd and 3rd), pharynx (3rd and 4th), esophagus (5th to 7th), gizzard (8th or 8th and 9th), stomach (9th or 10th to 14th), intestine (15th up to the last segment except anus) and anus (Aryal 2020). They also passively absorb dissolved chemicals through their body wall (Shi et al. 2020). These absorbed and eaten substrates are mixed with intestinal fluid and enzymes from microbes. Earthworm's intestine acts as warehouse for microbes and enzymes such as lipase, amylase, nitrate reductase, protease, phosphatase, cellobiase, etc. that bioprocess disintegration of ingested foodstuffs.

Earthworms maintain and improve soil quality parameters (Bhadauria and Saxena 2009) and act as bioindicators of soil quality (Fründ et al. 2011). Abundance and species composition of earthworms, their behaviour in contact with the soil, assimilation of chemicals in their body parts and biochemical or cytological stress markers can indicate soil quality (Fründ et al. 2011). Earthworms produce pores and aggregates (biostructures) in soil, thus influencing soil's physical properties, nutrient cycling and plant growth (Lal 1999; Scheu 2003). Anecic species make permanent burrows in mineral soils; they drag surface organic materials into the soil for food. Endogeic species are the ecosystem engineers who make nonpermanent burrows in the upper surface mineral layer through which other organisms get accessibility to underground resources (Jones et al. 1994). No till or minimal disturbance to the soil, as in conservation agriculture, enhances organic residues, thus creating ideal conditions for earthworm habitat (Labenz 2021). Mucus production associated with water excretion by earthworms enhances the activity of soil beneficial microorganisms that help in improving soil structure and aggregate stability. Earthworm's excreta (cast) are rich in plant-available nutrients, thus concentration of N, P, K, Ca, Mg and many more trace elements in soil increases and toxic materials including heavy metals get accumulated in their gut (Usmani and Kumar 2017) which make them biologically potent for remediation of FA (Fig. 10.2).

Metal accumulation mostly occurs in the chloragogenous tissue at the posterior end of the alimentary canal of earthworm (Usmani and Kumar 2017; Morgan and Morris 1982). On exposure to metals, earthworms synthesize metallothioneins (MT) that have low-molecular weight, cysteine-rich proteins with high affinity towards Cd, Cu and Zn (Dallinger 1994). These proteins protect organisms against toxic metal stress and thus can be used as indicator of soil pollution. While dealing the unnecessary heavy metals, earthworms detoxify their effects through interaction with many chemicals in the metabolic processes. Bioaccumulation of metals and organocomplex formation results in decline in the availability of heavy metals in soil as part of enzyme antioxidant systems such as superoxide dismutase (SOD) and MT (Li et al. 2008).

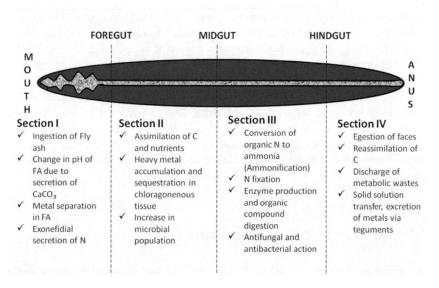

Fig. 10.2 Physicochemical transformations occurring in different compartments of earthworm illustrating heavy metal sequestration and nutrient assimilation on ingestion of FA [*source* Usmani and Kumar 2017 (Adapted and modified with the permission of the Publisher)]

As mentioned above, the highest metal accumulation occurs in the posterior alimentary canal (PAC) of the earthworm. Intracellular vesicles within PAC accumulate Pb and Zn and the superfluous metals interact with P ligands within the chloragosome matrix (Usmani and Kumar 2017; Morgan and Morgan 1990). The cation-exchange properties in chloragosomes (Fischer 1973, 1977) are considered as integral part for the physiological functioning of intracellular organelles (Morgan 1981; Fischer and Trombitts 1980). Microprobe X-ray analysis of air-dried chloragogenous tissue revealed Ca, Pb and Zn (in association with sulphur) accumulation in the chloragosomes while Cd was accumulated in an electron-lucent vesicular component called cadmosome (Usmani and Kumar 2017).

10.7 Process of Vermiremediation

Metal accumulation by earthworm (vermiremediation) may be in-situ or on-site treatment in the FA dumped sites (contaminated land), or it may be ex-situ through vermicomposting (Usmani and Kumar 2017). *Eisenia fetida* cannot tolerate 100% FA, thus addition of organic matter is essential (Niyazi and Chaurasia 2014). Considerable reduction in metal concentration occurs after vermiremedition. FA lacks N and C and thus organic matter addition is required to support microbial growth (Mupambw et al. 2015). Experiments on vermicomposting of cow dung with FA showed 30–50% reduction in heavy metal concentration up to 60% FA while 10–30% reduction was in 80% FA addition. Hence, 60% addition of FA with *E. fetida* was proposed to be a sustainable vermiremediation technique (Gupta et al. 2005). In another experiment, minimum mortality and maximum population growth were observed in 1:3 mixture of FA and cow dung. Significant reduction of heavy metals viz. Cu, Pb, Mn and Cr were also observed with vermiremediation at variable range of FA and cow dung mixtures. Vermistabilization resulted reduction in pH by 8–15.7%, EC by 16.2–53.6%, total organic carbon by 15.6–32.5% and C:N ratio by 43.2–97.4% (Singh et al. 2016b). A decline in heavy metal concentration in vermicompost was reported by Niyazi and Chaurasia (2014) like Anderson and Laursen (1982), Morgan and Morgan (1990) who observed variations in metal accumulation depending on inter-specific metal intake ability, worm age, their physiological utilization and transformation, season and many other factors (Usmani and Kumar 2017).

10.8 Strategies for Vermiremediation

Earthworm survival and mobility of contaminants are the two limiting factors in vermiremediation (Usmani and Kumar 2017). The performance of earthworms is affected by poor soil quality, environmental conditions and high concentration of pollutants (Sinha et al. 2008). Vermiremediation of FA-contaminated soils needs controlled mobility and bioavailability of toxicants and facilitated growth of earthworms under ameliorated soil environment. Nutrient and organic amendments and provisioning for better soil physical properties should be the prime management strategies for efficient and effective vermiremediation. The vermiremediation capacity of different earthworm species needs through assessment before their engagement in contaminated land reclamation. Suitability of crops to differential FA and organic residue amendments and bioaccumulation of toxic heavy metals across trophic levels need in-depth investigation for validation of the remediation technologies. Safe and timely evacuation of earthworms in vermiremediation is mostly lacking, which requires burning as specialized for hazardous waste (Sheoran et al. 2010; Ali et al. 2013). A brief account of different harvest methods of the earthworms used in vermiremediation is presented under Table 10.2. Vermiremediation can be facilitated through appropriate microbe-earthworm combined interactions as is evident in phosphorous solubilizing bacteria inoculated FA amendments (Lukashe

Table 10.2 Potential harvest methods of earthworms used in vermiremediation—based on the summary of earthworm sampling methods

Classifications		Expellant	Characteristics	References
Ethological methods	Chemical methods	Mustard or hot mustard	Non-destructive or 'environmental friendly'; more effective on anecic species; expensive	Chan and Munro (2001) Lawrence and Bowers (2002)
		Formalin	A standard method for the expulsion of earthworms; highly toxic to soil organism	Čoja et al. (2008)
		Detergent	Toxic	East and Knight (1998)
		Allyl isothiocyanate (AITC)	Environmental friendly; effective on deep-burrowing anecic species	Zaborski (2003)
		Onion solution	Environmental friendly	Steffen et al. (2013)
	Electrical method	Electroshocking	Little damage	Eisenhauer et al. (2008)
Hand-sorting	–	–	Physical disturbance of soil system; labour-intensive; time-consuming	Valckx et al. (2011)
Mechanical separation	–	–	Energy consuming	–

Source Shi et al. (2020)-reproduced with permission

et al. 2018). Harmonious integration of phytoremediation, vermiremediation and effective microorganisms has been far better option against any two of these remediation techniques to clean up residual contaminants (Deng and Zeng 2017). In heavily contaminated soils, vermiremediation can be used as polishing step after primary remedial treatment (Sinha et al. 2008). Another way of enhancement of vermiremediation is through quality food supplementation and optimization of the inoculation conditions (temperature, pH, aeration, moisture, etc.) that ultimately increase earthworm biomass and rate of uptake of contaminants as well (Curry and Schmidt 2007). Improvement of agronomic conditions such as soil texture, organic matter, hydraulic conductivity and homogenization of contaminants to avoid hotspots will certainly enhance vermiremediation (Gerhardt et al. 2017). Since it is impracticable and time-consuming to study individual species under all possible conditions, various models viz. empirical, rate, equilibrium-partition, mechanical and fugacity models predicting uptake and accumulation of toxic materials in earthworms need to be validated (Shi et al. 2020).

10.9 Conclusions and Prospects

Vermiremediation as an expanding, sustainable, ecofriendly and cost-effective technology available for treatment of polluted soils, including FA, has been well acknowledged widely. Unlike physiochemical remediation, vermiremediation is an environmental supportive and relatively cheaper, easier, effective and efficient technique that should be highlighted. Many researchers have studied vermiremediation of FA over past few decades thereby opening up an innovative scientific approach in remediating contaminated land. Vermiaccumulation and vermitransformation play important roles in vermiremediation of pollutants like heavy metals in FA. Furthermore, emphasis is to be given for enhancing bioavailability of organic residues and by providing congenial environment for optimum growth of earthworms. Integration of effective microorganisms, agronomic practices, phytoremediation, biomass enhancement, etc. has the potential to facilitate vermiremediation. Safe and timely harvest and disposal of contaminated earthworms could prevent biomagnification of pollutants in natural food chains which should be considered seriously. Available models for predicting uptake and accumulation in earthworms need to be validated so that the capacity, contribution and mechanism of different processes in vermiremediation are fully clarified.

References

Adriano DC, Page AL, Elseewi AA, Chang AC, Straughan I (1980) Utilization and disposal of fly ash and other coal residues in terrestrial Ecosystems: a review. J Environ Qual 9:333–344. https://doi.org/10.2134/jeq1980.00472425000900030001x

Ahmad A, Alam MM (1997) Utilization of fly ash and *Paecilomyces lilacinus* for the control of *Meloidogyne incognita*. Int J Nematol 7:162–164

Ali H, Khan E, Sajad MA (2013) Phytoremediation of heavy metals—concepts and applications. Chemosphere 91:869–881. https://doi.org/10.1016/j.chemosphere.2013.01.075

Anderson C, Laursen J (1982) Distribution of heavy metal in *Lumbricus terrestris, Aporrectodea longa* and *A. rosea* measured by atomic absoption and X ray fluoresecence spectrometry. Pedobiologia 24:347–356. https://inis.iaea.org/search/searchsinglerecord.aspx?recordsFor=SingleRecord&RN=14763698

Arthur MF, Zwick TC, Tolle DA, Voris PV (1984) Effects of fly ash on microbial CO_2 evolution from an agricultural soil. Water Air Soil Pollut 22(2):209–211

Aryal S (2020) Digestive system of earthworm. The biology notes. Retrieved from https://thebiologynotes.com/digestive-system-earthworm/

Basu M, Pande M, Bhadoria PBS, Mahapatra SC (2009) Potential fly-ash utilization in agriculture a global review. Prog Nat Sci 19:1173–1186. https://doi.org/10.1016/j.pnsc.2008.12.006

Berridge L (2020) Earthworm biology. Earthworm Soc Britain, 20 March 2020. https://www.earthwormsoc.org.uk/earthworm-biology

Bhadauria T, Saxena KG (2009) Role of earthworms in soil fertility maintenance through the production of biogenic structures. Appl Environ Sci 2010:7 pages. Article ID 816073. https://doi.org/10.1155/2010/816073

Bhaisare B, Matte DB, Badole WP, Deshmukh A, Pillewan S (1999) Effect of fly ash on physicochemical properties of Vertsoils and yield of green gram. J Soils Crops 9:255–257

Bhattacharya SS, Kim KH (2016) Utilization of coal ash: is vermitechnology a sustainable avenue? Renew Sustain Energy Rev 58:1376–1386. https://doi.org/10.1016/J.RSER.2015.12.345

Bhattacharya T, Pandey SK, Pandey VC, Kumar A (2021) Potential and safe utilization of Fly ash as fertilizer for Pisum sativum L. Grown in phytoremediated and non-phytoremediated amendments. Environ Sci Pollut Res. https://doi.org/10.1007/s11356-021-14179-9

Bouché MB (1975) Action de la faune sur les états de lamatière organique dans les ecosystemes. In: Kilber-tius G, Reisinger O, Mourey A, Cancela da Fon-seca JA (eds) Humification et biodégradation. Pier-ron, Sarreguemines, France, pp 157–168

Brown GG, Barois I, Lavelle P (2000) Regulation of soil organic matter dynamics and microbial activity in the drilosphere and the role of interactions with other edaphic functional domains. Eur J Soil Biol 36:177–198

Brown KD, Sherlock E (2021) Earthworm ecology. The Earthworm Society of Britain: digging into the world of earthworms. Retrieved on 20 April 2021 from https://www.earthwormsoc.org.uk/earthworm-ecology

Butt KR, Lowe CN (2011) Controlled cultivation of endogeic and anecic earthworms. In: Karaca A (ed) Biology of earthworms. Springer, Berlin, Heidelberg, pp 107–121. https://doi.org/10.1007/978-3-642-14636-7_7

Carlson CL, Adriano DC (1993) Environmental impact of coal combustion residues. J Environ Qual 22:227–247. https://doi.org/10.2134/jeq1993.00472425002200020002x

Chan KY, Munro K (2001) Evaluating mustard extracts for earthworm sampling. Pedobiologia 45:272–278. https://doi.org/10.1078/0031-4056-00084

Chang AC, Lund LJ, Page AL, Warneke JE (1977) Physical properties of fly ash-amended soils. J Environ Qual 6:267–270

Čoja T, Zehetner K, Bruckner A, Watzinger A, Meyer E (2008) Efficacy and side effects of five sampling methods for soil earthworms (Annelida, Lumbricidae). Ecotoxicol Environ Saf 71:552–565. https://doi.org/10.1016/j.ecoenv.2007.08.002

Curry JP, Schmidt O (2007) The feeding ecology of earthworms—a review. Pedobiologia 50:463–477. https://doi.org/10.1016/j.pedobi.2006.09.001

Dallinger R (1994) Invertebrate organisms as biological indicators of heavy metal pollution. Human Press Inc 48(a):27–31. https://doi.org/10.1007/BF02825356

Deng S, Zeng D (2017) Removal of phenanthrene in contaminated soil by combination of alfalfa, white-rot fungus, and earthworms. Environ Sci Pollut Res 24:7565–7571. https://doi.org/10.1007/s11356-017-8466-y

Dhindsa HS, Sharma RD, Kuma R (2016) Role of fly ash in improving soil physical properties and yield of wheat (*Triticum aestivum*). Agri Sci Dig 36(2):97–101. https://doi.org/10.18805/asd.v36i2.10626

Dwibedi SK, Sahoo SK (2017) Effect of fly ash and vermicompost on the performance of rice seedlings and physicochemical properties of the soil in West Central Table Land Zone of Odisha. M. Tech. (Environmental Science and Engineering) thesis, Sambalpur University, Jyoti Vihar, Odisha, India

Dwibedi SK, Sahu SK, Pandey VC, Rout KK, Behera M (2021) Seedling growth and physicochemical transformations of rice nursery soil under varying levels of coal fly ash and vermicompost amendment. Environ Geochem Hlth. https://doi.org/10.1007/s10653-021-01074-y

Dwibedi SK (2018) Wetland management through agro-geo-special technology in Odisha. 11th science communicator's meet. 105th Indian Science Congress, 19–20 March, Manipur University-795003, pp 27–28

East D, Knight D (1998) Sampling soil earthworm populations using household detergent and mustard. J Biol Educ 32:201–206. https://doi.org/10.1080/00219266.1998.9655621

Edwards CA (2004) Earthworm ecology. CRC Press, Boca Ratoon, FL

Edward CA, Arancon NQ (2006) The science of vermiculture: the use of earthworms in organic waste manangement. In: Guerrero RDIII, Guerrero-del Castillo MRA (eds) Vermi technologies for developing countries. Proceedings of the international symposium-workshop on vermi technologies for developing countries. Nov. 16–18, 2005

Eisenhauer N, Straube D, Scheu S (2008) Efficiency of two widespread non-destructive extraction methods under dry soil conditions for different ecological earthworm groups. Eur J Soil Biol 44(1):141–145. https://doi.org/10.1016/j.ejsobi.2007.10.002

Fischer E (1973) The chloragosomes of Lumbricidae as cation exchangers in vitro investigations. Acta Biol Acad Sci H 24:157–163

Fischer E (1977) The function of chloragosomes, the specific age pigment granules of annelids—a review. Exp Geront 12:69–74. https://doi.org/10.1016/0531-5565(77)90035-3

Fischer E, Trombitts K (1980) X-ray microprobe analysis of chloragosomes of untreated and EDTA-treated *Lumbricus terrestris* by using fresh air-dried smears. Acta Histochem 66:237–242. https://doi.org/10.1016/S0065-1281(80)80008-0

Fischer GL, Chang DPY, Brummer M (1978) Fly ash collected from electrostatic precipitator. Microcrystalline structures and the mystery of the spheres. Science 19:553. https://doi.org/10.1126/science.192.4239.553

Fründ HC, Graefe U, Tischer S (2011) Earthworms as bioindicators of soil quality. In: Karaca A (ed) Biology of earthworms. Soil Biol, vol 24. Springer, Berlin, Heidelberg. https://doi.org/10.1007/978-3-642-14636-7_16

Gerhardt KE, Gerwing PD, Greenberg BM (2017) Opinion: taking phytoremediation from proven technology to accepted practice. Plant Sci 256:170–185. https://doi.org/10.1016/j.plantsci.2016.11.016

Ghodrati M, Sims JT, Vasilas BS (1995) Evaluation of flyash as a soil amendment for the Atlantic coastal plain. I. Soil hydraulic properties and elemental leaching. J Water Soil Air Pollut 81:349–361. https://doi.org/10.1007/BF01104020

Goyal V, Angar MR, Srivastava DK (2002) Studies on the effect of fly ash treated soil on the increased protein contents in the seeds of *Glycine max* (soyabean). Asian J Chem 14:328–332

Gupta R, Garg VK (2009) Vermiremediation and nutrient recovery of non-recyclable paper waste employing *Eisenia fetida*. J Haz Mater 162(1):430–439. https://doi.org/10.1016/j.jhazmat.2008.05.055

Gupta SK, Tewari A, Srivastava R, Murthy RC, Chandra S (2005) Potential of *Eisenia fetida* for sustainable and efficient vermicomposting of fly ash. Water Air Pollut 163(1/4):293–302. https://doi.org/10.1007/s11270-005-0722-y

International Energy Agency (2020) Share of coal in global energy mix touches lowest level in 16 years, drops to 27 percent. IEA Clean Copal Centre. https://www.iea-coal.org/share-of-coal-in-global-energy-mix-touches-lowest-level-in-16-years-drops-to-27-per-cent/

Jones CG, Lawton JH, Shachak M (1994) Organisms as ecosystem engineers. Oikos 69(3):373–386. https://doi.org/10.1007/978-1-4612-4018-1_14

Kalra N, Jain M, Joshi H, Choudhry R, Harit R, Vasta B, Sharma S, Kumar V (1998) Fly ash as a soil conditioner and fertilizer. Biores Technol 64:163–167. https://doi.org/10.1016/S0960-8524(97)00187-9

Kalra N, Harit R, Sharma S (2000) Effect of fly ash incorporation on soil properties on texturally variant soils. Biores Technol 75:91–93. https://doi.org/10.1016/S0960-8524(00)00036-5

Khan MR, Khan MW, Singh K (1997) Management of root-knot disease of tomato by the application of fly ash in soil. Plant Pathol 46:33–43. https://doi.org/10.1046/j.1365-3059.1997.d01-199.x

Kuchawar OD, Matte DB, Kene DR (1997) Effect of fly ash application on physico-chemical properties of soil. J Soils Crops 7:73–75

Kumar V, Zacharia KA, Goswami G (2000) Fly ash use in agriculture: a perspective Proceedings of the 2nd international conference on fly ash disposal and utilization, vol 2, pp 1–13

Kuzyakov Y, Blagodatskaya E (2015) Microbial hotspots and hot moments in soil: concept & review. Soil Biol Biochem 83:184–199. https://doi.org/10.1016/j.soilbio.2015.01.025

Labenz AT (2021) Earthworm activity increases soil health. Natural Resources Conservation Service Hutchinson, Kansas. Retrieved on 20 March 2021. https://www.nrcs.usda.gov/wps/portal/nrcs/detail/ks/newsroom/features/?cid=stelprdb1242736

Lal R (1999) Soil conservation and biodiversity. In: Hawksworth DL (ed) The biodiversity of microorganisms and invertebrates: its role in sustainable agriculture. CAB International, Wallingford, UK, pp 89–103

Lawrence AP, Bowers MA (2002) A test of the 'hot' mustard extraction method of sampling earthworms. Soil Biol Biochem 34:549–552. https://doi.org/10.1016/S0038-0717(01)00211-5

Lazcano C, Arnold J, Tato A, Zaller JG (2009) Dominguez J (2009) Compost and vermicompost as nursery pot components: Effects on tomato plant growth and morphology. Span J Agric Res 7:944–951. https://doi.org/10.5424/sjar/2009074-1107

Lee H, Ha HS, Lee CH, Lee YB, Kim PJ (2006) Fly ash effect on improving soil properties and rice productivity in Korean paddy soils. Biores Technol 97:1490–1497. https://doi.org/10.1016/j.biortech.2005.06.020

Li LZ, Zhou DM, Wang P, Luo XS (2008) Subcellular distribution of Cd and Pb in earthworm *Eisenia fetida* as affected by Ca^{2+} ions and Cd-Pb interaction. Ecotoxicol Environ Saf 71:632–637. https://doi.org/10.1016/j.ecoenv.2008.04.001

Lukashe NS, Mupambwa HA, Green E, Mnkeni PNS (2018) Inoculation of fly ash amended vermicompost with phosphate solubilizing bacteria (*Pseudomonas fluorescens*) and its influence on vermi-degradation, nutrient release and biological activity. Waste Manage 83:14–22. https://doi.org/10.1016/j.wasman.2018.10.038

Morgan AJ (1981) A morphological and electron-microprobe study of the inorganic composition of the mineralized secretory products of the calciferous gland and chloragogenous tissue of the earthworm, *Lumbricus terrestris* L. The distribution of injected strontium. Cell Tissue Res 200:829–844. https://doi.org/10.1007/BF00210465

Morgan AJ, Morris B (1982) The accumulation and intracellular compartmentation of cadmium, lead, zinc and calcium in two earthworm species (*Dendrobaena rubida* and *Lumbricus rubellus*) living in highly contaminated soil. Histochem Cell Biol 75:269–285. https://doi.org/10.1007/BF00496017

Morgan JE, Morgan AJ (1990) The distribution of cadmium, copper, lead, zinc and calcium in the tissues of earthworm *Lumbricus rubellus* sampled from one uncontaminated and four polluted soils. Oecologia, (Berlin) 84:559–566. https://doi.org/10.1007/BF00328174

Mupambw HA, Dube E, Mnkeni PNS (2015) Fly ash composting to improve fertiliser value—a review. S Afr J Sci 111:7–8. https://doi.org/10.17159/SAJS.2015/20140103

Niyazi R, Chaurasia S (2014) Vermistabilization of fly ash amended with pressmud by employing *Eisenia foetida*. Int J Pharma Chem Biol Sci 4(1):85–95. http://www.ijpcbs.com/files/volume4-1-2014/13.pdf

Page AL, Elseewi AA, Straughan IR (1979) Physical and chemical properties of fly ash from coal-fired power plants with reference to environmental impacts. Residue Rev 71:83–120. https://link.springer.com/chapter/. https://doi.org/10.1007/978-1-4612-6185-8_2

Panda D, Tikadar P (2014) Effect of fly ash incorporation in soil on germination and seedling characteristics of rice (*Oryza sativa* L.). Biolife J 2(3):800–807

Pandey VC, Singh N (2010) Impact of fly ash incorporation in soil systems. Agric Ecosyst Environ 136:16–27. https://doi.org/10.1016/j.agee.2009.11.013

Pandey VC, Abhilash PC, Singh N (2009a) The Indian perspective of utilizing fly ash in phytoremediation, phytomanagement and biomass production. J Environ Manage 90:2943–2958. https://doi.org/10.1016/j.jenvman.2009.05.001

Pandey VC, Abhilash PC, Upadhyay RN, Tewari DD (2009b) Application of fly ash on the growth performance and translocation of toxic heavy metals within *Cajanus cajan* L.: implication for safe utilization of fly ash for agricultural production. J Hazard Mater 166:255–259. https://doi.org/10.1016/j.jhazmat.2008.11.016

Pandey VC (2020a) Phytomanagement of fly ash. Elsevier (Authored book), p 334. ISBN: 9780128185445. https://doi.org/10.1016/C2018-0-01318-3

Pandey VC (2020b) Fly ash properties, multiple uses, threats, and management: an introduction. In: Phytomanagement of fly ash. Elsevier, Amsterdam, pp 1–34. https://doi.org/10.1016/B978-0-12-818544-5.00001-8

Pandey VC (2020c) Scope of fly ash use in agriculture–prospects and challenges. In: Phytomanagement of fly ash. Elsevier, Amsterdam, pp 63–101. https://doi.org/10.1016/B978-0-12-818544-5.00003-1

Pandey VC (2020d) Opportunities and challenges in fly ash–aided paddy agriculture. In: Phytomanagement of fly ash. Elsevier, Amsterdam, pp 103–139. https://doi.org/10.1016/B978-0-12-818544-5.00004-3

Pani NK, Samal P, Das R (2015) Effect of fly ash on soil properties, changes in bio-chemical parameters and heavy metal uptake in sunflower (*Helianthus annus* L.). Int J Sci Technol Manage 4(7):1–11. http://www.ijstm.com/images/short_pdf/1438321607_P174-184.pdf

Parab N, Mishra S, Bhonde SR (2012) Prospects of bulk utilization of fly ash in agriculture for integrated nutrient management. Bull Nat Inst Ecol 23:31–46. https://nieindia.org/Journal/index.php/niebull/article/view/162

Priatmadi BJ, Saidy AR, Septiana M (2015) Soil properties and growth performance of rice (*Oryza sativa* L.) grown in a fly-ash amended soil. Trop Wetland J 1(1):19–24

Rautaray SK, Ghosh BC, Mittra BN (2003) Effect of fly ash, organic wastes and chemical fertilizers on yield, nutrient uptake, heavy metal content and residual fertility in a rice-mustard cropping sequence under acid lateritic soils. Biores Technol 90:275–283. https://doi.org/10.1016/S0960-8524(03)00132-9

Reddy TP, Umadevi M, Rao PC (2010) Effect of fly ash and farm yard manure on soil properties and yield of rice grown on an inceptisol. Agric Sci Dig 30(4):281–285. https://www.indianjournals.com/ijor.aspx?target=ijor:asd&volume=30&issue=4&article=012

Rodriguez-Campos J, Dendooven L, Alvarez-Bernal D, Contreras-Ramos SM (2014) Potential of earthworms to accelerate removal of organic contaminants from soil: a review. Appl Soil Ecol 79:10–25. https://doi.org/10.1016/j.apsoil.2014.02.010

Sarkar A, Singh A, Agrawal SB (2012) Utilization of fly ash as soil amendments in agricultural fields of North-Eastern Gangetic plains of India: potential benefits and risks assessments. Bull Nat Inst Ecol 23:0-00. https://nieindia.org/Journal/index.php/niebull/article/view/150

Scheu S (2003) Effects of earthworms on plant growth: patterns and perspectives. Pedobiologia 47(5–6):846–856. https://doi.org/10.1078/0031-4056-00270

Sharma K, Garg VK (2018) Chapter 17—Solid-state fermentation for vermicomposting: a step toward sustainable and healthy soil. In: Pandey A, Larroche C, Soccol CR (eds) Current developments in biotechnology and bioengineering. Elsevier, pp 373–413. https://doi.org/10.1016/B978-0-444-63990-5.00017-7

Sharma SK, Kalra N (2006) Effect of fly ash incorporation on soil properties and plant productivity—a review. J Sci Ind Res 65:383–390. https://www.researchgate.net/publication/224319633_Effect_of_flyash_on_soil_properties_and_productivity_of_crops_A_review#:~:text=Fly ash%20can%20be%20used%20for,macro%20and%20micronutrients%20for%20plants

Sheoran VA, Sheoran S, Poonia P (2010) Role of hyperaccumulators in phytoextraction of metals from contaminated mining sites: a review. Crit Rev Environ Sci Technol 41(2):168–214. https://doi.org/10.1080/10643380902718418

Shi ZM, Xu L, Hu F (2014) A hierarchic method for studying the distribution of phenanthrene in *Eisenia fetida*. Pedosphere 24:743–752. https://doi.org/10.1016/S1002-0160(14)60061-8

Shi Z, Tang Z, Wang C (2017) A brief review and evaluation of earthworm biomarkers in soil pollution assessment. Environ Sci Pollut Res 24:13284–13294. https://doi.org/10.1007/s11356-017-8784-0

Shi Z, Tang Z, Wang C (2019) Effect of phenanthrene on the physicochemical properties of earthworm casts in soil. Ecotoxicol Environ Saf 168:348–355. https://doi.org/10.1016/j.ecoenv.2018.10.032

Shi Z, Liu J, Tang Z, Zhao Y, Wang C (2020) Vermiremediation of organically contaminated soils: concepts, current status, and future perspectives. Appl Soil Ecol [Internet]. https://doi.org/10.1016/j.apsoil.2019.103377

Sikka R, Kansal BD (1995) Effect of fly ash application on yield and nutrient composition of rice, wheat and on pH and available nutrient status of soil. Biores Technol 51:199–203. https://doi.org/10.1016/0960-8524(94)00119-L

Singh SN, Kushreshtha K, Ahmed KJ (1997) Impact of fly ash soil amendment on seed germination, seedling growth and metal composition of Vicia faba L. Ecol Eng 9:203–208. https://doi.org/10.1016/S0925-8574(97)10004-0

Singh JS, Pandey VC (2013) Fly ash application in nutrient poor agriculture soils: impact on methanotrophs population dynamics and paddy yields. Ecotoxicol Environ Saf 89:43–51. https://doi.org/10.1016/j.ecoenv.2012.11.011

Singh S, Gond D, Pal A, Tewary BK, Sinha A (2011a) Performance of several crops grown in fly ash amended soil. In: World coal ash conference—May 9–12, 2011, Denver, CO, USA. http://www.flyash.info/

Singh JS, Pandey VC, Singh DP (2011b) Coal fly ash and farmyard manure amendments in dry-land paddy agriculture field: effect on N-dynamics and paddy productivity. Appl Soil Ecol 47:133–140. https://doi.org/10.1016/j.apsoil.2010.11.011

Singh K, Pandey VC, Singh B, Patra DD, Singh RP (2016a) Effect of fly ash on crop yield and physico-chemical, microbial and enzyme activities of sodic soils. Environ Eng Manage J 15(11):2433–2440

Singh S, Bhat SA, Singh J, Kaur R, Vig AP (2016b) Vermistabilization of thermal power plant fly ash using *Eisenia fetida*. J Ind Pollut Cont. Retrieved on 20 March 2021 from https://www.icontrolpollution.com/articles/vermistabilization-of-thermal-power-plant-fly-ash-using-eisenia-fetida-.php?aid=81970

Sinha RK, Bharambe G, Ryan D (2008) Converting wasteland into wonderland by earthworms-a low-cost nature's technology for soil remediation: a case study of vermiremediation of PAHs contaminated soil. Environmentalist 28:466–475. https://doi.org/10.1007/s10669-008-9171-7

Sinha RK, Herat S, Valani D, Chauhan K (2010) Earthworms-The environmental engineers: review of vermiculture technologies for environmental management and resource development. Int J Glob Environ 10:265–292. https://doi.org/10.1504/IJGENVI.2010.037271

Steffen GPK, Antoniolli ZI, Steffen RB, Jacques RJS, dos Santos ML (2013) Earthworm extraction with onion solution. Agric Ecosyst Environ Appl Soil Ecol 69:28–31. https://doi.org/10.1016/j.apsoil.2012.12.013

Stromberger ME, Keith AM, Schmidt O (2012) Distinct microbial and faunal communities and translocated carbon in *Lumbricus terrestris drilospheres*. Soil Biol Biochem 46:155–162. https://doi.org/10.1016/j.soilbio.2011.11.024

Tarannum A, Khan AA, Diva I, Khan B (2001) Impact of fly ash on hatching, penetration and development of root-knot nematode, *Meloidogyne javanica* on chickpea (*Cicer arietinum* L.). Nematol Mediterr 29:215–218

Tiunov AV, Scheu S (1999) Microbial respiration, biomass, biovolume and nutrient status in burrow walls of *Lumbricus terrestris* L. (Lumbricidae). Soil Biol Biochem 31:2039–2048. https://doi.org/10.1016/S0038-0717(99)00127-3

Tripathy A, Sahu RK (1997) Effect of coal fly ash on growth and yield of wheat. J Environ Biol 18(2):131–135. https://www.osti.gov/etdeweb/biblio/488690

Usmani Z, Kumar V (2017) The implications of fly ash remediation through vermicomposting: a review. Nature Environ Pollut Technol 16(2):363–374. http://www.neptjournal.com/upload-images/NL-60-5-(3)B-3395com.pdf

Valckx J, Govers G, Hermy M, Muys B (2011) Optimizing earthworm sampling in ecosystems. In: Karaca A (ed) Biology of earthworms. Springer, Berlin, Heidelberg, pp 19–38

West TO, McBride AC (2005) The contribution of agricultural lime to carbon dioxide emissions in the United States: dissolution, transport and net emissions. Agric Ecosys Environ 108:145–154. https://doi.org/10.1016/j.agee.2005.01.002

Wong JWC, Wong MH (1990) Effects of fly ash on yields and elemental composition of two vegetables, *Brassica parachinensis* and *B. chinensis*. Agric Ecos Environ 30:251–264. https://doi.org/10.1016/0167-8809(90)90109-Q

Yeledhalli N, Prakash S, Gurumurthy S, Ravi M (2007) Coal fly ash as modifier of physico-chemical and biological properties of soil. Karnataka J Agric Sci 20(3):531–534. http://14.139.155.167/test5/index.php/kjas/article/viewFile/899/892

Zaborski ER (2003) Allyl isothiocyanate: an alternative chemical expellant for sampling earthworms. Appl Soil Ecol 22:87–95. https://doi.org/10.1016/S0929-1393(02)00106-3

Zhang S, Hu F, Li H, Li X (2009) Influence of earthworm mucus and amino acids on tomato seedling growth and cadmium accumulation. Environ Pollut 157:2737–2742. https://doi.org/10.1016/j.envpol.2009.04.027

Zibilski JJ, Alva AK, Sajwan KS (1995) Fly ash. In: Reckcigi JE (ed) Soil amendments and environmental quality. CRC press Inc., Boca Raton, FL, pp 327–363. https://journals.lww.com/soilsci/Citation/1996/07000/Soil_Amendments_and_Environmental_Quality.7.aspx

Chapter 11
Management of Biomass Residues Using Vermicomposting Approach

Suman Kashyap, Seema Tharannum, V. Krishna Murthy, and Radha D. Kale

Abstract Currently, it is evident that the utilization of biomass as a feedstock for production of energy ultimately has deleterious environmental effects like soil disturbance, nutrient depletion and impaired water quality. There is a need to prioritize biomass to protect environmental quality. Vermicomposting process could be an eco-biotechnological approach that provides useful organic compost that addresses both the quality and eco-friendliness, besides rightly managing solid waste that innately contains even non-biodegradable metallic components. Treatises are provided about major biomass types and brief details of their features. Details of the vermicomposting process including subjecting both industrial and domestic solid wastes such as fly-ash and human excreta for vermicomposting are discussed. What and how are the vermicomposting agents that work, perform and yield the useful product is highlighted. A critical discussion on how vermicompost performs better compared to chemical fertilizers and consequently how the earthworms, the principal change agents, are to be protected to sustain the process is discussed. The chapter emphasizes that vermicomposting is a preferred, useful, simple yet effective bioconversion process that is responsible for not only to manage the biomass but also to upkeep environmental quality.

Keywords Biomass · vermicomposting · Solid waste · MSW

S. Kashyap (✉)
Biosciences, Dayanand Sagar University, Bengaluru, India
e-mail: k.suman@jainuniversity.ac.in

S. Tharannum
Department of Biotechnology, PES University, Bengaluru, India

V. Krishna Murthy
Department of Chemistry, Dayanand Sagar University, Bengaluru, India

R. D. Kale
Mount Carmel College, Bengaluru, India

© The Author(s), under exclusive license to Springer Nature Switzerland AG 2023
V. C. Pandey (ed.), *Bio-Inspired Land Remediation*, Environmental Contamination Remediation and Management, https://doi.org/10.1007/978-3-031-04931-6_11

11.1 Introduction

Population growth is leading to the increased industrialization, agricultural, forestry and cattle farming activities generating significant volume of biomass residues. This waste untreated biomass can contaminate the environment, consequently leading to ecological problems (i.e. nitrate and phosphate lixiviation of underground water conservation, salinization of soil and resins, aromatic oils and lignin accumulation). Biomass residues generated from forestry and agricultural activities act as a potential substrate for the contaminants, leading to health issues in humans. So the environmental impact can be minimized on using organic waste materials along with residual biomass by vermicomposting where earthworms aid the decomposition of organic materials by creating ideal environment for mixing and aerating of the organic materials (Santamaria Romero et al. 2001; Dominguez et al. 2003). vermicomposting is a process where earthworms, organic waste residue, microorganisms, carbon dioxide, mineral ions, water and environmental factors are held responsible for the conversion of organic waste into the vermicompost, nutrient-rich with humic substances (Ulle et al. 2004). Research on vermicomposting of hog manure when mixed with bovine manure and fruit fibres of palm oil aided the process of biomass conversion into nutrient-rich vermicompost (Hernández et al. 2008). Vermicomposting being an excellent biotechnological process and the vermicompost produced is physically, chemically and biologically enriched material which proves to be the best agricultural amendment (Nogales et al. 2005). Finally, organic biomass residue is transformed into stable end products viz., vermicompost that is capable of improvising the fertility of the soil in a safer and efficient way (Soto and Muñoz 2002; Hernández et al. 2002).

Plant and animal residues used in the production of electricity, heat and other forms of energy including in industries as raw substances to produce a range of products is termed as biomass. Renewable biomass resources which could either directly be used as fuel or be transformed into an alternative form of energy product are generally termed as feedstocks. The committed and lengthen energy crops like miscanthus, switchgrass; agricultural crop residues and the waste generated from crops viz., bagasse and wheat straw waste; horticultural waste generated from yard; corn cobs waste obtained after the food processing; manure with rich nitrogen and phosphorus content obtained from the animal farming; algae from forestry waste; waste residues obtained from processing of wood; municipal solid waste and wet waste as well includes waste from crops, waste from forest, purpose-grown grasses, waste from industries, segregated municipal solid waste (MSW), wood waste from urban areas and remnants food waste. The major constituents that form biomass can not only be purposely grown as energy crops but also the majority of natural materials also could be converted into useful biomass.

11.2 Vermicomposting

Vermicomposting involves physical, chemical and biological conversion of organic waste matter (agricultural residues of plant and animal origin) into nutrient-rich vermicompost by the action of mesophilic microorganisms and earthworms (Pizl and Novakova 2003; Garg and Gupta 2009).

11.3 Agents of Vermicompost

11.3.1 Microorganisms

Bio-oxidative and eco-biotechnological vermicomposting procedure that transforms solid organic waste material into valuable biological end product-vermicompost in an ecologically friendly manner. The process also involves mutual interaction of microorganisms and earthworms. The microbial biomass along with feedstock, present within the gut of earthworms also contributes to both chemical and biological decomposition of organic waste matter. While earthworms act as mediators and help in increasing the surface area for microbial action, thereby enhancing the enzymatic activities and directly held responsible for the physical status and indirectly for chemical status of organic waste (Fornes et al. 2012). Vermicast (excreta of earthworms) produced as the end product supported the microbial growth and action by acting as an organic substrate (Williams et al. 2006).

11.3.2 C/N Ratio

Considerable decrease in protein, carbohydrate, aliphatic, C/N ratio, lignocellulosic composition, volatile solids and increase in acid phosphatase activity, aromaticity and humic acid content is clearly denoted by the proper mineralization and maturity of vermicompost (Lv et al. 2013). Vermicompost as the end product is found to be consisting of Nitrogen, Phosphorus, Carbon, Potassium content, revealed enhanced activity of the enzymes and inhibition of plant-based pathogens (Pramanik et al. 2007; Yasir et al. 2009; Bhattacharya et al. 2012). The humic acid comprises of functional-carboxyl and phenolate groups that have shown an excellent chelating capacity, and being a ligand has capacity to form coordinate organic compounds with heavy metals such as zinc and copper complexes (Hsu and Lo 2000; Kang et al. 2011). Supplementary to these the earthworms; excretory products like mucus, body fluids including some enzymes enrich to the nitrogen content of the vermicompost.

11.3.3 Heavy Metals

Several studies have reported that vermicompost is free of toxic chemicals and pathogens. Earthworms biodegrade many organic and inorganic chemical residues of recalcitrant environmental pollutants present in the soil such as persistent organochlorine pesticides, heavy metals and polycyclic aromatic hydrocarbons (PAHs). Researchers have exemplified that earthworms assemble considerable quantities of toxic heavy metals like manganese, copper, iron and zinc in the tissue (Yadav and Garg 2009; Hait and Tare 2012). Studies have indicated that trace toxic heavy metals were found to be drawn up from the contaminated organic waste by the epithelial gut cells of the earthworms and consequently leading to the reduction of bioavailability of these elements (Dominguez and Edwards 2004).

11.4 Types of Earthworms Employed for Vermicomposting in India

A study reports that the earthworms being invertebrates and terrestrial organisms belonging to the Annelida family, originated and have existed since 600 million years dating to pre-Cambrian period (Piearce et al. 1990). They can live in diverse habitats and demonstrate good changes of soil, enhancing the fertility of soils by inducing physical, chemical changes in the texture of soils (Darwin 1881; Edwards et al. 1995; Kale 1998). However, among many species, *Eisenia foetida, Eudrilus eugeniae, Perionyx excavatus and P. sansibaricus* are largely employed in vermicomposting (Oyedele et al. 2005; Suthar 2007, 2009, 2010). Species, *Eudrilus eugeniae* and *Perionyx excavatus* have been reported to be suitable for vermicomposting in tropical and sub-tropical environmental conditions.

Eudrilus eugeniae or the 'African night crawler', is a tropical earthworm. *E. eugeniae* can tolerate higher temperatures with ample humidity. Earthworm compost produced by *Eseinia fetida* vigorously affects the soil fertility by its characteristic features like providing nourishment, recuperating soil structure and water-holding capacity. Worldwide *Eisenia foetida (Savigyn), Eudrilus eugeniae (Kinberg), Perionyx excavatus (Perrier),* are the fundamental species of earthworms evaluated for vermicomposting. These three-earthworm species have been considered as the most important species in organic waste recycling (Edwards 1998; Kale and Bano 1988; Giraddi 2000; Chaudhuri et al. 2001; Reddy and Ohkura 2004).

Soil health is indicated by the earthworms (Ismail 1997) hence exhibits a significant role in solid waste management and soil fertility. Earthworms are capable of stabilizing and transforming domestic waste and sewage sludge and converting them into soil nutrients thus they also are agents of significance in the public health field (Ismail 1997). In agriculture and horticulture vermicompost has been of immensely useful material (Edwards and Bohlen 1996; Ismail 2005; Ansari and Ismail 2008).

Studies on consumption of organics from waste by most earthworms are found to be half of their body weight in 24 h (Visvanathan et al. 2005). *Eisenia fetida* is capable of organic consumption equivalent to their whole-body weight in 24 h. Relating to the rate of degradation of organic waste by earthworms range between 60 and 80% more compared to conventional and natural methods of decomposition. When provided with optimum conditions like temperatures ranging between 20 and 30 °C; moisture content of 60–70% and approximately around 10,000 earthworms of about 5 kg can vermicompost one tonne of organic waste residue in one month. Significant reduction in volume of solid waste is noticed upon vermicomposting. And approximately 1–0.5 cum of vermicompost is produced revealing 50% conversion rate and the leftover becomes worm biomass. Vermicompost is effective in improving soil quality. It also remains to be an efficient as well as an effective process that is associated with production of higher biodiversity of useful microbial organisms (Ismail 1993; Vivas et al. 2009). Studies have also shown that pre-composting is necessary to avoid mortality rate of earthworms during vermicomposting (Kaushik and Garg 2003).

11.4.1 Care of Vermicomposting Earthworms

Earthworms transfer and receive sperms in the act of copulation, they are simultaneous hermaphrodites. Earthworm exhibits both male and female sex organs that produce sperm and egg, respectively and are dependent on another earthworm to reproduce, while mating two individual worms line up inverted so that the sperm is exchanged. Each earthworm will have two sperm receptacles and two male openings, which take in the sperm from another mate. Earthworms will have a pair of ovaries that can produce eggs. A mature earthworm within 7–10 days can produce many egg capsules, accommodating around one dozen hatchlings and capsules (cocoons or eggs). Newly hatched earthworm reaches maturity within 60–90 days period, meaning earthworm population multiplication happening each month. However, earthworm mortality rate is also high during harvesting of vermicompost and also because of predators. Successive breeding rate signifies that the earthworm population can easily adapt to environmental conditions and the food supply provided.

11.5 Vermicomposting on Ground Heaps

- Ground heaps could be a better option for vermicomposting than open pits
- Bed of organic biomass residues in dome shape are prepared for vermicomposting
- The dimension of heap with a length of 10 feet, width of 3 feet and height of 2 feet is appropriate one.

Depending on the conditions in the surroundings, above-ground tanks of the size mentioned above can be constructed using locally available materials like mud or cement bricks or stone slabs to keep away the rodents. Appropriate mesh covers are provided to the tanks for full-proof protection. Sheet or thatched roof to the tanks protects the vermicomposting unit from flooding during heavy rains.

11.6 Raw Materials for Degradation by Vermicomposting Process

- Biomass waste residues like:

    Agricultural waste that includes farm wastes like—soybean, mustard, chickpea, cotton seed meal, straw from wheat crops and other oil seed residues,

    Animal waste includes dead fish, waste generated from fish processing industry, night soil that is human faecal dry waste mixed with soil, activated sludge of microorganisms generated in sedimentation tanks of sewage water treatment and other such sources, pig manure, droppings of sheep and scrap poultry manure from poultry farming activities, meat scraps from slaughter and meat industry and forest waste,

    Industrial solid waste, municipal solid waste, etc.

- Fresh cow dung
- Biomass wastes residues were mixed in equal ratios with cow dung (1:1 on dry weight basis)
- One kilogram of earthworms was introduced into the waste biomass for vermicomposting, i.e. approximately around 1000–1200 adult earthworms were used for every one quintal (100 kg) of waste biomass.
- Providing sufficient water to restore 60–70% moisture.

The major types of biomass can be classified basing on their origins such as sourced from agriculture, animal waste, forests and municipal solid waste arising from human habitation. Table 1 provides the type of biomass with a few examples in each type.

11.6.1 Agricultural biomass

Protocols followed in agriculture practices generate huge loads of biomass which has to be processed before dumping in fields. If left untreated, agriculture wastes result in increased salinity of soils (salinization), nitrate leaching and phosphates reaching to aquifers resulting in contamination (Flotats and Solé 2008). Besides, accumulation of resins and aromatic oils in soil could result in land pollution (Achten and Hofmann

Table 1 Types of biomass with a few examples

Agricultural Waste Biomass
Biomass residual waste obtained from agricultural crops like leaves, branches, stalks, straw, husk and waste generated after pruning that was utilized in energy production becomes biomass residues. Crop residues are abundantly generated during crop cultivation, quite often posing disposal problems.
Animal Waste Biomass
Animal waste generated from intensive livestock operations, large scale poultry, pig, cattle farms and waste generated from slaughter houses; animal wastes and their byproducts such as dung, urine, bones, fish processing wastes become complex constituents of biomass sourced from animal wastes. In a few places like at Greece, raising- goats, sheep and lambs is done exclusively, spread over a large area stretched as pastures. However, in such instances, the waste is dispersed, becomes difficult to accumulate and utilize it for energy production.
Forest Waste Biomass
Forest waste biomass is used in energy production process. Waste mainly includes forestry residues generated from thinning and logging processes; firewood, etc. Waste material like fallen tree and its branches, tree leaves, dead and decayed entities has to be cleared off periodically from the forests as to prevent forest fires; substantial amounts of biomass is produced as byproducts from large-scale wood industries.
Human Habitation Waste/Municipal Waste
The biodegradable portion of municipal waste.

2009). Agri waste can also promote pest formation, proliferation of weeds, become a cause for various diseases and resulting in phytotoxicity issues. To overcome these risks it is recommended that the agricultural wastes be managed through proper decomposition methods (Fornes et al. 2012). Such degradation of agri biomass by vermicomposting process results in the formation of stable polymerized and useful biomolecules that enhance soil fertility (Bernal et al. 2009). Various other decomposition protocols produce organic fertilizers with profound requisite variations in their characteristics (Fornes et al. 2012).

11.6.1.1 Vermicomposting of Grape Marc

Vermicomposting of grape marc has been manifested to be effective that yields vermicast as an organic fertilizer. The seeds of grapes are considered to be a source of bioactive components. This conversion procedure reduces the agricultural biomass by transforming the unstable parts of grape marc into a product rich in nutrients, beneficial microbes, polyphenol-free and high-quality organic fertilizer. Vermicompost can be separated from the residues on sieving, while the residues contain grape seeds and also helps in dissolving the phytotoxicity of the polyphenols from vermicompost. Also facilitates vermicompost to attain various bioactive components like fatty acid-rich seed oil and polyphenol-rich extracts as a result of grape seeds processing. Coproducts produced find applications in food, pharmaceutical and cosmetic industries. Grape marc vermicomposting aided the reproduction rate of earthworms and noted applications like manufacture of fish bait, animal feed and as accumulators of soil pollutants, vermiceuticals (pharmaceutically active principle components derived from earthworms) and human food. Thus vermicompost derived from processing of grape marc have resulted to be rich-source of enzymes and was used in the improvement of biochemical performance of soil and in detoxification of pesticide-contaminated soils. This vermicomposting technique has proven to be very simple, eco-friendly, sustainable and effective which is easily scaled up and finds broad applications at various industries, resulting in a variety of value-added products (Domínguez et al. 2017).

11.6.1.2 Vermicomposting of Floral and Herbal Waste

Vermicomposting of floral and herbal waste in the aromatic oil extraction unit that creates disposal problems can be easily considered as the material for getting good quality vermicompost.

11.6.1.3 Vermicomposting of Spent Coffee Grounds (SCGs)

Large volume of spent coffee grounds (SCGs) has been generated worldwide annually. Researchers have foreseen vermicomposting as a rapid, economical, environment friendly and considered to be safe in the sizeable amount of spent coffee grounds conversion. Studies on the transformation process of SCGs into biologically degradable materials that have proven to be useful in energy, food, cosmetic and pharmaceutical industries. The coffee ground waste deserves a greater attention as it acts as a potential biostimulatory amendment in polluted soil that addresses bioremediation. Vermiprocess decreased substantially the biomass residue, resulting in an enzymatically active, nutrient-rich vermicompost in very less time. Carboxyl esterases (CbEs) in the vermicompost produced were assessed for its magnitude, reactivity and contaminant transformation as laccases. In animals, esterases have been perceived as enzymes that detoxify pesticides. In vermicompost, extracellular detoxifying enzymatic activity of esterases is unspecified. Therefore, bioddegradationegradation and chlorpyrifos (model organophosphorus pesticide) detoxification was demonstrated in both solid and liquid conditions derived from SCGs. The reaction of pesticides was a first-order kinetic reaction ($t_{1/2} = 4.74$ day^{-1}). Esterases acted as biological scavengers in response to chlorpyrifos contamination which resulted in inhibition of CbE activity. Vermicompost derived from SCGs has shown that this revalorized product has greater potential with regard to the biological remediation of pesticide-contaminated agricultural soils (Sanchez-Hernandez and Domínguez 2017).

11.7 Animal Waste Biomass

Animal manure or animal biomass is a well-studied source of zoonotic pathogens (Pell 1997). Its predisposing factor is responsible for outspread of diseases among humans and animals if the biomass is left untreated (Albihn and Vinnerås 2007). In urban regions, domesticated animals like cows, goats, pigs and chickens generate a considerable amount of manure, i.e. 59% is left behind without any kind of treatment or just dumped in storm channels while the remaining 32% is spread across the fields untreated as fertilizers (Komakech et al. 2014). Studies have reported that organic biomass waste and animal manure or animal biomass is rich in valuable plant nutrients and organic matter that can reinstate the physical, chemical and biologically degraded soil and fortify sustainable agricultural activity for longer duration (Diacono and Montemurro 2010). Conventional management of biomass waste can reduce the impact on the environment by avoiding the harmful emission of greenhouse gases from landfills (Hoornweg and Bhada-Tata 2012) thereby confronting with chemical fertilizers from causing further destruction (Pimentel et al. 2005).

11.8 Forest Biomass

Forest ecosystem has the maximum biomass, as it includes organisms of all trophic levels as compared to pond, lake or grassland ecosystems. In forest ecosystems productivity also is high which contributes to huge quantities of biomass. Trees are the most massive and complex ecosystems of the earth. Total volume of plant biomass produced per hectare of forest area is considerably high. Three-quarters and more of the total plant biomass in a mature forest is contained in the form of tree trunks that generally exceeds when compared to that present in the canopy and roots combined. However, the contribution of the component parts of the trees to the total biomass is variable. It depends on the age of the individual trees, as well as its developmental stage in the forest ecosystem. Herbivores consume a relatively small, but variable proportion of the annual photosynthetic production. Up to 75% of this may enter the detrital or decomposer food chain, taking the form of litter. This supports a diverse population of soil organisms. The volume of this dead organic material (DOM) can be twice that of the fresh leaf tissue. The type of stratification and degree of structural development of the component plant life forms of the forest ecosystem depends on the amount of tree biomass and, especially, on the density and depth of tree canopy that has been shaped by the particular environmental conditions. If we take tropical rain forests for example the multi-layered tree canopies reduce the penetration of sunlight. Consequently, the growth of small non-arboreal forms beneath it becomes limited. Forestry-livestock organic waste biomass residues is an alarming ecological challenge if left unnoticed and can become burdened in agricultural activities.

11.8.1 Vermicomposting of Mixed Leaves Waste

Research on vermicomposting of processed mixed leaves litter amended with cured cow dung in 50:50, 60:40 and 70:30 ratios, respectively resulted in high-quality, nutrient-rich vermicompost by earthworm *Eudrilus eugeniae*. The conversion rate was the same in all three treatments and vermicompost obtained carried desired plant nutrients (Jayanthi et al. 2010). Study carried out on vermicomposting of leaf litter amended with cow dung in 3:1 ratio by earthworm *Eisenia fetida* for 90 days resulted in nutrient-rich vermicompost with increase in NPK levels upto 17.90%, 44.73% and 18.24%, respectively while there were decreased levels in organic carbon of 13.130% to 10.780% and C:N ratio upto 32.60% was as well recorded. This signifies the enhanced nutrient content in vermicompost due to activity of earthworms (Sandeep et al. 2017).

Studies on vermicomposting of silver oak, bamboo leaves waste and cow dung by earthworm *Eisenia fetida* resulted in value-added manures for sustainable fertility of soil. The study also recorded reduced levels of total organic carbon by 2.4–11.8; K-exch by 4–10.07: C:N ratio by 13.4–45.2: 2.18–4.13 folds, respectively while there was increase in electrical conductivity of 12–142, ash content of 1.07–1.17, total Nitrogen of 2.1–3.72, total Phosphorus of 111–117, total Calcium of 182–215 and N-NO3 of 9.47–17.59, respectively. Highest microbial populations and cocoon numbers in vermibeds were reported. This indicates the successful conversion of urban forest leaf litter into value-added vermicompost for fertility of soil (Suthar and Gairola 2014).

Management of forest leaf biomass by vermicomposting will help reduce air pollution by burning up of the leaf litter and contamination caused by landfilling of leaf litter waste. The vermicomposting of forest leaf waste biomass can be used in organic farming which results in increased productivity of crops.

11.9 Human Habitation Waste

11.9.1 Vermicomposting of Industrial Sludge

During the process of biological industrial wastewater treatment, a considerable amount of industrial sludge is released annually, as a by-product. Most of the industries like paper and pulp, chemical, cement, power plants, tanneries, oil refineries, food processing and many more. Release of industrial sludge as a secondary pollutant is another challenge faced during the coagulation-flocculation process, by the treatment plant. Industrial sludge contains. This sludge is either semisolid or solid material, contains colloidal matter, particulate sand, all the other compounds separated from the industrial wastewater and the substances which were added during the biological and chemical treatment process.

Industrial sludge is found to be combined with following contaminants:

1. Organic compounds and secondary pollutants such as polychlorinated biphenyls (PCBs), polyaromatic hydrocarbons (PAHs), polychlorinated dibenzodioxins/furans (PCDD/Fs) and surfactants.
2. Inorganic compounds like trace elements and metallic components.
3. Microbes like viruses, pathogenic bacteria, parasitic helminths and protozoa.
4. At several places, the industrial sewage system is let into the municipal sewage system, ultimately the sludge is found to be mixed with organic matter and higher heavy metal content.

Ultimately, the composition of industrial sludge varies, based on the quality and treatment protocols followed. Industrial sludge generated from industries exhibits distinct characteristic features, so sludge has to be managed, treated distinctively and disposed of in eco-friendly manner. Therefore, management of industrial sludge is most challenging and intricate environmental worry. Industrial sludge could be tempered, transformed and subjected to digestion by proteolytic enzymes inside the digestive system of earthworms (Joo et al. 2015).

11.9.2 Vermicomposting of Paper and Pulp Industry Sludge

Demand for pulp production has surged quantum of waste sludge. Sludge release, management and disposal have raised a challenge because of inflexible environmental regulations imposed on solid waste disposal. The sludge has varied amounts of wood fibres, cellulose, hemicellulose and lignin, micronutrients, carbohydrates, macronutrients, trace metals mixed in water. The structural polysaccharides and lesser nitrogen content (<0.5%) of the sludge make the degradation process of the sludge hard. Research has paved a way to overcome this problem, mixing up of municipal biosolids rich in nitrogen content with the solids of pulp mills is considered ideal as it can act as inoculant to microbes and can later be subjected to vermicomposting (Quintern 2014). Paper mill sludge added to the debris of tomato plants in 1:2 ratio has been reported to be best feed for earthworm *E. fetida*, for their optimal growth and reproduction during vermicomposting. On using the high volumes of tomato plant debris to the paper mill sludge resulted in vermicompost rich in humic acid content (Fernández-Gómez et al. 2013). Studies have also shown that earthworm *Perionyx excavatus* were used in vermicomposting, equal ratios of paper and pulp mill sludge when amended with cow dung and waste from food processing industry, resulted in vermicompost with increased total Nitrogen and Phosphorus content by 58.7% and

76.1%, respectively, whereas reduction of around 74.5% total organic carbon content was reported (Sonowal et al. 2013). Another study reported that vermicomposting protocol of paper industry sludge and cow dung gave rise to vermicompost, significantly accounting for low levels of heavy metals, i.e. 95.3–97.5% of Pb, 32–37% Cd, 68.8–88.4% Cu and 47.3–80.9% of Cr (Suthar et al. 2014). Heavy metal bioremediation could be processed by vermicomposting of the industrial sludge showed safer disposal of sludge.

11.9.3 Vermicomposting of Agro-Based/Sugar Industry Sludge

Sugar processing industries release a considerable quantity of by-products like bagasse, sludge from fermentation of yeast, cane trash and pressmud. Pressmud sludge comprises fibre, sugar, soil particles, cane wax, inorganic salts and albuminoids. Management of pressmud sludge is most challenging because of its longer decomposition time and foul odour when dumped in open places. Researchers have reported that pressmud sludge when mixed with *Jeevamirtham Azospirillum* and cow dung in the vermicomposting process by earthworm species *Eudrilus eugeniae* successfully transformed pressmud sludge into vermicompost which was found to be odour free, a stabilized nutrient-rich product, which was rich in NPK and low on C:N ratio and organic Carbon components (Vasanthi et al. 2014). Another study reported the biological transformation of pressmud sludge and cow dung in various ratios resulted in final vermicompost product with increased phosphorus, nitrogen, sodium contents, pH and electrical conductivity while potassium content and the genotoxicity of pressmud sludge were found to be reduced (Bhat et al. 2014).

11.9.4 Vermicomposting of Sludge from Food Industry

Food industry generates semisolid waste and liquid waste in large volumes, which contains organic matter, proteins, sugars, enzymes, organic carbon and micro and macronutrients. Disposal of such waste onto land might cause variations in pH, foul odour and secondary salinization which could be because of presence of heavy metals (Yadav and Garg 2013). Further, application of such organic matter as fertilizers inhibits plant growth because of toxic metabolite production (Zucconi et al. 1981). Research reported that vermicomposting of bakery sludge when mixed with cow dung, affected the enzymatic and microbial parameters which resulted in growth and multiplication of earthworm *E. fetida* (Yadav et al. 2015). Vermicomposting of sludge released from food industry amended with the with the cow dung, slurry from biogas plant and poultry droppings resulted in enhanced earthworm biomass which in turn produced nutrient-rich vermicompost and as a conclusion vermicomposting proved to be suitable food industry waste management technique (Garg et al. 2012).

11.9.5 Vermicomposting of Wastewater Sludge from Milk Processing Industry

Management of the solid or sludge produced from the milk processing industry is the greatest challenge faced again. The wastewater sludge has an impact on the environment, as conventional disposal like landfilling have led to contamination of soil and groundwater resources whereas emission of greenhouse gases leading to pollution of air (Lim et al. 2016). Research has shown that vermicomposting process of 60% sludge from milk industry with 40% cow dung by the earthworm, *E. fetida* resulted in vermicompost rich in minerals, total N, P, K+, Ca^{2+}, Fe, Mn and Zn, as well showed low pH, C:N ratio and organic carbon content (Suthar 2012). Investigation on the vermicomposting of 60% dairy industry wastewater sludge, 10% cattle dung and 30% agricultural trash like sugarcane and wheat straw proved to be the appropriate feed for flawless growth of earthworms and nutrient-rich vermicompost.

11.9.6 Vermicomposting of Sludge from Tanning Industry

Tanning, the chemical process in which leather and its products are produced using animal hides and skin, such industries release huge amounts of waste as there is increasing demand for leather products because of increasing population growth. Tannery sludge released from the tanning industry comprises organic materials rich in nutrients, chromium, volatile organic compounds, sulfide, pathogens and various chemicals used in this process has a significant impact on the environment if left unmanaged. Vermicomposting of sludge from tannery industry and cow dung resulted in value-added, nutrient-rich vermicompost with low C:N ratio (Vig et al. 2011). Another study on vermicomposting of liming and primary tannery sludge amended with cattle dung in different proportions resulted into a vermicompost with increased N, K, Ca, Mg, Na while total organic carbon and C:N ratio was found to be lowered, proved to be better soil conditioner (Malafaia et al. 2015). Alternatively, another study on vermicomposting of tannery sludge along with sawdust resulted in decreased pH, C:N ratio, organic carbon and organic vermicompost matter with increased total nitrogen and cation exchange, while Cr (VI) was found to be negligible after 135 days of vermicomposting. Studies have reported that vermicomposting of tannery sludge by earthworms aided in Chromium bioremediation, i.e. biotransformation of Cr (VI) to Cr (III), a much stable form (Nunes et al. 2016).

11.9.7 Vermicomposting of Sludge from Textile Industries

Vermicomposting of textile mill sludge for 75 days with cattle dung in 1:3 ratio resulted in enhanced reproduction rate of earthworms, cocoon formation and earthworm biomass, as the sludge comprises of organic matter, N, P, K and some micronutrients in increased amounts. They also reported that increased concentration of sludge is attributed to sludge toxicity consequently leading to death of earthworms (Bhat et al. 2013).

11.9.8 Vermicomposting of Distillery Industry Waste

Distillery industry uses enormous raw materials for fermentation and distillation processes, which generates high volumes of waste or the sludge that is attributed with unpleasant odour, high COD viz., chemical oxygen demand, BOD viz., biochemical oxygen demand and organic materials. Disposal of untreated sludge is not recommended as it will have a negative impact on the environment. Research on vermicomposting of sludge from distillery industry when mixed with cow dung in 1:9 and 1:3 ratios, respectively, resulted in high-quality, nutrient-rich vermicompost later to be used as an eco-friendly fertilizer. They also reported that sludge consisting of low C:N ratio resulted in growth and development of earthworms and increased production of cocoons (Singh et al. 2014).

11.9.9 Vermicomposting of Carbide Sludge

In the production of acetylene gas, a significant quantity of carbide sludge is produced. Research on vermicomposting of carbide waste sludge of 1.5–2% when mixed with cow dung, sawdust and vegetable waste in 4:1:5 ratios along with 0.27 kg of dried leaves aided in the process of successful transformation of carbide waste sludge into nutrient-rich vermicompost (Varma et al. 2015).

11.9.10 Vermicomposting of Contaminated Groundwater

Chemicals like ferric sulphate and precipitated lime were introduced into groundwater treatment plants containing arsenic-contaminated water, while these chemicals aid in adjusting the pH and precipitate the contaminants which form sludge. Direct disposal of this sludge can cause a negative impact on the environment by releasing heavy metals. Research on vermicomposting of waste sludge when amended with horse manure and grass in 3:6:1 ratio resulted in nutrient-rich organic fertilizer with reduced arsenic content of about 1/3 by the end of 90-day trial. Results obtained concluded that earthworms can consume heavy metals like arsenic from sludge through passive diffusion and metamorphose into an inorganic or less toxic component (Maňáková et al. 2014).

11.10 Municipal Waste

The uncontrolled human population, especially in urban areas creates a surge in demand for basic necessities. This trend naturally forces local governments to manage the obvious surge in overproduction of solid wastes of several types. In India, the problem is assumed to be uncontrolled as in the processes of municipal solid waste collection, segregation, transfer and allocation. The solid waste management thus has become an insurmountable problem since there will be negligence on the part of both by the local governing bodies as well as of the producers not complying to source-segregation of wet (biodegradable) and dry (recyclable) wastes in India. This has resulted often in an unmanageable imbalance in the proportion of production wastes and its proper treatments to manage. India being the second most populated country in the world, the trend shows an increase in urbanization by almost 4% in the last decade (Plecher 2020). Increased population on a rapid rate lays stress on basic necessities of life besides depleting the environmental resources both in quality and quantity (Manser and Keeling 1996; Cointreau 2006; Kathivale and Muhd Yunus 2008). Increase in population also is leading to expeditious industrialization and urbanization processes which will directly or indirectly influence the amount of MSW being produced (Minghua et al. 2009). The management of solid waste is, therefore a daunting task in Indian metro cities. This scenario is also a common feature in counties of under-developed and developing stages of the countries of the world, as well.

Studies have been reported of vermicompost produced from earthworms by the degradation of wastes of sewage treatment plants and industries like paper cardboard, breweries, pulp, sericulture, distillery, vegetable oils, potato and corn chips, sugarcane, aromatic oil extraction, wood (Kale 1998; Kale et al. 1992; Seenappa et al. 1995; Gunathilagraj and Ravignanam 1996; Lakshmi and Vizaylakshmi 2000). The vermicomposting was resorted to by managing fixed proportion of toxic waste of the mining industry such as those with sulphurous residues with other organic wastes, else the disposal problems would arise owing to toxicity issues (Kale and Sunitha 1995).

11.10.1 Palm Oil Mill Waste

The palm oil mill waste (POMW) is a major form of solid waste in Malaysia. The oil effluent has the palm seed shells, the palm tree twigs and the fibre from oil seed covers. Normally waste generated is either thrown out into open dumps or used as plant fertilizers and also as animal feed. Solid waste used in land applications is a regular practice. But untreated POMW directly into agricultural soil results in water contamination and leaching of pollutants in soils. Vermicomposting of palm oil mill waste, therefore, was considered to address the land, water contamination. Unfortunately, only a few researchers are focussing on this prospective area. Vermicomposting of residual biomass from the oil industry could prove to be a useful alternative (Singh et al. 2011).

11.11 Vermicomposting of Human Excreta

Vermicomposting of human faeces or human excreta was studied successfully (Bajsa et al. 2004). The conversion of faeces to vermicompost was completed in six months. The vermicompost showed better texture that is free from odour and pathogens. Sawdust, therefore, was a better material for covering and also for the vermicomposting process in toilets to enhance the earthy smell, texture and colour of the compost.

11.12 Vermicomposting of Fly-Ash

Fly-ash from the coal-based power plants is considered as a hazardous waste and its disposal is a huge problem. It contains a few metallic components. Fly-ash happens to contain rich amounts of nitrogen and microbial biomass. The earthworms ingest the fly-ash while converting them into vermicompost complex. It is reported that 25% of fly-ash mixed with sisal green pulp, parthenium and green grass cuttings form a feed for *Eisenia fetida* species the earthworms. The vermicompost had higher NKP contents (Saxena et al. 1998).

11.13 Vermiremediation of Contaminated Soils

Earthworms are said to have an active role in removing Polycyclic aromatic hydrocarbons (PAHs) like benzo (a) pyrene is highly resistant to degradation biologically, hydrocarbons and several harmful chemicals from contaminated soils. The process involves earthworms, catabolically active microorganisms in vermiremediation. Earthworms are resistant to toxic PAHs and capable of tolerating toxic concentrations normally not found in soils. Study conducted in Brisbane during winter showed that around seven PAH (approx. 80%) compounds were removed successfully by 500 earthworms in twelve weeks. It is reported that rate of vermiremediation can be boosted by increasing the earthworm number (100 mature earthworms in a kilogram of soil) and the time taken for complete (100%) removal of the PAH compounds from the contaminated soil was 16 weeks (Sinha et al. 2008). Earthworms enlarge the pores in the soil by their burrowing actions. This action facilitates microorganisms to enter the pores and thus they could act and degrade soil contaminants. As the soil pore size is 20 nm or even less in diameter and the chemical contaminants are held inside such tiny pores. Earthworms stimulate multiplication of the decomposer microbial community for enhanced processes of biological degradation. The grinding process in the gut of earthworms makes chemical contaminants sequestered and thus 'bio-available' to decomposer microorganisms for biological degradation in the soil. Vermiremediation saves cost and is safe to the environment. It is also one of the safe ways to treat contaminated soil sites in weeks to months. The remediated soil and the land become suitable well as productive facilitating agriculture, horticulture or even for construction. Thus, both environmental and economic benefits could be derived from vermicomposting.

11.14 Properties of Vermicompost

11.14.1 The Physical Properties

Vermicompost is granular, porous material that helps in good aeration and in moisture-retaining capacity. The quality of the end products obtained has been considered for evaluation for its safe use in agricultural lands following standard stability criteria viz., C/N ratio, pH, Phytotoxicity of biologically transformed organic matter, temperature and enzymatic activity (Bernal et al. 2009).

11.14.2 Chemical and Biochemical Properties

The nutrient content of vermicompost was much higher for most elements except Magnesium (Magnesium sulphate can be used to rectify this deficiency). The nutrients in waste materials like Calcium, Magnesium, Nitrogen, Potassium and Phosphorus when processed by earthworms were simplified to more readily available forms to plants (Kale et al. 1992). It is reported to be rich in macronutrients, micronutrients, growth hormones, vitamins, enzymes viz., amylases, chitinase, proteases, cellulase, lipase and immobilized microflora. The enzymes help in the process of the organic matter degradation and further continue the process of degradation even after they have been ejected out by the earthworms (Kale et al. 1982).

11.14.3 Microbial Populations

The fragmentation of fresh organic matter, achieved by earthworms, provides the surface area for microbial colonization (Domínguez et al. 2010). There is an importance of fungi in the process of vermicomposting of cattle manure (Pizl and Novakova 2003). Earthworm species *Eiseniaandrei* feeding demonstrates microfungi addition to vermiculture resulted in increase in growth rates of fungi such as *Aspergillus flavus*. There is a prospect of seeding commercial vermiculture substrates with particular fungi to enhance the vermicomposting process (Roupas and Ferguson 2007). It is also a source of antibiotics, actinomycetes help in increasing biological resistance among the crop plants. Thus, pesticide spray could be minimized wherever earthworms and vermicomposts were utilised (Singh 1993).

11.14.4 Humus

Earthworms produce effective humic substances naturally (Masciandaro et al. 1997). Humus acts as a natural soil conditioner. Humic acid is one of the components of humus that has attachment sites for Calcium, Potassium, Iron, Sulphur and Phosphorus. Humic acid molecule is a storehouse for many of these nutrients in a readily available form to plants. Humic substances are humified dark-coloured organic matter, soluble in acid and alkali (Schnitzer 1991). These humic substances can improve plant nutrition and growth and are reminiscent of hormones or enzymes (Chen and Aviad 1990).

The vermicompost is rich in organic acids in the humus region such as humic and fulvic acids. Humic acids act as a great chelating agent. Humic substances can complex transition metal cation (such as iron, zinc, magnesium, calcium and manganese), which can result in enhanced uptake of macronutrients and micronutrients required by plants. Addition of humic acids increases water retention capacity of plant and also increases the cell membrane permeability thus showing hormone-like activity (Arancon 2006).

11.15 Quality of Vermicompost

Depending on the raw materials used, their decomposition stages, vermicomposting environment, maturity and storage time of vermicompost signifies the quality or the nutritional status of vermicompost (Durán and Henríquez 2007).

Quality of Vermicompost is related to the earthworm species used in vermicomposting. Economically high-value crops definitely deserve high-quality vermicompost. Important criteria to be followed in implementing the guidelines for the assessment of good quality vermicompost production includes—the organic feedstocks and its characteristic features, time taken for production of vermicompost by earthworms and other parameters which were used as maturity indicators. For the expansion of production, utilization and marketing of vermicompost, the vermicompost producing industry expects necessary compost quality indicators.

The high-quality vermicompost will significantly be best as it is enriched with essential minerals and microbes which were reported to be beneficial to soil and plants. Earthworms during the process of vermicomposting releases anti-pathogenic body fluid/coelomic fluid into the organic waste biomass. Release of coelomic fluid acts as a disinfectant and keeps vermicompost free from any kind of pathogens (Pierre et al. 1982). Earlier studies have proven that the compost produced by conventional methods contains high ammonium while the compost (vermicompost) produced by earthworms is considered to be rich in nitrates (biologically available form of nitrogen) which is readily available for plants. Vermicompost is said to contain higher available nitrogen (N) on the basis of weight and also said to supply several essential plant nutrients elements viz., potassium (K) magnesium (Mg), phosphorus (P) and sulphur (S) when compared with conventional compost. Several studies have determined that feedstock influences the earthworm populations and also the quality of resulting vermicompost.

The vermicompost is a form of stabilized organic fertilizer that possesses enhanced market value, good soil conditioning properties, low production cost, increased retail profits are a few notable features.

Vermicompost being the earthworm castings favours the soil and environment by reducing the use of synthetic chemical fertilizers and also minimizing the amount of waste getting piled up in landfills. Production of vermicompost is popularly increasing worldwide and is finding its application in Western, Asia–Pacific and Southeast Asian countries.

11.16 Advantages of Vermicompost

Vermicomposting has enumerated applications. In crop improvement, they aid in destruction of pathogens infecting the plants, retain water of the soil and improve crop yield. Extensive research has been carried out to examine the effect of vermicompost on plant growth, improvement in physical, chemical and biological properties of the soil (Edwards and Burrows 1988; Kale et al. 1987). Vermicompost aids the process of production of plant growth hormones/regulators by the microbial population.

Earthworm castings are rich in NPK in bio-available form which is gradually released spanning up to a month of application. Vermicompost or earthworm castings progress the plant growth, suppressing pathogenic diseases in plants.

Research on earthworms and vermicompost has reported that using vermicompost can result in 30–40% higher yield of crops when compared to use of chemical fertilizers. Chemical fertilizers need more water for irrigation whereas vermicompost uses less amounts of water because of its innate water retention capacity.

Vermicompost process supports life and population of the earthworms. Economically significant environmental advantage of using vermicompost, in successive years, in agriculture. They maintain prolonged soil fertility. Cocoons improve the fertility of soil and consequently minimize the amount of vermicompost use. This plays a key role in maintaining the health, quality, yield and productivity of the crop plants. Organically obtained plant products in turn help to maintain the health and wealth of the people, country and the entire universe. Naturally healthy fertile soil and chemical-free environment is protected and preserved for the future generations by using earthworms and its vermicompost. On the contrary, farmers are compelled to use chemical or synthetic fertilizers to get faster yield and to increase crop productivity for revenue generation. Continuously amending the soil with chemicals over the years may result in destruction of the naturally available fertile soil and this may result in soil getting habituated to synthetic chemicals. Consequently, this increases the demand for more volumes of chemicals to maintain or to increase the productivity and yield of previous years. Increased use of agrochemicals to meet the demand of food production is a self-defeating proposition.

Solid waste management is the biggest challenge in the present day. Vermicomposting is the better solution to overcome this challenge. Being environment friendly, it is proved to be a simple, less costly solid waste management technique. The earthworm castings preserve our environment in several ways. Vermicomposting helps solid waste degradation and the vermicompost is a resourceful fertilizer. This further produces healthy plants and high yields. Vermicomposting not only provides occupation for the farmers and other freelancers but also helps in revenue generation. Potentially earthworms aid the process of transformation of garbage to gold.

Vermicompost helps in producing healthy, disease-free plants by increasing the size, colour, smell, taste and flavour of the plant products. Vermicompost maintains the quality (storage value) of flowers, fruits, food grains and vegetables. Vermicompost saves our earth, water, energy, landfills and helps in rebuilding or retaining the fertility of soil. The socio-economic significance is that the plant and its products produced using earthworm compost is completely organic, safe and chemical-free.

Minerals or nutrients present in the vermicompost are easily and continuously supplied with bio-available forms to plants. Complex chemicals are broken down into simple water-soluble forms for the absorption by plants.

Research on earthworms and vermicompost has documented the production of healthy plants and its products on mixing top layers of soil with mature vermicompost. Vermicompost has a direct impact on the growth parameters like root formation and time taken for formation of roots; inflorescence; development of leaf and its area and elongation of internode. Mature vermicompost is nutritional, consisting of biologically stable growth-regulating substances which is responsible for the development and plant free from diseases.

The need is to establish, standardize and implement the applications of earthworms and vermicompost to build up efficient, eco-friendly and affordable environment for the future generation to use low-cost organic manure like vermicompost and its extracts along with coelomic fluid of the earthworms to achieve higher yields. This is an attempt based on the information available at field level for different crops as suggested by many researchers.

Earthworm compost is proving to be a highly nutritive plant 'growth promoter' compared to several other synthetic fertilizers. It has positive effects on the soil, improves fertility and reduces the levels of soil contamination. It is beneficial to the microbes in soil and tends to retain the necessary nutrients for plant growth.

Some of the Significant and concluding attributes of the vermicompost are listed below:

- Vermicompost can help restore the microbial population that helps in nitrogen fixation and stabilizing phosphates, etc.
- Vermicompost supplies soil nutrients to the plants.
- Help the soil texture, water-holding capacity, good soil aeration, improving root growth, preventing soil erosion and vermicast provides surface area for proliferation and colonization of soil microbes.
- reduces the pesticide application and helps in control plant pathogen.
- enhances the quality of grains/fruits due to increased sugar content.
- Vermicomposting process can hasten with proper management practices like providing, large surface area for earthworm activity, optimum temperature, good aeration in the processing material and moisture regulation.

11.17 Conclusion

In India, the most important cause for significant increase in mass production of waste is because of increasing population, rapid urbanization, industrialization and economic development. Ever since 1996, there has been a change in the composition owing to the economic growth. There is a dire need to address solid waste management as an entire process. When preparing long-term solutions, priority should be focused mainly on fixing existing problems like segregation and disposal. Waste reduction, waste to energy and recovery potential from wastes are to be combined with Integrated Solid waste management (ISWM).

Earthworms that are chief agents that yield vermicompost from waste organic material are easy to multiply. They double every 60–70 days. Thus, on a commercial scale, the vermicompost production is a possibility. The entire allied infrastructure is rather easy to install, maintain and run the process of vermicomposting. Being a 'one-time-investment'; the entire process economical, feasible yielding vermicompost a value product and an input in agriculture. This far weighs in resulting in soil fertility that is most beneficial and helps environmental protection and thus a sustainable practice.

The body fluid of earthworms is associated with a host or valuable bioactive components. They possess medicinal properties. The three versatile species *E. fetida, E. eugeniae* and *P. excavatus* perform wide social, economic and environmental functions almost everywhere.

The vermicomposting process has a few disadvantages—it can be quicker, but labour-intensive; It requires space because earthworms are surface feeders, tanks that hold waste material needs a depth of over metre; It is more vulnerable to environmental conditions like temperature, cold and drought; It requires more start-up resources, either in cash (to buy the worms) or in time and labour. However, the advantages outweigh the disadvantages and so vermicomposting is a welcome technology.

References

Achten C, Hofmann T (2009) Native polycyclic aromatic hydrocarbons (PAH) in coals—a hardly recognized source of environmental contamination. Sci Total Environ 407:2461–2473

Albihn A, Vinnerås B (2007) Biosecurity and arable use of manure and biowaste—treatment alternatives. Livestock Sci 112(3):232–239

Ansari AA, Ismail SA (2008) Reclamation of sodic soils through vermitechnology. Pak J Agric Res 21:92–97

Arancon NQ, Edwards CA, Lee S, Byrne R (2006) Effects of humic acids from vermicomposts on plant growth. Eur J Soil Biol 42:S65–S69

Bajsa O, Nair J, Mathew K, Ho GE (2004) Pathogen die-off in vermicomposting process. Paper presented at the international conference on small water and wastewater treatment systems, Perth

Bernal MP, Alburquerque JA, Moral R (2009) Composting of animal manures and chemical criteria for compost maturity assessment: a review. Bioresour Technol 100(22):5444–5453

Bhat SA, Singh J, Vig AP (2013) Vermiremediation of dying sludge from textile mill with the help of exotic earthworm Eisenia fetida Savigny. Environ Sci Pollut Res 20:5975–5982

Bhat SA, Singh J, Vig AP (2014) Genotoxic assessment and optimization of pressmud with the help of exotic earthworm Eisenia fetida. Environ Sci Pollut Res 21:8112–8123

Bhattacharya SS, Iftikarb W, Sahariaha B, Chattopadhyay GN (2012) Vermicomposting converts fly ash to enrich soil fertility and sustain crop growth in red and lateritic soils. Resour Conserv Recycl 65:100–106

Chaudhuri PS, Pal TK, Bhattacharjee G, Dey SK (2001) Suitability of rubber leaf litter (Hevea brasiliensis var. PRIM 600) as substrate for epigeic earthworms, Perionyx excavatus, Eudrilus eugeniae and Eisenia fetida. In: Proceedings of VII Nat. Symp. Soil. Biol. Ecol., GKVK, Bangalore, 7–9 Nov, 2001, pp 18–26

Chen Y, Aviad T (1990) Effects of humic substances on plant growth. In: MacCarthy P, Clapp CE, Malcolm RL, Bloom PR (eds) Humic substances in soil and crop sciences: selected readings. SSSA, Madison, pp 161–186

Cointreau S (2006) Occupational and environmental health issues of solid waste management; World Bank Urban Sector Board. Urban paper series no. UP-2, p 48

Darwin C (1881) The formation of vegetable moulds through the action of worms. Murray Publications, London

Diacono M, Montemurro F (2010) Long-term effects of organic amendments on soil fertility. A review. Agron Sustain Dev 30(2):401–422

Dominguez J, Edwards CA (2004) Vermicomposting organic wastes: a review. In: Hanna SHS, Mikhail WZA (eds) Soil zoology for sustainable development in the 21st century. Eigenverlag, Cairo, pp 369–395

Dominguez J, Edwards CA (2010) Biology and ecology of earthworm species used for vermicomposting. CRC Press Taylor & Francis Group, Boca Raton, USA, pp 28–38

Domínguez J, Hernandez JCS, Lores M (2017) Vermicomposting of winemaking by-products. In: Galanakis CM (ed) Handbook of grape processing by-products: sustainable solutions. Academic Press, Elsevier, London, pp 55–78 978-0-12-809870-7

Domínguez JR, Parmelee W, Edwards CA (2003) Interactions between Eisenia andrei (Oligochaeta) and nematode populations during vermicomposting. Pedobiologia 47:53–60. https://doi.org/10.1078/0031-4056-00169

Durán L, Henríquez C (2007) Caracterización química, física y microbiológica de vermicompostes producidos a partir de cinco sustratos orgánicos. Agron Costarricense 31(001):41–51

Edwards CA (1998) The use of earthworms in the breakdown and management of organic wastes. In: Earthworm ecology. CRC Press LLC, pp 327–354

Edwards CA, Bohlen PJ (1996) Biology and ecology of earthworms. Chapman and Hall, London, pp 35–55

Edwards CA, Burrows I (1988) The potential of earthworms composts as plant growth media. In: Edward CA, Neuhauser EF (eds) Earthworms in waste and environmental management. SPB Academic Publishing, The Hague, p 2132

Edwards CA, Bohlen PJ, Linden DR, Subler S (1995) Earthworms in agroecosystems. In: Hendrix PF (ed) Earthworm ecology and biogeography in North America. Lewis Publisher, pp 185–213

Fernández-Gómez MJ, Díaz-Ravina M, Romero E, Nogales R (2013) Recycling of environmentally problematic plant wastes generated from greenhouse tomato crops through vermicomposting. Int J Environ Sci Technol 10:697–708

Flotats JR, Solé FM (2008) Situación actual en el tratamiento de los residuos orgánicos: aspectos científicos, económicos y legislativos. In: Moreno J, Moral R (eds) Compostaje. Mundiprensa, Madrid, España, pp 44–73

Fornes F, Mendoza-Hernańdez D, Garcı́a-de-la-Fuente R, Abad M, Belda RM (2012) Composting versus vermicomposting: a comparative study of organic matter evolution through straight and combined processes. Bioresour Technol 118:296–305

Garg V, Gupta R (2009) Vermicomposting of agro-industrial processing waste. In: Singh nee' Nigam P, Pandey A (eds) Biotechnology for agro-industrial residues utilisation. Springer, Dordrecht. https://doi.org/10.1007/978-1-4020-9942-7_24

Garg VK, Surthar S, Yadav A (2012) Management of food industry waste employing vermicomposting technology. Bioresour Technol 126:437–443

Giraddi RS (2000) Influence of vermicomposting methods and season on the biodegradation of organic waters. Indian J Agric Sci 70:663–666

Gunthilingaraj K, Ravignanam T (1996) Vermicomposting of sericulture wastes. Madras Agric J 83:455–457

Hait S, Tare V (2012) Transformation and availability of nutrients and heavy metals during integrated composting vermicomposting of sewage sludges. Ecotoxicol Environ Saf 79:214–224

Hernández A, Guerrero F, Mármol L, Bárcenas J, Salas E (2008) Caracterización física según granulometría de dos vermicompost derivados de estiércol bovino puro y mezclado con residuos de fruto de la palma aceitera. INCI 33(9):668–671

Hernández JA, Mayarez L, Romero E, Ruiz J, Contreras C (2002) Efecto de la altura del cantero sobre el comportamiento de la lombriz roja (Eisenia spp). Lombricultura y abonos orgánicos Memorias. UAEM. México

Hoornweg D, Bhada-Tata P (2012) What a waste—a global review of solid waste management. Tech. rep., World Bank. http://documents.worldbank.org/curated/en/2012/03/16537275/waste-global-review-solid-waste-management

Hsu JH, Lo SL (2000) Characterization and extractability of copper, manganese, and zinc in swine manure composts. J Environ Qual 29:447–453

Ismail SA (1993) Keynote papers and extended abstracts. In: Proceedings of Congress on traditional sciences and technologies of India, vol 10. IIT, Mumbai, pp 27–30

Ismail SA (1997) Vermicology: the biology of earthworms. Orient Longman, India, p 92

Ismail SA (2005) The earthworm book. Other India Press, Mapusa, p 101

Jayanthi B, Ambiga G, Neelanarayanan P (2010) Utilization of mixed litter for converting into vermicompost by using an epigenic earthworm Eudrilus eugeniae. Nat Environ Pollut Technol 9:763–766

Joo SH, Monaco FD, Antmann E, Chorath P (2015) Sustainable approaches for minimizing biosolids production and maximizing reuse options in sludge management: a review. J Environ Manage 158:133–145

Kale RD (1998) Earthworm: Cinderella of organic farming. Prism Book, Bangalore, p 88

Kale RD, Bano K (1988) Earthworm cultivation and culturing techniques for production of Vee. Comp. 83 UAS. Vee. Meal 83P UAS. Mysore J Agric Sci 22:339–344

Kale RD, Sunitha NS (1995) Efficiency of earthworms (E. Eugeniae) in converting the solid waste from aromatic oil extraction industry into vermicompost. J IAEM 22:267–269

Kale RD, Bano K, Krishnamoorthy RV (1982) Potential of Perionyx excavatus for utilising organic wastes. Pedobiologia 23:419–425

Kale RD, Vinayaka K, Bano K, Srinivasa MN, Bagyaraj DJ (1987) Influence of Worm cast (Vee. E. UAS 83) on the growth and mycorrhizal colonization of two ornamental plants. South Indian Hort 35:433–437

Kale RD, Mallesh BC, Kubra B, Bhagyaraj DJ (1992) Influence of vermicompost application on available micronutrients and selected microbial populations in paddy field. Soil Biol Biochem 24:1317–1320

Kang J, Zhang Z, Wang JJ (2011) Influence of humic substances on bioavailability of Cu and Zn during sewage sludge composting. Bioresour Technol 102:8022–8026

Kathiravale S, Muhd Yunus MN (2008) Waste to wealth. Asia Europe J 6(2):359–371

Kaushik P, Garg VK (2003) Vermicomposting of mixed solid textile mill sludge and cow dung with the epigeic earthworm Eisenia foetida. Bioresour Technol 90:311–316

Komakech AJ, Banadda NE, Gebresenbet G, Vinnerås B (2014) Maps of animal urban agriculture in Kampala City. Agron Sustain Dev 34(2):493–500

Lakshmi BL, Vizaylakshmi GS (2000) Vermicomposting of sugar factory filter pressmud using African earthworms species (Eudrillus eugeniae). Pollut Res 9:481–483

Lim SL, Lee LH, Wu TY (2016) Sustainability of using composting and vermicomposting technologies for organic solid waste biotransformation: recent review, greenhouse gases emissions and economic analysis. J Clean Prod 111:262–278

Lv B, Xing M, Yang J, Qi W, Lu Y (2013) Chemical and spectroscopic characterization of water extractable organic matter during vermicomposting of cattle dung. Bioresour Technol 132:320–326

Malafaia G, Estrela DDC, Guimarães ATB, Araújo FGD, Leandro WM, Rodrigues ASDL (2015) Vermicomposting of different types of tanning sludge (liming and primary) mixed with cattle dung. Ecol Eng 85:301–306

Maňáková B, Kuta J, Svobodová M, Hofman J (2014) Effects of combined composting and vermicomposting of waste sludge on arsenic fate and bioavailability. J Hazard Mater 280:544–551

Manser AGR, Keeling AA (1996) Processing and recycling municipal waste. CRC, Boca Raton

Medina M (2000) Scavenger cooperatives in Asia and Latin America. Resour Conserv Recycl 31(1):51–69

Masciandaro G, Ceccanti B, Garcio C (1997) Soil agroecological management: fertirrigation and vermicompost treatments. Biores Technol 59:199–206

Minghua Z, Xiumin F, Rovetta A, Qichang H, Vicentini F, Bingkai L, Giusti A, Yi L (2009) Municipal solid waste management in Pudong New Area, China. Waste Manage 29:1227–1233

Nogales R, Cifuentes C, Benítez E (2005) Vermicomposting of winery wastes: a laboratory study. J Environ Sci Health 34:659–673

Nunes RR, Bontempi RM, Mendonça G, Galetti G, Rezende MOO (2016) Vermicomposting as an advanced biological treatment for industrial waste from the leather industry. J Environ Sci Health 51:271–277

Oyedele DJ, Schjonning P, Amussan AA (2005) Physicochemical properties of earthworm casts and uningested parental soil from selected sites in southwestern Nigeria. Ecol Eng 20:103–106

Pell AN (1997) Manure and microbes: public and animal health problem? J Dairy Sci 80(10):2673–2681z

Piearce TG, Oates K, Carruthers WJ (1990) A fossil earthworm embryo (Oligochaeta) from beneath a late bronze age midden at Potterna, Wiltshire, UK. J Zool Land 220:537–542

Pierre V, Phillip R, Margnerite L, Pierrette C (1982) Anti-bacterial activity of the haemolytic system from the earthworms Eisinia foetida Andrei. Invertebr Pathol 40:21–27

Pimentel D, Hepperly P, Hanson J, Douds D, Seidel R (2005) Environmental, energetic, and economic comparisons of organic and conventional farming systems. Bioscience 55(7):573–582

Pizl V, Novakova A (2003) Interactions between micro fungi and Eisenia Andrei (Oligochaeta) during cattle manure vermicomposting. Pedobiologia 47:895–899

Plecher H (2020) Population growth in India 2019, Statista, Oct 20

Pramanik P, Ghosh GK, Ghosal PK, Banik P (2007) Changes in organic—C, N, P and K and enzyme activities in vermicompost of biodegradable organic wastes under liming and microbial inoculants. Bioresour Technol 98:2485–2494

Quintern M (2014) Full scare vermicomposting and land utilization of pulpmill solids in combination with municipal biosolids (sewage sludge). J Ecol Environ 18:65–76

Reddy VM, Okhura K (2004) Vermicomposting of ricestraw and its effects on sorghum growth. Trop Ecol 45:327–331

Roupas P, Ferguson A (2007) Handbook of waste management and co-product recovery in food processing, vol 1

Sanchez Hernandez JC, Domínguez J (2017) Vermicompost derived from spent coffee grounds: assessing the potential for enzymatic bioremediation. In: Galanakis CM (ed) Handbook of coffee processing by-products sustainable applications. Academic Press, Elsevier, London, pp 369–398. ISBN 978-0-12-811290-8

Sandeep SD, Yadav J, Urmila (2017) Assessment of nutrient status of vermicompost of leaf litter using Eisenia foetida. J Entomol Zool Stud 5(2):1135–1137

Santamaría-Romero S, Ferrera CR, Almaraz SJ, Galvis SA, Barois BI (2001) Variación y relaciones de microorganismos, C-orgánico y N-total durante el composteo y vermicomposteo. Agrociencia 35(4):377–384

Saxena M, Chauhan A, Asokan P (1998) Flyash vermicompost from nonfriendly organic wastes. Pollut Res 17:5–11

Schnitzer M (1991) Soil organic matter—the next 75 years. Soil Sci 151:41–58

Seenappa SN, Rao J, Kale R (1995) Conversion of distillery wastes into organic manure by earthworm *Eudrillus euginae*. J IAEM 22:244–246

Singh RD (1993) Harnessing the earthworms for sustainable agriculture. Institute of National Organic Agriculture, Pune, India, pp 1–16

Singh R, Embrandiri A, Ibrahim M, Esa N (2011) Management of biomass residues generated from palm oil mill: Vermicomposting a sustainable option. Resour Conserv Recycl 55:423–434. https://doi.org/10.1016/j.resconrec.2010.11.005

Singh J, Kaur A, Vig AP (2014) Bioremediation of distillery sludge into soil-enriching material through vermicomposting with the help of Eisenia fetida. Appl Biochem Biotechnol 174:1403–1419

Sinha RK, Bharambe G, Ryan D (2008) Converting wasteland into wonderland by earthworms-a low-cost nature's technology for soil remediation: a case study of vermiremediation of PAHs contaminated soil. Environmentalist 28:466–475

Sonowal PKD, Khwairkpam M, Kalamdhad AS (2013) Feasibility of vermicomposting dewatered sludge from paper mills using Perionyx excavatus. Eur J Environ Sci 3:17–26

Soto G, Muñoz C (2002) Consideraciones teóricas y prácticas sobre el Compost, y su empleo en la Agricultura. Manejo Integrado De Plagas y Agroecología 65:123–125

Suthar S (2007) Vermicomposting potential of Perionyx sansibaricus (Perrier) in different waste materials. Bioresour Technol 97:2474–2477

Suthar S (2009) Potential of Allolobophora parva (Oligochaeta) invermicomposting. Bioresour Technol 100:6422–6427

Suthar S (2010) Recycling of agro-industrial sludge through vermitechnology. Ecol Eng 36:703–712

Suthar S, Gairola S (2014) Nutrient recovery from urban forest leaf litter waste solids using Eisenia fetida. Ecol Eng 71:660–666

Suthar S, Sajwan P, Kumar K (2014) Vermiremediation of heavy metals in wastewater sludge from paper and pulp industry using earthworm Eisenia fetida. Ecotoxicol Environ Saf 109:177–184

Suthar S (2012) Vermistabilization of wastewater sludge from milk processing industry. Ecol Eng 47:115–119

Ulle J, Fernandez F, Rendina A (2004) Evaluación analítica del vermicompost de estiércoles y residuos de cereales y su efecto como fertilizante orgánico en el cultivo de lechugas mantecosas. Hortic Bras 22(2):434–438

Varma VS, Yadav J, Das S, Kalamdhad AS (2015) Potential of waste carbide sludge addition on earthworm growth and organic matter degradation during vermicomposting of agricultural wastes. Ecol Eng 83:90–95

Vasanthi K, Chairman K, Ranjit Singh AJA (2014) Sugar factory waste (vermicomposting with an epigeic earthworm, Eudrilus eugeniae). Amer J Drug Disc Devel 4:22–31

Vig AP, Singh J, Wani SH, Dhaliwal SS (2011) Vermicomposting of tannery sludge mixed with cattle dung into valuable manure using earthworm Eisenia fetida (Savigny). Bioresour Technol 102:7941–7945

Visvanathan C et al (eds) (2005) Vermicomposting as an eco-tool in sustainable solid waste management. Asian Institute of Technology, Anna University, India

Vivas A, Moreno B, Garcia-Rodriguez S, Benitez E (2009) Assessing the impact of composting and vermicomposting on bacterial community size and structure, and microbial functional diversity of an olive-mill waste. Bioresour Technol 100(3):1319–1326

Williams AP, Roberts P, Avery LM, Killham K, Jones DL (2006) Earthworms as vectors of *Escgerichia coli* O157:H7 in soil and vermicomposts. FEMS Microbiol Ecol 58:54–64. https://doi.org/10.1111/j.1574-6941.2006.00142.x

Yadav A, Garg VK (2009) Feasibility of nutrient recovery from industrial sludge by vermicomposting technology. J Hazard Mater 168:262–268

Yadav A, Garg VK (2013) Nutrient recycling from industrial solid wastes and weeds by vermicomposting using earthworms. Pedosphere 23:668–677

Yadav A, Suthar S, Garg VK (2015) Dynamics of microbiology parameters, enzymatic activities and worm biomass production during vermicomposting of effluent treatment plant sludge of bakery industry. Environ Sci Pollut Res 22:14702–14709

Yasir M, Aslam Z, Kim S et al (2009) Bacterial community composition and chitinase gene diversity of vermicompost with antifungal activity. Bioresour Technol 100:4396–4403

Zucconi F, Pera V, Forte M, De Bertoldi V (1981) Evaluating toxicity of immature compost. Biocycle J Composting Organics Recycl 22:54–57

Chapter 12
Vermiremediation of Agrochemicals, PAHs, and Crude Oil Polluted Land

Shivika Datta, Simranjeet Singh, Praveen C. Ramamurthy, Dhriti Kapoor, Vaishali Dhaka, Deepika Bhatia, Savita Bhardwaj, Parvarish Sharma, and Joginder Singh

Abstract Earthworms, the 'ecological engineers of the earth' have a unique capability to significantly influence the dynamics of the medium they are present in. Vermiremediation is an eco-technology that involves earthworms for remediation of contaminated soils or another medium in which they are present. The earthworms form a natural bioreactor for the decomposition of organic matter and help in nutrient recycling. It is an expanding technology which is gaining worldwide attention because of its results and cost-effectiveness. Intensification of agriculture by the indiscriminate use of agrochemicals has led to soil infertility. Plants absorb nutrients from the soil in the form of free metal ions. The agrochemicals chelate with metal ions forming stable complexes, rendering them unavailable for plant absorption. The presence of earthworms in contaminated soils is an indication that they have an ability to survive in a wide range of different organic contaminants like pesticides, herbicides, polycyclic aromatic hydrocarbons (PAHs), Polychlorinated biphenyls (PCBs), crude oil. However, earthworm survival depends at first on the concentration of contaminants. The negative connotation and cost incurred in using physical and chemical techniques for remediation of PAHs and crude oil contaminated sites have amplified

---

Shivika Datta, Simranjeet Singh = Equal Contribution

S. Datta
Department of Zoology, Doaba College, Jalandhar, India

S. Singh · P. C. Ramamurthy
Interdisciplinary Centre for Water Research (ICWaR), Indian Institute of Sciences, Bangalore 560012, India

D. Kapoor · S. Bhardwaj
Department of Botany, Lovely Professional University, Phagwara, India

V. Dhaka · J. Singh (✉)
Department of Microbiology, Lovely Professional University, Phagwara, India
e-mail: joginder.15005@lpu.co.in

D. Bhatia
Department of Biotechnology, Baba Farid Group of Institutions, Deon, Bathinda, India

P. Sharma
School of Pharmaceutical Sciences, Lovely Professional University, Phagwara, India

the involvement of vermiremediation. Vermiremediation improves the quality of soil in terms of pH, electrical conductivity, metal concentration, porosity, and aeration. Earthworms mortify and aerate the substrate by acting as mechanical blenders. This splintering of organic matter alters the microbial activity, amends its physical and chemical nature by progressively reducing the C/N ratio and increasing the surface area, making it more encouraging for microbial activity and decomposition further. The earthworms thereby contribute to accelerated decomposition of contaminants; however, sometimes the pollutants get adsorbed to the vermicast due to which their dissipation is delayed, which is a huge limitation to this technology.

Keywords agrochemicals · Contaminants decomposition · Organic matter · Vermiremediation

12.1 Introduction

A wide range of terrestrial and aquatic habitats is contaminated by anthropogenic activities (Mohee and Mudhoo 2012). The ecological equilibrium is skewed due to industrialization and urbanization, increasing population pressure, and the problem is compounded by the limited stock of natural resources (Hanafi 2012). The magnitude and nature of the concern are dynamic, bringing new challenges and creating a constant lacuna in need for developing appropriate and effective technologies. Sustainable development requires environmental management and a constant search for green technologies to restore ecological equilibrium. The harmful effects of chemical fertilizers and pesticides have abstracted the interests of researchers toward organic amendments like vermicompost or use of plant growth-promoting bacteria or by the degradation of wastes by bacteria or maybe by using complexes of humic acids and metals (Scotti 2015). Large-scale industrialization has led to inappropriate, indiscriminate, and untimed disposal of wastes in agricultural fields and water bodies (Goel 2006). This leads to a sudden and massive contribution of toxic trace metals, inorganic salts, pathogens; emission of harmful gases like hydrogen sulphide, ammonia, etc.; nutrient loss in the form of nitrogen and phosphorus by leaching, runoff or erosion and several other environmental problems (Hutchinson et al. 2005). Moral et al. (2009) suggested that proper handling of organic wastes could create a new source of nutrients for agriculture, thus can alternate costly mineral fertilizers and for the production of renewable energy. Vermicompost is one such cost-effective means for the conversion of highly toxic waste into value-based products (Bhat et al. 2014). Vermicomposting is a process that is known to convert the biodegradable matter into vermicast by the help of earthworms (Fig. 12.1). This process of vermicomposting has taken the credit to increase the bioavailability of a major part of nutrients in the organic matter (Gajalakshmi and Abbasi 2008).

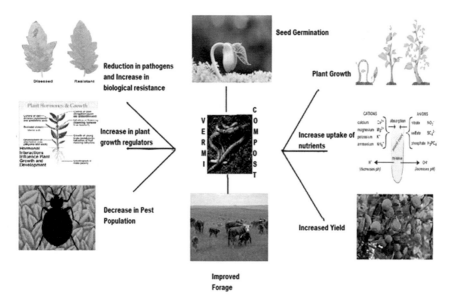

Fig. 12.1 Overview of vermicomposting

12.2 Agrochemicals: Classification, Effect on Environment, Health Hazards

The chemical products comprised of growth hormones, fertilizers, pesticides, and chemicals for the plant protection used in the field of agriculture are known as agrochemicals (Mandal et al. 2020). They involved a variety of chemicals extensively used in the field of agriculture to aid the growth of crops. To provide protection from the pests, these were manufactured as well as for the enhancement of crop yield (Ren et al. 2020). Agrochemicals concern to pesticides which consist of nematicides, herbicides, insecticides, and fungicides (Sparks 2013). Based on their mode of action and chemical structure, they are further classified into pyrethroids, organophosphorus, neonicotinoids, carbamates, and organochlorines (Xiao et al. 2017).

Although it provides many benefits in the agriculture field, agrochemicals are also found as a major pollutant widely detected in the soil (Tsatsakis et al. 2008). The numerous agrochemicals such as rodenticides, nematicides, fungicides, and other chemical fertilizers adversely affect the beneficial microbiota of the soil (Meena et al. 2016). The group of pesticides involves fungicides, insecticide, and herbicides are which function as to repel, control, or kill the life of a plant. The demand of these pesticides is constantly increasing. The pesticides have a good impact on the profit margin as well as on the crop yield, which causes them to be a significant component for agriculture (Meena et al. 2020). Though, the excessive use of agrochemicals leads to ecosystem degradation of soil microbiota (Önder et al. 2011). In agriculture, the main decreasing biotic factors are insects and weeds, which hamper the productivity and yield of the crop (Oliveira et al. 2014). Most of the insecticides and herbicides reach non-target microbes, which disturb the biodiversity of soil (Lo et al. 2015). This affects directly the microbiota of soil and soil fertility, which is a biological indicator (Santos and Flores 1995; Hussain et al. 2009). The herbicides namely 2,4-D, 2,4,5-T, Methomyl, Bensulfuron methyl, glyphosate adversely affect the *Rhizobium* species activities (Fabra et al. 1997), reduce the activity of purple non-sulphur bacteria like phosphatase and nitrogenase (Chalam et al. 1997). Which disrupts the signaling of *Rhizobium* and influences the nitrifying process (Fox et al. 2001), reduces the oxidation of methane to carbon dioxide (Arif et al. 1996), deceases the nitrogen mineralization (Subhani et al. 2000) and also suppresses the activity of phosphatase (Sannino and Gianfreda 2001).

Some fertilizers are rich in cadmium and copper heavy metals that cause toxicity in the environment of soil (Chen and Pu 2007; Kabata-Pendias and Pendias 1992). The soil consists of various enzymes. The persistent use of pesticides inhibits, decreases or increases the activity of soil enzymes viz. dehydrogenase, oxidoreductases, and hydrolases (Riah et al. 2014; Mayanglambam et al. 2005; Megharaj et al. 1999). It also alters the catabolic metabolism of the microbes (Niewiadomska 2004; Yale et al. 2017; Ortiz-Hernández et al. 2013). The major concern associated with this is water and soil pollution, causing the toxicity to animals and humans. The phosphate and nitrate compounds lead to groundwater contamination which is harmful to organisms (Aktar et al. 2009; Lamichhane et al. 2016). It also becomes a reason for the aquatic animal's death by increasing the algae growth in lakes and streams due to the overflow of fertilizers.

The exposure of agrochemical and its harmful effects on animals and humans are unavoidable (Sparks and Lorsbach 2017). It causes toxicity in the immune system, neurons, reproductive system, and also disrupts the endocrine system in humans (Mostafalou and Abdollahi 2017). Agrochemicals also play a role in disrupting the endocrine compounds of animals and humans (Luque and Muñoz-de-Toro 2020). They can mimic the interaction among the nuclear receptors (thyroid hormone, estrogen, aryl hydrocarbon, and androgen receptors) and endogenous hormones. It also interferes with the epigenetic changes and synthesis of amino acids, steroids, and peptide (Warner et al. 2020). There are different ways that agrochemicals can affect the signaling of estrogen (Vandenberg et al. 2020). A pesticide named organochlorine has properties of disrupting the endocrine in fish (Martyniuk et al. 2020). In amphibians, agrochemicals disrupt the multiple axes of endocrine, delay the metamorphosis, and influence sexual development (Trudeau et al. 2020). The endosulfan and atrazine pesticides impact the reproductive system of crocodiles (sentinels) belonging to the wetland ecosystem (Tavalieri et al. 2020). In Argentina, pesticides cause alterations of neurogenesis in hippocampal by way of disrupting the endocrine, i.e. agrochemical alters the function of the brain and hormone synthesis (Florencia and Cora 2020). Glyphosate herbicides influence female reproductive fertility. It may alter the functions of uterine and ovarian (Ingaramo et al. 2020). The agrochemical also causes obesogenic effects, for example, effects on transgenerational and development (Ren et al. 2020). They have also shown the epidemiological evidence of human obesity due to the exposure of agrochemicals.

The health risk is high to untrained farmers and their children during the usage of pesticides (Akbar et al. 2010). Many non-targeted organisms, such as small mammals, birds, and bees, suffer obliteration directly or due to the remaining traces left behind the utilization of agrochemicals (Paoli et al. 2015). Exposure of pesticides causes several health issues, for example, deformities of foetal, skin disorders, cancers, and acute poisoning (de Araujo et al. 2016).

12.3 PAHs (Classification, Effect on Environment, Health Hazards)

Polycyclic aromatic hydrocarbons (PAHs) having concerns because of their extensive occurrence, persistence, and carcinogenic characteristics in the ecosystem and human health. They are released into the environment both naturally and anthropogenically, due to the partial burning of biological resources, for instance, petroleum, tar, fossil fuels, debris, vehicular emission or other substances like plant material (Kim et al. 2013). PAHs found ubiquitously in the air, soil, sediments, aquatic ecosystems and are highly mobile in the ecosystem due to their physicochemical features and also used as air quality indicators (Baklanov et al. 2007). Sixteen PAHs are recognized as ecosystem contaminants by US EPA in accordance with PAHs abundance and harmfulness (Ghosal et al. 2016). High molecular weight PAHs, i.e. chrysene, fluoranthene, and pyrene, consisting of 4 or more rings are mostly recognized as genotoxic while low molecular weight PAHs like naphthalene, acenaphthene, fluorene, phenanthrene, consisting of 2–3 aromatic rings, are severely noxious (Abdel-Shafy and Mansour 2016). PAHs involve only C and H atoms however in some cases N, S, and O atoms added in the benzol to make heterocyclic aromatic compounds, that are generally congregated with PAHs (Alegbeleye et al. 2017). There exist various other PAHs by-products, like oxygenated PAHs (OPAHs) or nitrated PAHs (NPAHs), in addition to basic PAHs which consist of only C and H (Nováková et al. 2020).

Agroecosystems polluted with PAHs alter the agricultural soil properties, which dramatically result in a severe threat to ecosystem organisms found in that range. PAHs enter in mammals' body through breathing, skin contact, and ingestion, whereas in plants bioaccumulated via absorption from soils to roots and then transfer to various plant tissues (Veltman and Brunner 2012). PAHs are toxic to organisms when present in higher amounts in comparison to the effects range median (ERM) and harmless when lower than effects range low (ERL) (He et al. 2014). PAHs, i.e. naphthalene (NAP), fluorene (FLU), and pyrene (PYR), significantly affected the N_2-fixing bacterial organisms, which ultimately lead to a severe threat to the vigor of mangrove ecosystem by lowering the accessibility of N_2 in the mangrove regions (Sun et al. 2012). Soils which are extensively polluted with PAHs caused ecotoxic action in different plant species where the severity of the toxicity differs with the concentrations of PAHs, soil physiognomies, and the plant genotype involved (Maliszewska-Kordybach and Smreczak 2000). 1-nitronaphthalene and 1-nitropyrene affected the reproductive ability, i.e. hatchability of fish, *Fundulus heteroclitus* via lowest observed effect concentration (LOEC) of 447 μg/g and 958 ng/g, respectively (Onduka et al. 2015). Contamination of soil with PAHs caused alterations in soil characteristics, leaching and erosion, which dramatically lead to declined agricultural yield (Nwaejije et al. 2017).

Transfer of PAHs in food chains via intake of several foodstuffs is a foremost aspect of rapidly increasing concentration of PAHs in the ecosystem (Bansal and Kim 2015), and release of PAHs into the ecosystem and their toxicity to human have turned out to be a major subject of concern for researchers (Balcıoğlu 2016). The health hazards of PAHs rely upon the duration of exposure, amounts of PAHs, and the way of its intake, i.e. via inhalation, ingestion, or skin contact (ACGIH 2005; Kim et al. 2013). PAHs cause many short- and long-term health effects, carcinogenesis and also cause disruptions in the metabolism process because of their continual transfer into the food chain (Bansal and Kim 2015; Fig. 12.2). Yerba mate leaves and its hot and cold mate infusions showed carcinogenic activity, chiefly attributable to the presence of greater amounts of PAHs found in them (Kamangar et al. 2008). Acute oncogenic hazard due to PAHs has been observed in children and adults in Isfahan urban zone, entered through both PAHs dust ingestion and dermal contact (Soltani et al. 2015). PAHs cause not only cancers but various other non-genotoxic diseases also, for instance, diabetes mellitus, heart diseases (Hu et al. 2015). Nitro-PAHs caused genotoxicity by triggering severe micronuclear and nuclear aberrations in erythrocytes in comparison to control in *Pleuronectes yokohamae* fish (Bacolod et al. 2013).

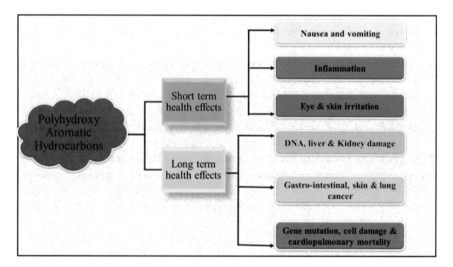

Fig. 12.2 Schematic representation of health hazards induced by PAHs (modified from Kim et al. 2013)

12.4 Crude Oil Polluted Land (Classification, Effect on Environment, Health Hazards)

Crude oil is a composite combination of organic substances which consists of hydrocarbons that differ in molecular weight and structural compounds and also contains heterocyclic molecules and some heavy metals. These biological substances include CH_4 gas, oils, crude wax, single or condensed rings and aromatic cycles like monocyclic and polycyclic aromatic hydrocarbons (Saadoun 2015). Crude oil pollution is recognized as an environmental stressor, and ecological pollutant which is released into the environment through anthropogenic activities, technical errors, transportation and storage faults, exploration and processing practices and has now become a major concern for ecosystem communities (Ivshina et al. 2015). Crude oil spills caused the devastation of fish territories in the mangrove ecosystem of the Niger Delta and also polluted the marshes and streams heavily eventually, converting them into an inappropriate habitat for fishing (Moses and Tami 2014). Total Hydrocarbon Content (THC) released from crude oil refineries resulted in altered soil chemical characteristics which ultimately lead to high noxiousness in the ecosystem (Yabrade and Tanee 2016).

In the aquatic environments, crude oil spill constructs a viscous surface slick, and H_2O-in-oil emulsion and accumulated in the aquatic habitat where it remains undecayed by microbes for a longer duration. This oil in H_2O diminished the level of O_2 in H_2O, because of the conversion of the organic moieties into inorganic substances, which dramatically lead to a decline in the biodiversity and hence, eutrophication (Onwurah et al. 2007). Waste released by crude oil refining practices discharged into marshes and the adjacent areas where it interrupts the quality of groundwater and also disrupts the health of the aquatic ecosystem (Amangabara and Njoku 2012). Crude oil adversely affected the health of various ecological niches via polluting the watercourses, canals, ponds, lakes, rivers, and mangroves (Sam and Zabbey 2018) and also resulted in deprived vigor of fish and their extinction due to abnormalities in reproductive abilities ultimately, higher mortality (Udotong et al. 2017). Petroleum hydrocarbons discharged into the water bodies and harm fish and other aquatic creatures (Clinton et al. 2014).

Wide-ranging crude oil toxicity relies upon several assets such as oil constituents and properties, weathered or un-weathered condition, contact pathway, i.e. via ingestion, skin contact, or inhalation and the oil accumulation, ultimately causes severe and long-term health problems (Ordinioha and Brisibe 2013). Crude oil pollutants disposed of refineries released into the soil and accumulated in the human body through the food chain and severely noxious to humans due to their noxious, mutagenic, and cancer-causing characteristics (Chikere and Fenibo 2018). Soil polluted with crude oil at Romanian field site resulted in high toxicity and produced carcinogenic effects where average calculated oncogenic hazard is about 1.07×10^{-5} for children and 6.89×10^{-6} for adults (Cocârță et al. 2017). Crude oil hydrocarbons can disturb genomic stability of many creatures which eventually lead to cancer, cellular mutations, and reproductive aberrations (Short and Heintz 1997). Crude oil caused substantial upsurges in the time occurrence for numerous diseases and exerted their toxic effects by inhibiting the protein synthesis, synaptic activity, obstructs the membrane transfer process and disrupts the plasma membrane (Ordinioha and Sawyer 2010).

12.5 Global Regulations on Use of Agrochemicals, PAHs, and Crude Oil

The U.S. Environmental Protection Agency (USEPA) fixed the maximum level of contaminant (MCLs) for community water supplies to decrease the probability of undesirable health impacts from contaminated water. The maximum level of contaminant limit meets by the public supply systems. These standards are lower than levels at which diverse health effects can occur. USEPA has not fixed the maximum contaminant level limit for individual aromatic hydrocarbons, but has established MCL for total PAHs of 0.2 ppb. Currently, there are no standards for regulating levels of these toxic chemicals in private wells. USEPA requires the data of any releases of PAHs into the environment that exceed 1 pound. There are no regulations fixed for the PAH content in food items.

Different countries have set different regulations and standards for agrochemicals. They include maximum limits for pesticide residues on food, product registration requirements, and restriction for using pesticide (Fenner-Crisp 2010). Regulatory agencies have fixed the standard values for pesticide residues in air, drinking water, soil, and agricultural goods for years. Currently, more than 19,000 pesticides soil regulatory guidance values (RGV's) and approximately 5000 pesticide drinking water maximum concentration levels (MCL's) have been established by 50 and 100 nations, respectively (Fenner-Crisp 2010). Over 100 nations have provided pesticide agricultural goods maximum residue limits (MRL's) for at least one of the twelfth most commonly consumed agricultural foods. A total of twenty pesticides have been regulated with more than 99 soil RGV's, and 20 pesticides have more than 95 drinking water MCL's (Li and Jennings 2017). This research indicates that those RGV's and MCL's for an individual pesticide could differ over 7 (DDT in drinking water MCL's), 8 (Lindane in soil RGV's), or even 10 (Dieldrin in soil RGV's) (Li and Jennings 2017).

12.6 Strategies to Overcome the Harmful Effects of Agrochemicals, PAHs, Crude Oil

These compounds incite carcinogenic effects, mutagenic effects, and other toxic effects reason being they are considered as hazardous pollutants. Various strategies employed to overcome the harmful effects are divided into Chemical and Biological.

12.6.1 Chemical

They can be originated from numerous natural (such as forest fire, volcanic eruptions, oil seeps, smoke from burning of woods) and anthropogenic activities (such as the burning of fossil fuels, coal tar, oil spillage, oil leakage, petroleum refinery effluents, and automotive emissions). They are hydrophobic in nature and show low water solubility, thus can bind to the organic matter present in soil (Bourceret et al. 2018). So, it is herculean to degrade them in non-toxic form. However, there is an emergence to degrade them since they engender pernicious effects. Selection of the method to be followed is determined by the type of soil, site of contamination, the associated risk with techniques and type of contaminants. It has also been seen, epoxides and dihydrodiols produced through the process of degradation of PAHs. Epoxides and dihydrodiols are even more harmful than their parent PAHs. Owing to which, it is essential to identify intermediate PAH metabolites. So, a number of physical methods (thermal desorption, microwave heating, vitrification, air sparging), chemical methods (oxidation using ozone, Fenton's reagent) and biological methods (phytoremediation, land farming, composting) can be followed to do so. Among all these strategies, it has been found that physical and chemical methods are efficient and effective but require a high amount of energy and are cost-effective. Besides all this, they also produce secondary pollutants. These limitations of chemical and physical methods are inevitable and are the main reason for the popularity of biological methods (Redfern et al. 2019). Biological methods are eco-friendly and convert toxic pollutants in non-toxic form without producing any other harmful secondary by-product.

12.6.2 Biological Method

In the biological method, bioremediation is being carried out for over two decades. Bioremediation is a method in which microorganisms (bacteria, fungi, yeast, and algae) are employed for the degradation of PAHs (Redfern et al. 2019). Microorganisms convert the contaminants in less toxic forms by producing numerous enzymes, water, and carbon dioxide along with it as a by-product. This technique has gained so much interest all over the world because of its eco-friendly nature. Type of microorganism to be used for degradation and end products are the most challenging task for effective contaminant degradation. Factors like temperature, nutrients, metabolites, and pH also play a vital role in the process (Haleyur et al. 2019). It is of two types: in situ and ex-situ.

12.6.2.1 In Situ Bioremediation

Bioaugmentation

This technique is carried out in the soils where the number of microorganisms is less in number. So, the addition of microorganisms either exogenous or indigenous is done to the contaminated site in the bioaugmentation process. Microbes to be added are appointed on the basis of their aptness to degrade the contaminant (Haleyur et al. 2019). Both aerobic and anaerobic type of microorganisms can be used for the bioaugmentation of PAH. Factors like microorganism survival, enzymatic activity, and pollutant bioavailability are foremost for the bioaugmentation. Various studies documented the degradation of PAH in soil by bioaugmentation by using fungi and bacteria. A fungal strain *S. brevicaulis* PZ-4 has been reported to remove more than 75% of polycyclic aromatic hydrocarbon where benzo-(a)-pyrene (70–75%) and phenanthrene (more than 85%) being highest to be removed, when isolated from an aged PAH contaminated soil and incubated for 25–28 days (Mao and Guan 2016). Concurrently, *Penicillium* sp. 06, when isolated and incubated for 25–28 days, showed an oxidation effect on the petroleum-contaminated soil. It was able to oxidize 88–89% of phenanthrene in waste residues originating from the petrochemical refining industry located in Singapore. If incubated for more than thirty days, it can also oxidize more than 70% of acenaphthene, fluorine, and fluoranthene (Zheng and Obbard 2003).

Biostimulation

It involves the environmental modification by adding oxygen and nutrients such as phosphorus, carbon, and nitrogen for the stimulation of oil/contaminant degrading activity by an indigenous microorganism. These nutrients inaugurate the allowance of synthesis of required enzymes for degradation of contaminants. It has been shown that there was an increase in the microbial biomass and activity when nutrients were added in PAH contaminated soils (Roy et al. 2018), when Zucchi et al. (2003) studied biostimulation by utilizing nutrients and surfactant solution in the hydrocarbon-degrading bacterial community for crude-oil contaminated soil. They noted 40% of reduction in hydrocarbon content. Similarly, when Abed et al. (2015) conducted the study and used ammonium chloride and sodium phosphate as N and P sources during the biostimulation of oil-contaminated desert soil, investigated 15–20% increase in the oil removal efficiency. One of the most efficient organic biostimulants is inactive biomass of *S. platensis*, phycocyanin or ammonium sulfate. These biostimulants were used for the biostimulation of soil contaminated with 3–4% of diesel for sixty days. The results concluded that 64% of biostimulation of 3–4% diesel by inoculation biomass of *S. platensis* for sixty days and extracted phycocyanin of *S. platensis* was found to be most effective as it biostimulated 89% of biodiesel in sixty days (Decesaro et al. 2017).

Bioventing

In the process of bioventing, air or oxygen is endowed through wells to prompt the growth of indigenous microorganisms as growth is the most indispensable factor of microorganisms to perform remediation. This technique has been extensively used for the remediation of the soils that are contaminated by petroleum hydrocarbons (Singh and Haritash 2019). A pilot-scale experiment was performed at Reilly Tar and chemical corporation site in St. Louis Park, Minnesota for the bioventing of 15.3 m^2 area, which included pyrene, benzo-(a)-anthracene, and fluoranthene. Ensue the completion of bioventing the results indicated, 20–24% reduction in six-membered ring PAHs, 15–20% reduction in five-membered ring, 30% reduction in four-membered ring, 45–50% reduction in three-membered ring, and 60% reduction in two membered rings PAHs (Alleman et al. 1995). When bioventing treatment of artificially contaminated soil by phenanthrene was done for seven months, 90–95% of phenanthrene was removed. Under conditions of carbon/nitrogen/phosphorus = 100:20:1 and humidity = 55–60% (Rodriguez et al. 2017).

12.6.2.2 Ex-Situ Bioremediation

Land-Farming

Land-Farming comprise the excavation, transportation of contaminated soil to the land-Farming site and then spreading over the prepared bed. Later tilling is done in order to provide aeration. It is the simplest technique for remediation of contaminated soil. Microorganisms degrade the contaminants by oxidation, a metabolic process. In South Africa, land-Farming of creosote contaminated soil was done and in six months and found that naphthalene, phenanthrene, fluorine, and anthracene (low molecular mass) were degraded. On the other hand, for the remediation of high molecular mass PAHs land farming was done for another ten months, and 75–88% of four-five membered rings PAHs were remediated (Atagana 2004). Not only PAHs but petroleum hydrocarbons such as trimethyl benzenes and diesel range organics can also be degraded by land farming (Katsivela et al. 2005).

Composting

Composting is a process in which both thermophilic and mesophilic microorganisms are used to degrade the organic contaminants at elevated temperature (60 °C). Microorganisms release the heat amidst the process which ameliorates the solubility of the contaminants. Scrutinization on spent mushroom compost was conducted for bioremediation of soil contaminated with PAHs in which the degradation of phenanthrene, naphthalene, and benzo-(a)-pyrene was observed after 48 h at 75–80 °C (Lau et al. 2003). The thermally insulated chamber was used for the remediation of soil contaminated with PAHs. Mushroom compost, consisting wheat straw, gypsum, and chicken manure was also used in it. This experiment was carried out for fifty days. In the end, 50–60% of compost was noticed. After another hundred days, 40–80% of aromatic hydrocarbons were eliminated (Sasek et al. 2003). For intensification of PAH's bioavailability coal tar, diesel, coal ash contaminated soil was mixed with compost (Wu et al. 2013).

Phytoremediation

It is a process that involves the employment of green plants and ally microorganisms to remove or degrade the PAHs or any type of contaminant. It is a cheaper and more convenient way to remediate the contaminants. Techniques like phytoextraction, phytostabilization, phytotransformation, rhizodegradation can opt for the remediation of soil (Kathi and Khan, 2011). Plants have the aptitude to secrete enzymes such as dioxygenase, monooxygenase, dehydrogenase, and hydrolase that assist in the remediation of contaminants (Cristaldi et al. 2017). It has been concluded that different types of plants such as *T. repensm* (white clover), yellow sweet clover (*M. officinalis*), *F. arundinacea,* and ryegrass (*L. multiflorum*) have the capability to degrade the PAHs such as fluoranthene, chrysene, naphthalene, and anthracene (Rezek et al. 2008). Contaminant nature, soil properties, type of plant, and bioavailability of the contaminants are various factors that can affect the phytoremediation process. A pot culture experiment was conducted for the remediation of petroleum hydrocarbons by using different species of plants, i.e. *M. sativa, E. purpurea, F. arundinacea.* All these plants removed the TPH, including polar compounds, aromatic hydrocarbons, and saturated hydrocarbons consequently (Liu et al. 2012). Various techniques, such as electrokinetic treatment or bioremediation, can be followed for the enhancement of phytoremediation. It is possible to boost up the remediation of anthracene or phenanthrene from the soil by electro-phytoremediation with *B. rapa* (Camesselle and Gouveia 2019). It took the choice of best remediation technology, environmental conditions, type of soil, toxicity to achieve highly efficient phytoremediation.

12.6.3 Remediation by Chemical Methods

It is a productive way to remove lethal waste from the soil at the location of oil spillage. Soil matrix mainly determines the efficiency of this method. In this method, Fenton's reagent is employed, which is a mixture of ferric ions and hydrogen peroxide. It carries out the oxidation as hydrogen peroxide is a strong oxidizing agent; it produces hydroxyl ions (Goi et al. 2006). On the other hand, ferric ions act as a catalyst. The effect of hydroxyl ions destroys contaminants. Fenton's reagent helps in remediating oil from the soil by lowering the pH of soil. It is the simplest and most efficient method for remediation of oil from the soil, but it has some drawbacks, i.e. very costly, time-consuming. Besides this, the transfer of contaminated soil to the disposal site is also a big issue.

12.6.4 Remediation by Bioremediation

Bioremediation is a biological and traditional method to remediate the harmful contaminants by using living organisms (bacteria, plants, and fungi). Employing this method for remediation of crude oil is efficient because it is environment friendly and cheap at the same time (Siles and Margesin 2018). Hydrocarbon concentration, soil characteristics, and pollutant constituents determine the efficiency of this method.

12.6.5 Remediation by Rhizoremediation

It is a method which assists plant microbes for remediation. Microorganisms that are present in the soil enhance the tendency of a plant to remediate crude oil by forming a cooperative nexus with one another. This cooperative nexus between soil microbe and plant is called rhizoremediation in which plants give space, and other required environments to the microbes and microbes degrade the contaminants in return. Lately, rhizoremediation is being the most efficient and cost-effective technique to remediate crude oil from the soil. Mainly it occurs naturally but can also be initiated by the addition of specific microbes (Kang et al. 2020). In a study, conducted on the wheat plant under hydroponic conditions concluded that more than 20% of the oil was eliminated by wheat seedlings from media and when associated with *Azospirilum* this ability increased by 25–30%. Bioremediation of soil contaminated by oil can be done by using yellow alfalfa in the association with *Acinetobcter sp.* Strain SS-33 which improved efficiency of remediation by 35% as compared to alone alfalfa which was 30–34% and *Acinetobacter* sp. S-33, which was 30–33%. Thus it was analyzed by fractional contaminants that plant–microbe association is very efficient technology in the clean-up of aromatic hydrocarbons from the soil (Muratova et al. 2018).

12.7 Vermicomposting in Bioremediation

12.7.1 Garden, Kitchen, and Agro Waste

According to Bouwman (2007), a number of tests have been developed to determine the effect of pollutants on earthworms, the reason being that earthworms are an essential ecological component of many soils (Bouwman 2007). That is why a lot of focus has deviated upon the possible role of vermiculture in solving the problem associated with waste disposal. Garden waste can be converted into manure by vermicomposting (Shah et al. 2015). Empty fruit bunches when mixed with cow dung and subjected to vermicomposting converts it into nutrient-rich organic fertilizer (Lim et al. 2014). According to another study, the cast from the earthworms produced by the ingestion of agricultural waste contains plant nutrients and growth-promoting substances in an assimilated form (Sinha et al. 2009). This increased level of nutrients can be attributed to the enzymatic and microbial activity of earthworms, and the results advocate that post-vermicompost samples derived from agricultural waste contain a fairly higher level of major and micronutrients in comparison to the initial levels of nutrients. The enzymatic and microbial activity of the earthworms contributes to this increased level of nutrients. Suthar (2009a) also reported the vermicomposting of post-harvest residues of some local crops like wheat, millets, and a pulse and concluded that agro waste could be converted to some value products like vermicompost which have the potential to be used for sustainable crop production (Suthar 2009a). Singh and Kalamdhad in 2013 also recycled temple waste that included floral offerings through vermicomposting through *Eisenia fetida* and a comparison was made with vermicompost from kitchen waste and farmyard waste (Singh and Kalamdhad in 2013). The maximum biomass was found in temple waste vermicompost, and also temple waste vermicompost showed an increase in the length of root, shoot; a number of secondary roots and total biomass when compared with kitchen waste and farmyard waste. Suthar (2007a, b) vermicomposted agriculture waste, farmyard manure, and urban solid waste with an earthworm, *Perionyx sansibaricus*. The decrease in organic carbon, C:N ratio; also the increase in NPK, plant metabolites in the end product and growth pattern of *P. sansibaricus* in different organic waste resources indicate that this species can be efficiently used for recycling of wastes with a low-cost input (Suthar 2007a, b).

12.7.2 Heavy Metal Reduction from Soil

Pathma and Sakthivel (2012) also suggested that vermicompost can potentially be used in sustainable agriculture and also can effectively manage wastes from agriculture, industrial, domestic, and medical sector which tends to be at high risk for both life and environment. The sewage sludge having high nutritive value for plants can be utilized as fertilizers after the elimination of heavy metals (Suthar 2009b;

Bettiol 2004). Singh and Kalamdhad (2013) reported the feasibility of earthworms in the reduction of metal toxicity and to increase the nutrient profile in water hyacinth vermicompost for sustainable land improvement practices. The bioavailability and leachability were marginally reduced by the vermicomposting of water hyacinth by *E. fetida*.

12.7.3 Municipal Sewage Waste

Earthworms are an important ecological part of many soils. Treatment of wastewater and sludge also utilize earthworms (Kaushik and Garg 2004). The municipal sewage waste by vermicomposting can be effectively converted to nutrient-rich, and eco-friendly biofertilizer (Mishra et al. 2014). Thus, if municipal sewage waste is managed in an appropriate manner, then it not only mitigates the negative effects, but it could help in meeting the demand of ecology and economy. Vermicomposting municipal biodegradable wastes at home are the best possible method for waste disposal. In terms of economy and impact on the environment, the most effective way to deal with solid waste is to reduce domestic waste at the source itself (Pirsaheb et al. 2013a, b, c). Vermicomposting of sewage sludge also resulted in a reduction in C:N ratio, total organic carbon (TOC) but increases in EC, total nitrogen, potassium, calcium, phosphorus, indicating sewage sludge could be converted to a good quality fertilizer. Khwairakpam and Bhargava in 2009 also vermicomposted sewage sludge and observed an increase in EC, N, K, Ca, Na, P; also, the heavy metals Cu, Mn, Pb, and Zn were now in permissible limits thus indicating that recycled sewage sludge through vermicompost can be used as an effective fertilizer.

12.7.4 Tannery Industry

The tanning industry is spread all over India and a major part in Tamil Nadu and Uttar Pradesh (Ravindran and Jindal 2008). It is one of the highly polluting and growth-oriented industries, and it generates tones of wastes in the form of rawhide (and skin) trimmings (Ozgunay et al. 2007). Tannery industries release not only organic material which forms a source of valuable nutrients on decomposition but also metals and pathogens and other toxin components which genuinely may put the environment to greater risks (Contreras-Ramos et al. 2004; Ganesh Kumar et al. 2009). The effluents from the tanning industry are high in organic and inorganic dissolved and suspended solids along with proclivity for high oxygen demand. The tanning activities lead to unpleasant odor that ensues from the decomposition of ammonia, solid protein waste, hydrogen sulphide, and volatile organic compounds. Most of the chemicals used in processing remain unabsorbed and thus are discharged into the environment. Tannery sludge can provide a nutrient supplement for crops after proper remediation as it contains plant nutrients like nitrogen, phosphorus, iron, zinc, and copper.

A successful attempt to recycle tannery sludge into manure through vermicomposting by *E. fetida* was made by Hemelata and Meenambal (2005). Another study evaluated the amendment of tannery sludge by vermicomposting. Results inferred an increase in nutrient content, lower C:N ratio and lower electrical conductivity which could be used as manure depicting that vermicompost could be considered as an effective technology for production of value-added products using tannery sludge as an input (Vig et al. 2011).

12.7.5 Improving Forage Quality

In a study reported in 2014, pre-composting prior to vermicomposting contributes for a powerful design for management of ruminant manure. This reduces the environmental pollution from ruminant production and can be effectively used as a feed supplement to ruminants (Nasiru et al. 2014). Further, the vermicast produced can be used as a good fertilizer. Vermicompost also increases the green fodder and dry matter in the case of sorghum (*Sorghum bicolor* (L.) Moench) (Sheoran and Rana 2005). Forage sorghum (*Sorghum bicolor* (L.) Moench) was produced by vermicompost and farmyard manure integrated with inorganic fertilizers (Sheoran and Rana 2005).

12.8 Vermicompost: Mechanism

Microorganisms primarily accomplish the task of biochemical decomposition of organic matter, but earthworms are the critical drivers of the process because they graze on microbes and stimulate their decomposer activity (Aira and Domínguez 2009; Monroy et al. 2009; Gomez-Brandon et al. 2011a, b), and in addition to this, they also increase the surface area available for microbes to act upon after decomposition of organic matter (Dominguez et al. 2010). The mechanism of converting 'garbage into gold' is very well studied and is comprised of the following steps (Fig. 12.3):

1. Ingestion of the substrate by the earthworms.
2. The grinding 'gizzard' located next to worm's mouth, helps in mincing of the ingested substrate and leads to an increase in the surface area of the substrate, which facilitates for microbes to act upon (Chan and Griffiths 1988).
3. Enzymes, along with the microflora of the worm's gut, digest the substrate further as it passes through the body.
4. The formation of the substrate as 'vermicast' which is microbially much more active than the ingested one.

Almost any industrial or agricultural organic material can be subjected to vermicompost, but out of the few may be toxic to be used directly for the earthworms

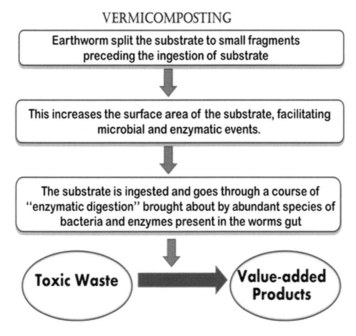

Fig. 12.3 Mechanism of conversion of toxic waste into value-added products by vermicomposting

and thus require a pre-processing. (Gajalakshmi and Abbasi 2008). This preliminary process can be in the form of washing, pre-composting, macerating or mixing. Precomposting facilitates vermicomposting (Tognetti et al. 2007) because it is a thermophilic phase and kills most pathogenic organisms which make sure that earthworms survive and grow well and also ensure pathogen-free vermicompost (Dominguez and Gomez-Brandon 2012).

Unlike composting where the substrate has to be tossed regularly to maintain aerobic conditions, in vermicomposting the earthworms take over the roles of both turning and maintaining the organics in an aerobic condition, eliminating the need for mechanical aeration (Misra et al. 2003; Sinha et al. 2010).

12.9 Conclusion

The waste material from various industries serves as a rich source of nutrients, proteins, and energy that should not be wasted by mere disposal in dumps or landfills. Rather, their immense energy should be utilized in one form or the other. Here, comes the role of vermicomposting which not only utilizes the wastes that would otherwise be problematic for the society but also converts and recycles that waste supplying the valuable nutrients back to the soil maintaining ecological sustainability. The post-vermicompost matter could be largely utilized as organic amendments in agriculture.

Thus, this solves the problem of waste disposal and equally benefits agro systems. Future research should be directed to increase the understanding of the impacts of organic fertilizers on soil microbial processes and nutrient cycling and disentangling the influence of different factors such as crop species, soil type, and compost properties, in order to increase crop yields under sustainable production systems. The combination of modern microbiological techniques with the knowledge of soil ecological processes would provide a unique opportunity to improve agronomical practices. The goal is to use and optimize the biological resources already existent in the soil and optimize fertilizer management to maximize yields while reducing environmental impacts.

References

Abdel-Shafy HI, Mansour MS (2016) A review on polycyclic aromatic hydrocarbons: source, environmental impact, effect on human health and remediation. Egypt J Pet 25(1):107–123

Abed RM, Al-Kharusi S, Al-Hinai M (2015) Effect of biostimulation, temperature and salinity on respiration activities and bacterial community composition in an oil polluted desert soil. Int Biodeterior Biodegradation 98:43–52

ACGIH (American Conference of Governmental Industrial Hygienists) (2005) Polycyclic aromatic hydrocarbons (PAHs) biologic exposure indices (BEI) Cincinnati. OH: American Conference of Governmental Industrial Hygienists

Aira M, Dominguez J (2009) Microbial and nutrient stabilization of two animal manures after the transit through the gut of the Earthworm *Eisenia fetida* (Savigny, 1826). J Hazard Mater 161:2–3

Akbar MF, Haq MA, Parveen F, Yasmin N, Sayeed SA (2010) Determination of synthetic and bio-insecticides residues during aphid, *Myzus persicae* (Sulzer) control on cabbage crop through high performance liquid chromatography. Pak. Entomol 32(2):155–162

Aktar W, Sengupta D, Chowdhury A (2009) Impact of pesticides use in agriculture: their benefits and hazards. Interdiscip Toxicol 2(1):1–12

Alegbeleye OO, Opeolu BO, Jackson VA (2017) Polycyclic aromatic hydrocarbons: a critical review of environmental occurrence and bioremediation. Environ Manage 60(4):758–783

Alleman BC, Hinchee RE, Brenner RC, McCauley PT (1995) Bioventing PAH contamination at the Reilly Tar site. In: Hinchee RE, Miller RN, Johnson PC (eds) In-situ aeration: air sparging, bioventing, and related remediation processes. Battelle Press, Columbus, pp 473–482

Amangabara GT, Njoku JD (2012) Assessing groundwater vulnerability to the activities of artisanal refining in Bolo and environs, Ogu/Bolo Local Government Area of Rivers State; Nigeria. Br J Environ Clim Change 2(1):28

Arif MS, Houwen F, Verstraete W (1996) Agricultural factors affecting methane oxidation in arable soil. Biol Fertil Soils 21(1–2):95–102

Atagana HI (2004) Bioremediation of creosote-contaminated soil in South Africa by landfarming. J Appl Microbiol 96(3):510–520

Azarmi R, Giglou MT, Hajieghrari B (2009) The effect of sheep manure vermicompost on quantitative and qualitative properties of cucumber (*Cucumis sativus* L.) grown in the greenhouse. Afr J Biotechnol 8:4953–4957

Bachman GR, Metzger JD (2008) Growth of bedding plants in commercial potting substrate amended with vermicompost. Bioresour Technol 99:3155–3161

Bacolod ET, Uno S, Tanaka H, Koyama J (2013) Micronuclei and other nuclear abnormalities induction in erythrocytes of marbled flounder, *Pleuronectes yokohamae*, exposed to dietary nitrated polycyclic aromatic hydrocarbons. Jpn J Environ Toxicol 16(2):79–89

Bahrampour T, Ziveh PS (2013) Effect of vermicompost on tomato (*Lycopersicum esculentum*) fruits. Int Agron Plant Prod 4(11):2965–2971

Baklanov A, Hänninen O, Slørdal LH, Kukkonen J, Bjergene N, Fay B et al (2007) Integrated systems for forecasting urban meteorology, air pollution and population exposure 7(3):855–874

Balcıoğlu EB (2016) Potential effects of polycyclic aromatic hydrocarbons (PAHs) in marine foods on human health: a critical review. Toxin Rev 35(3–4):98–105

Bansal V, Kim KH (2015) Review of PAH contamination in food products and their health hazards. Environ Int 84:26–38

Banu JR, Logakanthi S, Vijayalakshmi GS (2001) Biomanagement of paper mill sludge using two exotic and one indigenous earthworm species. J Environ Biol 22:181–185

Bettiol W (2004) Effect of sewage sludge on the incidence of corn stalk rot caused by *Fusarium*. Summa Phytopathologica 30:16–22

Beyer J, Jonsson G, Porte C, Krahn MM, Ariese F (2010) Analytical methods for determining metabolites of polycyclic aromatic hydrocarbon (PAH) pollutants in fish bile: a review. Environ Toxicol Pharmacol 30(3):224–244

Bhat SA, Singh J, Vig AP (2014) Genotoxic assessment and optimization of pressmud with the help of exotic earthworm *Eisenia fetida*. Environ Sci Pollut Res 21(13):8112–8123

Bhawalkar VU, Bhawalkar US (1993) Vermiculture: the bionutrition system. national seminar on indigenous technology for sustainable agriculture. Indian Agriculture Research Institute (IARI), New Delhi, 1–8

Blok WJ, Lamers JG, Termoshuizen AJ, Bollen GJ (2000) Control of soil-borne plant pathogens by incorporating fresh organic amendments followed by tarping. Phytopathology 90:253–259

Bombatkar V (1996) The miracle called compost. The Other India Press, Pune

Bourceret A, Leyval C, Faure P, Lorgeoux C, Cébron A (2018) High PAH degradation and activity of degrading bacteria during alfalfa growth where a contrasted active community developed in comparison to unplanted soil. Environ Sci Pollut Res 25(29):29556–29571

Bouwman H (2007) Modifications to a defined medium for the study of the biology and toxicology of the earthworm *Eisenia fetida* (Oligochaeta). Appl Soil Ecol 35:566–581

Burstyn I, Kromhout H, Partanen T, Svane O, Langård S, Ahrens W et al (2005) Polycyclic aromatic hydrocarbons and fatal ischemic heart disease. Epidemiology 744–750

Cameselle C, Gouveia S (2019) Phytoremediation of mixed contaminated soil enhanced with electric current. J Hazard Mater 361:95–102

Canellas LP, Olivares FL, Okorokova-Façanha AL, Façanha AR (2002) Humic acids isolated from earthworm compost enhance root elongation, lateral root emergence, and plasma membrane H+-ATPase activity in maize roots. Plant Physiol 130:1951–1957

Carvalho FP (2017) Pesticides, environment, and food safety. Food Energy Secur 6(2):48–60

Cavender ND, Atiyeh RM, Knee M (2003) Vermicompost stimulates mycorrhizal colonization of roots of *Sorghum bicolor* at the expense of plant growth. Pedobiologia 47:85–90

Chalam AV, Sasikala C, Ramana CV, Uma NR, Rao PR (1997) Effect of pesticides on the diazotrophic growth and nitrogenase activity of purple nonsulfur bacteria. Bull Environ Contam Toxicol 58(3):463–468

Chan PL, Griffiths DA (1988) The vermicomposting of pre-treated pig manure. Biol Wastes 24(1):57–69

Chandrakumar HL, Kumar CTA, Chakravarthy AK, Kumar NG, Puttaraju TB (2009) Influence of organic materials against shoot and fruit borer, *Leucinodes orbonalis* Guen. (Lepidoptera: Pyralidae) on brinjal (*Solanum melongena* L.). Current Biotica 2(4):495–500

Chavan RB (2001) Indian textile industry-environmental issues. Indian J Fibre Text Res 26(1/2):11–21

Chen F, Pu LJ (2007) Relationship between heavy metals and basic properties of agricultural soils in Kunshan County. Soils 39:291–296

Chhonkar PK, Datta SP, Joshi HC, Pathak H (2000) Impact of industrial effluents on soil health and agriculture -Indian experience: Pat-1, distillery and paper mill effluents. Sci Ind Es 59:350–361

Chikere CB, Fenibo EO (2018) Distribution of PAH-ring hydroxylating dioxygenase genes in bacteria isolated from two illegal oil refining sites in the Niger Delta, Nigeria. Sci Afr 1:e00003

Clinton EI, Ngozi ON, Ifeoma OL (2014) Heavy metals and polycyclic aromatic hydrocarbons in water and biota from a drilling waste polluted freshwater swamp in the mgbede oil fields of south-south Nigeria. J Bioremediat Biodegredation 5(7):1

Cocârță DM, Stoian MA, Karademir A (2017) Crude oil contaminated sites: evaluation by using risk assessment approach. Sustainability 9(8):1365

Contreras-Ramos SM, Alvarez-Bernal D, Trujillo-Tapia N, Dendooven L (2004) Composting of tannery effluent with cow manure an wheat straw. Bioresour Technol 94:223–228

Cristaldi A, Conti GO, Jho EH, Zuccarello P, Grasso A, Copat C, Ferrante M (2017) Phytoremediation of contaminated soils by heavy metals and PAHs. A brief review. Environ Technol Innov 8:309–326

de Araujo, Jessica SA, Delgado JF, Paumgartten FJR (2016) Glyphosate and adverse pregnancy outcomes, a systematic review of observational studies. BMC Public Health 16(1):472

Decesaro A, Rampel A, Machado TS, Thomé A, Reddy K, Margarites AC, Colla LM (2017) Bioremediation of soil contaminated with diesel and biodiesel fuel using biostimulation with microalgae biomass. J Environ Eng 143(4):04016091

Domestic scale vermicomposting for solid waste management

Dominguez J (2004) State-of-the-art and new perspectives on vermicomposting research. In: Edwards CA (ed) Earthworm ecology Ed. CRC Press, Boca Raton. pp 401–425

Dominguez J, Gómez-Brandón M (2012) Vermicomposting: composting with earthworms to recycle organic wastes. INTECH Open Access Publisher

Dominguez J, Aira M, Gómez-Brandón M (2010) Vermicomposting: earthworms enhance the work of microbes. In: Insam H; Franke-Whittle I, Goberna M (eds) Microbes at work: from wastes to resources. Springer, Heidelberg, Germany, pp 93–114. ISBN 978-3-642- 04042-9

Dong CD, Chen CF, Chen CW (2012) Determination of polycyclic aromatic hydrocarbons in industrial harbor sediments by GC-MS. Int J Environ Res Public Health 9:2175–2188

Edwards CA, Arancon NQ, Vasko-Bennett M, Askar A, Keeney G (2010) Effect of aqueous extracts from vermicomposts on attacks by cucumber beetles (*Acalymna vittatum*) (Fabr.) on cucumber and tobacco hornworm (*Manduca sexta*) (L.) on tomatoes. Pedobiologia 53:141–148

El Harti A, Saghi M, Molina JAE, Teller G (2001a) Production des composés indoliques rhizogénes par le ver de terre Lumbricus terrestris. Can J Zool 79:1921–1932

El Harti A, Saghi M, Molina J-AE, Téller G (2001b) Production d'une substance rhizogène à effet similaire à celui de l'acide indole acètique par le ver de terre *Lumbricus terrestris*. Can J Zool 79:1911–1920

El Harti A, Saghi M, Molina J-AE, Téller G (2001c) Production des composés indoliques rhizogénes par le ver de terre *Lumbricus terrestris*. Can J Zool 79:1921–1932

Espinoza MI, Meji SA, Estelle MV (2014) Phosphorus release kinetics in a soil amended with biosolids and vermicompost. Environmental Earth Sciences 71:1441–1451

Ezeji U, Anyadoh SO, Ibekwe VI (2007) Clean up of crude oil-contaminated soil. Terr Aquat Environ Toxicol 1(2):54–59

Fabra A, Duffard R, De Duffard AE (1997) Toxicity of 2, 4-dichlorophenoxyacetic acid to *Rhizobium* sp. in pure culture. Bull Environ Contam Toxicol 59(4):645–652

Fenner-Crisp PA (2010) Risk assessment and risk management: the regulatory process. In: Hayes' handbook of pesticide toxicology. Academic Press. pp 1371–1380

Fernández-Gómez MJ, Díaz-Raviña M, Romero E, Nogales R (2013) Recycling of environmentally problematic plant wastes generated from greenhouse tomato crops through vermicomposting. Int J Environ Sci Technol 10(4):697–708

Florencia RM, Cora S (2020) Agrochemicals and neurogenesis. Mol Cell Endocrinol 110820

Fox JE, Starcevic M, Kow KY, Burow ME, McLachlan JA (2001) Nitrogen fixation: endocrine disrupters and flavonoid signalling. Nature 413(6852):128–130

Gajalakshmi S, Abbasi SA (2008) Solid waste management by composting: state of the art. Crit Rev Environ Sci Technol 38(5):311–400

Gajalakshmi S, Ramasamy EV, Abbasi EV (2002) Vermicompositng of different forms of water hyacinth by the earthworm *Eudrilus eugeniae*, Kinberg. Biores Technol 82:165–169

Gajalakshmi S, Ramasamy EV, Abbasi SA (2005) Composting- vermicomposting of leaf litter ensuing from the trees of mango (*Mangifera indica*) Kinberg. Biores Technol 96:f1057-1061

Ganesh Kumar A, Venkatesan R, Prasad Rao B, Swarnalatha S, Sekaran G (2009) Utilization of tannery solid waste for protease production by *Synergistes* sp. insolid-state fermentation and partial protease characterization. Eng Life Sci 9:66–73

George J, Shukla Y (2011) Pesticides and cancer: insights into toxicoproteomic-based findings. J Proteomics 74(12):2713–2722

Ghosal D, Ghosh S, Dutta TK, Ahn Y (2016) Current state of knowledge in microbial degradation of polycyclic aromatic hydrocarbons (PAHs): a review. Front Microbiol 7:1369

Goel, P. K. (2006) *Water pollution: causes, effects and control*. New Age International.

Goi A, Kulik N, Trapido M (2006) Combined chemical and biological treatment of oil contaminated soil. Chemosphere 63(10):1754–1763

Gomez F, Aira M, Domínguez J (2009) Reduction of total coliform numbers during vermicomposting is caused by short-term direct effects of earthworms on microorganisms and depends on the dose of application of pig slurry. Sci Total Environ 407(20):5411–5416

Gómez-Brandón M, Aira M, Lores M, Domínguez J (2011a) Changes in microbial community structure and function during vermicomposting of pig slurry. Biores Technol 102(5):4171–4178

Gómez-Brandón M, Aira M, Lores M, Domínguez J (2011b) Epigeic earthworms exert a bottleneck effect on microbial communities through gut associated processes. PlosOne 6(9):1–9

Grant WB (2009) Air pollution in relation to US cancer mortality rates: an ecological study; likely role of carbonaceous aerosols and polycyclic aromatic hydrocarbons. Anticancer Res 29(9):3537–3545

Grappelli A, Galli E, Tomati U (1987) Earthworm casting effect on *Agaricus bisporus* fructification. Agrochimica 31(4–5):457–461

Haleyur N, Shahsavari E, Jain SS, Koshlaf E, Ravindran VB, Morrison PD et al (2019) Influence of bioaugmentation and biostimulation on PAH degradation in aged contaminated soils: Response and dynamics of the bacterial community. J Environ Manage 238:49–58

Hameeda B, Harini G, Rupela OP, Reddy G (2007) Effect of composts or vermicomposts on sorghum growth and mycorrhizal colonization. Afr J Biotech 6(1):9–12

Hanafi YS (2012) Ecology and agriculture in the Himalayan region: problems and prospects of agricultural development in North-western Himalaya. Concept Publishing Company

He X, Pang Y, Song X, Chen B, Feng Z, Ma Y (2014) Distribution, sources and ecological risk assessment of PAHs in surface sediments from Guan River Estuary, China. Mar Pollut Bull 80(1–2):52–58

Hemelata B, Meenabal T (2005) Reuse of industrial sludge alongwith yard waste by vermicomposting method. Nat Environ Pollut Technol 4(4):597–600

Hointik HAJ, Bohem MJ (1999) Biocontrol within the context of soil microbial communities: a substrate-dependent phenomenon. Annu Rev Phytopathol 37:427–446

Hu H, Kan H, Kearney GD, Xu X (2015) Associations between exposure to polycyclic aromatic hydrocarbons and glucose homeostasis as well as metabolic syndrome in nondiabetic adults. Sci Total Environ 505:56–64

Hussain S, Siddique T, Saleem M, Arshad M, Khalid A (2009) Impact of pesticides on soil microbial diversity, enzymes, and biochemical reactions. Adv Agron 102:159–200

Hutchinson ML, Walters LD, Avery SM, Munro F, Moore A (2005) Analyses of livestock production, waste storage, and pathogen levels and prevalences in farm manures. Appl Environ Microbiol 71(3):1231–1236

Ingaramo P, Alarcón R, Muñoz-de-Toro M, Luque EH (2020) Are glyphosate and glyphosate-based herbicides endocrine disruptors that alter female fertility? Mol Cell Endocrinol 110934

Ivshina IB, Kuyukina MS, Krivoruchko AV, Elkin AA, Makarov SO, Cunningham CJ et al (2015) Oil spill problems and sustainable response strategies through new technologies. Environ Sci Process Impacts 17(7):1201–1219

Joshi R, Vig AP (2010) Effect of vermicompost on growth, yield and quality of tomato (*Lycopersicum esculentum* L). Afr J Basic Appl Sci 2(3–4):117–123

Joshi R, Vig AP, Singh J (2013) Vermicompost as soil supplement to enhance growth, yield and quality of *Triticum aestivum* L.: a field study. Int J Recy Org Waste Agric 2:16

Kabata-Pendias A, Pendias H (1992) Trace elements in soils and plants, 2nd edn. CRC Press, Boca Raton, FL

Kamangar F, Schantz MM, Abnet CC, Fagundes RB, Dawsey SM (2008) High levels of carcinogenic polycyclic aromatic hydrocarbons in mate drinks. Cancer Epidemiol Prev Biomarkers 17(5):1262–1268

Kang CU, Kim DH, Khan MA, Kumar R, Ji SE, Choi KW et al (2020) Pyrolytic remediation of crude oil-contaminated soil. Sci Total Environ 713:136498

Karmegam N, Alagermalai K, Daniel T (1999) Effect of vermicompost on the growth and yield of greengram (*Phaseolus aureus* Rob.). Trop Agric 76(2):143–146

Kathi S, Khan AB (2011) Phytoremediation approaches to PAH contaminated soil. Indian J Sci Technol 4(1):56–63

Katsivela E, Moore ERB, Maroukli D, Strömpl C, Pieper D, Kalogerakis N (2005) Bacterial community dynamics during in-situ bioremediation of petroleum waste sludge in landfarming sites. Biodegradation 16(2):169–180

Kaushik P, Garg VK (2004) Dynamics of biological and chemical parameters during vermicomposting of solid textile mill sludge mixed with cow dung and agricultural residues. Bioresour Technol 94:203–209

Kavian MF, Ghatnekar SD, Kulkarni PR (1998) Conversion of coir pith into value added biofertiliser using *Lumbricus rubellus*. Indian J Environ Prot 18(5):354–358

Kesalkar VP, Khedikar IP, Sudame AM (2012) Physico-chemical characteristics of waste water from paper industry. Int J Eng Res Appl 2:137–143

Khwairakpam M, Bhargava R (2009) Vermitechnology for sewage sludge recycling. J Hazard Mater 161(2):948–954

Kim KH, Jahan SA, Kabir E, Brown RJ (2013) A review of airborne polycyclic aromatic hydrocarbons (PAHs) and their human health effects. Environ Int 60:71–80

Kumar R, Verma D, Singh BL, Kumar U, Shweta (2010) Composting of sugarcane waste by-products through treatment with microorganisms and subsequent vermicomposting. Bioresour Technol 101(17):6707–6711

Lamichhane JR, Dachbrodt-Saaydeh S, Kudsk P, Messéan A (2016) Toward a reduced reliance on conventional pesticides in European agriculture. Plant Dis 100(1):10–24

Latimer JS, Zheng J (2003) Fate of PAHs in the marine environment. PAHs Ecotoxicological Perspect 9

Lau KL, Tsang YY, Chiu SW (2003) Use of spent mushroom compost to bioremediate PAH-contaminated samples. Chemosphere 52(9):1539–1546

Lazarovits G, Tenuta M, Conn KL, Gullino ML, Katan J, Matta A (2000) Utilization of high nitrogen and swine manure amendments for control of soil-borne diseases: efficacy and mode of action. Acta Hortic 5:559–564

Lazcano C, Arnold J, Tato A, Zaller JG, Dominguez J (2009) Compost and vermicompost as nursery pot components: Effects on tomato plant growth and morphology. Span J Agric Res 7(4):944–995

Lazcano C, Sampedro L, Zas R, Dominguez J (2010) Vermicompost enhances germination of the maritime pine (*Pinus pinaster* Ait.). New 39:387–400

Lazcano C, Revilla P, Malvar RA, Domínguez J (2011) Yield and fruit quality of four sweet corn hybrids (*Zea mays*) under conventional and integrated fertilization with vermicompost. J Sci Food Agric 91(7):1244–1253

Li Z, Jennings A (2017) Worldwide regulations of standard values of pesticides for human health risk control: a review. Int J Environ Res Public Health 14(7):826

Lim PN, Wu TY, Clarke C, Daud NN (2014) A potential bioconversion of empty fruit bunches into organic fertilizer using *Eudrilus eugeniae*. Int J Environ Sci Technol 1–12

Lin C, Tjeerdema RS (2008) Crude oil, oil, gasoline and petrol. In: Jorgensen SE and Fath BD (eds) Encyclopedia of ecology. Volume 1: ecotoxicology. Elsevier, Oxford, UK, pp 797–805

Liu K, Han W, Pan WP, Riley JT (2001) Polycyclic aromatic hydrocarbon (PAH) emissions from a coal-fired pilot FBC system. J Hazard Mater 84:175–188

Liu R, Jadeja RN, Zhou Q, Liu Z (2012) Treatment and remediation of petroleum-contaminated soils using selective ornamental plants. Environ Eng Sci 29(6):494–501

Lo CC (2010) Effect of pesticides on soil microbial community. J Environ Sci Health B 45(5):348–359

Lo YH, Blanco JA, Welham C, Wang M (2015) Maintaining ecosystem function by restoring forest biodiversity: reviewing decision-support tools that link biology, hydrology and geochemistry. In: Lo YH, Blanco JA, Roy S (eds) Biodiversity in ecosystems: linking structure and function, pp 143–167

Lundestedt S (2003) Analysis of PAHs and their transformation products in contaminated soil and remedial processes. Dissertation, Umea University

Luque EH, Muñoz-de-Toro M (2020) Special Issue "Health effects of agrochemicals as endocrine disruptors"

Maliszewska-Kordybach B, Smreczak B (2000) Ecotoxicological activity of soils polluted with polycyclic aromatic hydrocarbons (PAHs)-effect on plants. Environ Technol 21(10):1099–1110

Mandal A, Sarkar B, Mandal S, Vithanage M, Patra AK, Manna MC (2020) Impact of agrochemicals on soil health. In: Agrochemicals detection, treatment and remediation. Butterworth-Heinemann, pp 161–187

Mao J, Guan W (2016) Fungal degradation of polycyclic aromatic hydrocarbons (PAHs) by *Scopulariopsis brevicaulis* and its application in bioremediation of PAH-contaminated soil. Acta Agriculturae Scandinavica Sect B-Soil Plant Sci 66(5):399–405

Marinescu M, Toti M, Tanase V, Carabulea V, Plopeanu G, Calciu I (2010) An assessment of the effects of crude oil pollution on soil properties. Ann Food Sci Technol 11:94–99

Martin P (1998) River pollution in India: an overview. In: Chari IK (ed) Employment news. March 28-April, 13. XXII (82), pp 1–2

Martyniuk CJ, Mehinto AC, Denslow ND (2020) Organochlorine pesticides: agrochemicals with potent endocrine-disrupting properties in fish. Mol Cell Endocrinol 110764

Mathivanan S, Kalaikandhan R, Chidambaram AA, Sundramoorthy P (2012) Effect of vermicompost on the growth and nutrient status in groundnut (*Arachis hypogaea* L). Asian J Plant Sci Res 3(2):15–22

Mayanglambam T, Vig K, Singh DK (2005) Quinalphos persistence and leaching under field conditions and effects of residues on dehydrogenase and alkaline phosphomonoesterases activities in soil. Bull Environ Contam Toxicol 75(6):1067–1076

Meena RS, Kumar S, Datta R, Lal R, Vijayakumar V, Brtnicky M, Pathan SI (2020) Impact of agrochemicals on soil microbiota and management: a review. Land 9(2):34

Meena H, Meena RS, Rajput BS, Kumar S (2016) Response of bio-regulators to morphology and yield of clusterbean [*Cyamopsis tetragonoloba* (L.) Taub.] under different sowing environments. J Appl Nat Sci 8(2):715–718

Megharaj M, Boul HL, Thiele JH (1999) Effects of DDT and its metabolites on soil algae and enzymatic activity. Biol Fertil Soils 29(2):130–134

Mishra K, Singh K, Tripathi CM (2014) Management of municipal solid wastes and production of liquid biofertilizer through vermic activity of epigeic earthworm *Eisenia fetida*. Int J Recycl Organ Waste Agric 3(3):1–7

Misra RV, Roy RN, Hiraoka H (2003) On-farm composting methods. UN-FAO, Rome, Italy

Mohee R, Mudhoo A (2012) Energy from biomass in Mauritius: overview of research and applications. In: Waste to energy. Springer, London, pp 297–321

Monroy F, Aira M, Domínguez J (2009) Reduction of total coliform numbers during vermicomposting is caused by short-term direct effects of earthworms on microorganisms and depends on the dose of application of pig slurry. Sci Total Environ 407(20):5411–5416

Moral R, Paredes C, Bustamante MA, Marhuenda-Egea R, Bernal MP (2009) Utilization of manure composts by high-value crops: safety and environmental challenges. Bioresource technology, vol 100, No. 22, pp. 5454–5460, ISSN: 0960-8524

Moses O, Tami AG (2014) Perspective: the environmental implications of oil theft and artisanal refining in the Niger delta region. Asian Rev Environ Earth Sci 1(2):25–29

Mostafalou S, Abdollahi M (2017) Pesticides: an update of human exposure and toxicity. Arch Toxicol 91(2):549–599

Mrema EJ, Rubino FM, Mandic-Rajcevic S, Sturchio E, Turci R, Osculati A et al (2013) Exposure to priority organochlorine contaminants in the Italian general population. Part 1. Eight priority organochlorinated pesticides in blood serum. Hum Exp Toxicol 32(12):1323–1339

Munroe G (2007) Manual of on-farm vermicomposting and vermi-culture. Publication of Organic Agriculture Centre of Canada, Nova Scotia

Muratova AY, Panchenko LV, Semina DV, Golubev SN, Turkovskaya OV (2018, January) New strains of oil-degrading microorganisms for treating contaminated soils and wastes. In: IOP conference series: earth and environmental science, pp 012066–012066

Nair J, Sekiozoic V, Anda M (2006) Effect of pre-composting of kitchen waste. Biores Technol 97:2091–2095

Najar IA, Khan AB (2013) Effect of vermicompost on growth and productivity of tomato (Lycopersicon esculentum) under field conditions. Acta Biol Malaysiana 2(1):12–21

Najar IA, Khan AB, Hai A (2015) Effect of macrophyte vermicompost on growth and productivity of brinjal (*Solanum melongena*) under field conditions. Int J Recycl Organ Waste Agric 1–11

Nasiru A, Ibrahim MH, Ismail N (2014) Nitrogen losses in ruminant manure management and use of cattle manure vermicast to improve forage quality. Int J Recycl Organ Waste 3:57

Neilson RL (1965) Presence of plant growth substances in earthworms, demonstrated by the paper chromatography and went pea test. Nature 208:1113–1114

Niewiadomska A (2004) Effect of Carbendazim, Imazetapir and thiram on nitrogenase activity, the number of microorganisms in soil and yield of red clover (*Trifolium pratense* L.). Pol J Environ Stud 13(4)

Noble R, Coventry E (2005) Suppression of soil-borne plant diseases with composts: a review. Biocontrol Sci Tech 15:3–20

Nováková Z, Novák J, Kitanovski Z, Kukučka P, Smutná M, Wietzoreck M et al (2020) Toxic potentials of particulate and gaseous air pollutant mixtures and the role of PAHs and their derivatives. Environ Int 139:105634

Nwaejije EC, Hamidu I, Obiosio EO (2017) Early to middle miocene sequence stratigraphy of well-5 (OML 34), Niger Delta, Nigeria. J Afr Earth Sc 129:519–526

Oliveira CM, Auad AM, Mendes SM, Frizzas MR (2014) Crop losses and the economic impact of insect pests on Brazilian agriculture. Crop Prot 56:50–54

Önder M, Ceyhan E, Kahraman A (2011, December) Effects of agricultural practices on environment. In: International conference on biology, environment and chemistry, Singapore, pp 28–30

Onduka T, Ojima D, Ito K, Mochida K, Koyama J, Fujii K (2015) Reproductive toxicity of 1-nitronaphthalene and 1-nitropyrene exposure in the mummichog. Fundulus Heteroclitus. Ecotoxicology 24(3):648–656

Onwurah INE, Ogugua VN, Onyike NB, Ochonogor AE, Otitoju OF (2007) Crude oil spills in the environment, effects and some innovative clean-up biotechnologies. Int J Environ Res 1(4):307–320

Ordinioha B, Brisibe S (2013) The human health implications of crude oil spills in the Niger delta, Nigeria: an interpretation of published studies. Nigerian Med J J Nigeria Med Assoc 54(1):10

Ordinioha B, Sawyer W (2010) Acute health effects of a crude oil spill in a rural community in Bayelsa State, Nigeria. Niger J Med 19(2)

Orlikowski LB (1999) Vermicompost extract in the control of some soil borne pathogens. In: International symposium on crop protection, vol 64, pp 405–410

Ortiz-Hernández ML, Sánchez-Salinas E, Dantán-González E, Castrejón-Godínez ML (2013) Pesticide biodegradation: mechanisms, genetics and strategies to enhance the process. Biodegradation-Life Sci 251–287

Ozgunay H, Çolak SELİME, Mutlu MM, Akyuz F (2007) Characterization of leather industry wastes. Pol J Environ Stud 16(6)

Pajon S (2007) The worms turn-Argentina. Intermediate Technology Development Group. Case Study Series 4; Quoted in Munroe

Pal R, Chakrabarti K, Chakraborty A, Chowdhury A (2010) Degradation and effects of pesticides on soil microbiological parameters-a review. Int J Agric Res 5(8):625–643

Pandya U, Maheshwari DK, Saraf M (2014) Assessment of ecological diversity of rhizobacterial communities in vermicompost and analysis of their potential to improve plant growth. Biologia 69(8):968–976

Paoli D, Giannandrea F, Gallo M, Turci R, Cattaruzza MS, Lombardo F, Gandini L (2015) Exposure to polychlorinated biphenyls and hexachlorobenzene, semen quality and testicular cancer risk. J Endocrinol Invest 38(7):745–752

Papathanasiou F, Papadopoulos I, Tsakiris I, Tamoutsidis E (2012) Vermicompost as a soil supplement to improve growth, yield and quality of lettuce. J Food Agric Environ 10(2):677–682

Pathma J, Sakthivel N (2012) Microbial diversity of vermicompost bacteria that exhibit useful agricultural traits and waste management potential. Springerplus 1:26

Pierre V, Phillip R, Margnerite L, Pierrette C (1982) Anti-bacterial activity of the haemolytic system from the earthworms *Eisenia foetida* Andrei. Invertebr Pathol 40(1):21–27

Pirsaheb M, Khosravi T, Sharafi K (2013b) Domestic scale vermicomposting for solid waste management. Int J Recycl Organ Waste Agric 2(1):1–5

Pirsaheb M, Khosravi T, Sharafi K, Babajani L, Rezaei M (2013c) Measurement of heavy metals concentration in drinking water from source to consumption site in Kermanshah—Iran. World Appl Sci J 21(3):416–423

Pirsaheb et al (2013a) Int J Recycl Organ Waste Agric 2:4

Prescott ML, Harley JP, Klan AD (1996) Industrial microbiology and biotechnology. Microbiology, 3rd edn. Wim C Brown Publishers, Chicago, pp 923–927

Qiao M, Wang C, Huang S, Wang D, Wang Z (2006) Composition, sources, and potential toxicological significance of PAHs in the surface sediments of the Meiliang Bay, Taihu Lake, China. Environ Int 32(1):28–33

Raguchander T, Rajappan K, Samiyappan R (1998) Influence of biocontrol agents and organic amendments on soybean root rot. Int J Trop Agri 16:247–252

Rao MA, Scelza R, Scotti R, Gianfreda L (2010) Role of enzymes in the remediation of polluted environments. J Soil Sci Plant Nutr 10(3):333–353

Ravindran N, Jindal N (2008) Limited feedback-based block diagonalization for the MIMO broadcast channel. IEEE J Sel Areas Commun 26(8):1473–1482

Reddy MV, Ohkura K (2004) Vermicomposting of rice-straw and its effects on sorghum growth. Trop Ecol 45(2):327–331

Reddy KS, Shantaram MV (2005) Potentiality of earthworms in composting of sugarcane by products. Asian J Microbiol Biotechnol Environ Sci 7:483–487

Redfern LK, Gardner CM, Hodzic E, Ferguson PL, Hsu-Kim H, Gunsch CK (2019) A new framework for approaching precision bioremediation of PAH contaminated soils. J Hazard Mater 378:120859

Reganold JP, Papendick RIP, James F (1990) Sustainable agriculture. Sci Am 262(6):112–120

Ren XM, Kuo Y, Blumberg B (2020) Agrochemicals and obesity. Mol Cell Endocrinol 110926

Rezek J, in der Wiesche C, Mackova M, Zadrazil F, Macek T (2008) The effect of ryegrass (*Lolium perenne*) on decrease of PAH content in long term contaminated soil. Chemosphere 70(9):1603–1608

Riah W, Laval K, Laroche-Ajzenberg E, Mougin C, Latour X, Trinsoutrot-Gattin I (2014) Effects of pesticides on soil enzymes: a review. Environ Chem Lett 12(2):257–273

Rodriguez J, García A, Poznyak T, Chairez I (2017) Phenanthrene degradation in soil by ozonation: effect of morphological and physicochemical properties. Chemosphere 169:53–61

Rostami M, Olia M, Arabi M (2014) Evaluation of the effects of earthworm *Eisenia fetida*-based products on the pathogenicity of root-knot nematode (*Meloidogyne javanica*) infecting cucumber. Int J Recycl Organ Waste 3:58

Roy A, Dutta A, Pal S, Gupta A, Sarkar J, Chatterjee A et al (2018) Biostimulation and bioaugmentation of native microbial community accelerated bioremediation of oil refinery sludge. Biores Technol 253:22–32

Saadoun IM (2015) Impact of oil spills on marine life. Emerging pollutants in the environment-current and further implications, pp 75–104

Sam K, Zabbey N (2018) Contaminated land and wetland remediation in Nigeria: opportunities for sustainable livelihood creation. Sci Total Environ 639:1560–1573

Sannino F, Gianfreda L (2001) Pesticide influence on soil enzymatic activities. Chemosphere 45(4–5):417–425

Santos A, Flores M (1995) Effects of glyphosate on nitrogen fixation of free-living heterotrophic bacteria. Lett Appl Microbiol 20(6):349–352

Sasek V, Bhatt M, Cajthaml T, Malachova K, Lednicka D (2003) Compost-mediated removal of polycyclic aromatic hydrocarbons from contaminated soil. Arch Environ Contam Toxicol 44(3):0336–0342

Scheuerell SJ, Sullivan DM, Mahaffee WF (2005) Suppression of seedling damping-off caused by *Pythium ultimum, P. irregulare*, and *Rhizoctonia solani* in container media amended with a diverse range of Pacific Northwest compost sources. Phytopathology 95:306–315

Sen B, Chandra TS (2007) Chemolytic and solid-state spectroscopic evaluation of organic matter transformation during vermicomposting of sugar industry wastes. Biores Technol 98:1680–1683

Shah RU, Abid M, Qayyum MF, Ulla R (2015) Dynamics of chemical changes through production of various composts/vermicompost such as farm manure and sugar industry wastes. Int J Recycl Organ Waste Agric 4:39–51

Sheoran RS, Rana DS (2005) Relative efficacy of vermicompost and farmyard manure integrated with inorganic fertilizers for sustainable productivity of forage sorghum (*Sorghum bicolor* (L.) Moench). Acta Agronomica Hungarica 53(3):303–308

Short JW, Heintz RA (1997) Identification of exxon valdez oil in sediments and tissue from prince william sound and the north western gulf of William based in a PAH weathering model. Environ Sci Technol 31:2375–2384

Siles JA, Margesin R (2018) Insights into microbial communities mediating the bioremediation of hydrocarbon-contaminated soil from an Alpine former military site. Appl Microbiol Biotechnol 102(10):4409–4421

Singh RD (1993) Harnessing the earthworms for sustainable agriculture. Institute of National Organic Agriculture, Pune, India, pp 1–16

Singh SK, Haritash AK (2019) Polycyclic aromatic hydrocarbons: soil pollution and remediation. Int J Environ Sci Technol 1–24

Singh J, Kalamdhad AS (2013) Reduction of bioavailability and leachability of heavy metals during vermicomposting of water hyacinth. Environ Sci Pollut Res 20(12):8974–8985

Singh R, Sharma RR, Kumar S, Gupta RK, Patil RT (2008) Vermicompost substitution influences growth, physiological disorders, fruit yield and quality of strawberry (Fragaria ananassa Duch.). Bioresour Technol 99:8507–8511

Singh J, Kaur A, Vig AP, Rup PJ (2010) Role of *Eisenia fetida* in rapid recycling of nutrients from bio sludge of beverage industry. Ecotoxicol Environ Saf 73(3):430–435

Singh A, Jain A, Sarma BK, Abhilash PC, Singh HB (2013) Solid waste management of temple floral offerings by vermicomposting using *Eisenia fetida*. Waste Manage 33(5):1113–1118

Singhai PK, Sarma BK, Srivastava JS (2011) Biological management of common scab of potato through Pseudomonas species and vermicompost Bio. Control 57:150–157

Sinha RK, Agarwal S, Chauhan K, Valani D (2010) The wonders of earthworms and its vermicompost in farm production: Charles Darwin's 'friends of farmers', with potential to replace destructive chemical fertilizers. Agric Sci 1(02):76

Sinha RK, Herat S, Valani D, Chauhan K (2009) Vermiculture and sustainable agriculture. Am-Eurasian J Agric Environ Sci (IDOSI Publication) 1–55. Sodh, SaAmiksha aur Mulyankan:307–311

Sivaram AK, Logeshwaran P, Lockington R, Naidu R, Megharaj M (2019) Low molecular weight organic acids enhance the high molecular weight polycyclic aromatic hydrocarbons degradation by bacteria. Chemosphere 222:132–140

Soltani N, Keshavarzi B, Moore F, Tavakol T, Lahijanzadeh AR, Jaafarzadeh N, Kermani M (2015) Ecological and human health hazards of heavy metals and polycyclic aromatic hydrocarbons (PAHs) in road dust of Isfahan metropolis, Iran. Sci Total Environ 505:712–723

Sparks TC (2013) Insecticide discovery: an evaluation and analysis. Pestic Biochem Physiol 107(1):8–17

Sparks TC, Lorsbach BA (2017) Perspectives on the agrochemical industry and agrochemical discovery. Pest Manag Sci 73(4):672–677

Subhani A, El-ghamry AM, Changyong H, Jianming X (2000) Effects of pesticides (Herbicides) on soil microbial biomass-a revievy. Pak J Biol Sci 3(5):705–709

Suhane RK (2007) Vermicompost. Publication of Rajendra Agriculture University, Pusa, Bihar, India, p 88

Sun FL, Wang YS, Sun CC, Peng YL, Deng C (2012) Effects of three different PAHs on nitrogen-fixing bacterial diversity in mangrove sediment. Ecotoxicology 21(6):1651–1660

Sunil K, Rawat CR, Shiva D, Suchit KR (2005) Dry matter accumulation, nutrient uptake and changes in soil fertility status as influenced by different organic and inorganic sources of nutrients to forage sorghum (*Sorghum bicolor*). Indian J Agric Sci 75(6):340–342

Surrage VA, Lafrenie C, Dixon M, Zheng Y (2010) Benefits of vermicompost as a constituent of growing substrates used in the production of organic greenhouse tomatoes. HortScience 45(10):1510–1515

Suthar S (2007a) Production of vermifertilizer from guar gum industrial wastes by using composting earthworm Perionyx sansibaricus (Perrier). Environmentalist 27(3):329–335

Suthar S (2007b) Vermicomposting potential of *Perionyx sansibaricus* (Perrier) in different waste materials. Biores Technol 98(6):1231–1237

Suthar S (2009a) Bioremediation of agricultural wastes through vermicomposting. Bioremediat J 13(1):21–28

Suthar S (2009b) Vermistabilization of municipal sewage sludge amended with sugarcane trash using epigeic *E. fetida* (Oligochaeta). J Hazard Mater 163:199–206

Sverdrup LE, Nielsen T, Krogh PH (2002) Soil ecotoxicity of polycyclic aromatic hydrocarbons in relation to soil sorption, lipophilicity, and water solubility. Environ Sci Technol 36(11):2429–2435

Szczech M (1999) Supressiveness of vermicompost against Fusarium wilt of tomato. J Phytopathol 147:155–161

Szczech M, Smolinska U (2001a) Comparison of suppressiveness of vermicompost produced from ani- mal manures and sewage sludge against *Phytophthora nicotianae* Breda de Haar var. nicotianae. J Phytopathol 149:77–82

Szczech M, Smolinska U (2001b) Comparison of suppressiveness of vermicompost produced from animal manures and sewage sludge against *Phytophthora nicotianae* Breda de Haar var. nicotianae. J Phytopathol 149:77–82

Tavalieri YE, Galoppo GH, Canesini G, Luque EH, Muñoz de Toro MM (2020) Effects of agricultural pesticides on the reproductive system of aquatic wildlife species, with crocodilians as sentinel species. Mol Cell Endocrinol 110918

Termorshuizen AJ, Van Rijn E, Van Der Gaag DJ, Alabouvette C, Chen Y, Lagerlöf J, Malandrakis AA, Paplomatas EJ, Rämert B, Ryckeboer J, Steinberg C, Zmora- NS (2006) Supressiveness of 18 composts against 7 pathosystems: variability in pathogen response. Soil Biol Biochem 38:2461–2477

Tognetti C, Mazzarino MJ, Laos F (2007) Improvingthequalityofmunicipal organic waste compost. Bioresour Technol 98:1067–1076

Tomati U, Galli E (1995) Earthworms, soil fertility and plant productivity. Acta Zool Fenn 196:11–14

Tomati U, Grappelli A, Galli E (1988) The hormone-like effect of earthworm casts on plant growth. Biol Fertil Soils 5:288–294

Tomati U, Grappelli A, Galli E (1983) Fertility factors in earthworm humus. In: Proceedings in international symposium on agriculture environment. Prospects in earthworm farming. Publication Ministero della Ricerca Scientifica e Technologia, Rome, pp 49–56

Tomati V, Grappelli A, Galli E (1987) The presence of growth regulators in earthworm-worked wastes. In: Proceeding of international symposium on 'Earthworms', pp 423–436

Trillas MI, Casanova E, Cotxarrera L, Ordovás J, Borrero C, Avilés M (2006) Composts from agricultural waste and the *Trichoderma asperellum* strain T-34 suppress Rhizoctonia solani in cucumber seedlings. Biol Control 39:32–38

Trudeau VL, Thomson P, Zhang WS, Reyaud S, Navarro-Martin L, Langlois VS (2020) Agrochemicals disrupt multiple endocrine axes in amphibians. Mol Cell Endocrinol 110861

Tsatsakis AM, Tzatzarakis MN, Tutudaki M, Babatsikou F, Alegakis AK, Koutis C (2008) Assessment of levels of organochlorine pesticides and their metabolites in the hair of a Greek rural human population. Hum Exp Toxicol 27(12):933–940

Udotong JIR, Udoudo UP, Udotong IR (2017) Effects of oil and gas exploration and production activities on production and management of seafood in Akwa Ibom State, Nigeria. J Environ Chem Ecotoxicol 9(3):20–42

Uma B, Malathi M (2009) Vermicompost as a soil supplement to improve growth and yield of Amaranthus species. Res J Agric Biol Sci 5:1054–1060

Uqab B, Mudasir S, Nazir R (2016) Review on bioremediation of pesticides. J Bioremediat Biodegrad 7(343):2

Vandenberg LN, Najmi A, Mogus JP (2020) Agrochemicals with estrogenic endocrine disrupting properties: lessons Learned? Mol Cell Endocrinol 110860

Veltman JA, Brunner HG (2012) De novo mutations in human genetic disease. Nat Rev Genet 13(8):565–575

Vig AP, Singh J, Wani SH, Dhaliwal SS (2011) Vermicomposting of tannery sludge mixed with cattle dung into valuable manure using earthworm *Eisenia fetida* (Savigny). Biores Technol 102:7941–7945

Vijaya KS, Seethalakshmi S (2011) Contribution of Parthenium vermicompost in altering growth, yield and quality of *Alelmoschus esculentus* (I) Moench. Adv Biotech 11(02):44–47

Warner GR, Mourikes VE, Neff AM, Brehm E, Flaws JA (2020) Mechanisms of action of agrochemicals acting as endocrine disrupting chemicals. Mol Cell Endocrinol 502:110680

World Health Organization (2017) Agrochemicals, health and environment: directory of resources

Wu G, Kechavarzi C, Li X, Sui H, Pollard SJ, Coulon F (2013) Influence of mature compost amendment on total and bioavailable polycyclic aromatic hydrocarbons in contaminated soils. Chemosphere 90(8):2240–2246

Xiao X, Clark JM, Park Y (2017) Potential contribution of insecticide exposure and development of obesity and type 2 diabetes. Food Chem Toxicol 105:456–474

Yabrade M, Tanee FG (2016) Assessing the impact of artisanal petroleum refining on vegetation and soil quality: a case study of warri South Wetland of Delta State, Nigeria. Res J Environ Toxicol 10(4):205–212

Yale RL, Sapp M, Sinclair CJ, Moir JWB (2017) Microbial changes linked to the accelerated degradation of the herbicide atrazine in a range of temperate soils. Environ Sci Pollut Res 24(8):7359–7374

Yardim EN, Arancon NA, Edwards CA, Oliver TJ, Byrne RJ (2006) Suppression of tomato hornworm (Manduca quinquemaculata) and cucumber beetles (*Acalymma vittatum* and *Diabotrica undecimpunctata*) populations and damage by vermicomposts. Pedobiologia 50:23–29

Zaller JG (2007) Vermicompost as a substitute for peat in potting media: effects on germination, biomass allocation, yields and fruit quality of three tomato varieties. Sci Hortic 112:191–199

Zeyer J, Ranganathan LS, Chandra TS (2004) Pressmud as biofertilizer for improving soil fertility and pulse crop productivity. ISCB—Indo-Swiss collaboration in Biotech

Zhang Y, Tao S (2009) Global atmospheric emission inventory of polycyclic aromatic hydrocarbons (PAHs) for 2004. Atmos Environ 43(4):812–819

Zheng Z, Obbard JP (2003) Oxidation of polycyclic aromatic hydrocarbons by fungal isolates from an oil contaminated refinery soil. Environ Sci Pollut Res 10(3):173–176

Zucchi M, Angiolini L, Borin S, Brusetti L, Dietrich N, Gigliotti C et al (2003) Response of bacterial community during bioremediation of an oil-polluted soil. J Appl Microbiol 94(2):248–257

Chapter 13
Biochar-Based Remediation of Heavy Metal Polluted Land

Abhishek Kumar and Tanushree Bhattacharya

Abstract The excessive use of heavy metals has led to the problem of pollution of land by heavy metals. The non-degradability, persistence, bioavailability and high mobility of heavy metals make them dangerous to human health and environment. In the previous decades, biochar has been suggested to remove the heavy metals from the soil effectively. Biochar is a carbonized material prepared by thermal treatment of a biomass feedstock. The variation in feedstock and thermal treatment affects the properties of the char produced. The properties of high sorption capacity, large surface area, high porosity, alkaline pH and remarkable oxygen-containing surface functional groups enable Biochar to minimize the mobility and bioavailability of the heavy metals. The high stability of biochar aids in removing the heavy metals for a long period of time. Mechanisms such as ion exchange, precipitation, diffusion, complex formation, electrostatic interaction and sorption, help in removal of heavy metals from the soil. Additionally, biochar could help in waste management, bioenergy production, crop production enhancement and climate change mitigation, which are indicative of the wide-ranging advantages associated with biochar production and its application. Keeping these things in mind, the chapter was conceptualized to review the developments in the field of biochar application for remediation of heavy metal polluted sites. The chapter has focussed upon its production, modification methods, physicochemical properties, and heavy metal removal mechanisms utilized by biochar. Additionally, the impact of biochar on mobility and bioavailability of heavy metals and case studies across the various parts of the world have been explored. Lastly, applications other than heavy metal removal, advantages and risks associated with biochar application and future scope for biochar production and application have been discussed.

Keywords Heavy metals · Biochar · Pyrolysis · Remediation · Sorption · Climate change

A. Kumar · T. Bhattacharya (✉)
Department of Civil and Environmental Engineering, Birla Institute of Technology, Mesra, Ranchi, Jharkhand 835215, India
e-mail: tbhattacharya@bitmesra.ac.in

13.1 Introduction

Our planet has seen emergence of numerous disasters inclusive of climate change, depletion of natural resources and pollution (Kumar et al. 2021b, c). Each of the issues is threatening for the survival of the planet and sustenance of the organisms thriving on it. Heavy metal pollution is one such significant issue that is undesirable for the twenty-first century and affects the socio-economic lives of people (Bhattacharya et al. 2021; Kumar et al. 2021a). The persistence and bioavailability of heavy metals make them toxic for the living organisms (Zhang et al. 2013). A number of remediation methods have been developed to remove the heavy metals from the environment (Pandey and Singh 2019). These methods could be physical, chemical or biological (Khalid et al. 2017; Pandey and Singh 2019). The physical methods include vitrification, isolation, soil replacement and electro-kinetic remediation. The chemical methods are inclusive of encapsulation, soil washing and chemical immobilization. The biological methods include phytoremediation (Pathak et al. 2020; Pandey and Bajpai 2019), bioremediation and biochar-based remediation (Dwibedi et al. 2022). Biochar is very optimistic technique for removing the heavy metals from soil and water (Dwibedi et al. 2022).

Biochar is a carbon–neutral recalcitrant substance obtained from the thermal treatment of a carbonaceous biomass (Manyà 2012; IBI 2015). Depending upon the type of biomass and thermal treatment technique used, properties of biochar vary (Tang et al. 2013). Biochar helps in reducing heavy metal pollution by decreasing their mobility and bioavailability (Kumar and Bhattacharya 2021,2022). Further, biochar improves the quality of soil, which helps in improving the soil and plant productivity (Lehmann et al. 2006). Additionally, biochar could help in waste management by consuming the waste materials for production of biochar; climate change mitigation by carbon sequestration and greenhouse gas emission reduction; fossil fuel management by biofuel production; and food security management by enhanced crop production (Lehmann et al. 2011; Titirici et al. 2012; Zhang et al. 2013; Mohan et al. 2014; Windeatt et al. 2014; Hossain 2016; Lee et al. 2018; Manyà et al. 2018). Therefore, production and application of biochar could be a sustainable solution for a number of threatening issues in addition to remediating heavy metal polluted soils.

13.2 Biochar and Its Production

Biochar is a stable carbonaceous residue (IBI 2015), obtained after thermal treatment of carbon-containing feedstock (Kumar et al.2022a, b; Shaikh et al. 2022b, a). Biochar is different from 'Amazonian dark earth', i.e. Terra preta, in structure and composition. Terra preta is produced by mixing low-temperature char with plant residues, bones, faeces and compost (Balée et al. 2016a, b). Identification of Terra preta's nutritional significance, promoted the production and use of biochar for various applications (Glaser et al. 2002).

A number of thermal treatment techniques have been used for biochar production (Kumar et al. 2020). These techniques include pyrolysis, combustion, torrefaction, gasification and carbonization (Meyer et al. 2011). Pyrolysis has been the most widely used method for producing biochar. It involves oxygen-deficient conditions and could be carried out in a kiln or furnace. The wide utilization of pyrolysis for biochar production is due to its efficiency and simplicity (Cha et al. 2016).

The properties of biochar vary depending upon the treatment method, conditions and the type of feedstock used (Sahota et al. 2018; Zhang et al. 2018a). Some of the properties significant for heavy metal removal are inclusive of large surface area, high porosity, high cation exchange capacity, a non-carbonized fraction and oxygen-containing surface functional groups (Mukherjee et al. 2011; Ahmad et al. 2014). Application of biochar for removing heavy metals from polluted lands has emerged in the recent times (Mohan et al. 2014).

13.2.1 Feedstock Variation

Theoretically, biochar could be produced by any type of biomass, but the costs of production and the applicability of biomass for compost and biofuel production, restrict the range of feedstock for biochar production (Kuppusamy et al. 2016; Tripathi et al. 2016). Additionally, feedstock composition and its calorific value are determined for biochar production. Some of the feedstock biomasses used for the production of biochar are crop residues, kitchen waste, animal litter, poultry litter, sewage sludge, rubber tyres and algae (Beesley and Marmiroli 2011; Cantrell et al. 2012; Lu et al. 2012; Ghani et al. 2013; Xu et al. 2013a; Zhao et al. 2013; Mazac 2016). Importantly, utilization of waste material for production of biochar would assist in waste management by decreasing generation of waste, which could decrease the pollution of soil and groundwater, increase the levels of sanitation and reduce the number of landfill sites.

Decreasing the moisture content in feedstock is necessary to increase the feasibility of the thermal treatment of biochar (Bryden and Hagge 2003; Lv et al. 2010). Moisture content in the feedstock above 30% depletes the rate of heating, thereby increases the time needed to achieve the conditions necessary for thermal treatment. Therefore, it is vital to decrease the moisture content in feedstock by drying it through natural or human-assisted means. Naturally, it could be dried under the sun or by the influence of wind. Feedstock drying through human assistance incorporates use of microwave ovens or instruments that generate heat. However, natural ways must be preferred to decrease the energy consumption burden, which could help in tackling energy security partially.

Thermal treatment of feedstock decomposes hemicellulose and cellulose at 200–315 °C and 315–400 °C, respectively (Sadaka et al. 2014). Lignin decomposition occurs beyond 400 °C. Therefore, feedstock rich in hemicellulose and cellulose could produce biochar at low-temperature thermal treatment. However, low-temperature chars are considered to be less efficient for heavy metal removal because of the low surface area, low porosity, less cation exchange capacity and less oxygen-containing surface functional groups obtained at lower temperatures (Igalavithana et al. 2017; Weber and Quicker 2018; Zhang et al. 2018b). Therefore, high-temperature chars are preferred for heavy metal removal purposes due to their high efficiency and efficacy. High lignin content in feedstock is necessary to increase the yield of biochar production at high-temperature thermal treatments (Angin 2013; Shivaram et al. 2013). Therefore, feedstocks with less moisture content and high lignin content are preferred for the production of biochar for remediating heavy metal-polluted soils.

13.2.2 Thermal Treatment

The thermal treatment processes involve thermal conservation of biomass feedstock. The different thermal treatment techniques are torrefaction, combustion, gasification, carbonization and pyrolysis (Meyer et al. 2011; Zhang et al. 2013). The various treatment methods have been summarized in Table 13.1. Low-temperature thermal treatment of feedstock in oxygen-depleted conditions is referred to as torrefaction. The temperatures are in the range of 200–300 °C. Torrefaction could be used for feedstock pre-treatment in gasification to enhance the quality of biochar produced.

Combustion involves direct burning of the feedstock to convert the stored chemical energy into thermal energy. However, combustion needs pre-treatment due to the low yield of biochar production (McKendry 2002). Gasification involves thermal treatment of feedstock at very high temperatures ranging from 700 to 900 °C. The feedstock is partially oxidized in gasification and the carbon content is transformed into a gaseous product apart from generation of soils and liquid products. Gasification results in 85% syngas, 10% biochar and 5% bio-oil as products (Neves et al. 2011; Asensio et al. 2013).

Carbonization is majorly of two types—flash carbonization and hydrothermal carbonization. In flash carbonization, feedstock is heated at 350–650 °C and elevated pressure for time less than 30 min. Flash carbonization yields syngas and biochar in equal amounts (Antal et al. 2003; Asensio et al. 2013). On the other hand, in hydrothermal carbonization, the wet biomass is thermally treated at elevated pressure and temperature. It results in conversion of wet biomass into hydrothermal carbon, i.e. hydrochar, along with the release of energy (Wang et al. 2018b).

Pyrolysis is the most widely used method for thermal treatment of feedstock. Pyrolysis involves thermal treatment of feedstock in oxygen-depleted conditions at 300–900 °C. Oxygen-deficit conditions allow feedstock to be heated above the thermal stability limits, resulting in formation of biochar with high stability. Additionally, bio-oil and syngas is also obtained in pyrolysis. As pyrolysis proceeds, the heat decomposes and devolatilizes the feedstock constituents. Oxygen-rich functional groups such as hydroxyl and carboxyl are formed on the surface after pyrolysis (Ekström et al. 1985). Pyrolysis could be divided into slow, intermediate or fast depending upon the heating rate.

Table 13.1 Various types of thermal treatment methods for production of solid (biochar), liquid (bio-oil) and gaseous (syngas) products

Treatment method	Feedstock used for production	Products obtained	References
Torrefaction	Rice husk, bagasse, peanut husk, sawdust, & water hyacinth	Solid	Pimchuai et al. (2010)
Combustion	Waste biomass	Solid and thermal energy	McKendry (2002), Caillat and Vakkilainen (2013)
Gasification	Lignocellulose rich plant biomass; Sedum alfredii	Gas	Pröll et al. (2007), Balat et al. (2009), Cui et al. (2018)
Flash carbonization	Woods (Oak & Leucaena); agricultural waste (corncob & macadamia nut shells)	Gas & solid	Antal et al. (2003), Asensio et al. (2013)
Hydrothermal carbonization	Agricultural waste; eucalyptus sawdust & barley straw	Solid (Hydrochars)	Sevilla et al. (2011), Titirici et al. (2012)
Slow pyrolysis	Softwood chip & grass; Crop residues	Solid	Onay and Kockar (2003), Windeatt et al. (2014), Behazin et al. (2016)
Fast pyrolysis	Corn cobs & Stover; Rice straw	Solid & liquid	Onay and Kockar (2003), Mullen et al. (2010), Eom et al. (2013)
Flash pyrolysis	Rapeseed; Sunflower oil cake	Liquid & gas	Yorgun et al. (2001), Onay and Kockar (2003)
Slow steam pyrolysis	Vegetal waste, switch grass	Solid & gas	Giudicianni et al. (2013)

Feedstock is heated with moderate heating rate at 400–500 °C in slow pyrolysis. Feedstock is heated at 500–650 °C in intermediate pyrolysis. Fast pyrolysis involves very rapid heating rate, where feedstock is heated up to 800–1200 °C. Slow pyrolysis yields the maximum amount of biochar (Tripathi et al. 2016). Apart from treatment temperature and heating rate, pyrolysis also depends upon vapour residence time, pressure, feedstock particle size and the technique used for production such as burning in a kiln or electrical heating in a furnace (Asensio et al. 2013). Rate of removal of volatile gases during pyrolysis affects the vapour residence time and the occurrence of secondary reactions on the surface of biochar, which consequently affects the properties of biochar produced (Meyer et al. 2011). It must be noted that pyrolysis is considered as the most efficient and cost-effective technique for biochar production (Cha et al. 2016).

13.3 Biochar Modification Methods

Application of biochar for removal of contaminants may need improvements for better remediation results. Recently, biochar modification has received attention for improving remediation performance in char (Alam et al. 2018; Shaikh et al. 2021). Some of the modification methods are digestion, oxidation, magnetization and activation. These methods affect the surface area, porosity, cation exchange capacity, pH and surface functional groups of biochar. These properties could be compared for the evaluation of heavy metal remediation efficiency in chars.

For activation of biochar, steam activation is an effective method. The pore volume is enhanced and the pore structure becomes complex after biochar activation. Hass et al. (2012) reported that steam activation increases the surface area and pH of biochar. They stated that steam activation of char prepared at 350 °C is similar in efficacy to char prepared at 700 °C in terms of their liming effect.

Magnetization is another efficient method reported for enhancing the sorption potential of biochar. Magnetization renders strong ferromagnetic capacity in biochar. Additionally, magnetization is beneficial in terms of the ability for its recollection by magnetic separation and reutilization. Chen et al. (2011) prepared magnetic biochar by chemical co-precipitation of orange peel powder with ferric and ferrous ions followed by their thermal treatment. They reported that the magnetic biochar had enhanced pore size and was more potent in removing contaminants. The ferric oxide particles on char surface aid in sorption enhancement by providing sites for electrostatic interaction. Wang et al. (2015) prepared magnetic biochar from pinewood and reported that the magnetized biochar could be used for removing metallic contaminants.

Oxidation is another method utilized efficiently for enhancing the sorption potential of biochar. Oxidation is achieved by the addition of oxidants in the pre- or post-treatment stages. Some of the oxidants used are hydrogen peroxide, potassium permanganate and nitric acid (Xue et al. 2012; Li et al. 2014). Oxidation facilitates acidic functional groups to the char surface after treatment. Li et al. (2014) reported that nitric acid is more effective for biochar modification by oxidation treatment in comparison to potassium permanganate.

Lastly, digestion is another method used effectively for enhancing the sorption capacity of biochar. Anaerobic digestion treatment of feedstock improves the sorption capacity of char in comparison to undigested feedstocks. Inyang et al. (2010) modified bagasse by anaerobic digestion and observed that the digested chars had greater cation exchange capacity, more surface area, surplus negative surface charges and higher pH than undigested chars. Similar results were reported by Yao et al. (2011) in the beetroot tailings-derived biochar and Inyang et al. (2012) in the dairy manure-derived biochar. These results are indicative of the enhanced char properties after modification by digestion treatment.

13.4 Properties of Biochar

13.4.1 Composition

Biochar composition depends on the composition of feedstock, rate of heating and treatment temperatures involved. Feedstock is generally composed of lignin and holocellulose, i.e. hemicellulose and cellulose. Thermal treatment decomposes the hemicellulose and cellulose in feedstock at 200–315 °C and 315–400 °C, respectively (Sadaka et al. 2014), while lignin decomposition occurs beyond 400 °C. Therefore, thermal treatment temperature could affect the biochar composition, which could affect the physical and chemical properties of biochar.

Thermal treatment of the biomass results in detachment of oxygen and hydrogen-containing surface functional groups, resulting in the decrease in their ratios with respect to carbon. Hydrogen, oxygen and nitrogen contents decrease with an increase in treatment temperature (Sun et al. 2014). Carbon contents increase in biochar at high treatment temperatures (Vassilev et al. 2010). The increase in treatment temperature increases the loss of volatile matter by enhancement of devolatilization and decomposition of the char matrix. Therefore, the volatile matter decreases at higher treatment temperatures in biochar (Pimchuai et al. 2010; Weber and Quicker 2016). Elements such as magnesium, calcium and potassium increase in biochar with an increase in the treatment temperature (Sun et al. 2014).

Addition of biochar to soil enhances dissolved organic carbon content. Such an enhancement stimulates the activity of micro-organisms in soils. Additionally, there is an alteration in redox processes and biochemical reactions. These changes affect the impact of biochar on soil contaminants (Beesley and Dickinson 2011; Choppala et al. 2012; Qian et al. 2016). Park et al. (2011b) reported that the dissolved organic matter increases the mobilization of copper, which could be indicative of a detrimental effect of biochar addition.

With an increase in the treatment temperatures, aromaticity in the char enhances. This could be due to thermodynamic stability of aromatic carbon at high treatment temperatures (Conti et al. 2014). The aromatic structures help in increasing the heavy metal remediation efficiency by enhancing the sorption potential of organic and inorganic contaminants (Wang et al. 2016).

13.4.2 pH and Ash Content

The removal of the acidic functional groups on the surface of char enhances its alkalinity (Fidel et al. 2017). The increase in alkalinity is accompanied by an increase in pH of the char. pH values as high as 10–12 are obtained for thermal treatment at high temperatures. A high pH enables the char to neutralize acidic soils, thereby increasing the availability of arable lands, which could be extremely significant in the present context, comprising of a rise in pollution and the need for more food crop production.

Increase in pH of the char could also be an outcome of the rise in the ash content at high treatment temperatures. Ash contributes in increasing the alkalinity in animal manure biochar. Volatilization of organic acids and removal of acidic functional groups contributes in the high pH in the agricultural waste biochar (Wang et al. 2019). In comparison to plant biomass-derived biochar, animal-derived biochar has a higher carbonate and ash content which could be responsible for the high pH (Rajkovich et al. 2012). Therefore, feedstock composition affects the pH of char (Wang et al. 2019). Ash could be composed of oxides of alkaline and alkali metals such as silicon, aluminium, potassium, calcium and magnesium (Vassilev et al. 2013b). High ash content could be detrimental for the applicability of the char in industrial applications due to the health problems related to ash.

Ash content regulatesion exchange in the soil matrix, while soil pH and alkalinity regulate co-precipitation (Wang et al. 2018a). Heavy metals are stable in an alkaline environment while unstable in an acidic environment. Addition of biochar to soil facilitates the carbonates and oxygen-containing functional groups, thereby increasing the pH in the soil making it alkaline. Such alkaline conditions enhance the stability of heavy metals. Further, the functional groups provide negative charges on the char surface, aiding in heavy metal removal (Yuan et al. 2011).

13.4.3 Cation exchange capacity

Majority of the functional groups on surface of the char provide a negative charge, indicating its anionic nature. It enables the char to attract the cations. Therefore, char produced at low treatment temperature has a high cation exchange capacity (Mukherjee et al. 2011). Rajkovich et al. (2012) reported that cation exchange capacity is greater in biochar derived from oak, corn stover, or manure than compared to biochar derived from hazelnut shells, paper mill waste, or food waste. Cow manure-derived char has a low cation exchange capacity in comparison to plant biomass-derived char due to their high ash content and low carbon/nitrogen content (Wang et al. 2019). Further, it helps in capturing the contaminants, thereby assisting in the remediation of polluted lands. It could also help in reducing the contaminant levels in the plants by reducing their availability for plant uptake (Cushman and Robertson-Palmer 1998; Liang et al. 2006).

13.4.4 Surface Area, Porosity and Pore Volume

The porosity and surface area of char depends on the feedstock used, the treatment temperature involved and the rate of heating (Manna et al. 2020). The thermal treatment of feedstock produces a porous biochar by releasing volatile gases and decomposition of the biomass matrix. With an increase in treatment temperature, porosity of the char increases. However, treatment temperatures above 800–1000 °C break the cell structures in the biomass, leading to a reduction in porosity at high treatment temperatures (Cetin et al. 2004). Treatment temperatures increase the pore volume in biochar (Fu et al. 2012). Furthermore, micropores (0.05–0.0001 μm) form volume above 80% in the char. The abundance of pores in biochar helps in sorption of heavy metals on the outer sphere and its transport to the inner sphere (Houben et al. 2013; Yin et al. 2016).

With an increase in treatment temperature, surface area of the char increases. Similar to the porosity, surface area of biochar decreases at temperatures above 800–1000 °C (Cetin et al. 2004). The decrease in surface area could be a result of shrinking solid matrix (Pulido-Novicio et al. 2001). Cao and Harris (2010) stated that the surface area of a char derived from dairy manure is less than a char obtained from plant biomass due to the abundance of carbon in its matrix. The abundance of organic carbon also enables the char derived from plant biomass to have a very high porosity. High porosity and surface area increase the heavy metal removal capability of biochar by enhancing the adsorption capacity (Rouquerol et al. 1999). A high surface area aids in increasing the cation exchange capacity, water holding potential and nutrient retention capacity of biochar (Weber and Quicker 2018). Surface area also plays a vital role in affecting the microbial community present in the soil matrix by providing pores to the microbes for survival (Igalavithana et al. 2017). The surface of biochar could develop both positive and negative charges, which could help in the sorption of both positively and negatively charged metal species such as chromium and arsenic. This is brought about by the stimulation of microbial processes which helps in the promotion of redox reaction in the soil (Solaiman and Anawar 2015).

13.4.5 *Mechanical Stability and Grindability*

Thermal treatment of feedstock decreases its mechanical stability due to an increase in the porosity and a decrease in structural complexity of char, i.e. the solid product formed after thermal treatment of feedstock (Byrne and Nagle 1997). Biochar becomes brittle and grindable due to the decrease in mechanical stability. High hemicellulose content in feedstock produces a highly grindable char. On the contrary, high lignin content in the feedstock produces biochar which is less brittle and has high mechanical stability (Emmerich and Luengo 1994).

High mechanical stability could assist the char in replacing coal for industrial applications. High stability is also significant for the carbon sequestration for a long period of time. The extended stability of biochar in the soil does not have any negative impact on the heavy metal removal. In a study by Li et al. (2016), biochar prepared from hardwood was applied to cadmium and copper contaminated soils and incubated for 3 years. The biochar application reduced the concentration of cadmium and copper by 58% and 64% in the 1st year, followed by a further decrease in the 2nd and 3rd years. These results are indicative of heavy metal removal from the soil and absence of negative impacts of biochar ageing on the soil.

Grindable nature of char affects its particle size distribution. Particle size affects the interaction between the char particles and the soil matrix (Liao and Thomas 2019). Smaller particle size enhances the surface area and micro-porosity of the char, thereby increasing the interaction between char particles and soil (Valenzuela-Calahorro et al. 1987; Sun et al. 2012a; Xie et al. 2015). It helps in increasing the nutrient availability for the plants grown in these soils (Xie et al. 2015).

13.4.6 Energy Content and Thermal Conductivity

The increase in carbon content in biochar helps in increasing the energy content (Weber and Quicker 2018). The energy content in char (30–35 MJ/kg) produced at 700 °C is nearly double the value of the energy levels of the feedstock (15–20 MJ/kg) from which the char is prepared. The high energy content could assist in its applicability as a source of bioenergy.

Thermal conductivity of biochar increases with a rise in its density. Increase in porosity decreases the thermal conductivity of the char by trapping air in the pores. The decrease in thermal conductivity helps the soil in providing soil insulation in colder areas. biochar could be used in the construction materials to assist in heat insulation and electromagnetic shielding (Usowicz et al. 2006).

13.4.7 Interaction with Water

Previous studies have reported contrasting results of the interaction between biochar and water. Chun et al. (2004) reported a rise in water-repelling tendency of biochar produced at high temperatures. This is seen due to the detachment of oxygen-containing surface functional groups, polar in nature. As a result of polar group detachment, hydrophobicity increases. However, Zornoza et al. (2016) reported that low-temperature chars are more hydrophobic in comparison to the high-temperature chars.

Rise in the treatment temperature increases the porosity of char which assists in enhancing its water holding capacity (Zhang and You 2013; Gray et al. 2014). Such an enhancement of water holding capacity increases the water retaining capability of soil. This helps in reducing the water lost due to leaching and increasing the water available to plant roots.

13.5 Heavy Metals and Their Removal

Metals or metalloids with a potential to affect human and environmental health negatively and possesses a specific density above 5 g/cm^3 are called heavy metals (Järup 2003). Lead, mercury, chromium, arsenic, cadmium, etc. are certain examples of heavy metals. They have been involved in vital processes such as cell division, redox reaction, enzymatic functioning, protein synthesis and regulation, etc. in the living organisms (Pilon-Smits et al. 2009). Additionally, they have been widely used by human beings for various domestic and industrial applications (Tchounwou et al. 2012).

The excessive use of the heavy metals has resulted in the pollution of land and water bodies. The non-biodegradability, high bioavailability and enhanced mobility of heavy metals make them toxic to the living beings (Zhang et al. 2013). Heavy metals enter the environment by natural (e.g. weathering) and anthropogenic routes (e.g. Industrial release, agricultural discharge, metal mining, etc.) and penetrate the soil and water bodies equally (Young 1995; Tchounwou et al. 2012). They are transported from soil to water and water to soil and do not get self-purified.

Heavy metals could be taken up by plants and microbes through which they enter the animal bodies upon ingestion (Mohammed et al. 2011; Tangahu et al. 2011). They trigger chlorosis, necrosis, growth stunting, enzymatic inhibition, photosynthetic stress and reactive oxygen species formation in plants (Stadtman 1990). Heavy metals affect the reproductive system, circulatory system, nervous system, digestive system and excretory system. Further, they damage the genetic material and could be mutagenic (Patra et al. 2006; Tchounwou et al. 2012).

The persistence and toxicity of the heavy metals have made it impertinent to look for remediation methods. The various methods developed are physical, chemical and biological in nature (Gunatilake 2015; Khalid et al. 2017). Biochar-based remediation has gained attention in the previous decades because of the low costs involved, simplicity, high efficacy and efficiency, to minimize the damage caused by the heavy metals (Ahmad et al. 2014; Dwibedi et al. 2022). The fate of heavy metals, their toxicity manifestations in plants and human beings and their removal from soil have been depicted in Fig. 13.1.

It is also important to distinguish the applicability of biochar for heavy metal removal in comparison to activated carbons. Activated carbons are prepared by oxygen activation of char, which renders them high porosity and surface area. However, the properties of porosity and surface area in activated carbons are comparable to biochar (Cao et al. 2011). Further, biochar does not need an additional treatment stage unlike activated carbons and contains surface functional groups rich in oxygen, possesses a non-carbonized fraction and a high cation exchange capacity for contaminant removal (Ahmad et al. 2012a). Additionally, biochar aids in soil quality enhancement, climate change mitigation, energy production and waste management (Atkinson et al. 2010; Sohi 2012; Lee et al. 2017b; Sophia Ayyappan et al. 2018). The advantages associated with biochar application are indicative for its preferability to activated carbons.

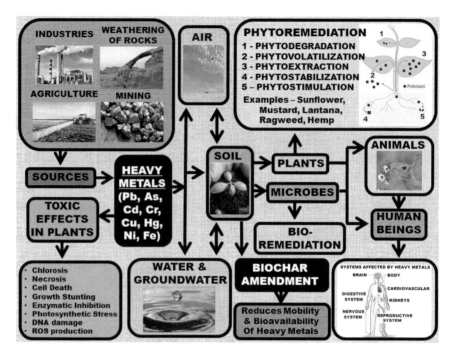

Fig. 13.1 Fate of heavy metals in the environment, their toxic effects in plants and human beings and their removal from soil by biochar amendment and phytoremediation

13.5.1 Heavy Metal Remediation Mechanisms

The properties of high porosity, adequate surface area, alkaline pH, aromaticity and oxygen-containing surface functional groups enable biochar remediating heavy metals from soil. The various mechanisms are summarized in Fig. 13.2. These mechanisms are elaborated in the following passages:

(1) Physical adsorption

It is also called as van der Waals adsorption and is an outcome of the intermolecular interaction between the adsorbent particles and the adsorbate. The heavy metals in soil get sorbed on the char surface (Yu et al. 2009; Lou et al. 2011). The process is reversible in general. High porosity, pore volume, large surface area, surface energy, high pH and adequate ionic strength affect heavy metal sorption (Zhang et al. 2009; Xie et al. 2011). A high surface area and large pore volume facilitate a greater contact between the heavy metals and biochar. An increase in pyrolysis temperatures increases the surface area and pore volume and consequentially contributes in a greater remediation of heavy metals. Liu et al. (2010) prepared chars from switchgrass and pine wood at 300 °C and 700 °C, respectively. They reported that these chars could immobilize uranium

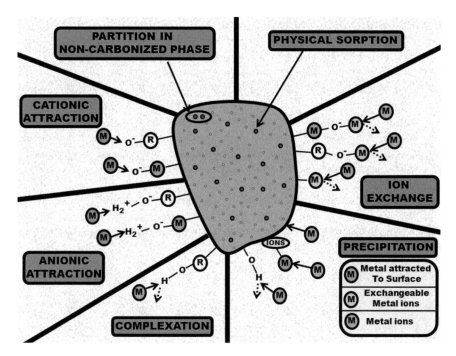

Fig. 13.2 Various mechanisms involved in removal of heavy metals

and copper effectively. Beesley and Marmiroli (2011) stated that biochar could immobilize zinc, cadmium and arsenic by physisorption remarkably.

(2) ion exchange

Exchange of metal ions such as magnesium, potassium and sodium on char surface by heavy metal ions is called as ion exchange. It is dependent on the chemical properties of char surface. The high cation exchange capacity of biochar assists in the process of ion exchange. Cation exchange capacity decreases with an increase in pyrolysis temperature and maximum cation exchange capacity is seen in chars produced at 250–300 °C (Lee et al. 2010). El-Shafey (2010) prepared biochar from rice husks at 175–180 °C and reported that mercury and zinc were effectively removed by these chars via ion exchange mechanism. Liu et al. (2010) observed that char prepared by pyrolysis possess greater surface area in comparison to hydrothermal chars. The greater surface area assists in copper removal by ion exchange and sorption. They also stated that ion exchange removes heavy metals more effectively in comparison to sorption. Sánchez-Polo and Rivera-Utrilla (2002) demonstrated that ion exchange is related to soil pH. When soil pH is less than biochar pH at point of zero charge, greater amount of heavy metals are removed via ion exchange.

(3) Electrostatic attraction/repulsion

Electrostatic interaction between cations (metal pollutants) and anionic char surface is involved in heavy metal remediation (Xu et al. 2011). Metal exchange between cations on char surface and heavy metals results in electrostatic outer sphere complex formation thereby aiding in heavy metal remediation (Ahmad et al. 2014). Electrostatic interaction depends on factors such as soil pH, point of zero charge of biochar, valency and ionic radius of the metallic contaminant (Dong et al. 2011; Mukherjee et al. 2011). Qiu et al. (2008) reported that chars derived from rice and wheat straw is more effective remediator than activated carbon as a result of the electrostatic interaction between lead ions and the char surface. Peng et al. (2011) stated that the increase in soil pH and cation exchange capacity after biochar addition results in enhanced electrostatic interaction consequentially boosting heavy metal remediation.

(4) diffusion

A significant distinguishing feature between biochar and activated carbon is presence of non-carbonized phase in biochar. The contaminants diffuse not only into the non-carbonized portions of the char but also in the carbonized portions (Xu et al. 2012).

(5) Complexation

Biochar surface has abundant oxygen-containing functional groups such as hydroxyl and carboxylic groups. These functional groups form surface complexes with heavy metals (Park et al. 2011a; Tong et al. 2011). Biochar prepared at lower treatment temperatures consists of greater number of these functional groups. Further, oxidation of the char surface could result in an increase in the surface functional groups (Harvey et al. 2011). Stable complexes could be formed between lead ions and hydroxyl/carboxyl groups (Cao et al. 2011; Lu et al. 2012). Dong et al. (2011) reported surface complexation as the main mechanism in chromium removal by biochar derived from sugar beet tailings. Further, smaller ionic radius of the metals aid in the enhancement of remediation (Wan Ngah and Hanafiah 2008).

(6) precipitation

precipitation is another mechanism through which biochar immobilizes the heavy metals and insoluble precipitates such as carbonates and phosphates are formed (Shen et al. 2015, 2017). Cao and Harris (2010) prepared biochar by thermal treatment above 300 °C and observed that these chars could be used for heavy metal removal by precipitate formation. In the study, lead formed lead-phosphate-silicate precipitates in the alkaline biochar. Cao et al. (2011) investigated lead immobilization by cow manure-derived biochar. These chars have high ash content, which is rich in magnesium, silicon, potassium, phosphorus and sodium. The phosphates could form insoluble precipitates with heavy metals, such as pyromorphite is formed with lead. Xu et al. (2013b) investigated cadmium, zinc, copper and lead removal by biochar derived from cow manure and rice husk and observed that precipitation, in the form of carbonate and phosphate precipitates, is the main mechanism involved in their removal.

(7) Hydrogen bond formation

Formation of hydrogen bonds could also be involved in the removal of heavy metals. Contaminants form hydrogen bonds with the oxygen-containing functional groups present in abundance on the surface of biochar. Some of these functional groups are phenol, hydroxyl and carboxyl. Sun et al. (2011, 2012b) stated that organic contaminants could form hydrogen bonds with the surface functional groups available on the char.

The properties of biochar are dependent upon the feedstock type and thermal treatment conditions, as previously stated. Different types of biochar could be used for different remediation performances and it would be difficult to pinpoint a biochar for universal heavy metal removal. Some of the biochars used for removal of heavy metals have been represented in Table 13.2. Further, biochar incorporates different types of mechanisms for removal of heavy metals from contaminated soil as previously discussed, and a universal specific mechanism cannot be pointed. Biochar could affect the mobility and bioavailability of different heavy metals when amended to the contaminated soils. Therefore, heavy metal type and biochar properties must be considered before using the biochar for soil amendment to remove heavy metals. The impact of biochar on mobility and bioavailability of heavy metals in soil is discussed in the following sections.

13.6 Impact of Biochar on Mobility of Heavy Metal

Applying biochar to contaminated soils decreases the mobility of heavy metals present in these soils. This helps in decreasing the metal taken up by the plants grown in contaminated soils. Previously, it has been reported that bamboo-derived biochar could help in adsorption of heavy metals such as cadmium, copper, chromium, mercury and nickel from contaminated soils (Skjemstad et al. 2002; Cheng et al. 2006). In a study by Cao et al. (2009), biochar prepared from dairy manure at 200 °C was more effective in lead sorption when compared to biochar prepared from dairy manure at 350 °C. They stated that this could be an outcome of higher soluble phosphate concentration in biochar prepared at 200 °C.

Beesley et al. (2010) investigated the impact of biochar prepared from hardwood on mobility of cadmium and zinc in contaminated soils and reported that the chars reduced the heavy metals in pore water. In another study by Beesley and Marmiroli (2011), biochar amendment immobilized zinc and cadmium in the contaminated soils. The concentration of zinc and cadmium decreased by 300 and 45 times, respectively in the pore water. Namgay et al. (2010) investigated impact of biochar application on mobility of heavy metals and reported an increase in zinc and arsenic concentration, a decrease in lead concentration, an irregular trend in cadmium concentration and an absence of change in copper concentration.

There could be involvement of redox processes between biochar and heavy metals, which could help in decreasing leaching of the heavy metals. Choppala et al. (2012) prepared biochar using chicken manure as feedstock and applied the chars to chromium-contaminated soils. They reported that chromium (III) ions are sorbed on cation exchange sites on biochar. Additionally, chromium precipitates as chromium hydroxides which help in chromium reduction. Therefore, biochar helps in

Table 13.2 Variation in feedstock and treatment temperature for removal of heavy metals

Feedstock	Treatment temperature (°C)	Remarks	References
Bamboo	500	Cadmium, Lead, Zinc, Copper (maximum removed—49%)	Lu et al. (2014)
Broiler litter	700	Cadmium, Nickel, Zinc, Copper (maximum removed—75%)	Uchimiya et al. (2011a)
Chicken manure	550	Cadmium, Lead (maximum removed—94%)	Park et al. (2011a)
Dairy manure	450	Lead (sorption capacity—132.81 mg/g)	Cao et al. (2011)
Miscanthus	600	Cadmium, Zinc, Lead (maximum removed—92%)	Houben et al. (2013)
Rice straw	500	Zinc, Copper, Cadmium, Lead (maximum removed—71%)	Lu et al. (2014)
Sewage sludge	500–550	Lead, Nickel, Cobalt, Chromium, Arsenic immobilization; Cadmium, Zinc, Copper mobilization	Khan et al. (2013)
Cottonseed hulls	200–800	Cadmium, Lead, Nickel, Copper removed by sorption, complex formation, precipitation and electrostatic interaction	Uchimiya et al. (2011b)
Hard wood	NA	Cadmium and Zinc removal; Arsenic mobilization	Beesley and Marmiroli (2011)
Oak wood	400	Bioavailability reduction of Lead by 76%	Ahmad et al. (2012b)

reducing chromium (VI) ions to chromium (III) ions, thereby resulting in a decrease in chromium leaching (Bolan et al. 2013). The long-term existence of biochar in soil as a result of its excellent stability triggers changes in physicochemical properties of the char. Biochar ageing results in the oxidation of its surface, thereby increasing in the presence of oxygen-containing functional groups. Such a process could be accompanied by an increase in the cation exchange capacity and surface negative charges in biochar. These processes help in heavy metal immobilization (Wang et al. 2019).

Biochar application to soil could need amendments in certain cases. For example, arsenic (V) could get reduced to arsenic (III) by biochar application, consequentially increasing its mobility (Ahmad et al. 2014). Therefore, such a scenario could ask for amendments in biochar. Warren et al. (2003) stated that magnetization of biochar by iron oxide treatment could help in anion exchange thereby reducing the arsenic mobility in soil. Interestingly, reduction of heavy metals by biochar addition could be helpful in decreasing its toxicity in most of the cases. Choppala et al. (2016) reported that chromium (VI) is reduced to chromium (III) by biochar addition, which helps in decreasing their toxicity and bioavailability. The study also observed an increase in mobility of arsenic by its reduction from arsenic (V) to arsenic (III) when biochar was added to the soil.

Furthermore, efficiency and efficacy of biochar application could be affected by the soil type. In a study by Shen et al. (2016a, b), biochar prepared from hardwood was applied to contaminated sandy soil and lead-contaminated kaolin. They reported that the biochar application reduced zinc and nickel concentrations in sandy soil. However, no major effect was observed on lead mobility in kaolin.

13.7 Impact of Biochar on Bioavailability of Heavy Metal

The bioavailability of heavy metals regulates its potential to cause toxicity in soil the risks associated with its entry in food chain and its accessibility by the organisms thriving in the soils (Naidu et al. 2008). Additionally, the bioavailability of heavy metals determines their degradation potential and ecotoxicology (Zhang et al. 2013).

Application of biochar aids in immobilization of heavy metals in soils, which decreases their phytotoxicity and bioavailability. In a study by Fellet et al. (2011), biochar was prepared from orchard prune residues and applied at rates varying from 1 to 10% to decrease the toxicity caused due to heavy metals in the mine tailings. They reported that there was an increase in water retention, cation exchange capacity and pH in the soils. Further, there was a decrease in bioavailability of cadmium, zinc and lead, with maximum decrease in cadmium. Zhou et al. (2008) prepared biochar using cotton stalks and applied them in contaminated soils to reduce cadmium uptake in cabbage plants. They reported that bioavailability of cadmium in soil was reduced by using co-precipitation and sorption.

Méndez et al. (2012) prepared biochar using sewage sludge and used them to decrease the solubility and bioavailability of heavy metals in soils. They reported that biochar diminished the bioavailable nickel, zinc, cadmium and lead in the agricultural soils. Park et al. (2011a, b) prepared biochar using green waste and chicken manure and reported that they decreased copper, lead and cadmium uptake in mustard plants. In a study by Jiang et al. (2012), biochar prepared using rice straw immobilized copper and lead more efficiently than cadmium. It is, therefore, clear that biochars prepared from different feedstock at different treatment temperatures are differently potent in immobilization of heavy metals. Namgay et al. (2010) prepared biochar using activated wood and applied them to heavy metal contaminated soils. They observed that there is a decrease in arsenic, cadmium and copper concentrations in the shoots of maize plants. However, the results on lead and zinc removal were inconclusive in the study.

pH of soil has been reported to be correlated to heavy metal bioavailability. Uchimiya et al. (2010) investigated the impact of biochar amendment in soils and reported that biochar increases the pH and cation exchange capacity of soil, thereby increasing heavy metal immobilization in soils. Ahmad et al. (2012b) reported a decrease in bioavailability of lead by 76% from contaminated soils in military shooting ranges by biochar application. They stated that biochar increases the pH of soil and the sorption potential, thereby aiding in heavy metal remediation. Beesley and Marmiroli (2011) investigated the impact of biochar prepared from fruit trees to remediate a naturally contaminated soil. They stated that biochar effectively decreased the heavy metal concentrations in soil and organic carbon content could have an important impact on decreasing heavy metals bioavailability.

13.8 Remediation of Polluted Sites by Application of Biochar

Studies have been conducted in various parts of the world to determine the efficiency and efficacy of biochar amendment for heavy metal removal from polluted soils. Koetlisi and Muchaonyerwa (2019) prepared biochar from different feedstocks such as pine bark and human faecal products. They reported that these chars could be used to effectively remove copper, chromium and zinc from industrial effluents in South Africa so that soil contamination could be reduced. Gwenzi et al. (2016) prepared biochar by using sewage sludge to study their impact on soil properties, plant growth, nutrient uptake and heavy metal removal from tropical clayey soils in Zimbabwe. They reported that biochar could decrease the copper, lead and zinc concentrations in these soils.

In a study by von Gunten et al. (2019), biochar was prepared from Tibouchina wood and applied to ferralsol in Brazilian forests. They observed that mobility of magnesium, calcium, potassium, barium and zinc concentrations in soil increased after biochar application. Puga et al. (2015) prepared biochar using sugar cane straw at 700 °C for amending Brazilian mine soils contaminated with heavy metals. They reported that biochar application reduced cadmium, lead and zinc concentrations in the pore water and the plants grown on these soils. Rodriguez et al. (2019) prepared biochar from corncobs for utilization as a lead-contaminated soil amendment. They observed that the biochar could immobilize lead in these Colombian soils. However, the immobilization is not that effective due to the extreme contamination of the soils.

Rees et al. (2014) investigated the short-term impact of biochar produced in Germany on heavy metal mobility in French soils. They concluded that biochar could immobilize lead, copper, zinc and cadmium in soils by increasing the pH of the soil and intra-particle diffusion in the biochar matrix. In a study by Beesley et al. (2014), biochar was prepared from orchard prunings at 500 °C and applied to contaminated mine soils in Spain. They reported that biochar could effectively remediate heavy metals from the soils. Further, they stated that mixing biochar with compost could enhance the efficacy of heavy metal immobilization and toxicity reduction.

In a review by He et al. (2019), it was concluded that biochar could be applied to Chinese soils to effectively minimize heavy metal contamination. The remediation potential is dependent on the properties of biochar and soil used. Further, biochar application could reduce the heavy metal accumulation in plants. Mohan et al. (2018) prepared biochar using corn stover and rice husk at 550 °C and 650 °C and observed that biochar could be applied in Indian soils to improve their productivity and remove heavy metals from the soil sustainably. Choudhary et al. (2017) prepared char using eucalyptus bark at 500 °C and highlighted their potential in effective chromium remediation from groundwater, wastewater and soil in India. Hina et al. (2019) prepared biochar using rice husk and plant waste as feedstock to immobilize arsenic from soils in Pakistan. Rice husk char was more effective for lower arsenic contamination, while plant waste char was more efficient in higher arsenic concentrations. Mazhar et al. (2020) reported that biochar could be applied to soils in Pakistan to improve the plant growth parameters and effective removal of chromium. Bandara et al. (2017) prepared biochar using wood as feedstock and applied them to soils in Sri Lanka to effectively remove chromium, nickel and manganese from the soils.

Samsuri et al. (2013) prepared biochar from rice husk and empty oil palm fruit bunch and used them to remove arsenic from Malaysian soils. Fahmi et al. (2018) used biochar derived from empty fruit bunch and demonstrated that they could remove cadmium and lead from soils in Malaysia. Mulder (2014) used biochar to remove heavy metals from Malaysian and Indonesian soils. Dang et al. (2019) prepared biochar from rice straws and applied them to contaminated soils in Vietnam. They reported that these chars could be used to remove zinc, cadmium, and lead from these soils. Saengwilai et al. (2020) used organic amendments to immobilize cadmium from the polluted soils in Thailand. Therefore, various studies across the globe have prepared biochar using different types of feedstocks at varying thermal treatment conditions. These chars have effectively decreased the mobility and bioavailability of heavy metals from polluted soils across the globe.

13.9 Applications of Biochar Other Than Heavy Metal Removal

The char properties of oxygen-containing surface functional groups, good porosity, surface area, high carbon content and remarkable energy content facilitate its wide-ranging applications, which could help in tackling the issues of climate change, energy security, food security and waste management simultaneously.

Waste biomass could be used as feedstock for biochar production. Examples of waste biomass include agricultural residues, food waste, kitchen waste, animal manure, sewage sludge, municipal solid waste and others (Cao and Harris 2010; van Zwieten et al. 2010; Yargicoglu et al. 2015; Kumar et al. 2016; Lee et al. 2017b). Further, the biochar could be used for decreasing the mobility and bioavailability of heavy metals consequently reducing their plant uptake and toxicity (Cui et al. 2012; Hmid et al. 2014). Thermal treatment of the waste biomass would also help in killing the prevalent microbes which could be harmful to the environment and human health (Dahal et al. 2018). Therefore, biochar production would help in waste management and risk reduction at the same time.

Biochar helps in carbon sequestration. The carbon content available in biomass is converted to stable forms by thermal treatment in biochar. Carbon captured in biochar could check carbon dioxide release by 0.3 billion tonnes every year (Liu et al. 2015). Biochar has a very high stability in soil (Singh et al. 2012). Further, biochar could capture methane and nitrous oxide thereby helping in their emission reduction (van Zwieten et al. 2010; Yaghoubi et al. 2014; Edwards et al. 2018). It has also been reported that biochar could stimulate the activity of micro-organisms and help in suppressing the greenhouse gas emissions (Castaldi et al. 2011; Liu et al. 2014a, b). In a study by Spokas et al. (2009) and Al-Wabel et al. (2013), it was observed that biochar prepared at thermal treatment temperatures of 500 °C and above decreases the greenhouse gas emissions, consequentially mitigating climate change.

Thermal treatment of biomass produces syngas, bio-oil and biochar in different concentrations depending on the feedstock variation and thermal treatment conditions. Bio-oil is produced in large quantities in fast pyrolysis, while gasification produces syngas in abundance (Mohan et al. 2006; Lombardi et al. 2015). Biochar could be utilized as catalyst for biodiesel production (Lee et al. 2017a). The presence of surface functional groups in char help in metal sorption and aid in the functioning of biochar as catalysts (Titirici et al. 2012; Cheng and Li 2018). The various sources of bioenergy could be used to replace fossil fuels, consequentially decreasing the greenhouse gas emissions and aid in climate change mitigation. Biochar production could, therefore, help in solving energy security issues to a certain extent.

The properties of high carbon content and remarkable water retention capacity in biochar promote its utilization as soil conditioner to tackle water deficit situations (Bryant 2015; Nichols 2015). Biochar application minimizes the nutrient loss from soil (Sohi et al. 2010). Alkaline conditions introduced by biochar into soil help in neutralizing the acidic conditions. Further, biochar application stimulates microbial communities in soil and the associated microbial activity (Lehmann et al. 2011). Microbes oxidize the char surface thereby increasing oxygen-containing functional groups and the cation exchange capacity of the soil matrix. These changes help in increasing nutrient retention by soil, correspondingly enhancing the growth in plants. Various studies have stated that biochar application increase crop yield by facilitating nutrients to the plant roots (Steiner et al. 2009; Vassilev et al. 2013a; Houben et al. 2014; Siebers et al. 2014). Biochar could be used to decrease the time needed for composting and increase the value of compost (Awasthi et al. 2017; Sanchez-Monedero et al. 2018). All of the aforementioned changes help in improving crop yield, consequently solving the problem of food security partially.

Apart from removal of heavy metals from soil, biochar could also help in removing organic contaminants from environment (Beesley et al. 2011; Ahmad et al. 2014). Soils contaminated with oil and petroleum could be treated by biochar amendment (Wang et al. 2017; Kandanelli et al. 2018). Biochar supports microbial population growth in its pores and on its surface, which assists in hydrocarbon degradation. Biochar could also be utilized for dye degradation and remediation (Nautiyal et al. 2016; Sophia Ayyappan et al. 2018). The surface area and high pH of biochar could help in removing hydrogen sulphide from biogas (Sahota et al. 2018).

13.10 Advantages and Risks Associated with Biochar Production and Application

Apart from the various applications, biochar production and its use has a number of advantages. Biochar is cheaper than activated carbons and does not require additional activation steps. Additionally, biochar has a rich surface oxygen-containing functional groups, a non-carbonized fraction and a great cation exchange capacity, as stated previously. These enhanced properties aid in enhanced contaminant removal (Cao and Harris 2010; Ahmad et al. 2012a; McCarl et al. 2012). Further, biochar supports the growth of microbial colonies, consequently enhancing food chain in the soil (Pietikäinen et al. 2000). Additionally, they enhance the water retention capacity in soil aiding in nutrient retention and crop growth (Ventura et al. 2013; Yu et al. 2013).

However, there could be risks associated with biochar production and application. There could be presence of contaminants such as heavy metals and polycyclic aromatic hydrocarbons in the feedstock used for biochar production (Hossain et al. 2007). Risks associated with these contaminants could, however, be removed by thermal treatment at 500 °C and above (Verheijen et al. 2010). Interestingly, in a study by Gong et al. (2018), it was observed that heavy metals in plants used for phytoremediation could be stabilized by charring. Further, chars prepared from such plants could be used for remediation of polluted sites. The ash content in chars could be a threat to human health (De Capitani et al. 2007). However, health safety guidelines, during production and application of char, could be enforced to minimize and remove the risks associated with ash. Biochar could sorb agro-chemicals, such as pesticides and herbicides, thereby decreasing their potential to increase the crop yield. However, such a sorption could help in immobilizing excess agrochemicals in soil (Sun et al. 2012b). In a few studies, biochar has been reported to negatively affect earthworms and increase nitrous oxide emissions (Topoliantz and Ponge 2003; Warnock et al. 2007; Angst et al. 2014; Verhoeven and Six 2014). However, wet biochar could be applied to minimize the damage to earthworms (Li et al. 2011). Therefore, risks associated with biochar production and application do prevail, but the risks could be minimized by appropriate steps taken and guidelines properly enforced. Further, the need for extensive research arises with regard to biochar production and their application.

13.11 Future Research

Although biochar has been used for remediation of heavy metals from a number of polluted sites, there is a lot of scope for future research. These opportunities are mentioned in the following points:

(1) Due to the variation in properties and performance of biochar produced from different feedstock and thermal treatment conditions, there is a need to establish a global standard for obtaining maximum advantage in terms of remediation of polluted sites.
(2) Most of the studies have been small-scale and limited to laboratories and tiny agricultural lands. Further, the experiments have been focussed on single heavy metal removal. However, real-time metal pollution involves multiple heavy metals and occurs on large areas of land. Therefore, there is a need for extensive research involving multiple heavy metal contamination.
(3) The complexity of soil systems brings about variation in biochar efficiency from remediation. The mechanisms involved in the metal removal could be studied extensively to bring clarity in remediation of polluted sites by biochar application.
(4) The dose and rate of biochar application in metal-polluted sites need further optimization. Additionally, the suitability of biochar could be determined for targeted and specific removal of heavy metals.
(5) Emergence of extreme weather events in the scenario of climate change, enquire for identification and confirmation of their impact on the efficacy and efficiency of biochar performance for heavy metal removal.

13.12 Conclusion

Biochar could be a sustainable alternative for effective and long-term removal of heavy metals from polluted lands. Biochar could be produced from wide-ranging biomass sources and a number of thermal treatment methods are employed for its preparation. Biomass type and thermal treatment conditions affect the properties of char produced. Properties of alkaline pH, high cation exchange capacity, high surface area, high porosity, abundant oxygen-containing surface functional groups and a non-carbonized fraction, enable the biochar to remove heavy metals from the soil. Biochar incorporates mechanisms such as ion exchange, precipitation, diffusion, complex formation, electrostatic interaction and sorption, for the removal of metal pollutants from soil. Biochar decreases the mobility and bioavailability of heavy metals, thereby minimizing their toxic effects. The potential of biochar for heavy metal removal has been tested in different studies conducted across the various parts of the globe and biochar was found to be effective in the remediation of heavy metal polluted sites.

Biochar could also be used for removing the organic contaminants from soil. Utilization of waste biomass assists in waste management and waste reduction. Enhancement of crop production can help in tackling issues of food security. Removal of contaminants from soil and water makes it safe for the animals and human beings. Extended stability of biochar in soils reduces the safety concerns. Biochar production could help in producing bioenergy which could be used as an alternative to fossil fuels. Biochar production would help in solving the problem of energy security and depleting fossil fuel reserves. Biochar would help in the mitigation of climate change by carbon sequestration and reduction in emission of greenhouse gases. Therefore, biochar could be a promising method for remediation of polluted sites and tackling the various problems endangering the environment and human health. Government of various countries could assist the scientists by providing them grants for research and they could commence policies to boost the production and application of biochar. Lastly, the risks associated with biochar production and application should be acknowledged and minimized for helping the society in the longer run.

Acknowledgements One of the authors (Abhishek Kumar) is thankful to the University Grants Commission, New Delhi, for providing NET-JRF Fellowship [Ref. No.—3635/(OBC)(NET-DEC.2015)].

References

Alam A, Shaikh WA, Alam O, Bhattacharya T, Chakraborty S, Show B, Saha I (2018) Adsorption of As (III) and As (V) from aqueous solution by modified Cassia fistula (golden shower) biochar. Appl Water Sci 8(7):198. https://doi.org/10.1007/s13201-018-0839-y

Ahmad M, Lee SS, Dou X et al (2012a) Effects of pyrolysis temperature on soybean stover- and peanut shell-derived biochar properties and TCE adsorption in water. Bioresour Technol 118:536–544. https://doi.org/10.1016/j.biortech.2012.05.042

Ahmad M, Soo Lee S, Yang JE et al (2012b) Effects of soil dilution and amendments (mussel shell, cow bone, and biochar) on Pb availability and phytotoxicity in military shooting range soil. Ecotoxicol Environ Saf 79:225–231. https://doi.org/10.1016/j.ecoenv.2012.01.003

Ahmad M, Rajapaksha AU, Lim JE et al (2014) Biochar as a sorbent for contaminant management in soil and water: a review. Chemosphere 99:19–33. https://doi.org/10.1016/j.chemosphere.2013.10.071

Al-Wabel MI, Al-Omran A, El-Naggar AH et al (2013) Pyrolysis temperature induced changes in characteristics and chemical composition of biochar produced from conocarpus wastes. Bioresour Technol 131:374–379. https://doi.org/10.1016/j.biortech.2012.12.165

Angin D (2013) Effect of pyrolysis temperature and heating rate on biochar obtained from pyrolysis of safflower seed press cake. Bioresour Technol 128:593–597. https://doi.org/10.1016/j.biortech.2012.10.150

Angst TE, Six J, Reay DS, Sohi SP (2014) Impact of pine chip biochar on trace greenhouse gas emissions and soil nutrient dynamics in an annual ryegrass system in California. Agric Ecosyst Environ 191:17–26. https://doi.org/10.1016/j.agee.2014.03.009

Antal MJ, Mochidzuki K, Paredes LS (2003) Flash carbonization of biomass. Ind Eng Chem Res 42:3690–3699. https://doi.org/10.1021/ie0301839

Asensio V, Vega FA, Andrade ML, Covelo EF (2013) Tree vegetation and waste amendments to improve the physical condition of copper mine soils. Chemosphere 90:603–610. https://doi.org/10.1016/j.chemosphere.2012.08.050

Atkinson CJ, Fitzgerald JD, Hipps NA (2010) Potential mechanisms for achieving agricultural benefits from biochar application to temperate soils: a review. Plant Soil 337:1–18

Awasthi MK, Wang M, Chen H et al (2017) Heterogeneity of biochar amendment to improve the carbon and nitrogen sequestration through reduce the greenhouse gases emissions during sewage sludge composting. Bioresour Technol 224:428–438. https://doi.org/10.1016/j.biortech.2016.11.014

Balat M, Balat M, Kirtay E, Balat H (2009) Main routes for the thermo-conversion of biomass into fuels and chemicals. Part 1: Pyrolysis systems. Energy Convers Manag 50:3147–3157. https://doi.org/10.1016/j.enconman.2009.08.014

Balée WL, Erickson CL, Graham E (2016a) 2. A neotropical framework for Terra Preta. Time complex. Hist Ecol

Balée WL, Erickson CL, Neves EG, Petersen JB (2016b) 9. Political economy and pre-columbian landscape transformations in Central Amazonia. Time complex. Hist Ecol

Bandara T, Herath I, Kumarathilaka P et al (2017) Role of woody biochar and fungal-bacterial co-inoculation on enzyme activity and metal immobilization in serpentine soil. J Soils Sediments 17:665–673. https://doi.org/10.1007/s11368-015-1243-y

Beesley L, Dickinson N (2011) Carbon and trace element fluxes in the pore water of an urban soil following greenwaste compost, woody and biochar amendments, inoculated with the earthworm Lumbricus terrestris. Soil Biol Biochem 43:188–196. https://doi.org/10.1016/j.soilbio.2010.09.035

Beesley L, Marmiroli M (2011) The immobilisation and retention of soluble arsenic, cadmium and zinc by biochar. Environ Pollut 159:474–480. https://doi.org/10.1016/j.envpol.2010.10.016

Beesley L, Moreno-Jiménez E, Gomez-Eyles JL (2010) Effects of biochar and greenwaste compost amendments on mobility, bioavailability and toxicity of inorganic and organic contaminants in a multi-element polluted soil. Environ Pollut 158:2282–2287. https://doi.org/10.1016/j.envpol.2010.02.003

Beesley L, Moreno-Jiménez E, Gomez-Eyles JL et al (2011) A review of biochars' potential role in the remediation, revegetation and restoration of contaminated soils. Environ Pollut 159:3269–3282. https://doi.org/10.1016/j.envpol.2011.07.023

Beesley L, Inneh OS, Norton GJ et al (2014) Assessing the influence of compost and biochar amendments on the mobility and toxicity of metals and arsenic in a naturally contaminated mine soil. Environ Pollut 186:195–202. https://doi.org/10.1016/j.envpol.2013.11.026

Behazin E, Ogunsona E, Rodriguez-Uribe A et al (2016) Mechanical, chemical, and physical properties of wood and perennial grass biochars for possible composite application. BioResources 11:1334–1348. https://doi.org/10.15376/biores.11.1.1334-1348

Bhattacharya T, Pandey SK, Pandey VC, Kumar A (2021) Potential and safe utilization of Fly ash as fertilizer for Pisum sativum L. Grown in phytoremediated and non-phytoremediated amendments. Environ Sci Pollut Res 28(36):50153–50166. https://doi.org/10.1007/s11356-021-14179-9

Bolan NS, Choppala G, Kunhikrishnan A et al (2013) Microbial transformation of trace elements in soils in relation to bioavailability and remediation. Rev Env Contam Toxicol 225:1–56. https://doi.org/10.1007/978-1-4614-6470-9_1

Bryant L (2015) Organic matter can improve your soil's water holding capacity. In: Nrdc. https://www.nrdc.org/experts/lara-bryant/organic-matter-can-improve-your-soils-water-holding-capacity

Bryden KM, Hagge MJ (2003) Modeling the combined impact of moisture and char shrinkage on the pyrolysis of a biomass particle. Fuel 82:1633–1644. https://doi.org/10.1016/S0016-2361(03)00108-X

Byrne CE, Nagle DC (1997) Carbonization of wood for advanced materials applications. Carbon N Y 35:259–266. https://doi.org/10.1016/S0008-6223(96)00136-4

Caillat S, Vakkilainen E (2013) Large-scale biomass combustion plants: an overview. In: Biomass combustion science, technology and engineering, pp 189–224

Cantrell KB, Hunt PG, Uchimiya M et al (2012) Impact of pyrolysis temperature and manure source on physicochemical characteristics of biochar. Bioresour Technol 107:419–428. https://doi.org/10.1016/j.biortech.2011.11.084

Cao X, Harris W (2010) Properties of dairy-manure-derived biochar pertinent to its potential use in remediation. Bioresour Technol 101:5222–5228. https://doi.org/10.1016/j.biortech.2010.02.052

Cao X, Ma L, Gao B, Harris W (2009) Dairy-manure derived biochar effectively sorbs lead and atrazine. Environ Sci Technol 43:3285–3291. https://doi.org/10.1021/es803092k

Cao X, Ma L, Liang Y et al (2011) Simultaneous immobilization of lead and atrazine in contaminated soils using dairy-manure biochar. Environ Sci Technol 45:4884–4889. https://doi.org/10.1021/es103752u

Castaldi S, Riondino M, Baronti S et al (2011) Impact of biochar application to a Mediterranean wheat crop on soil microbial activity and greenhouse gas fluxes. Chemosphere 85:1464–1471. https://doi.org/10.1016/j.chemosphere.2011.08.031

Cetin E, Moghtaderi B, Gupta R, Wall TF (2004) Influence of pyrolysis conditions on the structure and gasification reactivity of biomass chars. Fuel 83:2139–2150. https://doi.org/10.1016/j.fuel.2004.05.008

Cha JS, Park SH, Jung SC et al (2016) Production and utilization of biochar: a review. J Ind Eng Chem 40:1–15. https://doi.org/10.1016/j.jiec.2016.06.002

Chen B, Chen Z, Lv S (2011) A novel magnetic biochar efficiently sorbs organic pollutants and phosphate. Bioresour Technol 102:716–723. https://doi.org/10.1016/j.biortech.2010.08.067

Cheng CH, Lehmann J, Thies JE et al (2006) Oxidation of black carbon by biotic and abiotic processes. Org Geochem 37:1477–1488. https://doi.org/10.1016/j.orggeochem.2006.06.022

Cheng F, Li X (2018) Preparation and application of biochar-based catalysts for biofuel production. Catalysts 8:346. https://doi.org/10.3390/catal8090346

Choppala GK, Bolan NS, Megharaj M et al (2012) The influence of biochar and black carbon on reduction and bioavailability of chromate in soils. J Environ Qual 41:1175–1184. https://doi.org/10.2134/jeq2011.0145

Choppala G, Bolan N, Kunhikrishnan A, Bush R (2016) Differential effect of biochar upon reduction-induced mobility and bioavailability of arsenate and chromate. Chemosphere 144:374–381. https://doi.org/10.1016/j.chemosphere.2015.08.043

Choudhary B, Paul D, Singh A, Gupta T (2017) Removal of hexavalent chromium upon interaction with biochar under acidic conditions: mechanistic insights and application. Environ Sci Pollut Res 24:16786–16797. https://doi.org/10.1007/s11356-017-9322-9

Chun Y, Sheng G, Chiou GT, Xing B (2004) Compositions and sorptive properties of crop residue-derived chars. Environ Sci Technol 38:4649–4655. https://doi.org/10.1021/es035034w

Conti R, Rombolà AG, Modelli A et al (2014) Evaluation of the thermal and environmental stability of switchgrass biochars by Py-GC-MS. J Anal Appl Pyrolysis 110:239–247. https://doi.org/10.1016/j.jaap.2014.09.010

Cui L, Pan G, Li L et al (2012) The reduction of wheat Cd uptake in contaminated soil via biochar amendment: a two-year field experiment. BioResources 7:5666–5676. https://doi.org/10.15376/biores.7.4.5666-5676

Cui X, Shen Y, Yang Q et al (2018) Simultaneous syngas and biochar production during heavy metal separation from Cd/Zn hyperaccumulator (Sedum alfredii) by gasification. Chem Eng J 347:543–551. https://doi.org/10.1016/j.cej.2018.04.133

Cushman R, Robertson-Palmer K (1998) Protecting our children. Can J Public Heal 89:221–223. https://doi.org/10.1007/BF03403920

Dahal RK, Acharya B, Farooque A (2018) Biochar: a sustainable solution for solid waste management in agro-processing industries. Biofuels 1–9. https://doi.org/10.1080/17597269.2018.1468978

Dang VM, Joseph S, Van HT et al (2019) Immobilization of heavy metals in contaminated soil after mining activity by using biochar and other industrial by-products: the significant role of minerals on the biochar surfaces. Environ Technol (United Kingdom) 40:3200–3215. https://doi.org/10.1080/09593330.2018.1468487

De Capitani EM, Algranti E, Handar AMZ et al (2007) Wood charcoal and activated carbon dust pneumoconiosis in three workers. Am J Ind Med 50:191–196. https://doi.org/10.1002/ajim.20418

Dong X, Ma LQ, Li Y (2011) Characteristics and mechanisms of hexavalent chromium removal by biochar from sugar beet tailing. J Hazard Mater 190:909–915. https://doi.org/10.1016/j.jhazmat.2011.04.008

Dwibedi SK, Pandey VC, Divyasree D, Bajpai O (2022) Biochar-based land development. Land Degrad Dev. https://doi.org/10.1002/ldr.4185

Edwards JD, Pittelkow CM, Kent AD, Yang WH (2018) Dynamic biochar effects on soil nitrous oxide emissions and underlying microbial processes during the maize growing season. Soil Biol Biochem 122:81–90. https://doi.org/10.1016/j.soilbio.2018.04.008

Ekström C, Lindman N, Pettersson R (1985) Catalytic conversion of tars, carbon black and methane from pyrolysis/gasification of biomass. In: Overend RP, Milne TA, Mudge LK (eds) Fundamentals of thermochemical biomass conversion. Elsevier Applied Science Publishers Springer, Netherlands, pp 601–618

El-Shafey EI (2010) Removal of Zn(II) and Hg(II) from aqueous solution on a carbonaceous sorbent chemically prepared from rice husk. J Hazard Mater 175:319–327. https://doi.org/10.1016/j.jhazmat.2009.10.006

Emmerich FG, Luengo CA (1994) Reduction of emissions from blast furnaces by using blends of coke and babassu charcoal. Fuel 73:1235–1236. https://doi.org/10.1016/0016-2361(94)90266-6

Eom IY, Kim JY, Lee SM et al (2013) Comparison of pyrolytic products produced from inorganic-rich and demineralized rice straw (Oryza sativa L.) by fluidized bed pyrolyzer for future biorefinery approach. Bioresour Technol 128:664–672. https://doi.org/10.1016/j.biortech.2012.09.082

Fahmi AH, Samsuri AW, Jol H, Singh D (2018) Bioavailability and leaching of Cd and Pb from contaminated soil amended with different sizes of biochar. R Soc Open Sci 5:181328. https://doi.org/10.1098/rsos.181328

Fellet G, Marchiol L, Delle Vedove G, Peressotti A (2011) Application of biochar on mine tailings: Effects and perspectives for land reclamation. Chemosphere 83:1262–1267. https://doi.org/10.1016/j.chemosphere.2011.03.053

Fidel RB, Laird DA, Thompson ML, Lawrinenko M (2017) Characterization and quantification of biochar alkalinity. Chemosphere 167:367–373. https://doi.org/10.1016/j.chemosphere.2016.09.151

Fu P, Hu S, Xiang J et al (2012) Evaluation of the porous structure development of chars from pyrolysis of rice straw: effects of pyrolysis temperature and heating rate. J Anal Appl Pyrolysis 98:177–183. https://doi.org/10.1016/j.jaap.2012.08.005

Ghani WAWAK, Mohd A, da Silva G et al (2013) Biochar production from waste rubber-wood-sawdust and its potential use in C sequestration: chemical and physical characterization. Ind Crops Prod 44:18–24. https://doi.org/10.1016/j.indcrop.2012.10.017

Giudicianni P, Cardone G, Ragucci R (2013) Cellulose, hemicellulose and lignin slow steam pyrolysis: thermal decomposition of biomass components mixtures. J Anal Appl Pyrolysis 100:213–222. https://doi.org/10.1016/j.jaap.2012.12.026

Glaser B, Lehmann J, Zech W (2002) Ameliorating physical and chemical properties of highly weathered soils in the tropics with charcoal—a review. Biol Fertil Soils 35:219–230. https://doi.org/10.1007/s00374-002-0466-4

Gong X, Huang D, Liu Y et al (2018) Pyrolysis and reutilization of plant residues after phytoremediation of heavy metals contaminated sediments: for heavy metals stabilization and dye adsorption. Bioresour Technol 253:64–71. https://doi.org/10.1016/j.biortech.2018.01.018

Gray M, Johnson MG, Dragila MI, Kleber M (2014) Water uptake in biochars: the roles of porosity and hydrophobicity. Biomass Bioenerg 61:196–205. https://doi.org/10.1016/j.biombioe.2013.12.010

Gunatilake SK (2015) Methods of removing heavy metals from industrial wastewater. J Multidiscip Eng Sci Stud Ind Wastewater 1:13–18

Gwenzi W, Muzava M, Mapanda F, Tauro TP (2016) Comparative short-term effects of sewage sludge and its biochar on soil properties, maize growth and uptake of nutrients on a tropical clay soil in Zimbabwe. J Integr Agric 15:1395–1406. https://doi.org/10.1016/S2095-3119(15)61154-6

Harvey OR, Herbert BE, Rhue RD, Kuo LJ (2011) Metal interactions at the biochar-water interface: Energetics and structure-sorption relationships elucidated by flow adsorption microcalorimetry. Environ Sci Technol 45:5550–5556. https://doi.org/10.1021/es104401h

Hass A, Gonzalez JM, Lima IM et al (2012) Chicken manure biochar as liming and nutrient source for acid appalachian soil. J Environ Qual 41:1096–1106. https://doi.org/10.2134/jeq2011.0124

He L, Zhong H, Liu G et al (2019) Remediation of heavy metal contaminated soils by biochar: mechanisms, potential risks and applications in China. Environ Pollut 252:846–855. https://doi.org/10.1016/j.envpol.2019.05.151

Hina K, Abbas M, Hussain Q et al (2019) Investigation into arsenic retention in arid contaminated soils with biochar application. Arab J Geosci 12:671. https://doi.org/10.1007/s12517-019-4865-3

Hmid A, Al Chami Z, Sillen W et al (2014) Olive mill waste biochar: a promising soil amendment for metal immobilization in contaminated soils. Environ Sci Pollut Res 22:1444–1456. https://doi.org/10.1007/s11356-014-3467-6

Hossain MM (2016) Recovery of valuable chemicals from agricultural waste through pyrolysis. Electron thesis and dissertation repository University of West Ontario, Ontario, Canada

Hossain MK, Strezov V, Nelson P (2007) Evaluation of agricultural char from sewage sludge. Proc Int Agrichar Initiat 2007 Conf

Houben D, Evrard L, Sonnet P (2013) Beneficial effects of biochar application to contaminated soils on the bioavailability of Cd, Pb and Zn and the biomass production of rapeseed (Brassica napus L.). Biomass Bioenerg 57:196–204. https://doi.org/10.1016/j.biombioe.2013.07.019

Houben D, Sonnet P, Cornelis JT (2014) Biochar from Miscanthus: a potential silicon fertilizer. Plant Soil 374:871–882. https://doi.org/10.1007/s11104-013-1885-8

IBI (2015) Standardized product definition and product testing guidelines for biochar that is used in soil. Int Biochar Initiat 23

Igalavithana AD, Mandal S, Niazi NK et al (2017) Advances and future directions of biochar characterization methods and applications. Crit Rev Environ Sci Technol 47:2275–2330. https://doi.org/10.1080/10643389.2017.1421844

Inyang M, Gao B, Pullammanappallil P et al (2010) Biochar from anaerobically digested sugarcane bagasse. Bioresour Technol 101:8868–8872. https://doi.org/10.1016/j.biortech.2010.06.088

Inyang M, Gao B, Yao Y et al (2012) Removal of heavy metals from aqueous solution by biochars derived from anaerobically digested biomass. Bioresour Technol 110:50–56. https://doi.org/10.1016/j.biortech.2012.01.072

Järup L (2003) Hazards of heavy metal contamination. Br Med Bull 68:167–182. https://doi.org/10.1093/bmb/ldg032

Jiang J, Xu RK, Jiang TY, Li Z (2012) Immobilization of Cu(II), Pb(II) and Cd(II) by the addition of rice straw derived biochar to a simulated polluted Ultisol. J Hazard Mater 229–230:145–150. https://doi.org/10.1016/j.jhazmat.2012.05.086

Kandanelli R, Meesala L, Kumar J et al (2018) Cost effective and practically viable oil spillage mitigation: comprehensive study with biochar. Mar Pollut Bull 128:32–40. https://doi.org/10.1016/j.marpolbul.2018.01.010

Khalid S, Shahid M, Niazi NK et al (2017) A comparison of technologies for remediation of heavy metal contaminated soils. J Geochem Explor 182:247–268. https://doi.org/10.1016/j.gexplo.2016.11.021

Khan S, Chao C, Waqas M et al (2013) Sewage sludge biochar influence upon rice (Oryza sativa L) yield, metal bioaccumulation and greenhouse gas emissions from acidic paddy soil. Environ Sci Technol 47:8624–8632. https://doi.org/10.1021/es400554x

Koetlisi KA, Muchaonyerwa P (2019) Sorption of selected heavy metals with different relative concentrations in industrial effluent on biochar from human faecal products and pine-bark. Materials (Basel) 12. https://doi.org/10.3390/ma12111768

Kumar A, Bhattacharya T, Hasnain SM, Nayak AK, Hasnain MS (2020) Applications of biomass-derived materials for energy production, conversion, and storage. Mater Sci Energy Technol 3:905–920. S2589299120300665. https://doi.org/10.1016/j.mset.2020.10.012

Kumar A, Bhattacharya T (2021) Biochar: a sustainable solution. Environ Dev Sustain 23(5):6642–6680. https://doi.org/10.1007/s10668-020-00970-0

Kumar A, Bhattacharya T, Shaikh WA, Roy A, Mukherjee S, Kumar M (2021a) Performance evaluation of crop residue and kitchen waste-derived biochar for eco-efficient removal of arsenic from soils of the Indo-Gangetic plain: a step towards sustainable pollution management. Environ Res 200:111758. S0013935121010525. https://doi.org/10.1016/j.envres.2021.111758

Kumar A, Bhattacharya T (2022) Removal of arsenic by wheat straw biochar from soil. Bull Environ Contam Toxicol 108(3):415–422. https://doi.org/10.1007/s00128-020-03095-2

Kumar A, Bhattacharya T, Mukherjee S, Sarkar B (2022a) A perspective on biochar for repairing damages in the soil–plant system caused by climate change-driven extreme weather events. Biochar 4(1):22. https://doi.org/10.1007/s42773-022-00148-z

Kumar A, Bhattacharya T, Shaikh WA, Chakraborty S, Owens G, Naushad M (2022b) Valorization of fruit waste-based biochar for arsenic removal in soils. Environ Res 213:113710. S0013935122010374113710. https://doi.org/10.1016/j.envres.2022.113710

Kumar A, Nagar S, Anand S (2021b) Nanotechnology for sustainable crop production: recent development and strategies. In: Singh P, Singh R, Verma P, Bhadouria R, Kumar A, Kaushik M (eds) Plant-microbes-engineered nano-particles (PMENPs) nexus in agro-ecosystems. Advances in science, technology & innovation. Springer, Cham. https://doi.org/10.1007/978-3-030-669 56-0_3

Kumar A, Nagar S, Anand S (2021c) Climate change and existential threats. In: Singh S, Singh P, Rangabhashiyam S, Srivastava KK (eds) Global climate change. Elsevier, pp 1–31

Kumar A, Schreiter IJ, Wefer-Roehl A et al (2016) Production and utilization of biochar from organic wastes for pollutant control on contaminated sites. Environ Mater Waste Resour Recover Pollut Prev 91–116

Kuppusamy S, Thavamani P, Megharaj M et al (2016) Agronomic and remedial benefits and risks of applying biochar to soil: current knowledge and future research directions. Environ Int 87:1–12. https://doi.org/10.1016/j.envint.2015.10.018

Lee JW, Kidder M, Evans BR et al (2010) Characterization of biochars produced from cornstovers for soil amendment. Environ Sci Technol 44:7970–7974. https://doi.org/10.1021/es101337x

Lee J, Kim KH, Kwon EE (2017a) Biochar as a catalyst. Renew Sustain Energy Rev 77:70–79. https://doi.org/10.1016/j.rser.2017.04.002

Lee J, Yang X, Cho SH et al (2017b) Pyrolysis process of agricultural waste using CO2 for waste management, energy recovery, and biochar fabrication. Appl Energy 185:214–222. https://doi.org/10.1016/j.apenergy.2016.10.092

Lee HW, Kim YM, Kim S et al (2018) Review of the use of activated biochar for energy and environmental applications. Carbon Lett 26:1–10. https://doi.org/10.5714/CL.2018.26.001

Lehmann J, Gaunt J, Rondon M (2006) Bio-char sequestration in terrestrial ecosystems—a review. Mitig Adapt Strateg Glob Chang 11:403–427. https://doi.org/10.1007/s11027-005-9006-5

Lehmann J, Rillig MC, Thies J et al (2011) Biochar effects on soil biota—a review. Soil Biol Biochem 43:1812–1836. https://doi.org/10.1016/j.soilbio.2011.04.022

Li D, Hockaday WC, Masiello CA, Alvarez PJJ (2011) Earthworm avoidance of biochar can be mitigated by wetting. Soil Biol Biochem 43:1732–1737. https://doi.org/10.1016/j.soilbio.2011.04.019

Li Y, Shao J, Wang X et al (2014) Characterization of modified biochars derived from bamboo pyrolysis and their utilization for target component (furfural) adsorption. Energy Fuels 28:5119–5127. https://doi.org/10.1021/ef500725c

Li H, Ye X, Geng Z et al (2016) The influence of biochar type on long-term stabilization for Cd and Cu in contaminated paddy soils. J Hazard Mater 304:40–48. https://doi.org/10.1016/j.jhazmat.2015.10.048

Liang B, Lehmann J, Solomon D et al (2006) Black carbon increases cation exchange capacity in soils. Soil Sci Soc Am J 70:1719–1730. https://doi.org/10.2136/sssaj2005.0383

Liao W, Thomas S (2019) Biochar particle size and post-pyrolysis mechanical processing affect soil pH, water retention capacity, and plant performance. Soil Syst 3:14. https://doi.org/10.3390/soilsystems3010014

Liu Z, Zhang FS, Wu J (2010) Characterization and application of chars produced from pinewood pyrolysis and hydrothermal treatment. Fuel 89:510–514. https://doi.org/10.1016/j.fuel.2009.08.042

Liu J, Shen J, Li Y et al (2014a) Effects of biochar amendment on the net greenhouse gas emission and greenhouse gas intensity in a Chinese double rice cropping system. Eur J Soil Biol 65:30–39. https://doi.org/10.1016/j.ejsobi.2014.09.001

Liu L, Shen G, Sun M et al (2014b) Effect of biochar on nitrous oxide emission and its potential mechanisms. J Air Waste Manag Assoc 64:894–902. https://doi.org/10.1080/10962247.2014.899937

Liu WJ, Jiang H, Yu HQ (2015) Development of biochar-based functional materials: toward a sustainable platform carbon material. Chem Rev 115:12251–12285. https://doi.org/10.1021/acs.chemrev.5b00195

Lombardi L, Carnevale E, Corti A (2015) A review of technologies and performances of thermal treatment systems for energy recovery from waste. Waste Manag 37:26–44. https://doi.org/10.1016/j.wasman.2014.11.010

Lou L, Wu B, Wang L et al (2011) Sorption and ecotoxicity of pentachlorophenol polluted sediment amended with rice-straw derived biochar. Bioresour Technol 102:4036–4041. https://doi.org/10.1016/j.biortech.2010.12.010

Lu H, Zhang W, Yang Y et al (2012) Relative distribution of Pb^{2+} sorption mechanisms by sludge-derived biochar. Water Res 46:854–862. https://doi.org/10.1016/j.watres.2011.11.058

Lu K, Yang X, Shen J et al (2014) Effect of bamboo and rice straw biochars on the bioavailability of Cd, Cu, Pb and Zn to Sedum plumbizincicola. Agric Ecosyst Environ 191:124–132. https://doi.org/10.1016/j.agee.2014.04.010

Lv D, Xu M, Liu X et al (2010) Effect of cellulose, lignin, alkali and alkaline earth metallic species on biomass pyrolysis and gasification. Fuel Process Technol 91:903–909. https://doi.org/10.1016/j.fuproc.2009.09.014

Manna S, Singh N, Purakayastha TJ, Berns AE (2020) Effect of deashing on physico-chemical properties of wheat and rice straw biochars and potential sorption of pyrazosulfuron-ethyl. Arab J Chem 13(1):1247–1258. S1878535217301910. https://doi.org/10.1016/j.arabjc.2017.10.005

Manyà JJ (2012) Pyrolysis for biochar purposes: a review to establish current knowledge gaps and research needs. Environ Sci Technol 46:7939–7954. https://doi.org/10.1021/es301029g

Manyà JJ, González B, Azuara M, Arner G (2018) Ultra-microporous adsorbents prepared from vine shoots-derived biochar with high CO_2 uptake and CO_2/N_2 selectivity. Chem Eng J 345:631–639. https://doi.org/10.1016/j.cej.2018.01.092

Mazac R (2016) Assessing the use of food waste biochar as a biodynamic plant fertilizer. In: Departmental honors projects. https://digitalcommons.hamline.edu/dhp/43

Mazhar R, Ilyas N, Arshad M, Khalid A (2020) Amelioration potential of biochar for chromium stress in wheat. Pak J Bot 52:1159–1168. https://doi.org/10.30848/PJB2020-4(19)

McCarl BA, Peacocke C, Chrisman R et al (2012) Economics of biochar production, utilization and greenhouse gas offsets. In: Biochar for environmental management: science and technology, pp 341–357

McKendry P (2002) Energy production from biomass (part 1): overview of biomass. Bioresour Technol 83:37–46. https://doi.org/10.1016/S0960-8524(01)00118-3

Méndez A, Gómez A, Paz-Ferreiro J, Gascó G (2012) Effects of sewage sludge biochar on plant metal availability after application to a Mediterranean soil. Chemosphere 89:1354–1359. https://doi.org/10.1016/j.chemosphere.2012.05.092

Meyer S, Glaser B, Quicker P (2011) Technical, economical, and climate-related aspects of biochar production technologies: a literature review. Environ Sci Technol 45:9473–9483. https://doi.org/10.1021/es201792c

Mohammed AS, Kapri A, Goel R (2011) Heavy metal pollution: source, impact, and remedies. Environ Pollut 1–28

Mohan D, Pittman CU, Steele PH (2006) Pyrolysis of wood/biomass for bio-oil: a critical review. Energy Fuels 20:848–889. https://doi.org/10.1021/ef0502397

Mohan D, Sarswat A, Ok YS, Pittman CU (2014) Organic and inorganic contaminants removal from water with biochar, a renewable, low cost and sustainable adsorbent—a critical review. Bioresour Technol 160:191–202. https://doi.org/10.1016/j.biortech.2014.01.120

Mohan D, Abhishek K, Sarswat A et al (2018) Biochar production and applications in soil fertility and carbon sequestration-a sustainable solution to crop-residue burning in India. RSC Adv 8:508–520. https://doi.org/10.1039/c7ra10353k

Mukherjee A, Zimmerman AR, Harris W (2011) Surface chemistry variations among a series of laboratory-produced biochars. Geoderma 163:247–255. https://doi.org/10.1016/j.geoderma.2011.04.021

Mulder J (2014) Biochar on acidic agricultural lands in Indonesia and Malaysia. Cut Across Lands 1–27

Mullen CA, Boateng AA, Goldberg NM et al (2010) Bio-oil and bio-char production from corn cobs and stover by fast pyrolysis. Biomass Bioenerg 34:67–74. https://doi.org/10.1016/j.biombioe.2009.09.012

Naidu R, Semple KT, Megharaj M et al (2008) Chapter 3 Bioavailability: definition, assessment and implications for risk assessment. In: Developments in soil science, pp 39–51

Namgay T, Singh B, Singh BP (2010) Influence of biochar application to soil on the availability of As, Cd, Cu, Pb, and Zn to maize (Zea mays L.). Aust J Soil Res 48:638–647. https://doi.org/10.1071/SR10049

Nautiyal P, Subramanian KA, Dastidar MG (2016) Adsorptive removal of dye using biochar derived from residual algae after in-situ transesterification: alternate use of waste of biodiesel industry. J Environ Manage 182:187–197. https://doi.org/10.1016/j.jenvman.2016.07.063

Neves D, Thunman H, Matos A et al (2011) Characterization and prediction of biomass pyrolysis products. Prog Energy Combust Sci 37:611–630. https://doi.org/10.1016/j.pecs.2011.01.001

Nichols R (2015) A Hedge against drought: why healthy soil is 'Water in the Bank.' United States Dep Agric Nat Resour Conserv Serv

Onay O, Kockar OM (2003) Slow, fast and flash pyrolysis of rapeseed. Renew Energy 28:2417–2433. https://doi.org/10.1016/S0960-1481(03)00137-X

Pandey VC, Bajpai O (2019) Phytoremediation: from theory towards practice. In: Pandey VC, Bauddh K (eds) Phytomanagement of polluted sites. Elsevier, Amsterdam, pp 1–49. https://doi.org/10.1016/B978-0-12-813912-7.00001-6

Pandey VC, Singh V (2019) Exploring the potential and opportunities of recent tools for removal of hazardous materials from environments. In: Pandey VC, Bauddh K (eds) Phytomanagement of polluted sites. Elsevier, Amsterdam, pp 501–516. https://doi.org/10.1016/B978-0-12-813912-7.00020-X

Pathak S, Agarwal AV, Pandey VC (2020) Phytoremediation—a holistic approach for remediation of heavy metals and metalloids. In: Pandey VC, Singh V (eds) Bioremediation of pollutants. Elsevier, Amsterdam, pp 3–14. https://doi.org/10.1016/B978-0-12-819025-8.00001-6

Park JH, Choppala GK, Bolan NS et al (2011a) Biochar reduces the bioavailability and phytotoxicity of heavy metals. Plant Soil 348:439–451. https://doi.org/10.1007/s11104-011-0948-y

Park JH, Lamb D, Paneerselvam P et al (2011b) Role of organic amendments on enhanced bioremediation of heavy metal(loid) contaminated soils. J Hazard Mater 185:549–574. https://doi.org/10.1016/j.jhazmat.2010.09.082

Patra RC, Swarup D, Sharma MC, Naresh R (2006) Trace mineral profile in blood and hair from cattle environmentally exposed to lead and cadmium around different industrial units. J Vet Med Ser A Physiol Pathol Clin Med 53:511–517. https://doi.org/10.1111/j.1439-0442.2006.00868.x

Peng X, Ye LL, Wang CH et al (2011) Temperature- and duration-dependent rice straw-derived biochar: characteristics and its effects on soil properties of an Ultisol in southern China. Soil Tillage Res 112:159–166. https://doi.org/10.1016/j.still.2011.01.002

Pietikäinen J, Kiikkilä O, Fritze H (2000) Charcoal as a habitat for microbes and its effect on the microbial community of the underlying humus. Oikos 89:231–242. https://doi.org/10.1034/j.1600-0706.2000.890203.x

Pilon-Smits EA, Quinn CF, Tapken W et al (2009) Physiological functions of beneficial elements. Curr Opin Plant Biol 12:267–274. https://doi.org/10.1016/j.pbi.2009.04.009

Pimchuai A, Dutta A, Basu P (2010) Torrefaction of agriculture residue to enhance combustible properties. Energy Fuels 24:4638–4645. https://doi.org/10.1021/ef901168f

Pröll T, Aichernig C, Rauch R, Hofbauer H (2007) Fluidized bed steam gasification of solid biomass—performance characteristics of an 8 MWth combined heat and power plant. Int J Chem React Eng 5. https://doi.org/10.2202/1542-6580.1398

Puga AP, Abreu CA, Melo LCA, Beesley L (2015) Biochar application to a contaminated soil reduces the availability and plant uptake of zinc, lead and cadmium. J Environ Manage 159:86–93. https://doi.org/10.1016/j.jenvman.2015.05.036

Pulido-Novicio L, Hata T, Kurimoto Y et al (2001) Adsorption capacities and related characteristics of wood charcoals carbonized using a one-step or two-step process. J Wood Sci 47:48–57. https://doi.org/10.1007/BF00776645

Qian L, Zhang W, Yan J et al (2016) Effective removal of heavy metal by biochar colloids under different pyrolysis temperatures. Bioresour Technol 206:217–224. https://doi.org/10.1016/j.biortech.2016.01.065

Qiu Y, Cheng H, Xu C, Sheng GD (2008) Surface characteristics of crop-residue-derived black carbon and lead(II) adsorption. Water Res 42:567–574. https://doi.org/10.1016/j.watres.2007.07.051

Rajkovich S, Enders A, Hanley K et al (2012) Corn growth and nitrogen nutrition after additions of biochars with varying properties to a temperate soil. Biol Fertil Soils 48:271–284. https://doi.org/10.1007/s00374-011-0624-7

Rees F, Simonnot MO, Morel JL (2014) Short-term effects of biochar on soil heavy metal mobility are controlled by intra-particle diffusion and soil pH increase. Eur J Soil Sci 65:149–161. https://doi.org/10.1111/ejss.12107

Rodriguez A, Lemos D, Trujillo YT et al (2019) Effectiveness of biochar obtained from corncob for immobilization of lead in contaminated soil. J Heal Pollut 9. https://doi.org/10.5696/2156-9614-9.23.190907

Rouquerol F, Rouquerol J, Sing K (1999) Introduction. Adsorpt by Powders Porous Solids, pp 1–26

Sadaka S, Sharara MA, Ashworth A et al (2014) Characterization of biochar from switchgrass carbonization. Energies 7:548–567. https://doi.org/10.3390/en7020548

Saengwilai P, Meeinkuirt W, Phusantisampan T, Pichtel J (2020) Immobilization of cadmium in contaminated soil using organic amendments and its effects on rice growth performance. Expo Heal 12:295–306. https://doi.org/10.1007/s12403-019-00312-0

Sahota S, Vijay VK, Subbarao PMV et al (2018) Characterization of leaf waste based biochar for cost effective hydrogen sulphide removal from biogas. Bioresour Technol 250:635–641. https://doi.org/10.1016/j.biortech.2017.11.093

Samsuri AW, Sadegh-Zadeh F, Seh-Bardan BJ (2013) Adsorption of As(III) and As(V) by Fe coated biochars and biochars produced from empty fruit bunch and rice husk. J Environ Chem Eng 1:981–988. https://doi.org/10.1016/j.jece.2013.08.009

Sanchez-Monedero MA, Cayuela ML, Roig A et al (2018) Role of biochar as an additive in organic waste composting. Bioresour Technol 247:1155–1164. https://doi.org/10.1016/j.biortech.2017.09.193

Sánchez-Polo M, Rivera-Utrilla J (2002) Adsorbent-adsorbate interactions in the adsorption of Cd(II) and Hg(II) on ozonized activated carbons. Environ Sci Technol 36:3850–3854. https://doi.org/10.1021/es0255610

Sevilla M, Maciá-Agulló JA, Fuertes AB (2011) Hydrothermal carbonization of biomass as a route for the sequestration of CO2: chemical and structural properties of the carbonized products. Biomass Bioenerg 35:3152–3159. https://doi.org/10.1016/j.biombioe.2011.04.032

Shaikh WA, Islam RU, Chakraborty S (2021) Stable silver nanoparticle doped mesoporous biochar-based nanocomposite for efficient removal of toxic dyes. J Environ Chem Eng 9(1):104982. S2213343720313312. https://doi.org/10.1016/j.jece.2020.104982

Shaikh WA, Chakraborty S, Islam RU, Ghfar AA, Naushad M, Bundschuh J, Maity JP, Mondal NK (2022a) Fabrication of biochar-based hybrid Ag nanocomposite from algal biomass waste for toxic dye-laden wastewater treatment. Chemosphere 289:133243. S0045653521037176 133243. https://doi.org/10.1016/j.chemosphere.2021.133243

Shaikh WA, Kumar A, Chakraborty S, Islam RL, Bhattacharya T, Biswas JK (2022b) Biochar-based nanocomposite from waste tea leaf for toxic dye removal: from facile fabrication to functional fitness. Chemosphere 291:132788. S0045653521032604. https://doi.org/10.1016/j.chemosphere.2021.132788

Shen Z, Jin F, Wang F et al (2015) Sorption of lead by Salisbury biochar produced from British broadleaf hardwood. Bioresour Technol 193:553–556. https://doi.org/10.1016/j.biortech.2015.06.111

Shen Z, McMillan O, Jin F, Al-Tabbaa A (2016a) Salisbury biochar did not affect the mobility or speciation of lead in kaolin in a short-term laboratory study. J Hazard Mater 316:214–220. https://doi.org/10.1016/j.jhazmat.2016.05.042

Shen Z, Som AM, Wang F et al (2016b) Long-term impact of biochar on the immobilisation of nickel (II) and zinc (II) and the revegetation of a contaminated site. Sci Total Environ 542:771–776. https://doi.org/10.1016/j.scitotenv.2015.10.057

Shen Z, Zhang Y, Jin F et al (2017) Qualitative and quantitative characterisation of adsorption mechanisms of lead on four biochars. Sci Total Environ 609:1401–1410. https://doi.org/10.1016/j.scitotenv.2017.08.008

Shivaram P, Leong YK, Yang H, Zhang DK (2013) Flow and yield stress behaviour of ultrafine Mallee biochar slurry fuels: the effect of particle size distribution and additives. Fuel 104:326–332. https://doi.org/10.1016/j.fuel.2012.09.015

Siebers N, Godlinski F, Leinweber P (2014) Bone char as phosphorus fertilizer involved in cadmium immobilization in lettuce, wheat, and potato cropping. J Plant Nutr Soil Sci 177:75–83. https://doi.org/10.1002/jpln.201300113

Singh BP, Cowie AL, Smernik RJ (2012) Biochar carbon stability in a clayey soil as a function of feedstock and pyrolysis temperature. Environ Sci Technol 46:11770–11778. https://doi.org/10.1021/es302545b

Skjemstad JO, Reicosky DC, Wilts AR, McGowan JA (2002) Charcoal carbon in U.S. agricultural soils. Soil Sci Soc Am J 66:1249–1255. https://doi.org/10.2136/sssaj2002.1249

Sohi SP (2012) Carbon storage with benefits. Science (80-) 338:1034–1035. https://doi.org/10.1126/science.1225987

Sohi SP, Krull E, Lopez-Capel E, Bol R (2010) A review of biochar and its use and function in soil. In: Sparks DL (ed) Advances in agronomy. Academic Press, Burlington, pp 47–82

Solaiman ZM, Anawar HM (2015) Application of biochars for soil constraints: challenges and solutions. Pedosphere 25:631–638. https://doi.org/10.1016/S1002-0160(15)30044-8

Sophia Ayyappan C, Bhalambaal VM, Kumar S (2018) Effect of biochar on bio-electrochemical dye degradation and energy production. Bioresour Technol 251:165–170. https://doi.org/10.1016/j.biortech.2017.12.043

Spokas KA, Koskinen WC, Baker JM, Reicosky DC (2009) Impacts of woodchip biochar additions on greenhouse gas production and sorption/degradation of two herbicides in a Minnesota soil. Chemosphere 77:574–581. https://doi.org/10.1016/j.chemosphere.2009.06.053

Stadtman ER (1990) Metal ion-catalyzed oxidation of proteins: biochemical mechanism and biological consequences. Free Radic Biol Med 9:315–325. https://doi.org/10.1016/0891-5849(90)90006-5

Steiner C, Garcia M, Zech W (2009) Effects of charcoal as slow release nutrient carrier on N-P-K dynamics and soil microbial population: Pot experiments with ferralsol substrate. Amaz Dark Earths Wim Sombroek's Vis, pp 325–338

Sun K, Keiluweit M, Kleber M et al (2011) Sorption of fluorinated herbicides to plant biomass-derived biochars as a function of molecular structure. Bioresour Technol 102:9897–9903. https://doi.org/10.1016/j.biortech.2011.08.036

Sun H, Hockaday WC, Masiello CA, Zygourakis K (2012a) Multiple controls on the chemical and physical structure of biochars. Ind Eng Chem Res 51:1587–1597. https://doi.org/10.1021/ie201309r

Sun K, Gao B, Ro KS et al (2012b) Assessment of herbicide sorption by biochars and organic matter associated with soil and sediment. Environ Pollut 163:167–173. https://doi.org/10.1016/j.envpol.2011.12.015

Sun Y, Gao B, Yao Y et al (2014) Effects of feedstock type, production method, and pyrolysis temperature on biochar and hydrochar properties. Chem Eng J 240:574–578. https://doi.org/10.1016/j.cej.2013.10.081

Tang J, Zhu W, Kookana R, Katayama A (2013) Characteristics of biochar and its application in remediation of contaminated soil. J Biosci Bioeng 116:653–659. https://doi.org/10.1016/j.jbiosc.2013.05.035

Tangahu BV, Sheikh Abdullah SR, Basri H et al (2011) A review on heavy metals (As, Pb, and Hg) uptake by plants through phytoremediation. Int J Chem Eng 2011:1–31. https://doi.org/10.1155/2011/939161

Tchounwou PB, Yedjou CG, Patlolla AK, Sutton DJ (2012) Heavy metal toxicity and the environment. EXS 101:133–164

Titirici MM, White RJ, Falco C, Sevilla M (2012) Black perspectives for a green future: hydrothermal carbons for environment protection and energy storage. Energy Environ Sci 5:6796–6822. https://doi.org/10.1039/c2ee21166a

Tong XJ, Li JY, Yuan JH, Xu RK (2011) Adsorption of Cu(II) by biochars generated from three crop straws. Chem Eng J 172:828–834. https://doi.org/10.1016/j.cej.2011.06.069

Topoliantz S, Ponge JF (2003) Burrowing activity of the geophagous earthworm Pontoscolex corethrurus (Oligochaeta: Glossoscolecidae) in the presence of charcoal. Appl Soil Ecol 23:267–271. https://doi.org/10.1016/S0929-1393(03)00063-5

Tripathi M, Sahu JN, Ganesan P (2016) Effect of process parameters on production of biochar from biomass waste through pyrolysis: a review. Renew Sustain Energy Rev 55:467–481. https://doi.org/10.1016/j.rser.2015.10.122

Uchimiya M, Lima IM, Klasson KT, Wartelle LH (2010) Contaminant immobilization and nutrient release by biochar soil amendment: roles of natural organic matter. Chemosphere 80:935–940. https://doi.org/10.1016/j.chemosphere.2010.05.020

Uchimiya M, Klasson KT, Wartelle LH, Lima IM (2011a) Influence of soil properties on heavy metal sequestration by biochar amendment: 1. Copper sorption isotherms and the release of cations. Chemosphere 82:1431–1437. https://doi.org/10.1016/j.chemosphere.2010.11.050

Uchimiya M, Wartelle LH, Klasson KT et al (2011b) Influence of pyrolysis temperature on biochar property and function as a heavy metal sorbent in soil. J Agric Food Chem 59:2501–2510. https://doi.org/10.1021/jf104206c

Usowicz B, Lipiec J, Marczewski W, Ferrero A (2006) Thermal conductivity modelling of terrestrial soil media—a comparative study. Planet Space Sci 54:1086–1095. https://doi.org/10.1016/j.pss.2006.05.018

Valenzuela-Calahorro C, Bernalte-Garcia A, Gómez-Serrano V, Bernalte-García MJ (1987) Influence of particle size and pyrolysis conditions on yield, density and some textural parameters of chars prepared from holm-oak wood. J Anal Appl Pyrolysis 12:61–70. https://doi.org/10.1016/0165-2370(87)80015-3

van Zwieten L, Kimber S, Morris S et al (2010) Effects of biochar from slow pyrolysis of papermill waste on agronomic performance and soil fertility. Plant Soil 327:235–246. https://doi.org/10.1007/s11104-009-0050-x

Vassilev SV, Baxter D, Andersen LK, Vassileva CG (2010) An overview of the chemical composition of biomass. Fuel 89:913–933. https://doi.org/10.1016/j.fuel.2009.10.022

Vassilev N, Martos E, Mendes G et al (2013a) Biochar of animal origin: a sustainable solution to the global problem of high-grade rock phosphate scarcity? J Sci Food Agric 93:1799–1804. https://doi.org/10.1002/jsfa.6130

Vassilev SV, Baxter D, Andersen LK, Vassileva CG (2013b) An overview of the composition and application of biomass ash. Part 1. Phase-mineral and chemical composition and classification. Fuel 105:40–76. https://doi.org/10.1016/j.fuel.2012.09.041

Ventura M, Sorrenti G, Panzacchi P et al (2013) Biochar reduces short-term nitrate leaching from a horizon in an Apple Orchard. J Environ Qual 42:76–82. https://doi.org/10.2134/jeq2012.0250

Verheijen F, Jeffery S, Bastos AC et al (2010) Biochar application to soils: a critical scientific review of effects on soil properties, Processes and Functions. Environment 8:144. https://doi.org/10.2788/472

Verhoeven E, Six J (2014) Biochar does not mitigate field-scale N2O emissions in a Northern California vineyard: an assessment across two years. Agric Ecosyst Environ 191:27–38. https://doi.org/10.1016/j.agee.2014.03.008

von Gunten K, Hubmann M, Ineichen R et al (2019) Biochar-induced changes in metal mobility and uptake by perennial plants in a ferralsol of Brazil's Atlantic forest. Biochar 1:309–324. https://doi.org/10.1007/s42773-019-00018-1

Wan Ngah WS, Hanafiah MAKM (2008) Removal of heavy metal ions from wastewater by chemically modified plant wastes as adsorbents: a review. Bioresour Technol 99:3935–3948. https://doi.org/10.1016/j.biortech.2007.06.011

Wang S, Gao B, Zimmerman AR et al (2015) Removal of arsenic by magnetic biochar prepared from pinewood and natural hematite. Bioresour Technol 175:391–395. https://doi.org/10.1016/j.biortech.2014.10.104

Wang Z, Han L, Sun K et al (2016) Sorption of four hydrophobic organic contaminants by biochars derived from maize straw, wood dust and swine manure at different pyrolytic temperatures. Chemosphere 144:285–291. https://doi.org/10.1016/j.chemosphere.2015.08.042

Wang Y, Li F, Rong X et al (2017) Remediation of Petroleum-contaminated Soil Using Bulrush Straw Powder, Biochar and Nutrients. Bull Environ Contam Toxicol 98:690–697. https://doi.org/10.1007/s00128-017-2064-z

Wang M, Zhu Y, Cheng L et al (2018a) Review on utilization of biochar for metal-contaminated soil and sediment remediation. J Environ Sci (china) 63:156–173. https://doi.org/10.1016/j.jes.2017.08.004

Wang T, Zhai Y, Zhu Y et al (2018b) A review of the hydrothermal carbonization of biomass waste for hydrochar formation: Process conditions, fundamentals, and physicochemical properties. Renew Sustain Energy Rev 90:223–247. https://doi.org/10.1016/j.rser.2018.03.071

Wang Y, Wang H-S, Tang C-S et al (2019) Remediation of heavy-metal-contaminated soils by biochar: a review. Environ Geotech 1–14. https://doi.org/10.1680/jenge.18.00091

Warnock DD, Lehmann J, Kuyper TW, Rillig MC (2007) Mycorrhizal responses to biochar in soil—concepts and mechanisms. Plant Soil 300:9–20. https://doi.org/10.1007/s11104-007-9391-5

Warren GP, Alloway BJ, Lepp NW et al (2003) Field trials to assess the uptake of arsenic by vegetables from contaminated soils and soil remediation with iron oxides. Sci Total Environ 311:19–33. https://doi.org/10.1016/S0048-9697(03)00096-2

Weber K, Quicker P (2016) Eigenschaften von Biomassekarbonisaten. Biokohle 165–212

Weber K, Quicker P (2018) Properties of biochar. Fuel 217:240–261. https://doi.org/10.1016/j.fuel.2017.12.054

Windeatt JH, Ross AB, Williams PT et al (2014) Characteristics of biochars from crop residues: potential for carbon sequestration and soil amendment. J Environ Manage 146:189–197. https://doi.org/10.1016/j.jenvman.2014.08.003

Xie Z, Liu Q, Zhu C (2011) Advances and perspective of biochar research. Soils 43:857–861

Xie T, Reddy KR, Wang C et al (2015) Characteristics and applications of biochar for environmental remediation: a review. Crit Rev Environ Sci Technol 45:939–969. https://doi.org/10.1080/10643389.2014.924180

Xu RK, Xiao SC, Yuan JH, Zhao AZ (2011) Adsorption of methyl violet from aqueous solutions by the biochars derived from crop residues. Bioresour Technol 102:10293–10298. https://doi.org/10.1016/j.biortech.2011.08.089

Xu T, Lou L, Luo L et al (2012) Effect of bamboo biochar on pentachlorophenol leachability and bioavailability in agricultural soil. Sci Total Environ 414:727–731. https://doi.org/10.1016/j.scitotenv.2011.11.005

Xu X, Cao X, Zhao L et al (2013a) Removal of Cu, Zn, and Cd from aqueous solutions by the dairy manure-derived biochar. Environ Sci Pollut Res 20:358–368. https://doi.org/10.1007/s11356-012-0873-5

Xu X, Cao X, Zhao L (2013b) Comparison of rice husk- and dairy manure-derived biochars for simultaneously removing heavy metals from aqueous solutions: role of mineral components in biochars. Chemosphere 92:955–961. https://doi.org/10.1016/j.chemosphere.2013.03.009

Xue Y, Gao B, Yao Y et al (2012) Hydrogen peroxide modification enhances the ability of biochar (hydrochar) produced from hydrothermal carbonization of peanut hull to remove aqueous heavy metals: Batch and column tests. Chem Eng J 200–202:673–680. https://doi.org/10.1016/j.cej.2012.06.116

Yaghoubi P, Yargicoglu EN, Reddy KR (2014) Effects of biochar-amendment to landfill cover soil on microbial methane oxidation: initial results. Geotech Spec Publ 1849–1858

Yao Y, Gao B, Inyang M et al (2011) Biochar derived from anaerobically digested sugar beet tailings: characterization and phosphate removal potential. Bioresour Technol 102:6273–6278. https://doi.org/10.1016/j.biortech.2011.03.006

Yargicoglu EN, Sadasivam BY, Reddy KR, Spokas K (2015) Physical and chemical characterization of waste wood derived biochars. Waste Manag 36:256–268. https://doi.org/10.1016/j.wasman.2014.10.029

Yin D, Wang X, Chen C et al (2016) Varying effect of biochar on Cd, Pb and As mobility in a multi-metal contaminated paddy soil. Chemosphere 152:196–206. https://doi.org/10.1016/j.chemosphere.2016.01.044

Yorgun S, Ensöz S, Koçkar ÖM (2001) Flash pyrolysis of sunflower oil cake for production of liquid fuels. J Anal Appl Pyrolysis 60:1–12. https://doi.org/10.1016/S0165-2370(00)00102-9

Young S (1995) Toxic metals in soil-plant systems. In: Ross SM (ed) xiv + 469 pp. John Wiley & Sons, Chichester (1994). £55.00 (hardback). ISBN 0 471 94279 0. J Agric Sci 124:155–156. https://doi.org/10.1017/s0021859600071422

Yu XY, Ying GG, Kookana RS (2009) Reduced plant uptake of pesticides with biochar additions to soil. Chemosphere 76:665–671. https://doi.org/10.1016/j.chemosphere.2009.04.001

Yu OY, Raichle B, Sink S (2013) Impact of biochar on the water holding capacity of loamy sand soil. Int J Energy Environ Eng 4:1–9. https://doi.org/10.1186/2251-6832-4-44

Yuan JH, Xu RK, Zhang H (2011) The forms of alkalis in the biochar produced from crop residues at different temperatures. Bioresour Technol 102:3488–3497. https://doi.org/10.1016/j.biortech.2010.11.018

Zhang J, You C (2013) Water holding capacity and absorption properties of wood chars. Energy Fuels 27:2643–2648. https://doi.org/10.1021/ef4000769

Zhang W, Li G, Gao W (2009) Effects of biochar on soil properties and crop yield. Chinese Agric Sci Bull 25:153–157

Zhang X, Wang H, He L et al (2013) Using biochar for remediation of soils contaminated with heavy metals and organic pollutants. Environ Sci Pollut Res 20:8472–8483. https://doi.org/10.1007/s11356-013-1659-0

Zhang C, Liu L, Zhao M et al (2018a) The environmental characteristics and applications of biochar. Environ Sci Pollut Res 25:21525–21534. https://doi.org/10.1007/s11356-018-2521-1

Zhang S, Abdalla MAS, Luo Z, Xia S (2018b) The wheat straw biochar research on the adsorption/desorption behaviour of mercury in wastewater. Desalin Water Treat 112:147–160. https://doi.org/10.5004/dwt.2018.21850

Zhao X, Ouyang W, Hao F et al (2013) Properties comparison of biochars from corn straw with different pretreatment and sorption behaviour of atrazine. Bioresour Technol 147:338–344. https://doi.org/10.1016/j.biortech.2013.08.042

Zhou JB, Deng CJ, Chen JL, Zhang QS (2008) Remediation effects of cotton stalk carbon on cadmium (Cd) contaminated soil. Ecol Env 17:1857–1860

Zornoza R, Moreno-Barriga F, Acosta JA et al (2016) Stability, nutrient availability and hydrophobicity of biochars derived from manure, crop residues, and municipal solid waste for their use as soil amendments. Chemosphere 144:122–130. https://doi.org/10.1016/j.chemosphere.2015.08.046

Chapter 14
Soil Carbon Sequestration Strategies: Application of Biochar an Option to Combat Global Warming

Shweta Yadav, Vikas Sonkar, and Sandeep K. Malyan

Abstract Carbon dioxide (CO_2) is an important heat-trapping greenhouse gas (GHG) contributing substantially to global warming with the average annual emission of 409.8 ± 0.01 ppm (2019), increasing at the rate of 2.5 ppm/year. Transferring the atmospheric CO_2 and storing it in the long-lived natural carbon pools (i.e., oceans, biotic, pedologic, and fossil fuel) to avoid its reemission is often termed as carbon sequestration. The terrestrial carbon pool (pedologic pool and biotic pool) accounts for 3120 Petagram (Pg) of carbon (C) which is four times higher than the atmospheric carbon pool, contributing significantly to the global carbon cycle (GCC). Therefore, soil carbon sequestration is considered an important pathway to climate change mitigation by achieving the global target of <2 °C. Carbon sequestration in the soil can be achieved by several strategies based on the land use type such as conservation/reduced tillage, afforestation, restoration of peatlands, water conservation, and urban forests. However, adding soil amendments such as biochar is a promising strategy that considerably increases the positive carbon budget in the soil. Enhancing the soil organic carbon (SOC) using biochar provides multiple co-benefits such as increased soil fertility, improved water retention, nutrient retention thus provides food security, soil health, and can potentially reduce global warming. To maximize the carbon sequestration in soil, biochar application in conjunction with other carbon-capturing strategies such as crop rotation, no-tillage (NT), and reforestation should be followed vigorously.

Keywords Biochar · Carbon sequestration · Carbon dioxide · Soil organic carbon · Global carbon cycle

14.1 Introduction

Over the past few decades, global attention toward mitigating the effects of global warming by stabilizing the atmospheric CO_2 and other GHGs such as nitrous oxide

---

S. Yadav (✉) · V. Sonkar · S. K. Malyan
Research Management and Outreach Division, National Institute of Hydrology, Jalvigyan Bhawan, Roorkee 247667, India
e-mail: yadav.shweta.64s@kyoto-u.ac.jp; shwtdv@gmail.com

© The Author(s), under exclusive license to Springer Nature Switzerland AG 2023
V. C. Pandey (ed.), *Bio-Inspired Land Remediation*, Environmental Contamination Remediation and Management, https://doi.org/10.1007/978-3-031-04931-6_14

(N_2O) and methane (CH_4) has increased tremendously (Stocker et al. 2013; Malyan 2017; Malyan et al. 2019a, b, 2021b, c; Fagodiya et al. 2020; Kumar et al. 2020). The presence of GHGs increases the radiative forcing that alters the mean precipitation and temperature of the Earth. The global warming potential (GWP: relative greenhouse efficiency) of the atmospheric gases determines their contribution to climate change. CO_2 remains one of the most abundant GHG gases in the atmosphere with a GWP of 1, followed by CH_4 and N_2O with a 100-year GWP of 28 and 265, respectively (Gupta et al. 2016; Malyan et al. 2016; IPCC 2014). Since the early twentieth century, the Earth's climate is primarily driven by anthropogenic activities (mainly burning of fossil fuel) which increased the levels of heat-trapping GHG in the Earths' atmosphere. Carbon (C) is the fourth most abundant element which is present in all living organisms and is the main building block of life on planet Earth. Living cells are made up of complex macro-molecules that include carbohydrate, proteins, lipids, and nucleic acids. Carbon is present in various forms, predominantly as biomass (plant), CO_2, and soil organic matter (SOM). In the atmosphere, CO_2 plays a crucial role in maintaining Earth's temperature through global warming (Malyan et al. 2019a, b). The average Earth's temperature without global warming is expected to be $-19\ °C$, which is an unfavorable condition for the existence of life on earth. Therefore, the presence of various forms of C is essential for the existence of living beings on earth.

Under the Framework Convention on Climate Change (UNFCCC), "a process or activity which releases a greenhouse gas, or aerosol or a precursor of a GHGs into the atmosphere" is termed as the source. However, a sink is any activity or process which reduces or removes the GHGs and other heat-trapping gases from Earth's atmosphere. Carbon sequestration is the process in which carbon is captured and stored permanently or for a long time in Earth's carbon pool to prevent its reemission in the atmosphere (FAO 2000). The terrestrial biosphere facilitates carbon sequestration with the help of biomass, soil, and other vegetation in the removal of the atmospheric CO_2. From the plant tissue, the carbon is consumed by the animals and possibly returned to the soil as soil organic matter (SOM) when plants die or in the form of litter. SOM is a major pathway for storing carbon in the soil. SOM significantly influences the soil physicochemical and biological properties such as water retention and porosity, and therefore, it is often referred to as "black gold". Enhancing soil carbon in soil can substantially improve soil fertility, nutrient retention, and water retention. Apart from improving soil health, C storage in the soil helps in mitigating global warming. There are many conventional strategies through which the rate of C sequestration can be enhanced. Agroforestry, soil management, zero tillage, and afforestation are the most common conventional methods for C sequestration. Recently, biochar is gaining global attention due to its multiple applications in environmental management-related activities (Table 14.1). The chapter presents multiple strategies for carbon sequestration in soil emphasizing the application of biochar in increasing the positive carbon budget in the soil.

Table 14.1 Advantages of biochar in environmental management

S. No	Advantages of biochar	References
1	Methane emission mitigation	Huang et al. (2019), Wu et al. (2019a, b), Mona et al. (2021), Malyan et al. (2021a)
2	Nitrous oxide emission mitigation	Cayuela et al. (2014), Wu et al. (2019a), Liu et al. (2020)
3	Pollutants remediation	Zheng et al. (2019), Cheng et al. (2020), Khalid et al. (2020), Zand et al. (2020)
4	Soil health	Purakayastha et al. (2019), Zhang et al. (2019), Hue (2020)
5	Improve nutrient use efficiency	Zhang et al. (2016), Purakayastha et al. (2019)

14.2 Role of Carbon Dioxide (CO_2) in Global Warming

The climate of the earth is continuously changing due to various factors which is one of the most serious threats (Fig. 14.1). Climate change is directly associated with the anthropogenic GHGs (mainly CO_2) that leads to the warming of the surface of the earth and troposphere (Blunden and Arndt 2020). GHG absorbs and radiates the long wave infrared radiation. Like other gases in the atmosphere such as nitrogen and oxygen, GHGs are more transparent to incoming sunlight but unlike other gases, these are not transparent to long wave infrared radiation (heat). These GHGs act as a blanket insulating the earth (Montzka et al. 2011). The Sun heats the earth's surface and radiates day and night. Some of the heat escapes easily to space, but some of it

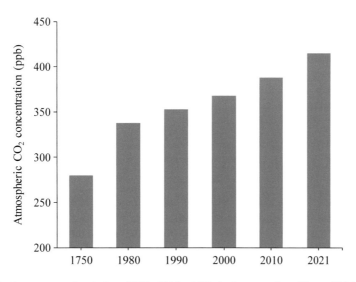

Fig. 14.1 Average annual emissions (1750–2021) of CO_2 in the atmosphere (*Source* NOAA 2021)

gets absorbed by the GHG molecules. These GHG molecules emit heat radiations back into their environments. Therefore, they are known as heat-trapping gases, and this phenomenon is termed as the greenhouse effect (Lindsey and Dlugokencky 2020). According to the Intergovernmental Panel on Climate Change (IPCC), since the mid-twentieth century the increase in the average temperature and CO_2 globally plays a major role in greenhouse effect (IPCC 2013). Since the Industrial Revolution, the atmospheric CO_2 concentration increased >40% (415 ± 0.01 ppm in 2021) (Fig. 14.1). It is higher than at any point in at least the past 8 million years (Moreira and Pires 2016; Lindsey and Dlugokencky 2020).

Among the various anthropogenic sources, fossil fuel (coal and oil contain carbon) burning for energy production is the major source for increasing CO_2 concentration in atmosphere (Lindsey and Dlugokencky 2020). According to State of the Climate in 2019 from NOAA and the American Meteorological Society, CO_2 emission due to burning of fossil fuel recorded from 1850 to 2018 is 440 ± 20 Pg C (1 Pg C = 1015 g C) (Friedlingstein et al. 2019). Almost half of the emitted CO_2 remains in the atmosphere since 1850. The rest of the CO_2 partially gets dissolved in the oceans and other natural sinks. Carbon dioxide increases 100 times faster in the past 60 years than the previous natural increase in the atmosphere (Friedlingstein et al. 2019).

CO_2 is an abundant and long-lived GHG on the earth. According to EPA, among the all GHGs such as CH_4, N_2O, and chlorofluorocarbons (CFCs), CO_2 absorbs unit heat per molecules. But CO_2, due to anthropogenic emission, is abundant, and its life span is much longer in the atmosphere than any other GHGs. The emitted CH_4 takes a decade to leave the atmosphere (converted to CO_2) and around a century for N_2O. While 40% of CO_2 will remain for century, and 20% will remain for 1000 years, and the last 10% takes 10,000 years to turn over in the atmosphere (Ucsusa.org 2017).

CO_2 is not as strong as water vapor in absorbing heat on a per molecule basis, whereas CO_2 has a property to absorb thermal energy that is not possible by water vapor. Because water vapor has a short life span (average 10 days) it falls to the earth before contributing to the climate change. Thus, CO_2 contributes uniquely to the global warming (Baldocchi and Wilson 2001; Al-Ghussain 2019). The GHGs contribute to the global warming defined as "the rise of average temperature of earth surface" which is evident by the increasing global mean temperature (Anderson et al. 2016).

14.2.1 Global Carbon Cycle (GCC)

The global carbon cycle (GCC) involves the storage and exchange of carbon (C) among several different reservoirs or pools of the planet Earth. The carbon cycle accounts for the removal of atmospheric CO_2 in the land and the ocean through the process of photosynthesis, along with the exchange of CO_2 by the natural processes such as respiration, ecological processes, biological growth, and atmospheric transport, gas solubility, and anthropogenic activities (e.g., land use change, biomass, and the combustion of fossil fuel) (Ito et al. 2020). Since industrialization, humans have moved various carbon reservoirs such as fossil fuel reservoirs (naturally fluctuate at geochemical time scale) in the form of greenhouse gases (e.g., CO_2) into the atmosphere largely due to anthropogenic activities. Consequently, the annual average atmospheric CO_2 increased from 280 ppm during the preindustrial era (1851–1900) (Ito et al. 2020) to 409.8 ± 0.01 ppm in 2019 (Arndt et al. 2020), on the short time scale (i.e., seconds to millennia). This exceeded the rate of increase of CO_2 to 2.5 ppm/year in 2017 and 2018 than the past century (~1.0 ppm/year) (Isson et al. 2020; Ito et al. 2020). The atmospheric CO_2 influences climate change by the greenhouse effect, altering the global carbon cycle (GCC). The emission of 4 petagrams (Pg) of carbon (1 Pg is equivalent to 1 billion metric ton or 1 gigaton) through various anthropogenic activities increases approximately 1 ppm of atmospheric CO_2 (Lal 2010; Arndt et al. 2020). The combined effect of several greenhouse gases and other halogenated gases on the radiative forcing was estimated to be 3.14 Wm^{-2} indicating a 45% increase from the year 1990 while CO_2 alone contributes 65% of this radiative forcing on the Earth's system (Arndt et al. 2020).

Globally, out of 9.9 Pg C yr^{-1} of the total anthropogenic emission (such as from fossil fuels, deforestation, agriculture, and land use alteration), the atmosphere and the oceans absorb 4.2 Pg C yr^{-1} and 2.3 Pg C yr^{-1}, respectively. The remaining emitted CO_2 is believed to be absorbed by the terrestrial sinks (Lal 2012). In general, the global carbon pools are divided into five major categories based on the carbon budget (Table 14.2), they are: (a) Oceanic pool with 38×10^3 Pg C, the rate of increase is 2.3 Pg C yr^{-1}; (b) Fossil fuels with 5×10^3 Pg C to 10×10^3 Pg C, mined/combusted at the rate of 8 Pg C yr^{-1}; (c) Pedologic pool, at 1 m depth comprised of 1.55×10^3 Pg of SOC and 950 Pg of soil inorganic carbon (SIC); (d) Atmospheric pool with 780 Pg C, the rate of increase is 4 Pg C yr^{-1}; and (e) Biotic pool with 560 Pg C of biomass along with detritus material of 60 Pg C (Lal 2010, 2012; FAO 2017). The terrestrial carbon pool comprises both pedologic pool and biotic pool, accounts for 3120 Pg C. The terrestrial C pool is four times higher than the atmospheric C pool, whereas the pedologic pool is 3.2 times higher (Lal 2010). Therefore, terrestrial C pool which through photosynthesis and respiration strongly interacts with the atmospheric pool is recognized as an important C sink.

Table 14.2 Global carbon budget and the major categories of global carbon pool (Lal 2010)

S. No	Categories of global carbon pool	Global carbon pool (Pg C yr^{-1})	Rate of increase (Pg C yr^{-1})	Global carbon pool (%)
1	Oceanic pool	38×10^3	2.3	77.4
2	Fossil fuels	5×10^3	8.0	14.9
3	Terrestrial pool	1.55×10^3 (SOC) + 950 (SIC) [Pedologic Pool]	–	5.0
		620 [Biotic Pool]	–	1.2
4	Atmospheric pool	780	4.0	1.5

14.3 Soil Organic Carbon (SOC)

Soil organic carbon (SOC) is a vital component of the much larger cycle (i.e., GCC). SOC sequestration can mitigate climate change by capturing the atmospheric CO_2 (Lehmann et al. 2020). The organic component of the soil such as plant tissues, microorganisms, and animals in different stages of breakdown (or decomposition) is referred to as the SOM. The turnover of the SOM influences the soil fertility, GCC, climate change, and the fate of soil pollutants. Moreover, SOM is also vital for nutrient retention in soil, water retention, and soil structure stabilization, thus contributing to the agricultural productivity and environment (FAO and ITPS 2015; Van der Wal and De Boer 2017). SOM comprises 55–60% of carbon (by mass), and most of this carbon stock represents SOC (except the inorganic C content of the soil). Based on the physical and chemical stability of the SOC, it is categorized into three different pools, and they are: (a) Fast pool—active pool with the turnover time of 1–2 years, (b) Intermediate pool—partially stabilized SOC with a turnover time of 10–100 years, and (c) Slow pool—stable pool with a turnover time of 100 to >1000 years (FAO and ITPS 2015). The amount of C in the soil largely depends on the rate of carbon losses and input in the soil system, controlled by several soil attributes, chemical, biological, and anthropogenic factors. SOC influencing factors such as soil attributes include soil texture and lithology; climatic variable includes, precipitation, and temperature, biotic factors include biomass production and microbial growth, while the anthropogenic factors include land use change and land management (Alidoust et al. 2018; Paustian et al. 2019). The processes influencing the dynamic equilibrium (carbon gains and losses) of the SOC are given in Fig. 14.2.

The dynamic equilibrium of SOC depends on the carbon gains and carbon losses. The inputs such as biomass (plants and animal residues), organic amendments (such as biochar, organic manure), and carbonaceous matter (inorganic and organic compounds) deposition act as the carbon gain. The added carbon stabilizes in the soil by the formation of structural aggregates, converting to substances which are resistant to breakdown or decomposition (recalcitration) and converting to humic

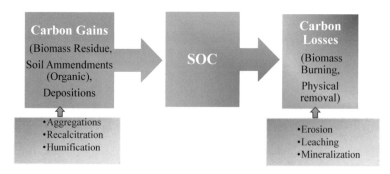

Fig. 14.2 Processes influencing the dynamic equilibrium (carbon gains and losses) of the SOC (Lal 2010)

substances (humification). Conversely, the removal of biomass by physical processes such as grazing, fire, and harvesting results in carbon losses in the soil. Other factors are soil erosion (water, gravity, and wind), mineralization, and leaching of organic compounds (dissolved form) further reduce the carbon content (Lal 2010, 2019). Soil is a potential sink and source of the C, influenced by biomass inputs, microclimatic conditions, and bioclimatic conditions. To mitigate climate change, the strategy is therefore to encourage carbon gain in the soil in comparison with carbon losses. The global agreement considers SOC sequestration a significant approach to mitigate climate change. Evidently, in 2015, at the United Nations Framework Convention for Climate Change-21st Conference of the Parties (UNFCCC-COP-21), the French Ministry of Agriculture officially launched a 4 per 1000 initiative (4p 1000) on soil for food security and climate change, with the aim to sequester 3.5 Gt C yr^{-1} (gigaton carbon) approximately in soils (as part of the Lima-Paris Action Plan). Consequently, the SOC sequestration was positively supported by scientists and several policymakers as an opportunity to mitigate the climate change (Zomer et al. 2017; Corbeels et al. 2019; Rumpel et al. 2020).

14.4 SOC Sequestration Strategies

SOC sequestration is a process that fixes the atmospheric CO_2 utilizing animal or plant residue and stores it in the soil. Enhancing soil carbon sequestration is an important pathway to climate change mitigation by achieving the global target of <2 °C (Paris Climate Agreement). Intensive agriculture and land use change over the past few decades has created a profound impact on C and nutrient (N, P) cycle as well as water quality (Yadav et al. 2019), worldwide. Land conversion to agriculture accounts for approximately 24% of the global GHG emission. Globally, 50% of the vegetated land was transformed to pastures, croplands, and other grasslands. Consequently, land use alteration and other agricultural activities have added 136 ± 55 Pg C since the industrial era (Zomer et al. 2017). Therefore, judicious management of soil and adopting the best management practices are crucial for enhancing the soil carbon pool. Several studies have suggested SOC sequestration strategies. The world soil can potentially sequester 0.4 to 1.2 Gt C yr^{-1} (Lal 2004). Crucial C sequestration approaches or strategies are discussed in this chapter based on the land use type (e.g., cropland, forest, grassland, peatland, and urban land) in this chapter (Fig. 14.3).

14.4.1 Cropland Soil Carbon Sequestration

In agriculture or cropland, cropping and tillage practice governs the plant residual input to the soil which promotes the SOC and aggregation. The carbon in the form of plant residue reenters the soil and released to the atmosphere, and these two processes control the SOC sequestration in croplands (McConkey et al. 2003; Blanco-Canqui

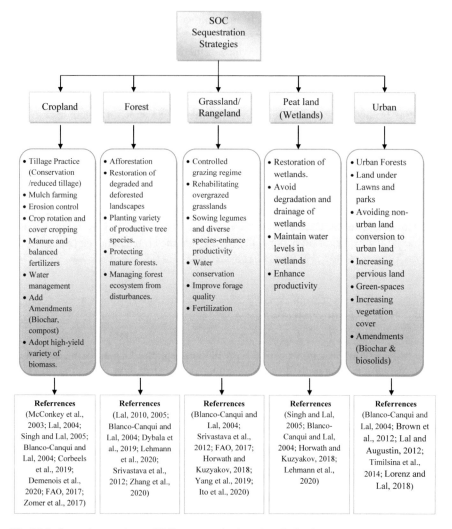

Fig. 14.3 Strategies to enhance SOC sequestration based on the land use type

and Lal 2004; Lal 2004; Singh and Lal 2005; FAO 2017; Zomer et al. 2017; Corbeels et al. 2019; Demenois et al. 2020). The cropland can potentially sequester 0.90–1.85 Pg C yr^{-1} which accounts for 25–53% of the 4 per 1000 initiative (Zomer et al. 2017).

14.4.1.1 Tillage

Reduced tillage or no-tillage (NT) prevents the breakdown or decomposition of the soil aggregates of stable form, thus protecting the organic matter from microbial degradation. In NT, the macroaggregates turnover remains low which aids in sequestering more carbon than conventional tillage (CT) practice. Studies have shown that NT soil sequesters more carbon (67–512 kg C ha^{-1} yr^{-1}) than CT (McConkey et al. 2003). In NT practice, leftover plant residues encourage the biological activity (microbial biomass such as bacteria and fungi), consequently promoting the macro and the micro-aggregates which shelter a significant amount of SOC than the CT. In some studies, GWP of NT was found to be 66% lower than CT, while the GHG emission/yield was 71% lower (Sainju 2016). Therefore, reduced tillage or no-tillage practice is recommended in agriculture to enhance SOC sequestration (Blanco-Canqui and Lal 2004; Horwath and Kuzyakov 2018).

14.4.1.2 Cover Crop and Crop Rotation

Crop rotation enhances the SOC by incorporating different crop residues. Crop residues act as an important amendment which improves the soil quality. Although crop rotation influences soil aggregates and improves soil carbon sequestration potential, however, factors such as tillage, quality of crop residue, crop type, and soil characteristics further affect the carbon sequestration during crop rotation. Legumes used for crop rotation alter the soil aggregates (micro and macroaggregates) for SOC sequestration. In some studies, the carbon sequestration by crop rotation was estimated to be 27–430 kg C ha^{-1} yr^{-1} using diverse crops (McConkey et al. 2003). For instance, Cha-un et al. (2017), found that crop rotation with sorghum-rice (RS) and corn-rice (RC) reduced the CO_2 emission by 68% to 78% and reduced the CH4 emission by 78% to 84% compared to the rice-rice (RR) cropping system during the dry and wet season (Cha-un et al. 2017). Crop rotation with NT further enhances the carbon sequestration in soils (McConkey et al. 2003; Blanco-Canqui and Lal 2004; Cha-un et al. 2017; Lorenz and Lal 2018).

14.4.1.3 Manure and Nutrient Management

Manuring and nutrient management are crucial for SOC sequestration in cropland. Improved soil fertility is beneficial for microbial biomass and carbon storages in soils. Crop rotation in conjunction with organic fertilizers substantially increases the SOC sequestration (Lal 2012, 2019). Organic manure increases the particulate organic matter (POM) in the soil, consequently stabilizes the macroaggregation, and thus enhances the carbon storage. Inorganic fertilization is crucial for C humification in residues and ultimately for SOC sequestration; however, in some cases (e.g., nitrogen fertilization), SOC sequestration is questionable. However, to improve the carbon budget in the soil, the combined application of green manure and organic fertilizers and crop rotation is recommended (Blanco-Canqui and Lal 2004; Paustian et al. 2019; Lehmann et al. 2020).

14.4.1.4 Irrigation

Well-managed irrigated land further aids in sequestering SOC in the soil predominantly in arid and semi-arid soil. In the long term, the mechanized irrigated cropland has highly stable aggregates in the soil than the land with limited irrigation or dry lands (Blanco-Canqui and Lal 2004; Lal 2004; Trost et al. 2013). Irrigated land supports biomass production by enhancing the net primary productivity in the agricultural land, thereby increasing the soil carbon content. Reduced tillage/no-tillage, in conjunction with the well-managed irrigation, further increases the SOC sequestration in the cropland. For instance, Campos et al. (2020), compared the rainfed and irrigated land SOC sequestration potential after the land use change and found that the irrigated sandy cropland sequester SOC at a much faster rate and has replenished SOC to a level similar to the native vegetation (by 20 years). In desert lands, irrigation significantly enhances the SOC sequestration from 90 to 500%, while in the semi-arid areas, the irrigation increases the SOC from 11 to 35% (Trost et al. 2013).

14.4.1.5 Soil Amendments (Biochar, Compost)

Biochar application in the soil is gaining considerable attention globally as a viable option for permanent carbon storage, with simultaneous co-benefits to the environment and society. Biochar is formed by the thermal decomposition (<700 °C) of the biomass (e.g., crop residue, and wood) under the anoxic condition, and thus, it is also a highly carbon-rich product less susceptible to microbial degradation. In croplands, the crop yield can be increased depending on the amount of biochar applied to the soil. The physicochemical properties of biochar enable it to influence the soil porosity, water retention, nutrient retention, and bulk density of soil and thus also used as a carrier of slow-release fertilizers. Biochar enhances crop productivity by providing balanced nutrient, fertilizers, and water, as well as by reducing the heavy metal absorption and pathogen infestation in soil (Lehmann et al. 2006; Matovic 2011; Sharma 2018; Chen et al. 2019; Singh and Singh 2020; Dwibedi et al. 2022).

14.4.2 Forest Soil Carbon Sequestration

Globally, forest land covers 4.03 B ha (billion hectares) which is equivalent to 30% of the Earth's surface. Out of total SOC stock, 85% of terrestrial carbon is concentrated in boreal forests, whereas, in temperate and tropical forests, the SOC stock is 60% and 50%, respectively (FAO 2017). In the semi-arid regions, the forest sequesters relatively higher SOC (~20 g/kg) than the other land uses (~10 g/kg) (Alidoust et al. 2018). Several studies indicated the significance of forest land in the SOC sequestration (Blanco-Canqui and Lal 2004; Lal 2005, 2010; Srivastava et al. 2012; Dybala et al. 2019; Lehmann et al. 2020; Zhang et al. 2020). A well-managed forest with a diverse and long-lived tree or plant species further enhances the carbon storage in forest soil. Activities such as afforestation and reforestation add to the SOC pool by photosynthesis in the biomass (Yadav et al. 2017) and by humification of the plant biomass (Pellis et al. 2019). However, the rate of C sequestration in the forest land is significantly influenced by soil age, carbon input by roots and litters, fertilization (N), moistures, forest management, and species type (McCarl et al. 2007; Xu et al. 2020a, b; Zhang et al. 2020).

14.4.3 Grasslands Soil Carbon Sequestration

Grasslands include pasturelands, rangelands, shrublands, and often time's cropland generally used for fodder crops or pasture. In 2000, grasslands cover 3.5 B ha of land, globally. However, much of the grasslands converted to other land uses such as for milk production, beef production, and cultivated crops (20%) (FAO 2017). Grasslands have high SOM content of 333 Mg ha^{-1}; however, 16% of it is degraded. In grasslands, particularly in pasturelands, the interaction between the organic carbon and soil is significant due to increased below-ground biomass (root biomass) and SOM content. The root biomass stabilizes the SOC in the aggregates and consequently enhances the sequestration (Blanco-Canqui and Lal 2004). Strategies such as fertilization, controlled grazing, sowing legumes, and water conservation further enhance the SOC sequestration in the grasslands (Blanco-Canqui and Lal 2004; Srivastava et al. 2012; FAO 2017; Horwath and Kuzyakov 2018; Yang et al. 2019; Ito et al. 2020).

14.4.4 Peatland (Wetlands) Soil Carbon Sequestration

Peatlands or wetlands are an important ecosystem that substantially contributes to mitigating climate change by sequestering the CO_2. Globally, peatland accounts for 5–8% of the terrestrial land which represents 20–30% of the carbon stock (Mitsch et al. 2013). Although wetlands are natural GHG emission sources such as CH_4 emission, yet they are an important carbon sink in the long term. The carbon sink and the net carbon retention of the world's wetlands are around 830 Tg yr^{-1} and 118 g C m^{-2} yr^{-1}, respectively (Mitsch et al. 2013; Xu et al. 2020a, b). The carbon sequestration rate in the wetlands is 0.75–3.1 Mg C ha^{-1} yr^{-1} (Horwath and Kuzyakov 2018). By preventing the degradation and drainage of the peatlands, the release of the C from the wetlands to the atmosphere can be avoided. Aerobic conditions in the peatlands result in SOC oxidation, and therefore, maintaining the water level particularly in the seasonal wetlands is crucial for SOC sequestration. Furthermore, restoration of peatland and highly productive wetland plant and animal biomass enhances the peatland productivity and thus increase their carbon capture potential. Limiting the cultivation of the wetlands further creates a positive C budget in the wetlands (Horwath and Kuzyakov 2018; Lal 2010; Lorenz and Lal 2018; Lehmann et al. 2020).

14.4.5 Urban Soil Carbon Sequestration

Urbanization increases the degraded soil surface (mainly during construction) and impermeable surface (Nachtergaele et al. 2016); as a result, the loss of soil C increases. In highly dense urban areas, urbanization can cover 80% of the land area. In urban areas, lawns, parks, recreational grounds, and urban forest can potentially enhance the carbon sequestration in urban soil. The SOC in the low-density urban area and the institutional area is 38–44% higher than the commercial land. Furthermore, longer-lived tree species with high wood density and tolerance to heat and urban stresses are significant for carbon sequestration. Besides, increasing the green spaces (green building and green roofs), vegetation cover, pervious land, and addition of amendments (biochar, municipal biosolids, and residuals) can substantially enhance the positive carbon budget in the urban soil (Brown et al. 2012; Lal and Augustin 2012; Blanco-Canqui and Lal 2004).

14.5 Carbon Sequestration in Soil Through Biochar

Biochar is an important soil amendment that significantly influences the soil physicochemical properties and aids in the carbon sequestration in soil (Duku et al. 2011; Qambrani et al. 2017; Li et al. 2018). Carbon sequestration potential of biochar in soil has been reported in both incubation (Maucieri et al. 2017; Walkiewicz et al. 2020; Wang et al. 2020) and field studies (Abagandura et al. 2019; Wang et al. 2019; Fan et al. 2020; Yang et al. 2020). Biochar application at the rate of 4% w/w soil in laboratory reduced CO_2 emission by 9.31% (Wang et al. 2020). In a similar study, Maucieri et al. (2017) found that the biochar application at the rate of 5% of w/w soil reduced 9.78% of cumulative CO_2 emission (Table 14.3). Walkiewicz et al. (2020) investigated the influence of biochar application on CO_2 flux from forest and orchard soil (Table 14.3). Furthermore, the water-holding capacity (WHC) of soil has a significant impact on CO_2 emission from soil under the biochar amendment (Walkiewicz et al. 2020). The soil with 100% WHC showed higher CO_2 fluxes over soil having 55% WHC under the same level of biochar application (Walkiewicz et al. 2020), which indicates that the biochar application at lower WHC enhances the carbon sequestration.

In China, Fan et al. (2020) investigated the impact of adding rice straw biochar and rice straw on carbon sequestration in the soil with a rice–wheat cropping system (Table 14.3). Biochar application reduced the 17.27% of annual CO_2 emission over rice straw application (Fan et al. 2020). Soil porosity substantially influences the SOC and soil CO_2 flux. Rice straw application reduced the macro-pores of the soil

Table 14.3 Impact of biochar application on carbon dioxide emission from soil

References	Location	Crop type	Type of soil	Treatment (Mg ha^{-1})	CO_2 emission (kg CO_2 ha^{-1})	Mitigation (%)
Zhang et al. (2012)	Jiangsu Province, China	Rice	Stagnic Anthrosol	BC (0)	4825	Control
				BC (10)	4558	5.54
				BC (20)	4506	6.61
				BC (40)	4847	−0.46
Fan et al. (2020)	Jiangsu Province, China	Rice–wheat	Alifisol	Rice straw	42,240	Control
				BC of straw	34,940	17.27
Wang et al. (2020)	Fujian, Southeastern, China	Rice	NM	BC (0)	59,880	Control
				Steel slag (8)	54,790	8.5
				BC (8)	48,950	18.25
				BC (8) + Steel slag (8)	47,830	20.12
Yang et al. (2020)	Linhe City, Mongolia, China	Corn	Sandy loam	BC (0^1)	5142	Control
				BC (15)	4007	22.07
				BC (30)	4035	21.53
				BC (45)	3350	34.85
Horak et al. (2020)	Slovak University of Agriculture, Slovakia	Barley-Corn-Wheat	Haplic Luvisol	BC (0)	94,490	Control
				BC (10)	90,090	4.66
				BC (20)	93,240	1.32
				BC (0) + N (100 kg ha^{-1})	101,820	Control

(continued)

Table 14.3 (continued)

References	Location	Crop type	Type of soil	Treatment (Mg ha^{-1})	CO_2 emission (kg CO_2 ha^{-1})	Mitigation (%)
				BC (10) + N (100 kg ha^{-1})	95,700	6.01
				BC (20) + N (100 kg ha^{-1})	102,300	−0.47
Abagandura et al. (2019)	South Dakota Stata, Brookings, USA	Corn-Soybean	Sandy Soil	BC (0)	11,084	Control
				Corn stover BC (10)	10,527	5.03
				Pinewood BC (10)	10,292	7.15
				Wsitchgrass BC (10)	10,366	6.48
			Clay soil	BC (0)	9999	Control
				Corn stover BC (10)	10,527	−0.99
				Pinewood BC (10)	10,784	−7.85
				Wsitchgrass BC (10)	11,073	−10.74
Wang et al. (2020)	Incubation study	NA	NM	BC (0% w/w of soil)	664.10 g CO_2 kg^{-1} soil	Control
				BC (4% w/w of soil)	602.38 g CO_2 kg^{-1} soil	9.31
				N + BC (0% w/w of soil)	744.14 g CO_2 kg^{-1} soil	Control
				N + BC (4% w/w of soil)	690.75 g CO_2 kg^{-1} soil	7.17
Walkiewicz et al. (2020)	Incubation study	Deciduous mixed forest	Slit laom	100% WHC + BC (0% w/w of soil)	3.80 mmol CO_2 kg^{-1} day^{-1}	Control

(continued)

Table 14.3 (continued)

References	Location	Crop type	Type of soil	Treatment (Mg ha^{-1})	CO_2 emission (kg CO_2 ha^{-1})	Mitigation (%)
				100% WHC + BC (4% w/w of soil)	3.40 mmol CO_2 kg^{-1} day^{-1}	10.53
				55%WHC + BC (0% w/w of soil)	6.77 mmol CO_2 kg^{-1} day^{-1}	Control
		Apple orchard	Slit laom	55%WHC + BC (4% w/w of soil)	4.09 mmol CO_2 kg^{-1} day^{-1}	39.59
				100% WHC + BC (0% w/w of soil)	1.68 mmol CO_2 kg^{-1} day^{-1}	Control
				100% WHC + BC (4% w/w of soil)	1.88 mmol CO_2 kg^{-1} day^{-1}	−11.9
				55%WHC + BC (0% w/w of soil)	2.75 mmol CO_2 kg^{-1} day^{-1}	Control
				55%WHC + BC (4% w/w of soil)	2.72 mmol CO_2 kg^{-1} day^{-1}	1.09
Maucieri et al. (2017)	Incubation study	NA	Clay-loam	BC (0% w/w of soil)	665.6 mg CO_2 kg^{-1} soil	Control
				BC (5% w/w of soil)	600.5 mg CO_2 kg^{-1} soil	9.78

Note BC—Biochar; NM—Not mention; NA—Not applicable; WHC—Water-holding capacity and increased micro-pores which results in higher CO_2 flux over biochar (Fan et al. 2020). Abagandura et al. (2019) investigate CO_2 flux from sandy and clay soil under a different type of the biochar amendments (Table 14.3). Biochar when applied in sandy soil enhanced SOC content, whereas the cumulative CO_2 emission was reduced; however, in clay soil, biochar increased CO_2 flux (Table 14.3). The study revealed that sandy soil has a positive correlation with carbon sequestration while clay soil has a negative correlation with carbon sequestration. Furthermore, in cornfields, under drip irrigation and mulching soil carbon sequestration increased significantly (Yang et al. 2020). Biochar application at the rate of 15 Mg ha^{-1}, 30 Mg ha^{-1}, and 45 Mg ha^{-1} reduced cumulative CO_2 emission by 22.07%, 21.53%, and 34.85%, respectively (Table 14.3). Xu et al. (2020a, b) quantified the carbon sequestration in Mosa Bamboo forest under 5 and 15 Mg biochar ha^{-1} amendment. Carbon sequestration under 0, 5, and 15 Mg ha^{-1} biochar is 3.36, 19.70, and 11.86 Mg CO_2-eq. ha^{-1}, respectively, which suggests that the biochar application at the rate of 5 Mg ha^{-1} is optimum for achieving the highest carbon sequestration in forest soil. Yan et al. (2020) investigated the effect of two types of biochar (*Spartina alterniflora* and *Phragmites communis*) in waterlogged soil and found that the biochar applied at a different rate (1% w/w to 10 w/w) reduced total CO_2 emission up to 36.27% (Yan et al. 2020). The application of optimum dose (up to 20 Mg ha^{-1}) of biochar in soil results in carbon sequestration, while excessive application of biochar ($\geq$40 Mg ha^{-1}) in soil does not show any positive correlation with carbon sequestration (Zhang et al. 2012; Horák et al. 2020). Soil texture also play important role in carbon sequestration and on this Abagandura et al. (2019) field experiment in can be easily concluded that biochar application in sandy soil show positive carbon sequestration while biochar application in clay soil show negative correlation with carbon sequestration in soil (Table 14.3). Water-holding capacity (WHC) of the soil has significant impact on carbon sequestration under biochar amended soils (Walkiewicz et al. 2020). At 100%, WHC application of biochar shows lower carbon sequestration rate as compared to soil having 55% WHC (Table 14.3) which clearly indicates that application of biochar in having lower WHC such as sandy soil has better carbon sequestration potential than the soil having higher WHC such as clay.

14.6 Conclusions

The chapter presents an overview of the role of carbon dioxide (CO_2) in global warming and highlighted various carbon sequestration strategies in soil with the emphasis on biochar application for climate change mitigation. Climate change is directly associated with anthropogenic greenhouse gas (GHGs) emissions such as CO_2, methane (CH_4), and nitrous oxide (N_2O). With the global warming potential of 1, CO_2 appears to be an important GHG contributing significantly to global warming. CO_2 is increasing at the rate of 2.5 ppm/year and reaches a level of 409.8 $\pm$ 0.01 ppm in the year 2019. Anthropogenic emissions from land use alteration and other agricultural activities have added 136 $\pm$ 55 Pg C since the industrial era. Out of the five natural carbon pools (oceans, biotic, pedologic, atmosphere, and fossils), the terrestrial pool, which is four times higher than the atmospheric pool, acts as an important carbon sink on the planet Earth and therefore recognized as an essential climate change mitigation strategy. The world soil can potentially sequester 0.4–1.2 Gt C yr^{-1}. The carbon sequestration strategies in the soil to enhance the soil organic carbon in various land uses are conservation/reduced tillage (cropland), biochar application (cropland and forest), afforestation (forest), restoration of peatlands, water conservation (wetlands), green spaces, and urban forests (urban land). Adding soil amendment such as biochar appears to be crucial strategy for long-term or permanent carbon sink in the terrestrial biosphere. The physicochemical properties of biochar enable it to influence the soil porosity, water retention, nutrient retention, and bulk density of soil and thus also used as a carrier of slow-release fertilizers. Biochar enhances crop productivity by providing balanced nutrients, fertilizers, and water, as well as by reducing the heavy metal absorption, and pathogen infestation in soil. The application of biochar in various types of land uses such as cropland and forest soil resulted in the CO_2 emission reduction between 1% and 39.5%. Biochar application in conjunction with other carbon sequestration strategies such as no-tillage and crop rotation is recommended for increased soil fertility, crop productivity, and enhanced carbon sequestration.

Acknowledgements We extend our gratitude toward the National Institute of Hydrology, Roorkee, for providing all the necessary resources to complete this study.

References

Abagandura GO et al (2019) Effects of biochar and manure applications on soil carbon dioxide, methane, and nitrous oxide fluxes from two different soils. J Environ Qual 48(6):1664–1674. https://doi.org/10.2134/jeq2018.10.0374

Al-Ghussain L (2019) Global warming: review on driving forces and mitigation. Environ Prog Sustain Energy 38(1):13–21. https://doi.org/10.1002/ep.13041

Alidoust E et al (2018) Soil carbon sequestration potential as affected by soil physical and climatic factors under different land uses in a semiarid region. CATENA 171(July):62–71. https://doi.org/10.1016/j.catena.2018.07.005

Anderson TR, Hawkins E, Jones PD (2016) CO2, the greenhouse effect and global warming: from the pioneering work of Arrhenius and Callendar to today's Earth System Models. Endeavour 40(3):178–187. https://doi.org/10.1016/j.endeavour.2016.07.002

Arndt DS, Blunden J, Dunn RJH (2020) State of the Climate in 2019. Bull Am Meteorol Soc 101(8):SI–S8. https://doi.org/10.1175/2020BAMSSTATEOFTHECLIMATE.1

Baldocchi DD, Wilson KB (2001) Modeling CO2 and water vapor exchange of a temperate broadleaved forest across hourly to decadal time scales. Ecol Model 142(1–2):155–184. https://doi.org/10.1016/S0304-3800(01)00287-3

Blanco-Canqui H, Lal R (2004) Mechanisms of carbon sequestration in soil aggregates. Crit Rev Plant Sci 23(6):481–504. https://doi.org/10.1080/07352680490886842

Blunden J, Arndt DS (2020) State of the climate in 2019. Bull Am Meteorol Soc 101(8):SI–S8. https://doi.org/10.1175/2020BAMSSTATEOFTHECLIMATE.1

Brown S, Miltner E, Cogger C (2012) Carbon sequestration in urban soil. In: Carbon sequestration in urban ecosystems, pp 1–385. https://doi.org/10.1007/978-94-007-2366-5

Campos R, Pires GF, Costa MH (2020) Soil carbon sequestration in rainfed and irrigated production systems in a new brazilian agricultural frontier. Agriculture (Switzerland) 10(5). https://doi.org/10.3390/agriculture10050156

Cayuela ML et al (2014) Biochar's role in mitigating soil nitrous oxide emissions: a review and meta-analysis. Agr Ecosyst Environ 191:5–16. https://doi.org/10.1016/j.agee.2013.10.009

Cha-un N et al (2017) Greenhouse gas emissions, soil carbon sequestration and crop yields in a rain-fed rice field with crop rotation management. Agr Ecosyst Environ 237:109–120. https://doi.org/10.1016/j.agee.2016.12.025

Chen W et al (2019) Past, present, and future of biochar. Biochar 1(1):75–87. https://doi.org/10.1007/s42773-019-00008-3

Cheng D et al (2020) Feasibility study on a new pomelo peel derived biochar for tetracycline antibiotics removal in swine wastewater. Sci Total Environ 720:137662. https://doi.org/10.1016/j.scitotenv.2020.137662

Corbeels M et al (2019) The 4 per 1000 goal and soil carbon storage under agroforestry and conservation agriculture systems in sub-Saharan Africa. Soil Tillage Res 188(December 2017):16–26. https://doi.org/10.1016/j.still.2018.02.015

Demenois J et al (2020) Barriers and strategies to boost soil carbon sequestration in agriculture. Front Sustain Food Syst 4(April). https://doi.org/10.3389/fsufs.2020.00037

Duku MH, Gu S, Hagan EB (2011) Biochar production potential in Ghana—a review. Renew Sustain Energy Rev 15(8):3539–3551. https://doi.org/10.1016/j.rser.2011.05.010

Dwibedi SK, Pandey VC, Divyasree D, Bajpai O (2022) Biochar-based land development. Land Degrad Dev. https://doi.org/10.1002/ldr.4185

Dybala KE et al (2019) Carbon sequestration in riparian forests: a global synthesis and meta-analysis. Glob Change Biol 25(1):57–67. https://doi.org/10.1111/gcb.14475

Fagodiya RK et al (2020) Global warming impacts of nitrogen use in agriculture: an assessment for India since 1960. Carbon Manag 11(3):291–301. https://doi.org/10.1080/17583004.2020.1752061

Fan R et al (2020) 'Straw-derived biochar mitigates CO2 emission through changes in soil pore structure in a wheat-rice rotation system. Chemosphere 243. https://doi.org/10.1016/j.chemosphere.2019.125329

FAO (2000) The state of food and Agriculture. Rome, Italy. https://www.fao.org/3/x4400e/x4400e.pdf

FAO (2017) Soil Organic Carbon: the hidden potential, Food and Agriculture Organization of the United Nations Rome, Italy

FAO and ITPS (2015) Status of the World's Soil Resources. s.n, Rome

Friedlingstein P et al (2019) Global carbon budget 2019. In: Earth system science data, pp 1782–1838. https://doi.org/10.3929/ethz-b-000385668 (Originally)

Gupta DK et al (2016) Mitigation of greenhouse gas emission from rice-wheat system of the Indo-Gangetic plains: Through tillage, irrigation and fertilizer management. Agr Ecosyst Environ 230(2016):1–9. https://doi.org/10.1016/j.agee.2016.05.023

Horák J et al (2020) Effects of biochar combined with n-fertilization on soil Co2 emissions, crop yields and relationships with soil properties. Pol J Environ Stud 29(5):3597–3609. https://doi.org/10.15244/pjoes/117656

Horwath WR, Kuzyakov Y (2018) The potential for soils to mitigate climate change through carbon sequestration. In: Climate change impacts on soil processes and ecosystem properties, pp 61–92. https://doi.org/10.1016/b978-0-444-63865-6.00003-x

Huang Y et al (2019) Methane and nitrous oxide flux after biochar application in subtropical acidic paddy soils under tobacco-rice rotation. Sci Rep 9(1):1–10. https://doi.org/10.1038/s41598-019-53044-1

Hue N (2020) Biochar for maintaining soil health. In: Soil health, pp 21–46. https://doi.org/10.1007/978-3-030-44364-1_2

IPCC (2013) Climate change 2013: the physical science basis. SE Asian J Trop Med Public Health

IPCC (2014) Climate change 2014: mitigation of climate change. Contribution of working group III to the fifth assessment report of the intergovernmental panel on climate change. Cambridge University Press, Cambridge, United Kingdom and New York, NY, USA. pp 1–1419

Isson TT et al (2020) Evolution of the global carbon cycle and climate regulation on earth. Global Biogeochem Cycles 34(2):1–28. https://doi.org/10.1029/2018GB006061

Ito G et al (2020) Global carbon cycle and climate feedbacks in the NASA GISS ModelE2.1. J Adv Model Earth Syst 1–44. https://doi.org/10.1029/2019ms002030

Khalid S et al (2020) A critical review of different factors governing the fate of pesticides in soil under biochar application. Sci Total Environ 711. https://doi.org/10.1016/j.scitotenv.2019.134645

Kumar SS et al (2020) Industrial wastes: fly ash, steel slag and phosphogypsum-potential candidates to mitigate greenhouse gas emissions from paddy fields. Chemosphere 241:124824. https://doi.org/10.1016/j.chemosphere.2019.124824

Lal R (2004) Soil carbon sequestration impacts on global climate change and food security. Science 304(5677):1623–1627. https://doi.org/10.1126/science.1097396

Lal R (2005) Soil carbon sequestration in natural and managed tropical forest ecosystems. J Sustain for 21(1):1–30. https://doi.org/10.1300/J091v21n01_01

Lal R (2010) Managing soils and ecosystems for mitigating anthropogenic carbon emissions and advancing global food security. Bioscience 60(9):708–721. https://doi.org/10.1525/bio.2010.60.9.8

Lal R (2012) Soil carbon sequestration: SOLAW background thematic report—TR04B

Lal R (2019) Eco-intensification through soil carbon sequestration: harnessing ecosystem services and advancing sustainable development goals. J Soil Water Conserv 74(3):55A–61A. https://doi.org/10.2489/jswc.74.3.55A

Lal R, Augustin B (2012) Urban trees for carbon sequestration. In: Carbon sequestration in urban ecosystems, pp 121–138. https://doi.org/10.1007/978-94-007-2366-5

Lehmann J, Gaunt J, Rondon M (2006) Bio-char sequestration in terrestrial ecosystems—a review. Mitig Adapt Strat Glob Change 11(2):403–427. https://doi.org/10.1007/s11027-005-9006-5

Lehmann J et al (2020) Persistence of soil organic carbon caused by functional complexity. Nat Geosci 13(8):529–534. https://doi.org/10.1038/s41561-020-0612-3

Li Y et al (2018) Effects of biochar application in forest ecosystems on soil properties and greenhouse gas emissions: a review. J Soils Sediments 18(2):546–563. https://doi.org/10.1007/s11368-017-1906-y

Lindsey R, Dlugokencky E (2020) Climate change: atmospheric carbon dioxide, pp 6–10

Liu Y et al (2020) Successive straw biochar amendments reduce nitrous oxide emissions but do not improve the net ecosystem economic benefit in an alkaline sandy loam under a wheat–maize cropping system. Land Degrad Dev 31(7):868–883. https://doi.org/10.1002/ldr.3495

Lorenz K, Lal R (2018) Carbon sequestration in wetland soils. In: Carbon sequestration in agricultural ecosystems, pp 1–392. https://doi.org/10.1007/978-3-319-92318-5

Malyan SK (2017) Reducing methane emission from rice soil through microbial interventions. ICAR-Indian Agricultural Research Institute, New Delhi-110012, India. Available at: http://krishikosh.egranth.ac.in/handle/1/5810074885

Malyan SK et al (2016) Methane production, oxidation and mitigation: a mechanistic understanding and comprehensive evaluation of influencing factors. Sci Total Environ 572:874–896. https://doi.org/10.1016/j.scitotenv.2016.07.182

Malyan SK, Bhatia A et al (2019a) Mitigation of greenhouse gas intensity by supplementing with Azolla and moderating the dose of nitrogen fertilizer. Biocatal Agric Biotechnol 20:101266. https://doi.org/10.1016/j.bcab.2019.101266

Malyan SK, Kumar A et al (2019b) Role of fungi in climate change abatement through carbon sequestration. In: Recent advancement in white biotechnology through fungi, pp 283–295. https://doi.org/10.1007/978-3-030-25506-0_11

Malyan SK, Kumar SS et al (2021a) Biochar for environmental sustainability in the energy-water-agroecosystem nexus. Renew Sustain Energy Rev 149(July 2020):111379. https://doi.org/10.1016/j.rser.2021.111379

Malyan SK, Bhatia A et al (2021b) Mitigation of yield-scaled greenhouse gas emissions from irrigated rice through Azolla, Blue-green algae, and plant growth–promoting bacteria. Environ Sci Pollut Res. https://doi.org/10.1007/s11356-021-14210-z

Malyan SK et al (2021c) Plummeting global warming potential by chemicals interventions in irrigated rice: a lab to field assessment. Agr Ecosyst Environ 319(February):107545. https://doi.org/10.1016/j.agee.2021.107545

Matovic D (2011) Biochar as a viable carbon sequestration option: global and Canadian perspective. Energy 36(4):2011–2016. https://doi.org/10.1016/j.energy.2010.09.031

Maucieri C et al (2017) Short-term effects of biochar and salinity on soil greenhouse gas emissions from a semi-arid Australian soil after re-wetting. Geoderma 307(August):267–276. https://doi.org/10.1016/j.geoderma.2017.07.028

McCarl BA, Metting FB, Rice C (2007) Soil carbon sequestration. Clim Change 80(1–2):1–3. https://doi.org/10.1007/s10584-006-9174-7

McConkey BG et al (2003) Crop rotation and tillage impact on carbon sequestration in Canadian prairie soils. Soil Tillage Res 74(1):81–90. https://doi.org/10.1016/S0167-1987(03)00121-1

Mitsch WJ et al (2013) Wetlands, carbon, and climate change. Landscape Ecol 28(4):583–597. https://doi.org/10.1007/s10980-012-9758-8

Mona S et al (2021) Towards sustainable agriculture with carbon sequestration, and greenhouse gas mitigation using algal biochar. Chemosphere 275:129856. https://doi.org/10.1016/j.chemosphere.2021.129856

Montzka SA, Dlugokencky EJ, Butler JH (2011) Non-CO_2 greenhouse gases and climate change. Nature 476(7358):43–50. https://doi.org/10.1038/nature10322

Moreira D, Pires JCM (2016) Atmospheric CO_2 capture by algae: negative carbon dioxide emission path. Biores Technol 215(March):371–379. https://doi.org/10.1016/j.biortech.2016.03.060

Nachtergaele FO, Petri M, Biancalani R (2016) Land degradation. In: World soil resources and food security, pp 471–498. https://doi.org/10.1017/cbo9780511622991.009

Paustian K et al (2019) Soil C sequestration as a biological negative emission strategy. Front Clim 1(October):1–11. https://doi.org/10.3389/fclim.2019.00008

Pellis G et al (2019) The ecosystem carbon sink implications of mountain forest expansion into abandoned grazing land: the role of subsoil and climatic factors. Sci Total Environ 672:106–120. https://doi.org/10.1016/j.scitotenv.2019.03.329

Purakayastha TJ et al (2019) A review on biochar modulated soil condition improvements and nutrient dynamics concerning crop yields: pathways to climate change mitigation and global food security. Chemosphere 227:345–365. https://doi.org/10.1016/j.chemosphere.2019.03.170

Qambrani NA et al (2017) Biochar properties and eco-friendly applications for climate change mitigation, waste management, and wastewater treatment: a review. Renew Sustain Energy Rev 79(November 2016):255–273. https://doi.org/10.1016/j.rser.2017.05.057

Rumpel C et al (2020) The 4p1000 initiative: opportunities, limitations and challenges for implementing soil organic carbon sequestration as a sustainable development strategy. Ambio 49(1):350–360. https://doi.org/10.1007/s13280-019-01165-2

Sainju UM (2016) A global meta-analysis on the impact of management practices on net global warming potential and greenhouse gas intensity from cropland soils. PLoS ONE 11(2). https://doi.org/10.1371/journal.pone.0148527

Sharma SP (2018) Biochar for carbon sequestration: bioengineering for sustainable environment. Omics technologies and bio-engineering: volume 2: towards improving quality of life, pp 365–385. https://doi.org/10.1016/B978-0-12-815870-8.00020-6

Singh BR, Lal R (2005) The potential of soil carbon sequestration through improved management practices in Norway. Environ Dev Sustain 7(1):161–184. https://doi.org/10.1007/s10668-003-6372-6

Singh JS, Singh C (2020) Biochar applications in agriculture and environment management. Biochar Appl Agric Environ Manag. https://doi.org/10.1007/978-3-030-40997-5

Srivastava P et al (2012) Soil carbon sequestration: an innovative strategy for reducing atmospheric carbon dioxide concentration. Biodivers Conserv 21(5):1343–1358. https://doi.org/10.1007/s10531-012-0229-y

Stocker TF et al (2013) Climate change 2013 the physical science basis: working Group I contribution to the fifth assessment report of the intergovernmental panel on climate change. https://doi.org/10.1017/CBO9781107415324

Trost B et al (2013) Irrigation, soil organic carbon and N2O emissions. a review. Agron Sustain Dev 33(4):733–749. https://doi.org/10.1007/s13593-013-0134-0

Ucsusa.org (2017) Why does CO2 get most of the attention when there are so many other heat-trapping gases?

Van der Wal A, De Boer W (2017) Dinner in the dark: illuminating drivers of soil organic matter decomposition. Soil Biol Biochem 105:45–48

Walkiewicz A et al (2020) Usage of biochar for mitigation of CO2 emission and enhancement of CH4 consumption in forest and orchard Haplic Luvisol (Siltic) soils. Appl Soil Ecol 156(May). https://doi.org/10.1016/j.apsoil.2020.103711

Wang C et al (2019) Effects of steel slag and biochar amendments on CO2, CH4, and N2O flux, and rice productivity in a subtropical Chinese paddy field. Environ Geochem Health 41(3):1419–1431. https://doi.org/10.1007/s10653-018-0224-7

Wang H et al (2020) Biochar mitigates greenhouse gas emissions from an acidic tea soil. Pol J Environ Stud 29(1):323–330. https://doi.org/10.15244/pjoes/99837

Wu Z, Zhang X et al (2019a) Biochar amendment reduced greenhouse gas intensities in the rice-wheat rotation system: six-year field observation and meta-analysis. Agric For Meteorol 278(July). https://doi.org/10.1016/j.agrformet.2019.107625

Wu Z, Song Y et al (2019b) Biochar can mitigate methane emissions by improving methanotrophs for prolonged period in fertilized paddy soils. Environ Pollut 253:1038–1046. https://doi.org/10.1016/j.envpol.2019.07.073

Xu L et al (2020a) Biochar application increased ecosystem carbon sequestration capacity in a Moso bamboo forest. For Ecol Manage 475(May):118447. https://doi.org/10.1016/j.foreco.2020.118447

Xu S, Sheng C, Tian C (2020b) Changing soil carbon: Influencing factors, sequestration strategy and research direction. Carbon Balance Manag 15(1):1–9. https://doi.org/10.1186/s13021-020-0137-5

Yadav S et al (2017) A satellite-based assessment of the distribution and biomass of submerged aquatic vegetation in the optically shallow basin of Lake Biwa. Remote Sens 9(9). https://doi.org/10.3390/rs9090966

Yadav S et al (2019) Land use impact on the water quality of large tropical river: Mun River Basin, Thailand. Environ Monit Assess 191(10). https://doi.org/10.1007/s10661-019-7779-3

Yan Z et al (2020) Waterlogging affects the mitigation of soil GHG emissions by biochar amendment in coastal wetland. J Soils Sediments 20(10):3591–3606. https://doi.org/10.1007/s11368-020-02705-0

Yang Y et al (2019) Soil carbon sequestration accelerated by restoration of grassland biodiversity. Nat Commun 10(1):1–7. https://doi.org/10.1038/s41467-019-08636-w

Yang W et al (2020) Impact of biochar on greenhouse gas emissions and soil carbon sequestration in corn grown under drip irrigation with mulching. Sci Total Environ 729(306):138752. https://doi.org/10.1016/j.scitotenv.2020.138752

Zand AD, Tabrizi AM, Heir AV (2020) Incorporation of biochar and nanomaterials to assist remediation of heavy metals in soil using plant species. Environ Technol Innov 20:101134. https://doi.org/10.1016/j.eti.2020.101134

Zhang A et al (2012) Effects of biochar amendment on soil quality, crop yield and greenhouse gas emission in a Chinese rice paddy: A field study of 2 consecutive rice growing cycles. Field Crop Res 127:153–160. https://doi.org/10.1016/j.fcr.2011.11.020

Zhang D et al (2016) Biochar helps enhance maize productivity and reduce greenhouse gas emissions under balanced fertilization in a rainfed low fertility inceptisol. Chemosphere 142:106–113. https://doi.org/10.1016/j.chemosphere.2015.04.088

Zhang C et al (2019) Biochar for environmental management: mitigating greenhouse gas emissions, contaminant treatment, and potential negative impacts. Chem Eng J 373(May):902–922. https://doi.org/10.1016/j.cej.2019.05.139

Zhang Z et al (2020) Carbon dynamics in three subtropical forest ecosystems in China. Environ Sci Pollut Res 27(13):15552–15564. https://doi.org/10.1007/s11356-019-06991-1

Zheng H et al (2019) Characteristics and mechanisms of chlorpyrifos and chlorpyrifos-methyl adsorption onto biochars: Influence of deashing and low molecular weight organic acid (LMWOA) aging and co-existence. Sci Total Environ 657:953–962. https://doi.org/10.1016/j.scitotenv.2018.12.018

Zomer RJ et al (2017) Global sequestration potential of increased organic carbon in cropland soils. Sci Rep 7(1):1–8. https://doi.org/10.1038/s41598-017-15794-8

Chapter 15
Remediation of Pharmaceutical and Personal Care Products in Soil Using Biochar

Amita Shakya, Sonali Swain, and Tripti Agarwal

Abstract Pharmaceuticals and personal care products (PPCPs) are relatively new, but a very diverse and unique group of emerging contaminants. These compounds can exert toxic effects to humans at very low doses (ng to μg). They are recalcitrant organic compounds with persistent nature. Regular addition of new compounds and upsurge in their discharge to the environment is of grave concern due to their ecological and environmental toxicity and human health risk. After use, the majority of active ingredients and metabolites of many pharmaceuticals excrete out from human and animal bodies and enter into the environment through various means. Sewage and wastewater treatment plants are the major sources for their contamination to the various environment mediums. PPCPs contamination to groundwater and surface water is well reported, and many techniques are available for their removal. However, the contamination of PPCPs to soil and their transport, translocation and bioaccumulation to various crops is relatively recent and raised the concern of food safety and human health with the development of antibiotic resistance. To address the problem, there is an urgent need for effective, efficient and rapid remediation methods of PPCPs from soil. Biochar is a pyrolyzed carbon material having a plethora of physicochemical properties. The biochar application as soil amendment for achieving high yield and soil nourishment is an age-old process. Use of biochar as an adsorbent for various organic and inorganic contaminants from soil and water is well reported in literature. However, application of biochar for PPCPs remediation from soil is relatively new. Due to its unique physicochemical properties such as high surface area, porosity, aromaticity and aliphatic polar surface functional groups, there are immense opportunities of biochar's use for PPCPs removal from soil. In this chapter, current research and future scope of biochar application for PPCPs removal from soil are discussed.

Keywords Biochar · Pharmaceuticals and personal care products · Soil contamination

---

A. Shakya · S. Swain · T. Agarwal (✉)
Department of Agriculture and Environmental Sciences, National Institute of Food Technology Entrepreneurship and Management, Kundli, Sonipat, Haryana, India
e-mail: tripti.niftem@gmail.com

15.1 Introduction

With modernization and civilization, mankind is facing the issues of population growth, industrialization, urbanization, environmental pollution and climate change which have created exceptional stress on natural resources. Pharmaceuticals and personal care products (PPCPs) are the unique group of emerging environmental contaminants, as they have the inherent ability to induce physiological effects in humans and animals even at very low doses. The presence of various PPCPs in different environmental compartments, including soil and water, has been confirmed with various studies all over the world, which raises concerns about the potential adverse effects to humans and wildlife (Ebele et al. 2017). The U.S. EPA has identified PPCPs as emerging contaminants of concern because little is known about their impact on the environment or risks to human health when they are released into ecosystems (Bastian and Murray 2012). Their persistence in the environment is of grave concern, as they cannot be removed from wastewater by the conventional methods.

The U.S. EPA defines pharmaceuticals and personal care products (PPCPs) as 'any product used by individuals for personal health or cosmetic reasons or used by agribusiness to enhance growth or health of livestock' (Daughton and Ternes 1999). The PPCPs include thousands of chemicals and compounds that make up fragrances, cosmetics, over-the-counter drugs and veterinary medicines. Since most of developing countries are still fighting for basic needs and more immediate problems such as clean water supply, sanitation, waste disposal and civil war, the long-term risk of PPCPs may not be seen as a pressing issue at the present time (Rahman et al. 2009). Typically, these PPCPs are present in the environment at very low concentration (ng/L to mg/L range), but their pseudo-persistence and transformation in the environment are of grave concern.

Figure 15.1 represents the nine classes of pharmaceuticals included in PPCPs and the five most important groups of personal care products of emerging concerns. Among them, many of the drugs such as acetaminophen, aspirin, codeine, diclofenac, ibuprofen, naproxen; antibiotics like β-lactams, macrolides, fluoroquinolones, aminoglycosides, tetracyclines and sulfonamides are considered 'pseudo-persistent'. The most common types of pharmaceuticals found in soil are antibiotics (trimethoprim, sulfadiazine and triclosan), analgesics (ibuprofen and diclofenac) and antiepileptic (carbamazepine) (Nikolaou et al. 2007). The extreme global PPCPs' usage, along with the escalating introduction of new pharmaceuticals to the market, is contributing substantially to the environmental presence of these chemicals and their active metabolites in the environment.

Most PPCPs are low molecular weight compounds and relatively hydrophilic in nature, which is a limitation for their complete removal by traditional treatment methods (Ebele et al. 2017). Studies have suggested that conventional wastewater treatment plants (WWTPs) can remove less than 75% of most PPCPs; however, chemical properties of PPCPs and treatment process widely influence the removal efficiency (Kandlakuti 2016). Unlike a number of methods available for PPCPs removal

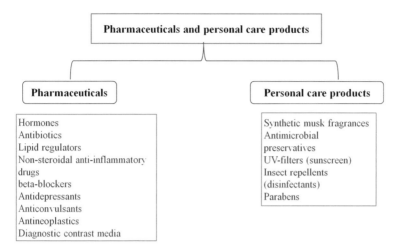

Fig. 15.1 Categories of products included in PPCPs

from wastewater, very few effective methods are available for PPCPs remediation from soil matrix.

Biochar is a carbon-rich material produced with thermochemical conversion of biomass under oxygen limited environment. Because of its unique properties such as high surface area, pore volume and cation exchange capacity, biochar can be used for applications such as soil improvement, fertility enhancement and carbon sequestration (Ahmad et al. 2013). Biochar amendment to agricultural soil is considered a win–win solution for its potential of sequestering carbon and improving soil fertility (Antal and Grønli 2003). In the present era, with distinctive characteristic properties, biochar has gained significant attention as an economical and sustainable adsorbent for abatement of various organic pollutants from soil and water. Aromaticity and functional groups are the features that make biochar an attractive adsorbent, even for many hydrophobic organic compounds. Due to its simple and hassle-free production, ease of scaling-up and environmental-friendly adsorption, it has been considered the most effective and sensitive method for contaminant removal (Xiang et al. 2019). The multidimensional benefits of biochar application on land made it an attractive choice for its use for organic contaminants, including PPCPs remediation.

15.2 Sources and Transport of PPCPs to the Soil Environment

The pharmaceuticals were prescribed to improve the health, while personal care products were introduced to improve the quality of life. Anthropogenic activities are the key source for constantly increasing level of PPCPs which could enter into the natural hydrological cycle of the environment. However, besides man-made PPCPs, various

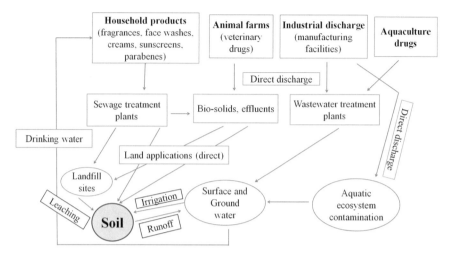

Fig. 15.2 Various sources and pathways of PPCPs release into the environment

hormones are naturally excreted by humans and animals. Unintentional presence of PPCPs in the environment has been increasing in the past few years due to their extensive use. The pharmaceuticals and related products are made with very high level of chemical compounds that human and animal bodies could not process (Lofrano et al. 2020). A large portion of these non-metabolized and dissolved compounds reach to the ecosystem via wastewater treatment plants (WWTPs) and surface leaching (Yang et al. 2017). Besides this, improper disposal of unwanted, expired, half used drugs, pharmaceuticals and many cosmetic compounds into sewer or trash is also a major cause for increasing environmental concentration of PPCPs.

PPCPs are reported in various aquatic as well as terrestrial ecosystems in many counties including India, China, and USA though at very low levels (Liu et al. 2020b; Xie et al. 2019). There are numerous ways that PPCPs can enter into the ecosystem (Fig. 15.2), and large animal farms have been identified as the major sources.

Municipal landfills, wastewater sources and septic tanks are point sources for PPCPs' pollution, whereas sewage sludge, artificial recharge of aquifers and groundwater surface interface are the secondary sources (Boxall et al. 2012). Domestic sewage, industrial services, release of PPCPs from manufacturing facilities, hospitals, aquaculture facilities, run-off from fields into surface waters, run-off into soil through animal farming and manure applications, run-off of veterinary medicines from farm-yards and disposal of the carcasses of treated animals are various sources of PPCPs into the environment (Lofrano et al. 2020).

Direct application of animal manure and dung, digest from manure-based biogas plants, is the major route of soil contamination with PPCPs. Further, translocation of such PPCPs to surface and groundwater resources increases the risk of environmental and health hazards. The main factors affecting the removal efficiency of PPCPs in WWTPs are affected by the physicochemical structure of drugs, the temperature during treatment, the redox status and hydraulic retention time. The hydraulic retention time is the most important factor influencing the removal efficiency of drugs in wastewater treatment plants (Jiang and Li 2020).

15.3 Environmental and Health Risk of PPCPs

Differences in the presence/absence and type of manufacturing sites, level and type of PPCPs usage, population demographics, cultural practices, environmental and climatic characteristics, dilution potential of receiving environments and infrastructure related to WWTPs are the factors influencing the risks due to the presence of PPCPs in the environment (Boxall et al. 2012). Surprisingly, there is a lack of research as well as governmental report data is scarce for the occurrence and possible toxic impacts of exposure to PPCPs. Use of bio-solids and reclaimed water for land application is again increasing the risk of accumulation of PPCPs into the soil environment which could easily be translocated to different parts of plants and enter into the food chain.

Dodgen et al. (2015) reported accumulation and translocation of many PPCPs and EDCs in roots, leaves of tomato, lettuce and carrot plants and observed increased accumulation due to transpiration. The major risks in the PPCPs are due to pharmaceuticals, especially various drugs, antibiotics and endocrine disruptors. Their direct or indirect uptake may cause increased risk of antibiotic resistance and various unknown health hazards which could pose possible lethal effects. Consumption of food and crops contaminated with PPCPs/EDCs is the most obvious and direct route of exposure to the living organisms. Many recent researches have shown accumulation of various PPCPs in different parts of the vegetable crops either treated with municipal bio-solids (Sabourin et al. 2012; Usyskin et al. 2015) or reclaimed water (Colon and Toor 2016; Liu et al. 2020b; Shenker et al. 2011).

The risk associated with exposure to various PPCPs can be estimated as hazard quotient (RQ/HQ) and cumulative health hazard index (HI). The HQ is determined by dividing the estimated daily intake (EDI) by the acceptable daily intake (ADI), which is the amount that can be consumed daily over a person's lifespan without adverse effects (Prosser and Sibley 2015).

$$\text{Hazard Quotient} = \frac{\text{Estimated daily intake}}{\text{Acceptable daily intake}}$$

$$\text{Estimated Daily Intake (EDI)} = \frac{C^{food} \times IR_{veg} \times \beta_{ww/dw}}{m}$$

where C^{food} is concentration of the studied pharmaceutical in crop ng/g dry weight, IR_{veg} is the intake rate of crops in units of wet weight, and $\beta_{ww/dw}$ w is the wet-to-dry conversion factor for plant tissue; m represents the average body weight of the person.

The PPPCPs exposure to humans can be estimated as follow (Wu et al. 2013):

$$\text{Human exposure} = C \times D \times m \times T$$

where C is the concentration of PPCPs in vegetables (ng/kg$_{\text{wet weight}}$), D is the average daily consumption of vegetables (g$_{\text{wet weight}}$/kg$_{\text{body weight/day}}$), m is the human body weight, and T is the exposure (time).

A negligible human risk is considered when the value of RQ is <0.01, and humans are considered to be exposed to a potential hazard if the HQ value is >0.1 rather than >1, when humans may be exposed through other pathways. An acceptable annual range of PPCPs for humans is considered to be 20–200 mg (Liu et al. 2020b).

Further the conversion of PPCPs into potentially toxic secondary metabolites during the sewage sludge treatment or wastewater treatment is of grave concern (Ferreira et al. 2017). The simultaneous consumption of more than one antibiotic can also result in their cross interaction effects in humans.

Acute embryonic developmental toxicity and reproductive toxicity due to exposure to various PPCPs was reported in fish. Besides this, effects of interactions of PPCPs with other contaminants such as metal and other organic compounds are still unknown (Jiang and Li 2020). Evolution of antibiotic resistance genes (ARGs) due to presence of various antibiotics in soil, water and sediments have arisen which reduce the therapeutic potential against various pathogens (Jiang and Li 2020). There are a number of reports about increasing unintended presence of PPCPs in the aquatic environment, and effluents from WWTPs are the key point source for this contamination. This leads to the continuous exposure of PPCPs to non-targeted aquatic organisms throughout their life cycle exerting detrimental impacts on aquatic life (Ebele et al. 2017). Death of vultures in the Indian subcontinent is the most prominent example of biological activity of pharmaceuticals when released into the environment. The population of three species of vultures, namely *Gyps bengalensis*, *Gyps indicus* and *Gyps tenuirostris,* continuously fell in the Indian subcontinent when they consumed the carcasses of animals who were earlier treated with drug diclofenac (Yang et al. 2017). The diclofenac residues remain in the body of animals even after their death, which bio-accumulate into the food chain and become a lethal reason for mortality of vultures.

15.4 Fate and Occurrence of PPCPs in the Soil Environment

The PPCPs are mostly not persistent, but they are called 'pseudo-persistent' since they are released continuously to the environment. The occurrence and fate of the PPCPs are highly dependent upon the properties of soil matrix and chemical nature of the PPCPs (Liu et al. 2020b).

Studies suggested poor adsorption and degradation of various PPCPs in soil resulting in their persistence under anaerobic conditions (Lin and Gan 2011). The nature of PPCPs also influences their fate in the environment. As reported by Dodgen et al. (2015) neutral and cationic PPCPs showed potential to accumulate in leaf and root tissues, while anionic PPCP/EDCs preferentially accumulated in root tissues. Herklotz et al. (2010) also reported symplastic transportation of many pharmaceutical products in the leaf and seed pods of the plants. Chemical nature of the PPCPs is the key factor which governs the accumulation and leaching of PPCPs in the soil matrix (Chen et al. 2013). Type of soil (sandy, loamy or clay) and presence of organic matter in the soil also affect the simultaneous translocation and bioaccumulation of the PPCPs (Usyskin et al. 2015). Following are the properties which control the fate and occurrence of the PPCPs in soil.

15.4.1 Transformation

Most of the PPCPs undergo transformation and the transformed products are more toxic than their parent compounds. The transformation of PPCPs in WWTPs depends on the physicochemical properties of compounds and condition of WWTPs (Boxall et al. 2012). PPCPs may be destroyed, partially transformed or left unchanged. Transformation products with unknown toxicity and persistence are found in effluent and in receiving water bodies (Keerthanan et al. 2020). Such secondary metabolites can also form due to the degradation of parent PPCPs in treatment plants during photolysis, advanced oxidation processes (AOPs), enzyme-based treatment processes and other treatment processes. The major problem with these secondary metabolites or transformed compounds is their identification and quantification due to their presence in trace amounts and lack of standard chemicals for the same. Research suggested that transformed products produced during AOPs could be much more biologically active and toxic than their parent compounds. Most PPCPs have relatively short environmental half-lives, compared with persistent organic pollutants (POPs) suggesting an easier formation of their transformed products (Yin et al. 2017). Besides this, incomplete removal and continuous discharge of PPCPs into the environment could result in continuous accumulation of PPCPs leading to their pseudo-persistent in the environment and hence possibly, resulting in the pseudo-persistence of their transformed products or metabolites. For example, acetaminophen yielded 13 transformation products, while sulfamethoxazole yielded nine transformation products by photodegradation. At least 30 carbamazepine metabolites, including pharmacologically active or genotoxic compounds, have been identified in humans, and nine transformed products of carbamazepine were reported in WWTPs (Yin et al. 2017). Variety of transformation products of bisphenol A and diclofenac were also found in soil (Dodgen et al. 2014). More focused research is required to understand and assess the toxicity and hazards of transformed products and degradation intermediates.

15.4.2 Bio-Adsorption and Accumulation

Bio-adsorption process affects the movement, uptake, bioavailability and bioaccumulation of the PPCPs in various organisms. Many biochemical and physiological parameters such as ionic strength, pH, temperature and difference in food matrix significantly affect the adsorption of the PPCPs into plants, humans and microbes. Reportedly, uptake was found higher in the hydroponic system than soil matrix (Herklotz et al. 2010). Chinese cabbage grown in organic matter-rich soil amend with the environmentally relevant concentration of the pharmaceuticals and soil amended with bio-solids from a local WWTP was found to be contaminated with all the reported pharmaceuticals (Holling et al. 2012). The adsorption of residual PPCPs in soil not only depends on the physicochemical properties of the PPCPs such as molecular structure, water solubility and hydrophobicity but also on soil properties (type, dissolved organic compounds, pH and liming) (Al-Farsi et al. 2017). According to the OECD, pharmaceuticals with log $K_{ow} > 3$, have a tendency to accumulate (OECD 2008).

The ability of plants to accumulate the PPCPs from exposed soil in their areal parts can be estimated using the bioconcentration factor (BCF_F), which is calculated as the ratio of PPCP concentration in the plant to the PPCP concentration in the soil (all on a dry weight basis) (Liu et al. 2020b):

$$BCF_F(L/kg) = \frac{\text{PPCP concentration in plant tissue } (\mu g/kg)}{\text{PPCP concentration in soil or soultion } (\mu g/kg \text{ or } \mu g/L)}$$

The Sorption and low mobility of various pharmaceuticals in soil have been reported which restrict their mobility into water limiting the chances of groundwater contamination (Yu et al. 2013).

15.4.3 Translocation

Movement of PPCPs from roots to other aerial parts of the plants is referred to as translocation and can be calculated as translocation factor (Dodgen et al. 2015).

$$\text{Translocation Factor (TF)} = \frac{\text{Concentration of PPCPs in aerial part}}{\text{Concentration of PPCPs in root}}$$

Plants are the first and the most vulnerable entities when grown in the soil contaminated with PPCPs. Biochemical properties of plants (molecular size, K_{ow}, and pKa, partition coefficient) and environmental conditions (concentration of DOM, ionic strength, etc.) are the key factors that influence the translocation of the PPCPs from roots to aerial parts of the plants (Al-Farsi et al. 2017). Uptake of PPCPs generally takes place by passive diffusion through plants roots along with many dissolved solutes. Once entered into plant roots, the translocation of these small organic molecules could take place through apoplastic movement, symplastic movement or vacuolar movement (Öztürk et al. 2015). Furthermore, the differences in plant lipid contents, detoxification, metabolic systems, growth rates and transpiration rates also influence PPCPs uptake and translocation behaviours in plants (Guasch et al. 2012).

15.4.4 Degradation

The diverse physical and chemical properties of PPCPs affect their degradability in the environment. In soil, the elimination of PPCPs takes place through biotic or abiotic medium. Photodegradation (Photolysis) and biodegradation by bacteria and fungi are the biotic means. Biodegradation of PPCPs by microbes takes place when they utilize PPCPs as substrate for carbon or energy source. However, increase in PPCPs concentration inhibits biodegradation and toxic to microorganisms (Ebele et al. 2017). The non-biotic pathways include sorption, hydrolysis, photolysis, oxidation and reduction processes. Many PPCPs can undergo natural photolysis and biodegradation, e.g. naproxen, gemfibrozil and ibuprofen, while gemfibrozil and ibuprofen can be degraded by biotransformation (Gurr and Reinhard 2006). Some of these processes take place in soil or surface water and some of them in treatment plants. Sorption, photodegradation and biodegradation are the most important processes occurring in the soil matrix (Kinney et al. 2006).

15.5 Strategies for Remediation of PPCPs from Soil

A significant number of techniques are being used to remove the pharmaceuticals and their metabolites. Combination of various advanced techniques like membrane filtration, ultrafiltration, membrane bioreactors, advanced oxidation processes (AOPs), nanofiltration, biological techniques, separation processes, adsorption, ozonization, reverse osmosis and combination of chemical and biodegradation process could be applied to improve the efficiency of WWTPs in removing PPCPs (Klatte et al. 2017; Sarkar et al. 2019).

Several factors govern the selection of specific treatment processes for the removal of PPCPs from effluents including the cost of the process, concentration of the specific pollutants, volume of the effluent to be treated, flow of the effluent to be treated, etc. (Xu et al. 2017). Among all, ease of operation, operational and functional flexibility, low cost, low maintenance and simplicity of design are the attractive features of the adsorption process. Various materials such as metal organic frameworks (MOFs), clay materials, zeolites, carbonaceous materials (e.g. activated carbon, carbon nanotubes (CNTs) and graphene oxide) and biosorbents like sludge and agriculture soil have been studied so far for the removal of PPCPs (Xu et al. 2017).

AOPs include ozonisation, Fenton oxidation, permanganate oxidation, photocatalytic technology, O_3/UV, UV/H_2O_2, electrochemical oxidation and ionizing radiation technology. They can be used to remove toxic or recalcitrant PPCPs but involve a high cost of operation due to the need of electrical energy to run the corresponding devices (Sarkar et al. 2018; Xu et al. 2017). Efficiency, treatment and utilizing processes such as sorption, plant uptake and biological degradation are the factors that affect the PPCPs removal by biological technique or microbial removal. The use of separation techniques including nanofiltration, microfiltration and ultrafiltration is also limited as many PPCPs could pass through the membrane. Such operations are generally employed in combination with other techniques for efficient results, such as combination of adsorption and AOPs (Kandlakuti 2016). Adsorption by activated carbon can remove pharmaceuticals more efficiently than coagulation and flocculation. High surface area, porous structure, cation/anion exchange capacity and variety of surface functional groups are the key features of an adsorbent which increase the removal of PPCPs in a more efficient manner. Use of carbon-based adsorbent material for adsorption of various organic and inorganic pollutants has been in use for a long time. Activated carbon (AC) has been explored in many studies for removal of various pollutants from soil as well as water (Ahmad et al. 2013, 2015).

Since activation process requires additional work and resources, this makes the process energy intensive and economically expensive. In the present scenario, biochar produced from abundantly available agro-processing residual biomass composed of cellulose, hemicellulose and lignin has come up as a sustainable and economically viable solution for remediation purposes. Though, production technology and treatment capability are the most sticking factors which decide the economic value of the biochar. Biochar produced from wood biomass with moderate capacity was reported to have larger environmental benefits for global warming, respiratory effects and noncarcinogenics than powdered activated carbon (Thompson et al. 2016). Biochar made from hemicellulosic biomass is able to remove various organic as well as inorganic contaminants (Anawar et al. 2015; Shakya and Agarwal 2019; Shakya et al. 2019).

15.6 Biochar for PPCPs Removal from Soil

Biochar was initially used as an amendment in soil to improve soil fertility, increase agricultural productivity, increase the soil nutrient and water holding capacity and to reduce GHGs emission (Lehmann and Joseph 2015). Biochar became an attractive adsorbent for the remediation of PPCPs due to its unique physicochemical properties. The possibility of surface modifications in biochars such as physical, chemical magnetic modification or biochar composite formation with impregnation of mineral sorbents is also an attractive advantage. A variety of factors including application time, biochar properties and process matrix influence the reduction of contaminant bioavailability. Along with this, the addition of biochar increases the C and N content in the soil resulting in the soil microbial growth (Yue et al. 2019).

Land application of biochar could be have a potential to be used as a shield against the leaching of PPCPs of particular concern into surface or groundwaters during application of reclaimed water for irrigation purposes (Yao et al. 2012). Reportedly, biochar was found to be more efficient than commercially available activated carbon for removal of various endocrine disrupting compounds (EDCs); (bisphenol A, 17 α-ethinylestradiol, 17 β-estradiol) and PPCPs (sulfamethoxazole, carbamazepine, ibuprofen, atenolol, benzophenone, benzotriazole, caffeine, gemfibrozil, primidone and triclocarban) (Kim et al. 2015). High temperature biochar (700 °C) derived from invasive plant was applied for the removal of most common veterinary drug sulfamethazine. Even small amount of biochar addition (5%) to the soil reduce the uptake of the drug to lettuce by 86% (Rajapaksha et al. 2014).

Forest pine wood biochar was applied to sandy loam soil, the uptake of fifteen different pharmaceuticals by radish was absorbed, and the accumulation of acetaminophen, carbamazepine, sulfadiazine, sulfamethoxazole, lamotrigine, carbadox, trimethoprim, oxytetracycline, tylosin, estrone and triclosan decreased by 33–83% in the soil amended with 1% biochar. Presence of biochar in soil acts as a barrier which lowered the uptake of drugs via lowering their concentrations in pore water (Li et al. 2020). Yue et al. (2019) amended the soils contaminated with a mixture of tetracycline, oxytetracycline and chlortetracycline and their corresponding intermediates epitetracycline, anhydrotetracycline, epianhydrotetracycline, epioxytetracycline, epichlortetracycline and demethylchlortetracycline with biochars derived from cow manure and plant materials. A 10% increase in removal rate of antibiotic was observed in the presence of biochar. Addition of biochar in soil increased the electrical conductivity that facilitates the accessibility of the microbes to the compounds. This elevates the microbe association with PPCPs and their secondary compounds, resulting in microbial degradation of pharmaceuticals (Shakya and Agarwal 2017; Yue et al. 2019).

Presence of antibiotics and the antibiotic resistance genes in soil can directly harm the human population through food chain. Addition of biochar to soil for reduction of ARGs could be a new dimension to explore about biochar use. Duan et al. (2017) suggested only 2% of biochar addition to the oxytetracycline contaminated soil not only limit the uptake of the drug but also reduce the relative abundance of ARGs from the crop by 50%. Further, disappearance of human pathogenic bacteria was also noticed with biochar application. Application of biochar to soil could led to the succession in bacterial population resulting in reduction in harmful pathogens (Duan et al. 2017). Li et al. (2019b) also reported decrease in total relative abundance of ARGs by 37.18% when clay composite biochar was added to the soil. Addition of rice straw and mushroom biochar as soil amendment not only limit the ARGs but also reduce the pathogenic bacteria biochar with 57% of removal rate (Cui et al. 2016).

Biochar can be used as the precursor material for the production of activated carbon, biochar-based metal or organic conjugates and designer or engineered biochars. All such modifications would increase the adsorption capacity of the biochars many folds. Presences of variety of surface functional groups on biochar surface provide ample possibilities of further such modifications. Use of compounds as activation/modification agent for biochar activation which are generally used in other treatment process could make the WWTPs more cost effective and hassle-free process. Persulfate is one of such agent which is used for AOP generally applied for soil and water treatment (Liu et al. 2020a). However, before use persulfate must be activated to generate sulphate radicals by means of various activation methods which extend the extra economic burden. Liu et al. (2020a) used persulfate as the activation agent for lychee branch biochar (600 °C) for efficient degradation of bisphenol A from soil as well as water. Also, biochar was found to have capability to activate persulfate to sulphate radicals (Jeon et al. 2017).

Kumar et al. (2017) synthesized nano-hetero assembly of superparamagnetic Fe_3O_4 and bismuth vanadate stacked on Pinus roxburghii derived biochar for thiophanate methyl removal. Use of biochar not only as adsorbent but as a catalyst in WWTPs for various treatment process could also be a possible approach.

Table 15.1 summarized the various studies conducted with biochar for PPCPs removal from soil. From table, it can be seen that application of biochar significantly limits the mobility of various pharmaceuticals and PCPs in the soil matrix. However, in some lab scale experiments, the studied concentration of PPCPs was higher than their actual presence in the environment, but performance of biochar for their removal is notably higher. Beside this, type of biochar and production conditions also affects the properties of biochars which significantly affects the adsorption potential of the biochars. All studies suggested that soil type and soil conditions also affect the performance of biochar and ionic state of the chemical compounds.

Though, the presence of PPCPs is generally in ng level in the environment, and their natural degradation could also happen in natural conditions. Still PPCPs elevated and unorganized usage of such compound mounting their concentration in the natural environment hence referred as 'emerging contaminants'. Promising results of biochar application of their removal suggests are as a boon for PPCPs remediation from soil. PPCPs removal from soil with biochar is still in initial stage and requires more extensive research for new dimensions about the use of biochar.

Table 15.1 Studies conducted for various PPCPs removal from soil with biochar

S. No	Biochar feedstock	Pyrolysis temperature	Retention time	Heating rate	Soil type	Pharmaceuticals or personal care product	Category of compound	Concentration of PPCPs to be studied	Removal mechanism	References
1	Grain and straw residues	650 °C	1 h (air atmosphere)	20 °C/min	Clay soil (5.27% organic matter), sand (0.07% organic matter) and reference soil (2.79% organic matter)	B-blockers, anti-inflammatory drugs, sulfonamides, 17-aethinylestradiol, carbamazepine and caffeine	Pharmaceutical	5 μg/mL	Π-π interactions, electrostatic interactions	Caban et al. (2020)
2	Forest pine wood	650 °C	30 min	–	Sand (74.6%), silt (24.6%), clay (0.78%), and organic matter (2.3%)	Acetaminophen, carbamazepine, sulfadiazine, sulfamethoxazole, lamotrigine, carbadox, trimethoprim, oxytetracycline, tylosin, estrone and triclosan	Pesticides, antibiotics	–	Π-π interactions, electrostatic interactions	Li et al. (2020)
3	Wheat straw	300 °C, 700 °C	6 h (Oxygen deficient condition)	–	Loessial soil	Ketoprofen, atenolol and carbamazepine	Pharmaceutical	10 μg/L	π - π interactions	Wu and Bi (2019)
4	Sugarcane harvest residues	550 °C	–	–	Sandy loam and clay soil	17α ethinylestradiol	Synthetic steroid estrogen	20 μg/L	Adsorption and abiotic transformation	Wei et al. (2019)
5	Mixed crop straw	500 °C	–	–	Sand (36%), silt (40%) and clay (24%)	Oxytetracycline Florfenicol	Antibiotics		Hydrophobic partitioning	He et al. (2019)

(continued)

Table 15.1 (continued)

S. No	Biochar feedstock	Pyrolysis temperature	Retention time	Heating rate	Soil type	Pharmaceuticals or personal care product	Category of compound	Concentration of PPCPs to be studied	Removal mechanism	References
6	Cow manure, peanut shell, *Salsolacollina* Pall., *Firmianaplatanifolia* sawdust and *Suaeda salsa* Pall	500 °C	2 h	10 °C/min	Soil organic carbon (7.2 g/kg), Total N_2 (0.86 g/kg), available N (174.13 mg/Kg), Olsen-P (46.67 mg/Kg), available K (469.82 mg/Kg)	Tetracycline and its derivatives	Antibiotics	200 µg/Kg	Electrostatic interactions	Yue et al. (2019)
7	Pine sawdust, bamboo sawdust, woodchips and pine cones	350 °C, 500 °C	2 h	10 °C/min	Organic carbon (0.755%), poorly grained sand with silt	Sulfamethoxazole	Antibiotics	–	Hydrophobic partitioning, hydrogen bonding, electrostatic interaction and π-π interaction, complexation	Yao (2018)
8	Vineyard wood	650 °C	–	–	Sand (90%), silt (8%), clay (2%)	Bisphenol A, caffeine, carbamazepine, clofibric acid, furosemide, methyl dihydrojasmonate, tonalide, triclosan and Tris(2-chloroethyl) phosphate	PPCP	–	π-π interactions (BPA & IBU), hydrophobic adsorption (CBZ)	Hurtado et al. (2017)

(continued)

Table 15.1 (continued)

S. No	Biochar feedstock	Pyrolysis temperature	Retention time	Heating rate	Soil type	Pharmaceuticals or personal care product	Category of compound	Concentration of PPCPs to be studied	Removal mechanism	References
9	Wheat, rice and corn stalks	400 °C	6 h (Oxygen deficient condition)	10 °C/min	Greenhouse soil	17β-estradiol	Steroids	–	Pore-filling	Zhang et al. (2017)
10	Pinus massoniana and Cunninghamia lanceolata trunks	300, 450, 700 °C	1 h	20 °C/min	Soil	Florfenicol	Antibiotics	5–20 mg/L	Hydrophobic interactions, π–π bonds hydrogen bonds and van der Waals forces	Jiang et al. (2017)
11	Spruce trees	450 °C(Softwood), 750 °C (Hardwood)	2.5 h to Few mins	–	Sand (92.2%), silt (4.3%), clay (3.5%) and organic matter (2.97%)	Progesterone	Steroids (endocrine disrupting compounds)	5 mg/Kg	Partitioning and hydrophobic interaction	Alizadeh et al. (2016)
12	Wood biochar	–	–	–	Calcareous purple soil	Ciprofloxacin, ofloxacin and enrofloxacin	Fluoroquinolones	–	–	Xuan et al. (2017)
13	Softwood	450 °C	–	–	Sand (92.2%), Silt (4.3%), Clay (3.5%)	17β-oestradiol (E2) and estrone (E1, primary metabolite	Steroids	–	–	Mann et al. (2016)
14	Corn stalks	500 °C	1.5 h	–	Agricultural soil	Bisphenol A (BPA) and 17a-ethynylestradiol (EE2)	EDCs	30 mg/Kg	Pore-filling, hydrogen bonding and π-π interactions	Xu et al. (2015)

(continued)

Table 15.1 (continued)

S. No	Biochar feedstock	Pyrolysis temperature	Retention time	Heating rate	Soil type	Pharmaceuticals or personal care product	Category of compound	Concentration of PPCPs to be studied	Removal mechanism	References
15	Wheat chaff, Eucalyptus	450 °C, 520 °C	–	–	Loamy sand soil	Carbamazepine and propranolol	Antibiotics	1.47 µg/L (CBZ), 2.13 µg/L (PRL)	Hydrophobic interactions and pore-filling	Williams et al. (2015)
16	Burcucumber plant and tea waste	700 °C	2 h	–	–	Sulfamethazine	Antibiotics	10 mg/L	Pore-filling	Rajapaksha et al. (2014)
17	Bur cucumber	300⁰C, 700⁰C	2 h	7 °C/min	Loamy sand and sandy loam soil	Sulfamethazine	Antibiotics	0.05 mg/L	π–π interactions and electrostatic cation exchange (pH 3), cation exchange (pH 5 & 7)	Vithanage et al. (2014)
18	Commercial hardwood biochars	–	–	–	Soil	Sulfamethazine	Veterinary antibiotic	–	Poe filling	Teixidó et al. (2013)
19	Bamboo waste	600 °C	–	–	Sand (26.9%), silt (22.7%), powder (40.4%), organic matter (46.9 g/kg)	Pentachlorophenol	Biocide	–	Diffusion and partitioning mechanism	Xu et al. (2012)

(continued)

Table 15.1 (continued)

S. No	Biochar feedstock	Pyrolysis temperature	Retention time	Heating rate	Soil type	Pharmaceuticals or personal care product	Category of compound	Concentration of PPCPs to be studied	Removal mechanism	References
20	Corn straw	100–600 °C	–	–	Agricultural soil	Tetracycline	Antibiotic	–	Partitioning, pore-filling	Zhang et al. (2012)
21	Hardwood (sweetgum and oak) Softwood (Yellow pines)	850 °C, 900 °C	–	–	Sandy loam forest soil, silt soil from corn field	Tylosin	Veterinary Antibiotic	–	Hydrophobic interactions	Jeong et al. (2012)
22	Corncob, pine saw dust and greenwaste	600 °C, 550 °C, 700 °C	–	–	Dairy farm soil	17β-oestradiol (E2) and estrone (E1, primary metabolite	Steroids	–	Pore-filling and hydrophobic interactions	Sarmah et al. (2010)

15.7 Factors Influencing the Removal of PPCPs Using Biochar

15.7.1 Biochar Properties

Biochar is always considered an efficient and effective adsorbent for the remediation of organic compounds. Biochar with small particle size and large surface area is considered to be more effective for remediation of organic compounds (Ahmad et al. 2013). When applied to the soil, biochar absorbs the organic compounds and alters their bioavailability to plants and microbes. When applied to the soil, biochar could affect the persistence and translocation of PPCPs. The alkaline nature, nearly 40–85% of C-content, and mineral fractions make biochar more recalcitrant in soil environment, hence hard for microbial degradation. Due to this, the bound pharmaceutical may not go for transformation process and cannot be easily released to the environment.

Biochars are produced from a number of widely available feedstocks including agriculture residues, food processing industries residue, forestry residues aquatic and invasive plants and various other non-conventional waste residues (Shakya and Agarwal 2018; Shakya and Agarwal 2020). This advantage makes biochar a multipurpose material which can be an adsorbent and also has carbon sequestration potential. Variation in feedstock has an obvious effect on the properties of the biochar. The compositional variation in the constituents of the feedstock (hemicellulose, cellulose, lignin and extractives) and the difference in their rate of decomposition significantly affect the physicochemical and surface properties of the biochar leading to the variation in adsorption potential of the biochars (Shakya and Agarwal 2019). Decomposition of cellulose and lignin not only affects the biochar yield but also influences the surface properties of the biochars.

Pyrolysis, gasification and hydrothermal carbonization are the most prominent thermochemical conversion techniques, which are used nowaday for biochar production. Among them, thermochemical treatment of biomass through slow pyrolysis yields highest amount of biochar. Variation in production process alters the properties of biochars (Aller 2016). Since, most of PPCPs are neutral in nature, high temperature biochar can be the more appropriate option for specifically PPCPs removal. The pyrolytic conditions highly influence the composition, surface properties and adsorption potential of the biochars. At low to medium range of temperature, most biochars are not fully carbonized (partial carbonization) and constituted with the carbonized organic matter (COM) and noncarbonized organic matter (NOM) with less aromatic fraction, less C-content but high polar hydroxyl and carbonyl functional groups (Chen et al. 2008; Shakya and Agarwal 2019). The COM is expected to behave as an adsorbent and the NOM as a partition (absorption) phase (Chen et al. 2008). Biochars produced at high temperature are nearly completely charred with relatively high surface area, low oxygen and hydrogen content, low polar surface functional groups and little organic fraction (Chun et al. 2004). It is suggested that carbonized surface possesses aromatic ring structure and high C-content having high affinity towards neutral organic contaminants through π-π interaction adsorbs, while the residual organic matter acts as a partition medium (Chun et al. 2004). Adsorption of neutral organic compounds on high temperature biochars was found to be exclusively on carbonized surface, while the sorption on low-temperature biochars resulted from the surface adsorption and the concurrent smaller partition into the residual organic matter phase (Chun et al. 2004). Chen et al. (2008) prepared biochars from 100 to 700 °C pyrolytic temperature with crop residue and suggested that the partition phase is evolved from an amorphous aliphatic domain to a condensed aromatic core with increasing pyrolytic temperature. The partition phase of (i) low-temperature biochar (100–300 °C) was suggested to be originated from an amorphous aliphatic fraction, which is enhanced with a reduction of the substrate polarity; (ii) for biochar prepared at 400–600 °C pyrolysis temperature, the partition occurs with a condensed aromatic core that diminishes with a further reduction of the polarity (Chen et al. 2008).

With the increase in pyrolysis temperature, the surface of biochar evolved simultaneously and makes transition in the adsorption components from a polarity-selective (low-temperature biochar—200–400 °C) to a porosity-selective (medium pyrolysis temperature, 500–600 °C) process and displays no selectivity with 700 °C biochar (Chen et al. 2008).

Surface properties of the biochars specifically polar surface functional groups are always the key players of the adsorption process which differentiate it from black carbon, soot or activated carbon. These polar surface functional groups electrostatically attract the contaminant and made complex with them resulting in their immobilization in soil matrix (Ahmad et al. 2013). Adsorption is the most inherent feature of biochar for which surface properties are the key responsible factor. The biochar prepared at low pyrolysis temperature is not fully carbonized and possesses various hydroxyl, phyenyl, carbonyl and carboxyl functional groups (Novak et al. 2009). This non-carbonized organic matter of biochar acts as primary adsorption phase, which interacts with neutral organic compounds more efficiently. The surface acidity and basicity of the biochars also influence their affinity towards polar and nonpolar compounds (Chun et al. 2004). Modification of biochar may create additional and abundant sorption sites on the surface of biochar by increasing the surface areas. This makes biochar surface more conducive to electrostatic attraction, surface complexation and/or surface precipitation, as well as enabling greater sorption affinity through stronger interactions with specific surface functional groups (Rajapaksha et al. 2016).

15.7.2 PPCPs Properties and Behaviour

The chemical nature and characteristic features of PPCPs significantly affect their presence, persistence, degradation, sorption and transformation in the environment. Ionization state of the pharmaceutical compound was found to be the key feature which directs the sorption of compound while lipophilicity had a negligible impact (Caban et al. 2020). The characteristic properties of PPCPs, such as solubility, polarity, acid dissociation constants (pKa values), hydrophobicity and hydrophilicity,

K_{ow}, and the distribution coefficient (KD), all depend upon the ionic state of the PPCPs (Rajapaksha et al. 2019). The nature of PPCPs could be acidic, basic and neutral or can be found as zwitterionic molecules depending on the environmental conditions and octanol/water partition coefficient (K_{ow}) values.

Acidic PPCPs may break down in the solution and form undissociated acid, and anions are released. On the other hand, basic PPCPs may dissociate to form both neutral and cationic molecules. Anions are difficult to be taken up by plants due to the negative electrical potential at the cell membrane of plant cells which repels the anions of negatively charged. Neutral molecules can pass through bio membranes of cells at a higher rate than ions with charge, thereby reducing bioaccumulation of chemicals by roots. Uptake of neutral chemicals from soil is largely governed by their hydrophobicity (log K_{ow}) (Guasch et al. 2012). The sorption behaviour of any PPCP was found to be highly dependent on the pKa and log K_{ow} values, and generally, higher removal capacities were noticed with larger log K_{ow} values of pharmaceuticals (Yoon et al. 2003).

15.7.3 Soil Properties

For PPCPs adsorption, soil properties, specifically natural soil organic matter (dissolved organic matter, humic acids and fulvic acids), play an important role. The natural organic matter can increase the adsorption potential of biochars by providing them with the more adsorption sites by superior dispensability. However, this could also block the active adsorption sites if natural organic matter consists of high molecular weight or by competing for the adsorption sites with the actual contaminant (Xiang et al. 2019). The microbes present in the soil have enzymatic activity, which could degrade the organic molecules, and limit their possible run-off and leaching to the groundwater.

It can be concluded that adsorption of PPCPs with biochar from soil is a synergistic process, which not only govern by properties of biochar but also soil physicochemical characteristics influence it. However, chemical nature of PPCPs has significant role especially the ionization state of the PPCPs.

15.8 Mechanism of PPCPs Removal from Soil with Biochar

The removal process of PPCPs with biochar is a synergistic action which involves physical as well as chemical interactions (Fig. 15.3). Pore-filling, electrostatic interaction, dipole–dipole interactions, van der Waals forces, π-π EDA interactions and hydrogen bonding are the possible interaction between biochar and the PPCP compounds (Li et al. 2019a; Rajapaksha et al. 2019). The most important factors which influence these interactions between biochar and PPCPs are the ionic state of the soil matrix. According to the ionic state and pH of the soil, the nature of

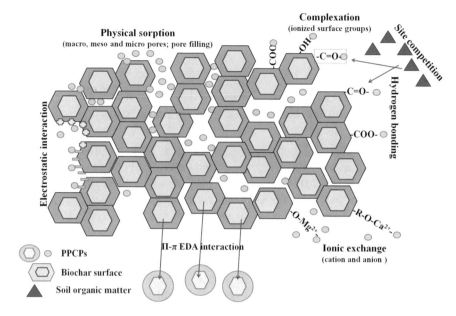

Fig. 15.3 Proposed mechanism of PPCPs removal from soil with biochar

pharmaceutical compound may change, leading to the impact on their adsorption through biochar. Further, the inorganic mineral content also plays important role for the adsorption of PPCPs. De-mineralization of the biochar surface will create the pockets or room for the interactions of ionic molecules of pharmaceuticals to get attached with. This would result in pore-filling or physical sorption of PPCPs on the biochar. However, large molecular size of the compounds would be a possible limitation to adsorption via pore-filling (Keerthanan et al. 2020). The released minerals in pore water of the soil could work as micronutrient for plants. The aromatic ring structure of biochar is the most attractive feature when it comes to the adsorption of organic compounds. The aromatic ring structure of biochar interacts with the aromatic rings of the pharmaceuticals via π-π EDA interactions and makes strong bond leading to the immobilization of the compound. The biochar surface is considered as graphitized surfaces and expected to have higher π-electron density. The π-π EDA interaction of the protonated aniline ring with the π-electron rich biochar surface referred to as π$^+$-π EDA interaction rather than ordinary electrostatic cation exchange (Li et al. 2019a). At high pyrolysis temperature, formation of more condensed aromatic ring structure takes place, while the polar hydroxyl and carboxyl groups decrease. This leads to the formation of strong interactions between high temperature biochar (π-donor) and PPCPs (π-acceptor) rather low-temperature biochars (Zhu et al. 2004). Part of trace metal or metal oxide is either present on the surface of biochar due to their presence in feedstock or their modifications to the biochar to stimulate the adsorption of antibiotics resulted in electrostatic interactions and surface complexation (Rajapaksha et al. 2019). The adsorption of organic molecules on biochar is suggested to be a collaborative behaviour of the carbonized organic matter and non-carbonized organic matter present in the biochar (Chen et al. 2008).

15.9 Possible Risk Factors Associated with Biochar Application for Removal of PPCPs from Soil

Land application of biochar is always considered beneficial for its advantage such as soil nourishment, improved fertility, recalcitrant organic carbon and nutrient retention. Occupational health hazards, environmental pollution, depletion in water quality and eutrophication could be the possible negative impacts when biochar applied to the soil. During the production of biochar, aggregation of recalcitrant minerals, possible formation of poly aromatic hydrocarbons, volatile organic compounds, dioxins and carbon nanoparticles takes place (Dutta et al. 2017). Soil application of biochar could possibly expose living organism to such inherent pollutants present in biochar. The major drawback of land application of biochar is that it cannot be removed from soil once applied. With time, the adsorption/immobilization potential of biochar decreased due to occupation of active sites with contaminant or other organic or inorganic compounds already present in the soil matrix. Besides this, PPCPs or contaminant immobilized biochars could pose possible negative and toxic effects on soil microbes and other organisms. Possibility of competitive behaviour of biochar towards soil micronutrient could not be ruled out, and it can react with essential soil micronutrients and immobilize them leading to their limited bioavailability to plants (Kavitha et al. 2018; Shakya and Agarwal 2020). Many PPCPs have complex structure and surface functional groups, and other essential compounds of similar structure may also get immobilized into soil matrix when come in contact with biochar. Since biochar has longer half-lives, the adsorbed PPCPs may remain in soil environment for longer time, and they could be transformed to more toxic metabolites. Continuous use of biochar for removal purpose could alter the soil environment. Thompson et al. (2016) reported relatively worse impacts of wood-based biochar than powdered activate in case of environmental performance due to energy use for bio-solids drying and the need for supplemental adsorbent.

15.10 Conclusion and Future Approaches

The knowledge gap, lack of economically sustained identification techniques and constant addition of various new compounds into the environment are the major challenges for PPCPs contamination management. Amalgamation of two or more processing techniques for WWTPs could be a possible solution for effective remediation of PPCPs, to limit their environmental discharge. Application of biochar for remediation of PPCPs could be an effective and economically viable substitute in comparison with other expansive process. Use of biochar for adsorption of PPCPs from soil not only provides clean environment but also nourishes the soil leading to the improved soil fertility. The characteristic diversity of the PPCPs and their chemical nature change with the change in the surrounding environment which makes their remediation more challenging and complex. The diversity of biochar's adsorption properties, such as porosity, surface area, surface functional groups, non-carbon fraction and carbon content, could be utilized in more effective ways against PPCPs in the real environment.

References

Ahmad M et al (2013) Biochar as a sorbent for contaminant management in soil and water: a review. Chemosphere 99:19–33. https://doi.org/10.1016/j.chemosphere.2013.10.071

Ahmed MB, Zhou JL, Ngo HH, Guo W (2015) Adsorptive removal of antibiotics from water and wastewater: progress and challenges. Sci Total Environ 532:112–126. https://doi.org/10.1016/j.scitotenv.2015.05.130

Al-Farsi RS, Ahmed M, Al-Busaidi A, Choudri B (2017) Translocation of pharmaceuticals and personal care products (PPCPs) into plant tissues: A review. Emerging Contaminants 3:132–137

Alizadeh S, Prasher SO, ElSayed E, Qi Z, Patel RM (2016) Effect of biochar on the fate and transport of manure-borne progesterone in soil. Ecol Eng 97:231–241

Aller MF (2016) Biochar properties: transport, fate, and impact. Crit Rev Environ Sci Technol 46:1183–1296. https://doi.org/10.1080/10643389.2016.1212368

Anawar HM, Akter F, Solaiman ZM, Strezov V (2015) Biochar: an emerging panacea for remediation of soil contaminants from mining. Ind Sewage Wastes Pedosphere 25:654–665. https://doi.org/10.1016/S1002-0160(15)30046-1

Antal MJ, Grønli M (2003) The art, science, and technology of charcoal production. Ind Eng Chem Res 42:1619–1640

Bastian R, Murray D (2012) Guidelines for water reuse. EPA Office of Research and Development, Washington, DC, USA

Boxall ABA et al (2012) Pharmaceuticals and personal care products in the environment: what are the big questions? Environ Health Perspect 120:1221–1229. https://doi.org/10.1289/ehp.1104477

Caban M et al (2020) Critical study of crop-derived biochars for soil amendment and pharmaceutical ecotoxicity reduction. Chemosphere 248:125976

Chen B, Zhou D, Zhu L (2008) Transitional adsorption and partition of nonpolar and polar aromatic contaminants by biochars of pine needles with different pyrolytic temperatures. Environ Sci Technol 42:5137–5143. https://doi.org/10.1021/es8002684

Chen W, Xu J, Lu S, Jiao W, Wu L, Chang AC (2013) Fates and transport of PPCPs in soil receiving reclaimed water irrigation. Chemosphere 93:2621–2630. https://doi.org/10.1016/j.chemosphere.2013.09.088

Chun Y, Sheng G, Chiou CT, Xing B (2004) Compositions and sorptive properties of crop residue-derived chars. Environ Sci Technol 38:4649–4655

Colon B, Toor GS (2016) Chapter three—a review of uptake and translocation of pharmaceuticals and personal care products by food crops irrigated with treated wastewater. In: Sparks DL (ed) Advances in agronomy, vol 140. Academic Press, pp 75–100. https://doi.org/10.1016/bs.agron.2016.07.001

Cui E, Wu Y, Zuo Y, Chen H (2016) Effect of different biochars on antibiotic resistance genes and bacterial community during chicken manure composting. Biores Technol 203:11–17. https://doi.org/10.1016/j.biortech.2015.12.030

Daughton CG, Ternes TA (1999) Pharmaceuticals and personal care products in the environment: agents of subtle change? Environ Health Perspect 107:907–938

Dodgen LK, Li J, Wu X, Lu Z, Gan JJ (2014) Transformation and removal pathways of four common PPCP/EDCs in soil. Environ Pollut 193:29–36. https://doi.org/10.1016/j.envpol.2014.06.002

Dodgen LK, Ueda A, Wu X, Parker DR, Gan J (2015) Effect of transpiration on plant accumulation and translocation of PPCP/EDCs. Environ Pollut 198:144–153

Duan M, Li H, Gu J, Tuo X, Sun W, Qian X, Wang X (2017) Effects of biochar on reducing the abundance of oxytetracycline, antibiotic resistance genes, and human pathogenic bacteria in soil and lettuce. Environ pollut 224:787–795. https://doi.org/10.1016/j.envpol.2017.01.021

Dutta T, Kwon E, Bhattacharya SS, Jeon BH, Deep A, Uchimiya M, Kim KH (2017) Polycyclic aromatic hydrocarbons and volatile organic compounds in biochar and biochar-amended soil: a review. Gcb Bioenergy 9:990–1004

Ebele AJ, Abdallah MA-E, Harrad S (2017) Pharmaceuticals and personal care products (PPCPs) in the freshwater aquatic environment. Emerg Contam 3:1–16

Ferreira AR, Ribeiro A, Couto N (2017) Remediation of pharmaceutical and personal care products (PPCPs) in constructed wetlands: applicability and new perspectives. In: Phytoremediation. Springer, pp 277–292

Guasch H, Ginebreda A, Geiszinger A (2012) Emerging and priority pollutants in rivers: bringing science into river management plans, vol 19. Springer Science & Business Media

Gurr CJ, Reinhard M (2006) Harnessing natural attenuation of pharmaceuticals and hormones in rivers. ACS Publications

He Y, Liu C, Tang X-Y, Xian Q-S, Zhang J-Q, Guan Z (2019) Biochar impacts on sorption-desorption of oxytetracycline and florfenicol in an alkaline farmland soil as affected by field ageing. Sci Total Environ 671:928–936

Herklotz PA, Gurung P, Vanden Heuvel B, Kinney CA (2010) Uptake of human pharmaceuticals by plants grown under hydroponic conditions. Chemosphere 78:1416–1421. https://doi.org/10.1016/j.chemosphere.2009.12.048

Holling CS, Bailey JL, Vanden Heuvel B, Kinney CA (2012) Uptake of human pharmaceuticals and personal care products by cabbage (Brassica campestris) from fortified and biosolids-amended soils. J Environ Monit 14:3029–3036. https://doi.org/10.1039/C2EM30456B

Hurtado C, Cañameras N, Domínguez C, Price GW, Comas J, Bayona JM (2017) Effect of soil biochar concentration on the mitigation of emerging organic contaminant uptake in lettuce. J Hazard Mater 323:386–393

Jeon P, Lee M-E, Baek K (2017) Adsorption and photocatalytic activity of biochar with graphitic carbon nitride (g-C3N4). J Taiwan Inst Chem Eng 77:244–249. https://doi.org/10.1016/j.jtice.2017.05.010

Jeong CY, Wang JJ, Dodla SK, Eberhardt TL, Groom L (2012) Effect of biochar amendment on tylosin adsorption–desorption and transport in two different soils. J Environ Qual 41:1185–1192

Jiang X, Li M (2020) Ecological safety hazards of wastewater. In: High-risk pollutants in wastewater. Elsevier, pp 101–123

Jiang C, Cai H, Chen L, Chen L, Cai T (2017) Effect of forestry-waste biochars on adsorption of Pb (II) and antibiotic florfenicol in red soil. Environ Sci Pollut Res 24:3861–3871

Kandlakuti VR (2016) Removal of target PPCPs from Secondary Effluent Using AOPs and Novel Adsorbents

Kavitha B, Reddy PVL, Kim B, Lee SS, Pandey SK, Kim K-H (2018) Benefits and limitations of biochar amendment in agricultural soils: a review. J Environ Manage 227:146–154

Keerthanan S, Jayasinghe C, Biswas JK, Vithanage M (2020) Pharmaceutical and personal care products (PPCPs) in the environment: plant uptake, translocation, bioaccumulation, and human health risks. Crit Rev Environ Sci Technol 1–38. https://doi.org/10.1080/10643389.2020.1753634

Kim E, Jung C, Han J, Son A, Yoon Y (2015) Removal of micro pollutants using activated biochars and powdered activated carbon in water. In: AGU Fall meeting abstracts, pp H33H-1701

Kinney CA, Furlong ET, Zaugg SD, Burkhardt MR, Werner SL, Cahill JD, Jorgensen GR (2006) Survey of organic wastewater contaminants in biosolids destined for land application. Environ Sci Technol 40:7207–7215. https://doi.org/10.1021/es0603406

Klatte S, Schaefer H-C, Hempel M (2017) Pharmaceuticals in the environment–a short review on options to minimize the exposure of humans, animals and ecosystems. Sustain Chem Pharm 5:61–66

Kumar A et al (2017) Facile hetero-assembly of superparamagnetic Fe3O4/BiVO4 stacked on biochar for solar photo-degradation of methyl paraben and pesticide removal from soil. J Photochem Photobiol A Chem 337:118–131. https://doi.org/10.1016/j.jphotochem.2017.01.010

Lehmann J, Joseph S (2015) Biochar for environmental management: science, technology and implementation. Routledge

Li L et al (2019a) Biochar as a sorbent for emerging contaminants enables improvements in waste management and sustainable resource use. J Clean Prod 210:1324–1342. https://doi.org/10.1016/j.jclepro.2018.11.087

Li Y et al (2019b) Effects of struvite-humic acid loaded biochar/bentonite composite amendment on Zn(II) and antibiotic resistance genes in manure-soil. Chem Eng J 375:122013. https://doi.org/10.1016/j.cej.2019.122013

Li Y, He J, Qi H, Li H, Boyd SA, Zhang W (2020) Impact of biochar amendment on the uptake, fate and bioavailability of pharmaceuticals in soil-radish systems. J Hazard Mater 122852

Lin K, Gan J (2011) Sorption and degradation of wastewater-associated non-steroidal anti-inflammatory drugs and antibiotics in soils. Chemosphere 83:240–246

Liu J, Jiang S, Chen D, Dai G, Wei D, Shu Y (2020a) Activation of persulfate with biochar for degradation of bisphenol A in soil. Chem Eng J 381:122637. https://doi.org/10.1016/j.cej.2019.122637

Liu X, Liang C, Liu X, Zhao F, Han C (2020b) Occurrence and human health risk assessment of pharmaceuticals and personal care products in real agricultural systems with long-term reclaimed wastewater irrigation in Beijing, China. Ecotoxicol Environ Safety 190:110022. https://doi.org/10.1016/j.ecoenv.2019.110022

Lofrano G et al (2020) 1—Occurrence and potential risks of emerging contaminants in water. In: Sacco O, Vaiano V (eds) Visible light active structured photocatalysts for the removal of emerging contaminants. Elsevier, pp 1–25. https://doi.org/10.1016/B978-0-12-818334-2.00001-8

Mann S, Qi Z, Prasher SO (2016) Transport and fate of estrogens from swine manure in a biochar amended sandy soil in a freeze-thaw environment. Can Biosyst Eng 58:1.1–1.10

Nikolaou A, Meric S, Fatta D (2007) Occurrence patterns of pharmaceuticals in water and wastewater environments. Anal Bioanal Chem 387:1225–1234

Novak JMLI, Xing B, Gaskin JW, Steiner C, Das KC, Ahmedna M, Rehrah D, Watts DW, Busscher WJ, Schomberg H (2009) Characterization of designer biochar produced at different temperatures and their effects on a loamy sand Annals of Environmental. Science 3:195–206

OECD (2008) OECD guidelines for the testing of chemicals Bioaccumulation in Sediment-dwelling. Benthic Oligochaetes OECD Guideline 315

Öztürk M, Ashraf M, Aksoy A, Ahmad MSA, Hakeem KR (2015) Plants, pollutants and remediation. Springer

Prosser RS, Sibley PK (2015) Human health risk assessment of pharmaceuticals and personal care products in plant tissue due to biosolids and manure amendments, and wastewater irrigation. Environ Int 75:223–233. https://doi.org/10.1016/j.envint.2014.11.020

Rahman M, Yanful E, Jasim S (2009) Endocrine disrupting compounds (EDCs) and pharmaceuticals and personal care products (PPCPs) in the aquatic environment: implications for the drinking water industry and global environmental health. J Water Health 7:224–243

Rajapaksha AU, Vithanage M, Lim JE, Ahmed MBM, Zhang M, Lee SS, Ok YS (2014) Invasive plant-derived biochar inhibits sulfamethazine uptake by lettuce in soil. Chemosphere 111:500–504. https://doi.org/10.1016/j.chemosphere.2014.04.040

Rajapaksha AU et al (2016) Engineered/designer biochar for contaminant removal/immobilization from soil and water: potential and implication of biochar modification. Chemosphere 148:276–291

Rajapaksha AU, Premarathna KSD, Gunarathne V, Ahmed A, Vithanage M (2019) Sorptive removal of pharmaceutical and personal care products from water and wastewater. In: Pharmaceuticals and personal care products: waste management and treatment technology. Elsevier, pp 213–238

Sabourin L, Duenk P, Bonte-Gelok S, Payne M, Lapen DR, Topp E (2012) Uptake of pharmaceuticals, hormones and parabens into vegetables grown in soil fertilized with municipal biosolids. Sci Total Environ 431:233–236. https://doi.org/10.1016/j.scitotenv.2012.05.017

Sarkar B, Mandal S, Tsang YF, Kumar P, Kim K-H, Ok YS (2018) Designer carbon nanotubes for contaminant removal in water and wastewater: a critical review. Sci Total Environ 612:561–581. https://doi.org/10.1016/j.scitotenv.2017.08.132

Sarkar B et al (2019) Sustainable sludge management by removing emerging contaminants from urban wastewater using carbon nanotubes. In: Industrial and municipal sludge. Elsevier, pp 553–571

Sarmah AK, Srinivasan P, Smernik RJ, Manley-Harris M, Antal MJ, Downie A, van Zwieten L (2010) Retention capacity of biochar-amended New Zealand dairy farm soil for an estrogenic steroid hormone and its primary metabolite. Soil Res 48:648–658

Shakya A, Agarwal T (2019) Removal of Cr(VI) from water using pineapple peel derived biochars: Adsorption potential and re-usability assessment. J Mol Liq 111497. https://doi.org/10.1016/j.molliq.2019.111497

Shakya A, Agarwal T (2020) Potential of biochar for the remediation of heavy metal contaminated soil. In: Biochar applications in agriculture and environment management. Springer, pp 77–98

Shakya A, Agarwal T (2017) Poultry litter biochar: an approach towards poultry litter management–a review. Int J Curr Microbiol App Sci 6(10):2657–2668

Shakya A, Agarwal T (2018) Green pea pod biochar as a low-cost adsorbent: an alternative approach for the removal of Cr (VI) from aqueous solution. Int J Pure Appl Biosci 6(4):375–386

Shakya A, Núñez-Delgado A, Agarwal T (2019) Biochar synthesis from sweet lime peel for hexavalent chromium remediation from aqueous solution. J Environ Manage 251:109570. https://doi.org/10.1016/j.jenvman.2019.109570

Shenker M, Harush D, Ben-Ari J, Chefetz B (2011) Uptake of carbamazepine by cucumber plants—a case study related to irrigation with reclaimed wastewater. Chemosphere 82:905–910. https://doi.org/10.1016/j.chemosphere.2010.10.052

Teixidó M, Hurtado C, Pignatello JJ, Beltrán JL, Granados M, Peccia J (2013) Predicting contaminant adsorption in black carbon (biochar)-amended soil for the veterinary antimicrobial sulfamethazine. Environ Sci Technol 47:6197–6205

Thompson KA, Shimabuku KK, Kearns JP, Knappe DRU, Summers RS, Cook SM (2016) Environmental comparison of biochar and activated carbon for tertiary wastewater treatment. Environ Sci Technol 50:11253–11262. https://doi.org/10.1021/acs.est.6b03239

Usyskin A, Bukhanovsky N, Borisover M (2015) Interactions of triclosan, gemfibrozil and galaxolide with biosolid-amended soils: Effects of the level and nature of soil organic matter. Chemosphere 138:272–280. https://doi.org/10.1016/j.chemosphere.2015.05.095

Vithanage M, Rajapaksha AU, Tang X, Thiele-Bruhn S, Kim KH, Lee S-E, Ok YS (2014) Sorption and transport of sulfamethazine in agricultural soils amended with invasive-plant-derived biochar. J Environ Manage 141:95–103

Wei Z et al (2019) Effect of biochar amendment on sorption-desorption and dissipation of 17α-ethinylestradiol in sandy loam and clay soils. Sci Total Environ 686:959–967

Williams M, Martin S, Kookana RS (2015) Sorption and plant uptake of pharmaceuticals from an artificially contaminated soil amended with biochars. Plant Soil 395:75–86

Wu L, Bi E (2019) Sorption of ionic and neutral species of pharmaceuticals to loessial soil amended with biochars. Environ Sci Pollut Res 26:35871–35881

Wu X, Ernst F, Conkle JL, Gan J (2013) Comparative uptake and translocation of pharmaceutical and personal care products (PPCPs) by common vegetables. Environ Int 60:15–22. https://doi.org/10.1016/j.envint.2013.07.015

Xiang Y et al (2019) Carbon-based materials as adsorbent for antibiotics removal: mechanisms and influencing factors. J Environ Manage 237:128–138

Xie H et al (2019) Pharmaceuticals and personal care products in water, sediments, aquatic organisms, and fish feeds in the Pearl River Delta: occurrence, distribution, potential sources, and health risk assessment. Sci Total Environ 659:230–239. https://doi.org/10.1016/j.scitotenv.2018.12.222

Xu T, Lou L, Luo L, Cao R, Duan D, Chen Y (2012) Effect of bamboo biochar on pentachlorophenol leachability and bioavailability in agricultural soil. Sci Total Environ 414:727–731

Xu N, Zhang B, Tan G, Li J, Wang H (2015) Influence of biochar on sorption, leaching and dissipation of bisphenol A and 17α-ethynylestradiol in soil. Environ Sci Process Impacts 17:1722–1730

Xu Y, Liu T, Zhang Y, Ge F, Steel RM, Sun L (2017) Advances in technologies for pharmaceuticals and personal care products removal. J Mater Chem A 5:12001–12014

Xuan P, Tang X, Xian Q, Liu C, Yang F, Huang Y (2017) Effects of biochar on adsorption-desorption of fluoroquinolones in purple soil. China Environ Sci 37:2222–2231

Yang Y, Ok YS, Kim K-H, Kwon EE, Tsang YF (2017) Occurrences and removal of pharmaceuticals and personal care products (PPCPs) in drinking water and water/sewage treatment plants: a review. Sci Total Environ 596:303–320

Yao W (2018) Removal of sulfamethoxazole by adsorption and biodegradation in the subsurface: batch and column experiments with soil and biochar amendments

Yao Y et al (2012) Adsorption of sulfamethoxazole on biochar and its impact on reclaimed water irrigation. J Hazard Mater 209:408–413

Yin L, Wang B, Yuan H, Deng S, Huang J, Wang Y, Yu G (2017) Pay special attention to the transformation products of PPCPs in environment. Emerg Contam 3:69–75. https://doi.org/10.1016/j.emcon.2017.04.001

Yoon Y, Westerhoff P, Snyder SA, Esparza M (2003) HPLC-fluorescence detection and adsorption of bisphenol A, 17β-estradiol, and 17α-ethynyl estradiol on powdered activated carbon water research 37:3530–3537. https://doi.org/10.1016/S0043-1354(03)00239-2

Yu Y, Liu Y, Wu L (2013) Sorption and degradation of pharmaceuticals and personal care products (PPCPs) in soils. Environ Sci Pollut Res 20:4261–4267

Yue Y, Shen C, Ge Y (2019) Biochar accelerates the removal of tetracyclines and their intermediates by altering soil properties. J Hazard Mater 380:120821. https://doi.org/10.1016/j.jhazmat.2019.120821

Zhang G, Liu X, Sun K, He F, Zhao Y, Lin C (2012) Competitive sorption of metsulfuron-methyl and tetracycline on corn straw biochars. J Environ Qual 41:1906–1915

Zhang F, Li Y, Zhang G, Li W, Yang L (2017) The importance of nano-porosity in the stalk-derived biochar to the sorption of 17β-estradiol and retention of it in the greenhouse soil. Environ Sci Pollut Res 24(10):9575--9584. https://doi.org/10.1007/s11356-017-8630-4

Zhu D, Hyun S, Pignatello JJ, Lee LS (2004) Evidence for π−π electron donor−acceptor interactions between π-donor aromatic compounds and π-acceptor sites in soil organic matter through pH effects on sorption. Environ Sci Technol 38:4361–4368. https://doi.org/10.1021/es035379e

Chapter 16
Biochar for Improvement of Soil Properties

Abhishek Kumar and Tanushree Bhattacharya

Abstract Anthropogenic activities have deteriorated the quality of soil all across the globe. It becomes extensively essential to improve the soil quality to support various life systems and sustain the planet. Several options, such as agro-chemicals, nanotechnology, phytoremediation, etc., have been sought for to boost the quality of soil. A simple, cheap, renewable, and sustainable material called biochar has been used by farmers for enhancing the soil quality since time immemorial. Biochar is a carbon-rich material obtained from thermal treatment of biomass in a limited supply of oxygen. Biochar facilitates nutrients to plants, increases soil pH, uplifts cation exchange capacity, supports soil microbial community, and remediates polluted soils. Additionally, biochar aids in waste management, crop productivity enhancement, clean energy production, and climate change mitigation. The chapter focuses on application of biochar for improving the physical, chemical, and biological parameters of soil which would be vital for sustainable development and is the need of the hour.

Keywords Biochar · Pyrolysis · Electrical conductivity · Cation exchange capacity · Soil organic matter · Porosity · Bulk density · Water holding capacity

16.1 Introduction

Soil is among the most important resources available on our planet. It supports life systems by retaining water and nutrients which promotes growth of plants and microorganisms. Soil helps in anchoring plant roots. Soil is a vital component of biogeochemical cycles. Soil helps in filtering rainwater and polluted waters. Soil is the basis of our agricultural systems. Growth of crops helps in food production which sustains the growing population. Additionally, soil helps in growing fodder crops for the cattle domesticated by human beings. Soil is also an important sink for

A. Kumar · T. Bhattacharya (✉)
Department of Civil and Environmental Engineering, Birla Institute of Technology, Mesra, Ranchi, Jharkhand 835215, India
e-mail: tbhattacharya@bitmesra.ac.in

greenhouse gases which aids in carbon sequestration, thereby supporting mitigation of climate change. Lastly, it provides raw materials for construction of infrastructure.

Due to the various anthropogenic activities, the quality of soil has deteriorated drastically all across the globe (Kumar et al. 2021b, c). Soil has been impinged with heavy metals by activities such as industrial dumping, waste disposal, nuclear action, and military operation (Young 1995; Tchounwou et al. 2012; Panagos et al. 2013). Mining is another such activity leading to heavy metal pollution (Al-Farraj et al. 2013). Agricultural activity is another such major reason responsible for release of heavy metals into the environment (Bhattacharya et al. 2021; Kumar et al. 2021a).

The human population has risen rapidly in the previous century, especially in the last few decades. Such an increase has boosted the demand for food, resources, and energy, which has ultimately led to over-exploitation of natural resources. The demand for food crops has given way to intensification of agriculture. Plants need macro- and micro-nutrients for appropriate and proper growth. Harvesting of crops triggers a decrease in nutrient level in the soil with time, since they are not returned to the soils (Pathak 2010). This enquires for addition of nourishment to soil in the form of chemical fertilizers for a high yield of crops. However, minimal amount of nutrients are absorbed by the crops, which makes it inevitable for regular addition of the chemical fertilizers. The large-scale application of these chemicals reduces nutrient content in the crops and initiates degradation of soil fertility (Hariprasad and Dayananda 2013; Yargholi and Azarneshan 2014).

Apart from fertilizers, enormous amounts of agro-chemicals such as pesticides, weedicides, and insecticides have been applied to agricultural fields. These chemicals affect the microbial community present in soil, which disturbs the health of the soil and decreases fungal and bacterial biomass (Wu 2012; Prashar and Shah 2016). The escalation in agricultural activities has also resulted in degradation of soil properties such as organic matter decomposition (Reynolds and Stafford 2002; Middleton 2004). Further, human activities have given way to harsh climatic conditions such as elevated atmospheric temperatures, which ultimately decomposes soil organic matter thereby degrading the soil (Díaz et al. 1997).

Such damaging effects of agro-chemicals, apart from the reduction in soil fertility, enhanced soil erosion, soil acidification, and depletion of soil organic matter, have necessitated the need for development of simple, cost-effective, and sustainable options in the field of agriculture (Jianping 1999; Annabi et al. 2011; De Meyer et al. 2011). Further, the extensive addition of heavy metals to soil asks for efficient removal via sustainable methods. Non-degradability and high bioavailability of the heavy metals is responsible for their immense toxicity, and it becomes essential to decrease their toxic manifestation (Zhang et al. 2013).

Various methods have been used for remediating the soils contaminated with heavy metals. These include chemical, physical, and biological alternatives (Khalid et al. 2017). The chemical methods include chemical immobilization, soil washing, and soil encapsulation; the physical alternatives are inclusive of soil replacement, soil isolation, and vitrification, and the biological techniques involve bioremediation, phytoremediation, and biochar-based remediation. Apart from application of biochar, most of the methods have some sort of drawbacks associated with them. For example, phytoremediation has been used to decrease the levels of heavy metals from polluted soils (Cristaldi et al. 2017).

When compared to chemical and physical methods, phytoremediation is eco-friendly and cost-effective. It uses plants to take up the heavy metals from the soil. The metals could be stabilized by the root systems, which could help in their spread by erosion and leaching (Khalid et al. 2017; Liu et al. 2018). The metals are also transported to the aerial parts of the plant, which later needs to be harvested (Gomes et al. 2016; Khalid et al. 2017). However, polluted soils are low in nutrient content, and organic matter content, and are alkaline, which make them low in fertility. Additionally, high metal content inhibits plant growth. These factors make phytoremediation unsuitable for heavy metal remediation. Use of soil amendments could overcome these disadvantages by immobilizing heavy metals, supplying nutrients to plants, and improving other soil properties. Manures and composts have been used to amend the soils, but pathogens and other harmful chemicals could be present in these amendments (Ding et al. 2016). These could be damaging to the agricultural soils. Additionally, they could boost global warming by releasing greenhouse gases. It is important to note that soil amendment should be safe for the environment, support soil structure, and enhance soil fertility (Paz-Ferreiro et al. 2014).

Lately, biochar has emerged as a promising amendment in the previous decades (Kumar and Bhattacharya 2021, 2022). Biochar is a renewable eco-friendly resource obtained after thermal treatment of biomass in a limited supply of oxygen (Lahori et al. 2017). The properties of biochar depend on the feedstock used for biochar production and the thermal treatment techniques. Amendment of soil with biochar would help in decreasing the mobility of heavy metals. The efficiency of removal of heavy metals is dependent on the soil type, the heavy metal under consideration, biochar type, and the application rate of biochar (Debela et al. 2012; Dwibedi et al. 2022).

Additionally, it would help in improving the soil properties by adding nutrients to the soil, increasing its pH, and enhancing cation exchange capacity (Bayu et al. 2016; Bonanomi et al. 2017). Further, it would boost soil biota, mitigate climate change, generate clean energy, and facilitate waste management. Improvement in soil properties would help in promoting growth of plants, thereby assisting in healing and protecting the ecosystem and environment from degradation (Placek et al. 2016; Dwibedi et al. 2022). Further, nutrient-enriched biochar has been used to ameliorate the polluted soils and enhance their fertility (Spokas et al. 2012; Kammann et al. 2015; Schmidt et al. 2015). This chapter explores the effects of biochar on the properties of soil. In the beginning, biochar is defined followed by its production. Later, some of the significant properties of biochar are discussed. It is followed by a discussion on the various physical, chemical, and biological properties of soil and the impact that biochar has on these properties.

16.2 Biochar Basics

Thermal treatment of any type of biomass, which contains carbon, results in the production of a stable, carbon–neutral, and recalcitrant substance is called biochar (Manyà 2012; IBI 2015). Biochar has been used by farmers since ages for increasing the quality of the soil (Kumar et al. 2022a, b; Shaikh et al. 2022a, b). The properties of biochar are dependent upon the type of feedstock used for biochar production and the thermal technique used for treatment of the feedstock (Tang et al. 2013). The origin of biochar dates back to the use of "Amazonian dark earth" by Amerindian population (Glaser et al. 2002; Forján et al. 2017). "Amazonian dark earth" is also referred to as *Terra preta*. However, biochar is different from *terra preta* in composition. *Terra preta* is a kind of dark-coloured soil produced by mixing plant residues, compost, manure, bones, and faeces with charcoal produced at low-temperature (Balée et al. 2016a, b). Recognition of nutritional status of *Terra preta* enabled people to produce and utilize biochar for wide ranging applications (Glaser et al. 2002).

Since the previous few decades, biochar has been used for remediating soils contaminated with heavy metals by reducing their bioavailability and mobility. Additionally, biochar boosts the soil quality which would enable in enhancement of the crop yield (Lehmann et al. 2006). Furthermore, biochar could be used for a number of applications (Lehmann et al. 2011; Titirici et al. 2012; Zhang et al. 2013; Mohan et al. 2014; Windeatt et al. 2014; Hossain 2016; Lee et al. 2018; Manyà et al. 2018). For example, the increase in crop yield would help in managing food security issues. Biochar could be produced by using waste substances as feedstock which would help in management of waste. Biochar could also enable carbon sequestration and reduction in emission of greenhouse gases. Biochar could be utilized for production of biofuel which would assist in management of fossil fuels (Kumar et al. 2020). It could be safely said that biochar production and its utilization would assist in tackling a number of threatening issues sustainably.

16.3 Production of Biochar

Thermal treatment of biomass results in production of biochar. Various thermal techniques have been used for producing biochar such as torrefaction, combustion, gasification, carbonization, and pyrolysis (Meyer et al. 2011; Zhang et al. 2013). Among these, pyrolysis is the most commonly used techniques for biochar production. Pyrolysis is a process where biomass is thermally treated in the range of 300–900 °C in a limited supply of oxygen in a furnace or a fabricated kiln to produce biochar. Pyrolysis results in maximum yield of biochar among the various mentioned techniques and its simplicity of operation favours its wide usage for production purposes (Cha et al. 2016). A limited supply of oxygen enables biomass to be heated above its thermal stability to form a stable biochar. Heat decomposes the feedstock and brings about devolatilization of its components. Thermal treatment results in formation of biochar, bio-oil, and syngas. A number of functional groups are formed which are rich in oxygen such as hydroxyl, carboxyl, and carbonyl (Ekström et al. 1985; Overend 1999). It is important to note that the treatment method and the treatment conditions affect the properties of biochar produced (Sahota et al. 2018; Zhang et al. 2018a).

Pyrolysis could be divided into slow or fast depending on the heating rate utilized during the process. Heating biomass at temperatures of 400–500 °C with a moderate kind of heating rate is slow pyrolysis. On the contrary, heating biomass at a temperature range of 800–1200 °C with a very rapid rate of heating is called fast pyrolysis. Among them, slow pyrolysis results in the maximum yield of biochar, and the yield could be as high as 40% (Peterson et al. 2012). However, there is scope to reduce the production time and increase the efficiency of production (Tripathi et al. 2016). Feedstocks such as crop residues and softwood chip have been used to produce biochar via slow pyrolysis (Onay and Kockar 2003; Windeatt et al. 2014; Behazin et al. 2016). Biomass materials such as corn cobs, corn stover, and rice straw have been used to produce biochar by fast pyrolysis (Onay and Kockar 2003; Mullen et al. 2010; Eom et al. 2013).

Variations of pyrolysis such as flash pyrolysis and slow steam pyrolysis have also been used for synthesis of biochar (Yorgun et al. 2001; Onay and Kockar 2003; Giudicianni et al. 2013). They utilized waste materials such as vegetable waste and sunflower oil cake in these processes. Pyrolysis is affected by heating rate, treatment temperature, pressure conditions, vapour residence time, and the heating technique, such as electrical heating or fuel burning (Asensio et al. 2013). Vapour residence time is governed by the rate of removal of volatile gases from the kiln as they bring about secondary reactions on the surface of biochar (Meyer et al. 2011). It could be stated as a conclusion that pyrolysis is simple, efficient, and cost-effective technique to produce biochar (Cha et al. 2016).

Among other thermal treatment techniques, torrefaction involves feedstock heating at temperatures of 200–300 °C. Feedstocks such as sawdust, peanut husk, bagasse, rice husk, and water hyacinth have been used for biochar production by torrefaction (Pimchuai et al. 2010). In combustion, feedstock is burnt directly to obtain thermal energy apart from biochar. Biomass need to be pre-treated to obtain a higher yield. Various waste biomasses have been used to produce biochar via combustion (McKendry 2002; Caillat and Vakkilainen 2013). Biomass is heated at temperatures of 700–900 °C during gasification. Gasification yields a low amount of biochar (Neves et al. 2011; Asensio et al. 2013). Lignocellulose-rich plant biomasses have been used to produce biochar and syngas through gasification (Pröll et al. 2007; Balat et al. 2009; Cui et al. 2018).

Temperatures of 350–650 °C for 30 min coupled with high pressure is flash carbonization and yields biochar and syngas as major products. Woody biomass such as Oak and agricultural waste material such as corncob have been used to produce biochar through flash carbonization (Antal et al. 2003; Asensio et al. 2013). When wet biomass is thermally subjected to high temperature and pressure, it is called hydrothermal carbonization and yields hydrochar (Wang et al. 2018). Waste materials such as barley straw and eucalyptus sawdust have been used to produce hydrochar (Sevilla et al. 2011; Titirici et al. 2012).

A wide variety of feedstocks could be used for production of biochar. Some of the feedstocks used for the production of biochar are kitchen waste, agricultural residues, sewage sludge, rubber tyres, wood biomass, leaf litter, animal litter, poultry litter, and algae (Beesley and Marmiroli 2011; Cantrell et al. 2012; Lu et al. 2012; Ghani et al. 2013; Xu et al. 2013; Zhao et al. 2013; Mazac 2016; Kumar and Bhattacharya 2020). Importantly, utilization of waste materials is a supporting factor for management of waste by aiding in reduction in generation of waste, reducing soil and groundwater pollution, and decreasing the number of landfill spots. However, it must be kept in mind that a lot of these biomasses are also used for biofuel production and composting, which makes it necessary to properly sieve out biomasses for biochar production (Kuppusamy et al. 2016; Tripathi et al. 2016). It has also been stated that moisture content in feedstocks should be decreased below 30% to increase the feedstock heating rate consequently reducing the time required to achieve thermal treatment conditions (Bryden and Hagge 2003; Lv et al. 2010). Feedstock could be dried under sun or through human-aided techniques, so that feasibility of thermal treatment enhances.

The nutritional content of biomass affects the presence of nutrients in biochar (Chan et al. 2007). Since biochar has been used as a soil amendment to improve its fertility status, it becomes a necessity to select the feedstock for production of biochar accordingly. Furthermore, composition of biomass feedstock could affect the temperature at which biochar would be synthesized. For example, holocellulose-rich biomass could be utilized to synthesize biochar at lower temperatures, while lignin-rich biomass would be ideal for biochar production at comparatively higher temperatures. Further, yield of biochar production could be increased by presence of a high lignin content in the biomass (Angin 2013; Shivaram et al. 2013). Interestingly, it would be ideal to determine the production temperature depending on the applicability of biochar. For example, biochar produced at high temperatures are beneficial for heavy metal removal purposes, when compared to biochars synthesized at low temperatures, because of their higher efficiency (Igalavithana et al. 2017; Weber and Quicker 2018; Zhang et al. 2018b).

16.4 Properties of Biochar

The properties of biochar are highly variant and are majorly dependent on the variation in biomass and the thermal treatment conditions. The composition of biochar varies according to the variation in biomass composition, heating rate, and treatment temperature. Feedstocks are mostly composed of hemicellulose, cellulose, and lignin. When a feedstock is thermally treated, hemicellulose, cellulose and lignin begin decomposing at 200 °C–315 °C, 315 °C–400 °C, and 400 °C respectively (Sadaka et al. 2014). Biochar is majorly composed of carbon. Carbon content rises with a rise in treatment temperature (Vassilev et al. 2010). On the contrary, nitrogen, oxygen, and hydrogen contents fall with a rise in treatment temperature (Sun et al. 2014). A rise in thermal treatment temperature triggers a decrease in surface functional groups containing hydrogen and oxygen. It results in a reduction in O/C and H/C atomic ratios. A rise in thermal treatment temperature also accentuates the processes of devolatilization and biomass decomposition during thermal conservation. It results in a decrease in volatile matter content in the char matrix, with an increase in treatment temperature (Pimchuai et al. 2010; Weber and Quicker 2016). Further, a rise in treatment temperatures results in an increase in elemental composition inclusive of potassium, calcium, and magnesium (Sun et al. 2014). Rise in treatment temperature increases the aromaticity in biochar as a result of aromatic carbon's greater stability at higher temperatures (Conti et al. 2014). Aromatic structures boost potential of biochar to remove pollutants (Wang et al. 2016).

Removal of oxygen-containing surface functional groups brings about a rise in pH and alkalinity in biochar (Fidel et al. 2017). The pH of biochar increases at higher treatment temperatures. A higher pH could aid in neutralizing acidic nature in soils. Similar to pH, electrical conductivity in biochar increases with an increase in treatment temperatures. A rise in pH and electrical conductivity could also be attributed to the fact that ash content in biochar rises with a rise in treatment temperatures. Ash is composed of oxides of metals like magnesium, calcium, silicon, and others (Vassilev et al. 2013). Further, organic acids get volatilized during thermal treatment, which could also be a contributing factor in pH rise, as seen in biochar derived from crop residues (Wang et al. 2019b). Additionally, it has been reported that biochar derived from animal remains have a higher pH when compared to biochar derived from plant biomass, and could be contributed to a greater carbonate and ash content (Rajkovich et al. 2012). Feedstock is, therefore, a significant factor in affecting the pH of biochar (Wang et al. 2019b).

In contrast to pH, cation exchange capacity of biochar is higher at lower treatment temperatures (Mukherjee et al. 2011). Higher cation exchange capacity is a result of the anionic nature furnished by negative charges present on the surface of biochar. The negative charges attract the positively charged cations, thereby aiding in pollutant removal. It also assists in decreasing pollutant uptake in plants, in particular crop species (Cushman and Robertson-Palmer 1998; Liang et al. 2006). The cation exchange capacity of animal waste-derived biochar such as cow manure-derived biochar is less in comparison to biochar derived from plant biomass (Wang et al. 2019b). The research team attributed it to the presence of greater ash content and lower content of carbon and nitrogen. Biochar obtained from thermal treatment of manure, corn stover, or oak, has higher cation exchange capacity when compared to biochar produced using food waste, paper mill waste, or hazelnut shells (Rajkovich et al. 2012).

Surface area and porosity of biochar are dependent on the biomass material used for production and conditions utilized during thermal treatment (Manna et al. 2020). Thermal treatment results in devolatilization and decomposition processes, thereby releasing volatile gases from the feedstock material and providing porosity to the matrix of biochar. Porosity enhances with a rise in treatment temperatures. In line with porosity, surface area increases with enhancement in treatment temperatures. Although it must be stated that porosity and surface area decreases at temperatures above 800–1000 °C due to the breakdown in cell structures in the feedstock matrix

(Cetin et al. 2004). Pulido-Novicio et al. (2001) stated that the decrease could also be an outcome of decreasing solid matrix. Thermal treatment of biomass material also brings about an enhancement in the pore volume of the biochar produced (Fu et al. 2012). More than four-fifth of the volume in biochar is occupied by micropores (0.05–0.0001 μm). The porous nature of biochar, in addition to high surface area, assists in pollutant removal from soil and water (Rouquerol et al. 1999; Houben et al. 2013; Yin et al. 2016). Surface area in biochar derived from plant material is more than biochar obtained from dairy manure. A greater amount of carbon in plant biomass-derived biochar is responsible for the larger surface area and porosity (Cao and Harris 2010). Higher surface area boosts nutrient and water retention potential apart from supporting microbial community and enhancing cation exchange capacity (Igalavithana et al. 2017; Weber and Quicker 2018). Microbial community bring about redox reactions in the soil matrix and assist biochar in contaminant removal (Solaiman and Anawar 2015).

As stated, thermal treatment reduces the porosity in biochar, which brings about a reduction in mechanical stability and enhancement in brittleness. Further, it enhances grindability of biochar (Byrne and Nagle 1997). Feedstock composition affects the grindability and brittle nature of biochar. A high lignin content decreases grindability while a high hemicellulose content increases grindability of biochar (Emmerich and Luengo 1994). A high mechanical stability assists in carbon sequestration and coal replacement in industries. Importantly, the high stability of biochar does not deter the pollutant removal efficiency, i.e. contaminant removal efficiency of biochar does not decrease with its aging (Li et al. 2016). Interestingly, stability of biochar has been reported to be greater than thousand years in soil (Mahimairaja and Shenbagavalli 2012). Further, biochar grindability also affects the particle size distribution, which plays a significant role in altering its interaction with soil matrix, apart from affecting porosity, surface area, and nutrient availability for plants (Valenzuela-Calahorro et al. 1987; Sun et al. 2012a; Xie et al. 2015; Liao and Thomas 2019).

Enhancement in carbon content is accompanied by increase in energy content of biochar (Weber and Quicker 2018). The high energy content in biochar (30–35 MJ/kg) could help in promoting its use for bio-energy. Thermal conductivity of biochar diminishes with a rise in its porosity, which traps more air, thereby providing an insulating effect and decreasing its conductivity (Usowicz et al. 2006). Such a property could be helpful in using biochar for heat insulation in construction materials and soil insulation in cold countries.

Interaction of biochar with water could be hydrophobic or hydrophilic, according to previous results. Chun et al. (2004) and Zornoza et al. (2016) have reported that removal of polar functional groups from surface of biochar results in hydrophobicity. They also stated that biochar produced at low thermal treatment temperatures are more hydrophobic. On the other hand, enhancement in porosity of biochar with rise in thermal treatment temperature, increases the potential to retain more water, and could be helpful for the plants grown on soils amended with biochar (Zhang and You 2013; Gray et al. 2014).

16.5 Impact of Biochar on Chemical Properties of Soil and Their Consequent Improvement

Amendment of soil with biochar changes the properties of soil such as pH, electrical conductivity, cation exchange capacity, soil organic matter, nutrient content, nutrient availability, and metals. The impact of biochar on some of these properties has been represented in Table 16.1.

16.5.1 Effect of Biochar on Soil pH

Soil pH is one of the most important parameters affecting the soil properties, nutrient status, and the presence of organic and inorganic pollutants. Soil pH in the range of 6.5–7.5 is considered to be adequate for the growth of plants (Marini Köpp et al. 2011). As regards the presence of contaminants in soil, cationic pollutants are found at lower pH while anionic pollutants are found at higher pH.

The pH of biochar generally lies above 7. Although Ahmad et al. (2014) and Khan et al. (2014) have reported acidic biochar. Decomposition of hemicellulosic contents forms propionic, formic, and acetic acids at low temperatures resulting in acidic pH. However, pH of biochar generally lies in the alkaline range. pH of biochar rises with rising thermal treatment temperatures (Igalavithana et al. 2017). Removal of acidic functional groups, such as hydroxyl and carboxyl, from surface during the thermal treatment results in alkaline pH of biochar. Presence of higher ash content causes higher pH (Weber and Quicker 2018). Formation of alkali carbonates on the surface of biochar also causes rise in pH (Sadaka et al. 2014). Biochar pH also affects the sorption potential of biochar. High pH increases the sorption potential of biochar (Zhang et al. 2018b).

Addition of biochar enhances the pH of soil. The increase in pH caused because of amendment of soil with biochar could be an outcome of a number of factors (Hmid et al. 2014; Xu et al. 2017; Meng et al. 2018), such as (i) a liming effect introduced by the alkaline nature of biochar, (ii) release of ions like oxide and carbonate after biochar addition which neutralizes the hydrogen ions, (iii) presence of functional groups on the surface of biochar which removes the hydrogen ions, and (iv) alkalinity provisioned by the ash contents in biochar. The alkaline metals and transition metals, such as Sodium, potassium, calcium, magnesium, copper, manganese, zinc and others form oxides and carbonates in biochar during the thermal treatment of biomass which forms the constituents in ash and is responsible for the alkaline nature of biochar (Sadaka et al. 2014; Weber and Quicker 2018). Vassilev et al. (2013) reported the presence of CaO, K_2O, and SiO_2 in biomass ash which is retained in biochar and contributes to alkaline nature. Biochar rich in ash content could act as a good fertilizer in the soils (Wu et al. 2012). Although, an ash could block pores in the biochar matrix, consequentially diminishing the available surface area for sorption and retention of water and nutrients (Manna et al. 2020). Interestingly, rise in alkaline conditions in

Table 16.1 Impact of biochar on chemical properties of soil

Biochar	Dose	pH	CEC (cmol/kg)	SOM/SOC	Nutrients	Reference
Peanut shells pyrolyzed at 400 °C	2%	5.37–6.05	–	12.8–250.8 mg/kg	N (68–65 mg/kg) K (138–211 mg/kg) P (49–47 mg/kg)	Zhang et al. (2019)
Hardwood pyrolyzed at 500 °C	10 t ha^{-1}	5.61–5.72	1.1–3.4	1.78–2.52%	N (0.16–0.17%) K (0.09–0.12 cmol/kg) P (8.6–10.1 mg/kg)	Adekiya et al. (2020)
Hardwood pyrolyzed at 500 °C	20 t ha^{-1}	5.61–5.88	1.1–5.3	1.78–2.75%	N (0.16–0.17%) K (0.09–0.13 cmol/kg) P (8.6–10.7 mg/kg)	Adekiya et al. (2020)
Hardwood pyrolyzed at 500 °C	30 t ha^{-1}	5.61–5.96	1.1–7.5	1.78–2.97%	N (0.16–0.18%) K (0.09–0.14 cmol/kg) P (8.6–14.6 mg/kg)	Adekiya et al. (2020)
Reed pyrolyzed at 500 °C	1%	4.67–4.7	–	16.4–13.9 mg/kg	K (211–156 mg/kg) P (34–38 mg/kg)	Dai et al. (2013)
Reed pyrolyzed at 500 °C	3%	4.67–4.77	–	16.4–18.2 mg/kg	K (211–228 mg/kg) P (34–52 mg/kg)	Dai et al. (2013)
Sow manure pyrolyzed at 500 °C	1%	4.67–5.95	–	16.4–124.5 mg/kg	K (211–310 mg/kg) P (34–129 mg/kg)	Dai et al. (2013)
Sow manure pyrolyzed at 500 °C	3%	4.67–6.76	–	16.4–13.9 mg/kg	K (211–687 mg/kg) P (34–175 mg/kg)	Dai et al. (2013)
Pineapple peel pyrolyzed at 500 °C	1%	4.67–5.58	–	16.4–157.1 mg/kg	K (211–745 mg/kg) P (34–62 mg/kg)	Dai et al. (2013)
Pineapple peel pyrolyzed at 500 °C	3%	4.67–6.83	–	16.4–89.1 mg/kg	K (211–1790 mg/kg) P (34–116 mg/kg)	Dai et al. (2013)
Wood pyrolyzed at 450 °C	50 t ha^{-1}	6.86–7.18	–	–	N (0.24–0.25%) P (0–16 mg/kg) K (62–82 mg/kg)	Jones et al. (2012)

(continued)

Table 16.1 (continued)

Biochar	Dose	pH	CEC (cmol/kg)	SOM/SOC	Nutrients	Reference
Peanut hull pyrolyzed at 400 °C	2%	5.6–7.3	2.2–2.7	2.78–18.8 g/kg	N (0.35–0.77 g/kg) K (37–319 mg/kg) P (28–47 mg/kg) Ca (131–173 mg/kg) Mg (24–46 mg/kg)	Novak and Busscher (2013)
Peanut hull pyrolyzed at 500 °C	2%	5.6–7.4	2.2–2.4	2.78–19.55 g/kg	N (0.35–0.75 g/kg) K (37–304 mg/kg) P (28–38 mg/kg) Ca (131–151 mg/kg) Mg (24–31 mg/kg)	Novak and Busscher (2013)
Hard wood from fast pyrolysis	2%	5.6–6.1	2.2–2.6	2.78–18.42 g/kg	N (0.35–0.35 g/kg) K (37–85 mg/kg) P (28–28 mg/kg) Ca (131–187 mg/kg) Mg (24–28 mg/kg)	Novak and Busscher (2013)

soil via biochar addition makes the heavy metals more stable thereby decreasing their toxic potential (Yuan et al. 2011).

In a study by Kelly et al. (2014), addition of beetle-killed lodge pine biochar led to an increase in pH in multi-contaminated soil arising from lead, zinc, silver, and gold extraction. Huang et al. (2018) reported an enhancement in pH in soils contaminated with lead, zinc, copper, cadmium and arsenic after amendment with chicken manure biochar. Mokarram-Kashtiban et al. (2019) stated an enhancement in pH in soils spiked with lead, copper, and cadmium after amendment with hornbeam biochar. Lebrun et al. (2017) found enhancement in pH in soils contaminated with lead and arsenic after amendment with pinewood biochar.

Interestingly, the particle size of biochar also affects the variation in pH increase. Finer particles in biochar bring about greater increase in pH in comparison to coarser particles (Lebrun et al. 2018b, c, d). Further, dose of biochar application also affects the changes in soil pH, and a higher dose increases the soil pH more than compared to lower dose of biochar application.

Furthermore, the increase in pH is also dependent on the initial pH of the soil to which biochar is added. Pinewood biochar was applied to different industrial and mine sites. The increase in pH was different at different sites. An increase in 2.2 and 2.9 units was observed in an acidic mine site when biochar was applied at 2 and 5%, while no effect was observed in an alkaline Issoudun industrial site, and an increase in 0.5 units was observed in a slightly acidic Mortagne-du-Nord industrial site when biochar was applied to the contaminated soil (Lebrun et al. 2017, 2018b; Lomaglio et al. 2018). These differences explain the variation in pH arising out of the biochar addition to the soils with varying pH.

It is also important to note that biochars produced from different biomasses bring about different increase in pH of the soil. In a study by Lebrun et al. (2020a, b, c), biochars were prepared form different feedstocks such as oak heartwood, oak sapwood, lightwood, oak bark, pinewood, and hardwood, and applied to a former mine site at Pontgibaud. The maximum increase was seen in biochars prepared from oak bark, while minimal decrease was seen in biochars prepared from heartwood and hardwood.

Further, biochar produced at different thermal treatment temperatures also alter the increase in pH of soils differently. The enhancement in thermal treatment temperatures increases the ash contents in biochar (Weber and Quicker 2018; Zhang et al. 2018b). It is caused because of the release of moisture and volatile matter from the feedstock and retention of biomass ash during thermal treatment. In a study by Al-Wabel et al. (2013, 2019), biochar was produced from date palm at 300, 500, and 700 °C, and applied to mine site contaminated with cadmium, copper, lead, iron, manganese, and zinc. They reported that biochar produced at 300 °C reduced the soil pH, while the ones produced at 500 and 700 °C increased the soil pH.

16.5.2 Effect of Biochar on Soil Electrical Conductivity

Electrical conductivity (EC) is the measure of the ability of any given solution to conduct electricity. It is dependent on the quantity and nature of dissolved salts or water-soluble ions. The EC of soil affects different processes prevailing in soil, particularly, biotic activities and nutrient supply to plants grown in the soils. Presence of an excess of salt affects the balance between water and soil and hampers growth of the plants. Soil EC below 100 μS/cm lies in the non-saline range which implies an absence of salt hindrances for growth of the plants (Smith and Doran 1996). EC of polluted soils is generally on the lower side (Namgay et al. 2010; Olmo et al. 2014).

EC of biochar rises with an increase in thermal treatment temperature. The increase in ash contents in biochar with an increase in thermal treatment temperature could be the reason behind increase in EC (Junna et al. 2014). EC of biochar affects physical properties in soil such as hydraulic conductivity and alters various biogeochemical processes in soil, by varying plant growth and microbial community thereby altering the nutrient cycling (Wang et al. 2015; Igalavithana et al. 2017).

The EC of soil rises with the application of biochar. In a study by Hossain et al. (2010), EC of soil was found to rise by 6 times after biochar amendment. Lebrun et al. (2017) reported a 2 time increase in EC of contaminated soil after biochar application. They stated that the elevated EC of biochar along with presence of soluble salts is responsible for the increase in soil EC after biochar amendment (Hmid et al. 2014; Melaku et al. 2020).

The increase in EC is also dependent on the rate and type of biochar application (Lebrun et al. 2017, 2018d, a, b; Lomaglio et al. 2017, 2018). Application of pinewood biochar to Pontgibaud mine soil and La Petite Faye soil increased the EC by 2 times and 1.5 times respectively. Application of pinewood biochar to Issoudun industrial soil did not increase the EC. They stated that such increase could be dependent on initial EC of soil. Application of lightwood biochar to Pontgibaud mine soil enhanced the EC twice when applied at 2% dose while it increased by 5 times when applied at 5%. Application of pinewood biochar increased soil EC 1.4 times when applied at 5% dose.

Further, particle size in biochar is significant in altering EC of soil. Lebrun et al. (2018d) applied hardwood biochars to Pontgibaud mine site at 2 and 5% dose. Finer sized particles increased EC by 3.3 and 1.5 times at 2% application rate and 4.9 and 2.7 times at 5% application rate. However, coarse sized biochar did not alter EC in the initial stages, but did increase EC after 46 days of application. Lebrun et al. (2018d) stated that a low surface area in coarse biochar could be the reason behind such a variation in EC in soil.

16.5.3 Effect of Biochar on Soil Cation Exchange Capacity

The capacity of soil to retain exchangeable cations is referred to as cation exchange capacity of the soil (Dai et al. 2017). It provides information about the metal sorption capability of the soil. Biochar has been reported to possess a high cation exchange capacity. The presence of negative charges on the surface of biochar coupled with a high surface area is responsible for the high cation exchange capacity. Further, abundance of carboxyl and hydroxyl groups helps in enhancement of the cation exchange capacity (Stella Mary et al. 2016). Consequently, amendment of soil with biochar helps in boosting its cation exchange capacity (Yuan et al. 2019).

Rafael et al. (2019) added baby corn peel biochar to an acidic soil and reported an increase in cation exchange capacity. Uzoma et al. (2011) reported an increase in cation exchange capacity of sandy soil after cow manure biochar addition. Nigussie et al. (2012) observed that addition of biochar derived from maize stalk to chromium contaminated soil enhanced the cation exchange capacity of the soil. Bandara et al. (2017) applied woody biochar to serpentine soil and stated an increase in cation exchange capacity. They also observed that the rise in cation exchange capacity was greater at higher application rate.

However, the increase in cation exchange capacity of soil with biochar addition was not universal. Nandillon et al. (2019a, b) stated that amendment of mine soil with hardwood biochar does not bring about changes in cation exchange capacity of soil. They observed such a neutral effect could be due to the low cation exchange capacity of the applied biochars. Biochars prepared at lower thermal treatment temperatures have a low cation exchange capacity (Janus et al. 2018).

16.5.4 Effect of Biochar on Soil Organic Matter and Soil Organic Carbon

Carbon containing compounds are present in soil. They are collectively called soil organic matter (SOM). Apart from carbonaceous compounds, minerals could also be present in SOM. The presence of these minerals in addition to carbon containing compounds makes SOM a facilitator of plant growth. The presence of 4–8% of SOM is considered to be fertile (Pettit 2014; Crouse 2018). Nevertheless, SOM is generally present in low quantities in contaminated soil. Apart from provisioning of minerals, SOM enables soil to retain nutrients and water and facilitates the growth of soil microbes by providing substrates (Agegnehu et al. 2015). Additionally, SOM provides energy to the soil microbes (Ramesh et al. 2019). Further, a high content of SOM in soil increases the metal retention capacity (Forján et al. 2016).

Carbon is retained in biochar during the process of thermal treatment. Rise in thermal treatment temperatures enhances the carbon contents in biochar (Weber and Quicker 2018). However, carbon content could decrease in certain cases as a result of low biomass density (Armynah et al. 2019). It was stated by Keiluweit et al. (2010) and Naeem et al. (2019) that the turbostratic structure in biochar could result in a graphitization effect. Further, Armynah et al. (2019) observed that presence of carbon in the form of graphene could be responsible for sheet-like structures on biochar surface.

Amendment of soil with biochar uplifts organic carbon content in soil. Such an increase triggers microbial activity in soils and changes and redox and biochemical processes in soils (Beesley and Dickinson 2011; Choppala et al. 2012; Qian et al. 2016). The elevated organic matter and organic carbon content in biochar enhances organic matter and organic carbon content in contaminated soils (Janus et al. 2015). In a study by Lebrun et al. (2018b), lightwood biochar and pinewood biochar in the particle size range of 0.2–0.4 mm was applied to Pontgibaud mine soil. The SOM in soil increased by 1.36–2.77 and 4.94% when lightwood biochar was applied at a dosage of 2 and 5%, while application of pinewood biochar enhanced SOM by 2.88–5.08%.

Contrastingly, Lomaglio et al. (2017, 2018) did not report any increase in dissolved organic carbon when pinewood biochar was applied to soils at Mortagne-du-Nord smelting site and La Petite Faye mine site. Interestingly, Lebrun et al. (2017) stated that dissolved organic carbon reduced in Pontgibaud mine site, and could be an outcome of improved microbial activity resulting in degradation of organic carbon (Hass et al. 2012). In a study by Li et al. (2018a), rice straw biochar was applied to soils contaminated with cadmium and lead and observed that SOM increased 2.3 times. Further, Cui et al. (2012) stated that biochar dose could affect SOM and soil organic carbon in soils. They reported no change when wheat straw biochar was applied at 10 t ha^{-1}, but remarkable increase was observed at 20 and 40 t ha^{-1} dosage. Lastly, Park et al. (2011b) observed that SOM enhancement boosted copper mobilization, which could be detrimental for soil.

16.5.5 Effect of Biochar on Soil Nutrients

A number of nutrients are necessary for growth of plants inclusive of nitrogen, phosphorus, and potassium. However, soils contaminated with organic and inorganic pollutants are low in nutritional status (Nandillon et al. 2019a). Biochar is generally rich in nutrient content and the nutritional status is dependent on the biomass used for its production. Further, the high nutrient retention capability also aids in provisioning of nutrients to plants. The alkaline nature of biochar boosts nutrient facilitation (Arienzo et al. 2009; Nigussie et al. 2012). Interestingly, the high sorption potential of biochar could help in reducing the nutrients lost due to leaching, but could simultaneously be detrimental for plant growth by sorbing the nutrients like phosphate, ammonium, and nitrate (Sarkhot et al. 2013; Zeng et al. 2013; Bakshi et al. 2014; Gai et al. 2014; Gao et al. 2015; Yang et al. 2017; Zhao et al. 2018).

The thermal treatment conditions could also affect the nutritional status of biochar. An increase in thermal treatment temperatures increases the carbon and sulphur contents while decreases the nitrogen contents in biochar (Sadaka et al. 2014; Li et al. 2018b; Kubier et al. 2019). Such trends are outcome of removal of cellulose, hemicellulose, and lignin contents from the biomass coupled with volatilization of nitrogen components in the form of ammonia and oxides of nitrogen. Chen et al. (2020) stated that presence of sulphur in biochar could enable it in contaminant removal. Nutrients such as magnesium, calcium, manganese, zinc, and iron are present in biochar, and their contents enhance with an increase in thermal treatment temperatures (Muhammad et al. 2017; Naeem et al. 2017; Zhang et al. 2018b).

Previous studies have reported an enhancement in nutrient content and availability with application of biochar to soils. Nandillon et al. (2019a, b) applied hardwood biochar to mine soils at 5% dose, and observed that potassium and phosphorus content multiplied by 2.7 and 3 times respectively. Olmo et al. (2014) amended farm soils with biochar derived from olive tree pruning, and stated a 1.6 times increase in nitrogen and phosphorus contents. Houben et al. (2013) amended multi-contaminated soil with biochar derived from Miscanthus straw and reported an enhancement in calcium, magnesium, potassium, and phosphorus contents.

16.6 Impact of Biochar on Physical Properties of Soil and Their Consequent Improvement

Amendment of soil with biochar changes the properties of soil such as water holding capacity, porosity, bulk density, and soil aggregation. The impact of biochar on some of these properties has been represented in Table 16.2.

Table 16.2 Impact of biochar on physical properties of soil

Biochar	Dose	Porosity	Bulk density	Water holding capacity	Reference
Peanut hull pyrolyzed at 500 °C	25%	0.5–0.55%	1.3 to 1.15 g cm^{-3}	17–25%	Githinji (2014)
Peanut hull pyrolyzed at 500 °C	50%	0.5–0.61%	1.3 to 0.85 g cm^{-3}	17–43%	Githinji (2014)
Peanut hull pyrolyzed at 500 °C	75%	0.5–0.69%	1.3 to 0.6 g cm^{-3}	17–56%	Githinji (2014)
Peanut hull pyrolyzed at 500 °C	100%	0.5–0.78%	1.3 to 0.38 g cm^{-3}	17–51%	Githinji (2014)
Corn stover pyrolyzed at 350 °C	1.13%	10% increase	1.01 to 0.94 g cm^{-3}	12–16% increase	Herath et al. (2013)
Corn stover pyrolyzed at 550 °C	1%	19% increase	1.01 to 0.91 g cm^{-3}	12–16% increase	Herath et al. (2013)
Pine pyrolyzed at 650 °C	12.5%	69.6–71.3%	–	90.14–93.34% (Soil A) 87.35–91.48% (Soil B)	Rehman et al. (2011)
Pine pyrolyzed at 650 °C	25%	69.6–72.1%	–	90.14–98.23% (Soil A) 87.35–92.98% (Soil B)	Rehman et al. (2011)
Birch pyrolyzed at 400 °C	1.2%	51–53%	1.3 to 1.25 g cm^{-3}	12.5% increase	Karhu et al. (2011)
Pecan shells pyrolyzed at 400 °C	2.1%	–	1.2 to 1.45 g cm^{-3}	10% decrease	Busscher et al. (2011)
Pondersoa Pine pyrolyzed at 450 °C	0.5%	–	–	11.9 to 12.4 g cm^{-3}	Briggs et al. (2012)
Pondersoa Pine pyrolyzed at 450 °C	1%	–	–	11.9 to 13 g cm^{-3}	Briggs et al. (2012)
Pondersoa Pine pyrolyzed at 450 °C	5%	–	–	11.9 to 18.8 g cm^{-3}	Briggs et al. (2012)
Hardwood pyrolyzed at 500 °C	10 t ha^{-1}	40–46%	1.58 to 1.44 Mg m^{-3}	40–70% increase	Adekiya et al. (2020)
Hardwood pyrolyzed at 500 °C	20 t ha^{-1}	40–51%	1.58 to 1.3 Mg m^{-3}	60–110% increase	Adekiya et al. (2020)

(continued)

Table 16.2 (continued)

Biochar	Dose	Porosity	Bulk density	Water holding capacity	Reference
Hardwood pyrolyzed at 500 °C	30 t ha^{-1}	40–59%	1.58 to 1.08 Mg m^{-3}	80–140% increase	Adekiya et al. (2020)

16.6.1 Effect of Biochar on Soil Water Holding Capacity

Amount/quantity of water that a material can retain is called as water holding capacity. Water holding capacity is an essential property of soil that enables plants to grow. Plants absorb water and the nutrients available in water, which helps them in their growth. Further, a high soil water holding capacity would decrease the irrigation frequency of crops, thereby aiding in sustainable agriculture. Interestingly, crops grow well in soils with a high soil water holding capacity. Contaminated soils have a lower water holding capacity mostly because of a low aggregation structure in the soils. Biochar has a high sorption potential, a high porosity, and boosts soil aggregation. These properties of soil enhance its potential to hold and retain water (Herath et al. 2013; Obia et al. 2016). Presence of hydrophilic functional groups on the pores and surface of biochar could assist in increasing water holding capacity (Uzoma et al. 2011). Consequently, addition of biochar to soil enhances its water holding capacity. Furthermore, water holding capacity of soil is also affected by the soil texture (Tryon 1948). Biochar application could increase water holding capacity in sandy soils more than clayey and loamy soils.

Biochar application could enhance the available soil water contents up to 97% and saturated water contents up to 56% (Uzoma et al. 2011). Biochar application could increase moisture retention in soil by 15% (Laird et al. 2010). In a study by Lebrun et al. (2018b), biochars prepared from pinewood and lightwood was added to mine soils at a rate of 5%, and observed an enhancement in water holding capacity. Similarly, Karhu et al. (2011) found a rise in water holding capacity in soil after amendment with biochar derived from birch. Fellet et al. (2011) observed that water holding capacity in mine soil contaminated with lead and zinc increased by 10% after amendment with biochar derived from prune residue. Molnár et al. (2016) observed an increase of about 5% in water holding capacity in agricultural sandy soil after amendment with biochar derived from paper fibre sludge and grain husk. Nevertheless, amendment of soil with biochar does not necessarily increase water holding capacity. Lebrun et al. (2018d) observed that fine sized hardwood biochars do not affect water holding capacity of soil while the coarser ones increase water holding capacity when amended at a rate of 5%. They observed that such a result could be an outcome of the pore size of biochar where more water is held by larger pores.

16.6.2 Effect of Biochar on Soil Porosity

Ratio of pore volume to soil volume is referred to as soil porosity. Soil porosity is a significant soil attribute that affects growth of the plants. The pores in soil are classified depending upon the pore size—macro pores, meso pores, and micro pores. Pores are crucial for aeration in addition to retention and movement of water and nutrients. Further, they provide refuge to the microbial community in the soils.

Amendment of soil with biochar helps in enhancing the overall porosity of soil (Masulili et al. 2010; Devereux et al. 2013; Burrell et al. 2016; Obia et al. 2016). The enhancement is dependent upon the soil and biochar type (Herath et al. 2013). The soil and biochar type also influences the percentage of pore types (Githinji 2014). The high porosity of biochar enables it to increase soil porosity (Mukherjee and Lal 2013). However, biochar could clog soil pores thereby decreasing the soil porosity. The interaction between biochar and soil minerals could also result in an increase in soil porosity with biochar addition (Blanco-Canqui 2017).

16.6.3 Effect of Biochar on Soil Bulk Density

The measurement of the extent to which the soil particles are tightly pressed together is called bulk density. In other words, mass of dry soil divided by volume of soil particles and pore spaces is called bulk density of soil. Soil bulk density massively affects soil properties and growth of plants. Soil bulk density above >1.6 Mg cm^{-3} decreases the water holding and water sorption capacity of soil. Additionally, a high bulk density provides penetration resistance to the roots of plants. This could ultimately affect soil properties and plant growth (Goodman and Ennos 1999).

Application of biochar could reduce the bulk density of soil (Masulili et al. 2010; Devereux et al. 2013; Burrell et al. 2016; Obia et al. 2016). It could be an outcome of the high porosity and pore volume of biochar (Mukherjee and Lal 2013). Enhancing the dose of biochar application could increase the reduction in bulk density of soil (Githinji 2014). The decrease in bulk density could also be a result of a dilution effect induced by the low density of biochar (Burrell et al. 2016; Blanco-Canqui 2017). The support to microbial activity through biochar addition enhances soil aeration which could decrease bulk density (Burrell et al. 2016).

16.6.4 Effect of Biochar on Soil Aggregation

The attractive forces in the soil system bind them together via adhesive and cohesive forces. These contribute in keeping the soil colloidal particles together. Such a property is significant from the point of view of structure of soil. A well aggregated soil possesses a good structure which facilitates a good medium for movement of water and nutrients in soil. Such an aggregated soil structure also helps in uptake of water and nutrients by plants (Borselli et al. 1996; Aslam et al. 2014). Interestingly, microbes present in soil could secrete some polysaccharides which might enhance the adherence between the soil colloidal particles, thereby boosting soil aggregation.

Amendment of soil with biochar boosts enhances soil aggregation in soil systems (Masulili et al. 2010; Devereux et al. 2013; Burrell et al. 2016; Obia et al. 2016). Further, biochar addition to soil pumps microbial growth by furnishing refuge to the soil microbes. Additionally, biochar also prevents desiccation of microbes and prevents the microbes from predators (Aslam et al. 2014). Increase in microbial activity and community structure would boost soil aggregation (Burrell et al. 2016).

16.7 Impact of Biochar on Other Properties of Soil and Their Consequent Improvement

Apart from changes in chemical and physical properties of soil, amendment of soil with biochar changes certain other properties of soil such as—metals and metalloids mobility and bioavailability; soil microbial community; and growth of plants.

16.7.1 Effect of Biochar on Metals and Metalloids

Soils could be contaminated with a number of metals and metalloids. Amendment of such polluted soils with biochar could have varying results depending on the type of metal or metalloid in consideration along with the type of biochar applied. Application of biochar to soil reduces their mobility and bioavailability. For example, lead is a metal with high mobility and bioavailability in acidic conditions. Application of biochar minimizes their mobility and bioavailability. In a study by Lebrun et al. (2017), lead concentrations decreased by 69 and 97% in Pontgibaud mine soil when biochar derived from pinewood was applied at 2% and 5% respectively. In a similar study, lead concentrations decreased by 86% when biochar derived from pinewood was applied and lead concentrations decreased by 69% when biochar derived from lightwood was applied Pontgibaud mine soil (Lebrun et al. 2018b).

Further, particle size of biochar could affect the immobilization and reduction in bioavailability of metals in polluted soils. Lebrun et al. (2018c) studied the impact of hardwood biochars on lead immobilization. They reported that finer biochars reduced lead mobility right after their addition, while coarse biochars reduced lead mobility after 46 days of their addition. Lu et al. (2014) studied the impact of rice straw and bamboo biochars on lead immobilization at a smelting site. They stated that rice straw biochar at 5% dose showed the best immobilization result. They also inferred that there was an increase in lead immobilization with increase in biochar dose. Sorption of lead on the surface of biochar is considered as the chief mechanism of lead immobilization (Lu et al. 2014; Lebrun et al. 2018c). Further, the abundance of oxygen containing functional groups, such as hydroxide, on the surface of biochar helps in removing metals and metalloids from soils via electrostatic interaction (Lebrun et al. 2018d). Interestingly, the high pH of biochar could help in lead immobilization via formation of metal(oxy)hydroxide precipitates at alkaline pH (Liang et al. 2016; Lebrun et al. 2018b). With regards to particle size, finer biochars, as compared to coarser biochars, are more efficient in metal removal via sorption. It could be due to the high surface area and porosity along with the cation exchange capacity of finer biochars (Liang et al. 2016; Lebrun et al. 2018c).

However, biochar is not considered capable for anion sorption. Amendment of arsenic-contaminated soil with biochar did not change arsenic concentrations in some studies (Lebrun et al. 2018b, c) and increased in others (Lomaglio et al. 2017). In a study by Yin et al. (2016), amendment of soil with biochar prepared from water hyacinth increased bioavailability of arsenic. Huang et al. (2018) used biochar prepared from sewage sludge to amend arsenic polluted soil and reported a rise in arsenic extractability. The repulsion between negatively charged biochar surface and negative arsenic ions could be the reason behind diminished arsenic removal from polluted soils (Lebrun et al. 2018d). Nevertheless, some studies reported a decrease in mobility of arsenic after biochar addition. Lebrun et al. (2020b) used sapwood biochars while Gregory et al. (2015) used willow biochar and reported a reduction in arsenic concentrations in the soils. The impact of biochar on metals and metalloids present in soil are summarized in Table 16.3.

16.7.2 Effect of Biochar on Soil Microbes

Microbes are present all around in the soils. Presence of metals and metalloids in soil are inhibitory for the growth and proliferation of soil microbes (Garau et al. 2017; Chen et al. 2018). Enhancement in metals and metalloids affects the diversity, richness, and bioactivity of the soil microbes (Xie et al. 2016). The increase in concentration of arsenic, copper, and lead in soil decreased metabolism and activity of microbes in soil (Boshoff et al. 2014).

The presence of a high porosity and pore volume in biochar facilitates habitat for soil microbes thereby protecting them from predators and resulting in proliferation of microbial community structure (Lu et al. 2015; Nie et al. 2018). Further, Purakayastha et al. (2015) stated that microbial growth could be supported by the labile fraction of biochar. Biochar furnishes carbon and nutrients to microbes for their enzymatic activity and metabolism, resulting in their growth (Nie et al. 2018; Xu et al. 2018). Additionally, enhancement of soil properties after biochar addition indirectly boosts microbial growth (Gul et al. 2015). A rise in pH with biochar addition supports growth of soil microbes (Huang et al. 2017). Furthermore, biochar addition enhances soil organic matter and water holding capacity and improves physical structure of soil. These contribute in a boosted microbial growth (Khadem and Raiesi 2017; Nie et al.

Table 16.3 Impact of biochar on metals and metalloids present in soil

Biochar	Impact on metals and metalloids in soil	Reference
Bamboo pyrolyzed at 500 °C	Removed about 49% of Cadmium, Lead, Zinc, Copper	Lu et al. (2014)
Broiler litter pyrolyzed at 700 °C	Removed about 75% of Cadmium, Nickel, Zinc, Copper	Uchimiya et al. (2011a)
Chicken manure pyrolyzed at 550 °C	Removed about 94% of Cadmium, Lead	Park et al. (2011a)
Dairy manure pyrolyzed at 450 °C	Removed Lead with sorption capacity of 132.81 mg/g	Cao et al. (2011)
Miscanthus pyrolyzed at 600 °C	Removed about 92% of Cadmium, Zinc, Lead	Houben et al. (2013)
Rice straw pyrolyzed at 500 °C	Removed about 71% of Zinc, Copper, Cadmium, Lead	Lu et al. (2014)
Sewage sludge pyrolyzed at 500–550 °C	Immobilized Lead, Nickel, Cobalt, Chromium, Arsenic	Khan et al. (2013)
Cottonseed hulls pyrolyzed at 200–800 °C	Removed Cadmium, Lead, Nickel, Copper	Uchimiya et al. (2011b)
Hard wood	Removed Cadmium and Zinc	Beesley and Marmiroli (2011)
Oak wood pyrolyzed at 400 °C	Reduced Bioavailability of Lead by 76%	Ahmad et al. (2012)

2018). Lastly, biochar removes metals and metalloids from soils which helps in microbe proliferation (Park et al. 2011a; Moore et al. 2018).

Amendment of soils with biochar has been shown to be advantageous for soil microbes. Biochar application influences growth, activity, and community structure of soil microbes. In a study by Ahmad et al. (2016), amendment of agricultural soil with biochars derived from pine needle and soybean stover resulted in enhancement in fungi, Gram negative bacteria, and Gram positive bacteria. In a study by Nie et al. (2018), biochar derived from corn straw was added to soil contaminated with lead. It was reported that there was a rise in Gram negative bacteria biomass and soil microbial diversity. Al-Wabel et al. (2019) applied biochar prepared from date palm to a mine soil and observed an increase in microbial activity and microbial biomass in the soil sample. Moore et al. (2018) prepared biochar using chicken manure and oat hull, and applied the biochars to polluted soils. They observed that there was an increase in soil microbes in both the cases. Amendment with chicken manure biochar demonstrated better results than oat hull biochar. Lu et al. (2015) amended contaminated soils with biochar prepared from poultry litter and eucalyptus and reported that there was a rise in the soil enzymatic and soil metabolic activity. Al Marzooqi and Yousef (2017) prepared biochar from a dwarf glasswort called *Salicornia bigelovii* was applied to non-contaminated soils and reported that soil enzymatic activity and soil biomass increased. Tian et al. (2016) amended paddy soil with biochar and stated that there was a rise in amines and amino acids utilization indicative of microbial way for compensating high carbon/nitrogen ratio.

Chen et al. (2016) and Xu et al. (2017) stated that biochar could also have an impact on composition of soil microbes by altering microbial diversity. In a study by Ahmad et al. (2016), biochar was applied to agricultural soils near a mine site and it was observed that there was an increase in *Actinobacteria* and a decrease in *Acidobacteria* and *Chloroflexi*. Huang et al. (2017) reported a change in community composition of soil microbes when biochar was added to contaminated river sediments. Wang et al. (2019a) observed a rise in *Proteobacteria* and *Verrucomicrobia* when biochar was added to polluted paddy soils. Xu et al. (2017) applied biochar to polluted industrial soils and reported a rise in *Cyanobacteria, Planctomycetes, Proteobacteria, Firmicutes,* and *Actinobacteria* and a decrease in *Gemmatimonadetes*. Xu et al. (2017) and Wang et al. (2019a) stated that a rise in soil nutrients, an increase in organic matter, and a rise in pH could be contributing towards growth of soil microbes. The impact of biochar on the microbes present in soil has been represented in Table 16.4.

16.7.3 Effect of Biochar on Plants

Soils are one of the greatest requirements for the growth of plants. Soils contaminated with organic and inorganic pollutants are deterrent for growth of plants. There was a reduction in growth of *Trifolium repens, Salix viminalis, Salix alba,* and *Salix purpurea*, when grown on mine soils polluted with lead and arsenic (Lebrun et al. 2017, 2018b, 2019; Nandillon et al. 2019b). The low fertility of polluted soils and toxicity caused by pollutants inhibits growth of plants. Amendment of polluted soils with biochar helps in boosting growth and establishment of different plants.

In a study by Lebrun et al. (2017), pinewood biochar was applied to Pontgibaud mine soils and Salix plant species (*S. viminalis, S. alba,* and *S. purpurea*) were grown. They observed that biochar addition helped the soils in increasing dry weight production of the plants. Amendment of soils with pinewood and lightwood biochars helped in increasing growth of *Populus euramericana* and *S. viminalis* (Lebrun et al. 2018b). Similarly, amendment with hardwood biochars helped in increasing growth of *Agrostis capillaris* and *T. repens* (Nandillon et al. 2019a).

Yu et al. (2017) reported an increase in growth of rice plants when biochar prepared from corn straw was applied to arsenic contaminated soils. Brennan et al. (2014) prepared biochar using olive tree prunings and applied them to multi-contaminated mine soil. They observed that there was an increase in shoot and root biomass, leaf surface area, and root length in maize plants.

Interestingly, plant growth is also affected by type of biochar, application dose, particle size of biochar, and properties of soil. Lebrun et al. (2018d) applied hardwood

Table 16.4 Impact of biochar on plant growth and microbes present in soil

Biochar	Impact on plant growth and microbes present in soil	Reference
Grass pyrolyzed at 250 and 650 °C	Increased bacterial population from 31.8 ± 1.4 CFUs to 118.7 ± 121.0 CFUs	Khodadad et al. (2011)
Oak pyrolyzed at 250 and 650 °C	Increased bacterial population from 31.8 ± 1.4 CFUs to 87.7 ± 4.4 CFUs	Khodadad et al. (2011)
Wood pyrolyzed at 450 °C	Increased bacterial growth by 80%	Jones et al. (2012)
Wood pyrolyzed at 450 °C	Increased fungal growth by 21%	Jones et al. (2012)
Soybean stover and pine needle pyrolyzed at 300 and 700 °C	Increased *Actinobacteria*; decreased *Acidobacteria* and *Chloroflexi*	Ahmad et al. (2016)
Bamboo chips pyrolyzed at 350 °C	Increased *Proteobacteria* and *Verrucomicrobia*	Wang et al. (2019a)
Wine lees pyrolyzed at 600 °C	Increased *Cyanobacteria, Planctomycetes, Proteobacteria, Firmicutes,* and *Actinobacteria*; decreased *Gemmatimonadetes*	Xu et al. (2017)
Savannah wood in "hot tail" oven	Increased biomass root yield by 35% and plant height by 10%	Yeboah et al. (2009)
Wood (mixed Deciduous)	Increased root length/plant biomass ratio by 17%	Prendergast-Miller et al. (2011)
Douglas fir pyrolyzed at 900 °C	Increased root biomass and shoot biomass by 33% and 20%	Bista et al. (2019)
Rice-husk obtained at 900–1100 °C	Increased stem length and above ground biomass by 50% and 900% respectively	Carter et al. (2013)

biochars of four different particle sizes to Pontgibaud mine soil and stated that there was maximum improvement in growth of *S. viminalis* when finer biochars were applied. However, coarser biochars increased plant growth at 5% application dose. Huang et al. (2018) prepared three different biochars from sewage sludge, Hibiscus cannabinus, and chicken waste. They applied them at different application dose to multi-contaminated soils and cultivated *Cassia alata*. All the amendments increased growth and there was a rise in root and shoot biomass.

There could be several reasons behind the positive impact of biochar amendment on plant growth. Amendment of soil with biochar enhances soil pH, water retention capacity, and organic matter in soil. Further, the nutritional enrichment furnished by biochar addition boosts plant growth. Correspondingly, the increase in plant biomass and leaf size would aid increasing transpiration and the consequent water uptake. Additionally, biochar decreases toxicity caused due to the presence of metals and metalloids, which aids in growth of plants. Correspondingly, the growth of plants would bring about a dilution effect assisting in a decrease in metals and metalloid content. However, there could be situations where biochar was not found to decrease metal or metalloid concentration from soil. For example, Lebrun et al. (2018b, d, c) concluded in their multiple studies that application of biochars derived from pinewood could not decrease uptake of arsenic and lead in *S. alba* and *S. viminalis* plants. Although, the arsenic and lead concentrations did decrease when biochars prepared from pinewood or lightwood was applied to *P. euramericana* and *S. viminalis* plants. Nevertheless, in most of the cases amendment of soil with biochar helps in decreasing metal and metalloid content from the soil, and helps in plant growth. In another study by Yu et al. (2017), arsenic concentrations in rice plants diminished when biochar made from corn straw was added to arsenic contaminated soils.

Metals and metalloids trigger the production of reactive oxygen species which are damaging to DNA and other molecules present in the cells (Ali et al. 2006). Application of biochar assists in decreasing the oxidative stress caused by metals or metalloids in plants. Abbas et al. (2018) applied biochar to wheat plants and reported a reduction in peroxide content and peroxidase activity and enhancement in superoxide dismutase and catalase activity in the plants. Gong et al. (2019) reported a reduction in oxidative damage in ramie seedlings when contaminated soils with amended with tea waste biochar. Biochar induces the reduction in oxidative stress by immobilizing metals and metalloids. Additionally, they also activate hydrogen peroxide and form superoxide ions on the surface that helps in depleting hydrogen peroxide in plants (Fang et al. 2014; Gong et al. 2019). Further, the presence of free radicals and functional groups on the surface of biochar participates in removing the reactive oxygen species formed in plants (Gong et al. 2019). Lastly, biochar could change the speciation of toxic metals such as arsenic, resulting in a reduction in the toxic manifestations. The impact of biochar on the plant growth has been represented in Table 16.4.

16.8 Amendments and Modifications of Biochar for Enhancing the Impact on Soil Properties

There could be situations where biochar did not decrease metal or metalloid concentrations from soil, as stated previously. However, properties of biochar could be modified by physical or chemical methods which could aid in removal of metals and metalloids (Alam et al. 2018; Shaikh et al. 2021). Physical modification or functionalization involves gas, steam, or magnetization. Chemical modification or functionalization involves methanol, amines, or acid/base treatments (Rajapaksha et al. 2016; Tan et al. 2017). These modifications enhance surface properties of biochar and increase sorption of metals and metalloids. Lebrun et al. (2018d) modified hardwood biochar by $FeCl_3$ treatment and reported an improvement in arsenic sorption. Zhou et al. (2013) modified bamboo biochar via chitosan which helped in enhancing its pH, nitrogen, oxygen, and hydrogen contents and improved removal capacity of metals and metalloids. Kwak et al. (2019) treated canola straw biochar with steam which improved its surface area, pH and metal removal efficiency. However, not all the modifications could be beneficial. Wu et al. (2016) stated that different modifications performed on coconut biochar did not necessarily improve metal removal efficiency. Further, modification methods could be costly and difficult to produce on a large-scale level for field application. Therefore, it becomes critical to select biochar modification method appropriately.

Apart from physical and chemical modifications of biochar, different amendments could be performed with biochar to improve its efficacy in improving the soil properties. For example, compost could be added to biochar. Compost is obtained by biological degradation of organic waste materials and is rich in nutrients, humic materials, and microorganisms (Huang et al. 2016). Addition of compost to biochar would aid in improving its nutritional status, enhancing its suitability as fertilizer for soils. Further, biochar could improve quality and humification in compost (Liang et al. 2017). Lebrun et al. (2019) reported that combined application of biochar and compost enhances water holding capacity, organic matter content, pH and electrical conductivity of soil, more than individual applications. Cao et al. (2017) stated that combined application of compost and biochar boosted the yield of water melons. Biochar could also be combined with other materials such as iron sulphate and iron grit. Presence of iron in these amendments helps in removing anions such as arsenic from soils and improving the nutrient content and overall properties of soils (Lebrun et al. 2019; Fresno et al. 2020).

16.9 Disadvantages of Biochar Application to Soil

Having discussed the various positive impacts that biochar could have on soil properties, it becomes necessary to mention the probable risks and disadvantages associated with biochar application. Simultaneously, it would be crucial to tackle these risks to maximize the advantages that biochar could offer.

Firstly, during the production process, improper thermal treatment methods could result in release of greenhouse gases and air pollutants (Smebye et al. 2017). Therefore, safer reactors could be fabricated which prevent the release of such gases. Additionally, biomass could consist of organic and inorganic pollutants which could be transferred to biochar during the production process. Application of contaminated biochars could be harmful to soil health and for the crops grown on such soils. Nevertheless, thermal treatment temperatures above 500 °C could prevent accumulation of such pollutants in biochar (Hossain et al. 2007), and prevent the consequent contamination of soils and crops. Such suitable thermal treatment techniques must be incorporated in the production process (Verheijen et al. 2010).

Utilization of crop residues for production of biochar has been criticized in terms of loosening of soil matrix and the consequent soil erosion (Verheijen et al. 2010).

Erosion could be carried out by forces of water and wind on the loosened soil. Interestingly, addition of fine biochars could decrease the overall particle size of soil, and initiate soil erosion (Verheijen et al. 2010). It becomes crucial to enforce suitable guidelines that help in maintaining soil quality and prevent soil erosion. Further, presence of ash in biochar could cause respiratory problems in the various stakeholders associated with production, transportation, and application stages pertaining to biochar (De Capitani et al. 2007). Compliance of procedural safety guidelines must be enforced to prevent health issues.

There is a possibility of sorption of fertilizers, insecticides, herbicides, and pesticides by biochar which could diminish the efficacy of the applied agro-chemicals. However, such sorption of agro-chemicals could help in reducing their off-site movement (Sun et al. 2012b). Additionally, biochar application could help in sorption of agro-chemicals thereby decreasing their uptake by crops (Saito et al. 2011).

Biochar could have negative impact on the earthworm community living in the soils (Topoliantz and Ponge 2003; Warnock et al. 2007). Nevertheless, use of wet biochar has been suggested to curtail the negative impacts (Li et al. 2011).

16.10 Research Gaps

Biochar holds immense potential for improving the properties of soil. However, there are abundant opportunities for future research that must be addressed. A standard needs to be charted out that re-affirms the sustainable production of biochar, adequate nutrition provisioning to crops via biochar, appropriate selection of biochar according to the soil type, and suitable biochar application rate. Extensive field studies must be carried out to confirm real-time applicability of biochars. Additionally, long-term impact of biochar on soils should be determined for safe application of biochar and identification of biochar ageing. Furthermore, there is a need to study the impact of climate change and extreme weather events on biochar application to soils. Lastly, modification methods pertaining to enhancement of biochar performance should be researched upon accordingly.

16.11 Conclusion

Soil is a vital resource that supports growth of plants by anchoring plant roots and facilitating water and nutrients. Growth of plants is critical in supporting our agricultural systems. Further, soil is involved in biogeochemical cycles, filtering polluted waters, and acts as a sink for greenhouse gases. Anthropogenic activities have deteriorated the soil quality in most parts of the world. Further, rise in human population needs food, energy, and other resources to sustain. A number of alternatives have been researched upon to minimize the damage done to soils and heal the ailing soil quality. Biochar has emerged as one of the most cost-effective and sustainable methods to improve the quality of soil and decrease the use of agro-chemicals such as fertilizers, insecticides, pesticides, and weedicides.

Biochar is a renewable substance obtained through thermal treatment of biomass feedstocks in absence or a limited supply of oxygen. Application of biochar helps in facilitating nourishment to the plants grown; increasing soil organic matter content in soil, enhancing the soil pH, electrical conductivity, and cation exchange capacity in soil; boosting the native soil microbes; and removing the organic and inorganic pollutants from contaminated soils. In addition to these advantages, biochar application helps in boosting crop productivity, mitigating climate change, producing clean energy, and facilitating waste management. It could be safely concluded that application of biochar would not only improve the soil quality sustainably, but also help in ameliorating the soils polluted with organic and inorganic contaminants and enhancing their fertility.

Acknowledgements One of the authors (Abhishek Kumar) is thankful to the University Grants Commission, New Delhi, for providing NET-JRF Fellowship [Ref. No.- 3635/(OBC)(NET-DEC.2015)].

References

Abbas T, Rizwan M, Ali S et al (2018) Biochar application increased the growth and yield and reduced cadmium in drought stressed wheat grown in an aged contaminated soil. Ecotoxicol Environ Saf 148:825–833. https://doi.org/10.1016/j.ecoenv.2017.11.063

Adekiya AO, Agbede TM, Olayanju A et al (2020) Effect of biochar on soil properties, soil loss, and cocoyam yield on a tropical sandy loam alfisol. Sci World J 2020:9391630. https://doi.org/10.1155/2020/9391630

Agegnehu G, Bass AM, Nelson PN et al (2015) Biochar and biochar-compost as soil amendments: effects on peanut yield, soil properties and greenhouse gas emissions in tropical North Queensland, Australia. Agric Ecosyst Environ 213:72–85. https://doi.org/10.1016/j.agee.2015.07.027

Ahmad M, Soo Lee S, Yang JE et al (2012) Effects of soil dilution and amendments (mussel shell, cow bone, and biochar) on Pb availability and phytotoxicity in military shooting range soil. Ecotoxicol Environ Saf 79:225–231. https://doi.org/10.1016/j.ecoenv.2012.01.003

Ahmad M, Rajapaksha AU, Lim JE et al (2014) Biochar as a sorbent for contaminant management in soil and water: a review. Chemosphere 99:19–33. https://doi.org/10.1016/j.chemosphere.2013.10.071

Ahmad M, Ok YS, Kim BY et al (2016) Impact of soybean stover- and pine needle-derived biochars on Pb and As mobility, microbial community, and carbon stability in a contaminated agricultural soil. J Environ Manage 166:131–139. https://doi.org/10.1016/j.jenvman.2015.10.006

Alam, Md A, Shaikh WA, Alam Md O, Bhattacharya T, Chakraborty S, Bibhutibhushan S, Saha I (2018) Adsorption of As (III) and As (V) from aqueous solution by modified Cassia fistula (golden shower) biochar. Appl Water Sci 8(7):198. https://doi.org/10.1007/s13201-018-0839-y

Al-Farraj AS, Usman ARA, Al Otaibi SHM (2013) Assessment of heavy metals contamination in soils surrounding a gold mine: comparison of two digestion methods. Chem Ecol 29:329–339. https://doi.org/10.1080/02757540.2012.735660

Ali MB, Singh N, Shohael AM et al (2006) Phenolics metabolism and lignin synthesis in root suspension cultures of Panax ginseng in response to copper stress. Plant Sci 171:147–154. https://doi.org/10.1016/j.plantsci.2006.03.005

Al Marzooqi F, Yousef LF (2017) Biological response of a sandy soil treated with biochar derived from a halophyte (*Salicornia bigelovii*). Appl Soil Ecol 114:9–15. https://doi.org/10.1016/j.apsoil.2017.02.012

Al-Wabel MI, Al-Omran A, El-Naggar AH et al (2013) Pyrolysis temperature induced changes in characteristics and chemical composition of biochar produced from conocarpus wastes. Bioresour Technol 131:374–379. https://doi.org/10.1016/j.biortech.2012.12.165

Al-Wabel MI, Usman ARA, Al-Farraj AS et al (2019) Correction to: date palm waste biochars alter a soil respiration, microbial biomass carbon, and heavy metal mobility in contaminated mined soil. Environ Geochem Health 41(4):1705–1722. https://doi.org/10.1007/s10653-017-9955-0). Environ Geochem Health 41:1809. https://doi.org/10.1007/s10653-017-0049-9

Angin D (2013) Effect of pyrolysis temperature and heating rate on biochar obtained from pyrolysis of safflower seed press cake. Bioresour Technol 128:593–597. https://doi.org/10.1016/j.biortech.2012.10.150

Annabi M, Le Bissonnais Y, Le Villio-Poitrenaud M, Houot S (2011) Improvement of soil aggregate stability by repeated applications of organic amendments to a cultivated silty loam soil. Agric Ecosyst Environ 144:382–389. https://doi.org/10.1016/j.agee.2011.07.005

Antal MJ, Mochidzuki K, Paredes LS (2003) Flash carbonization of biomass. Ind Eng Chem Res 42:3690–3699. https://doi.org/10.1021/ie0301839

Arienzo M, Christen EW, Quayle W, Kumar A (2009) A review of the fate of potassium in the soil-plant system after land application of wastewaters. J Hazard Mater 164:415–422. https://doi.org/10.1016/j.jhazmat.2008.08.095

Armynah B, Tahir D, Tandilayuk M et al (2019) Potentials of biochars derived from bamboo leaf biomass as energy sources: effect of temperature and time of heating. Int J Biomater 2019:12–18. https://doi.org/10.1155/2019/3526145

Asensio V, Vega FA, Andrade ML, Covelo EF (2013) Tree vegetation and waste amendments to improve the physical condition of copper mine soils. Chemosphere 90:603–610. https://doi.org/10.1016/j.chemosphere.2012.08.050

Aslam Z, Khalid M, Aon M (2014) Impact of biochar on soil physical properties. Sch J Agric Sci 4:280–284

Bakshi S, He ZL, Harris WG (2014) Biochar amendment affects leaching potential of copper and nutrient release behavior in contaminated sandy soils. J Environ Qual 43:1894–1902. https://doi.org/10.2134/jeq2014.05.0213

Balat M, Balat M, Kirtay E, Balat H (2009) Main routes for the thermo-conversion of biomass into fuels and chemicals. Part 1: pyrolysis systems. Energy Convers Manag 50:3147–3157. https://doi.org/10.1016/j.enconman.2009.08.014

Balée WL, Erickson CL, Graham E (2016a) 2. A neotropical framework for terra preta. Time Complex. Hist. Ecol.

Balée WL, Erickson CL, Neves EG, Petersen JB (2016b) 9. Political Economy and Pre-Columbian Landscape Transformations in Central Amazonia. Time Complex. Hist Ecol

Bandara T, Herath I, Kumarathilaka P et al (2017) Role of woody biochar and fungal-bacterial co-inoculation on enzyme activity and metal immobilization in serpentine soil. J Soils Sediments 17:665–673. https://doi.org/10.1007/s11368-015-1243-y

Bayu D, Tadesse M, Amsalu N (2016) Effect of biochar on soil properties and lead (Pb) availability in a military camp in South West Ethiopia. African J Environ Sci Technol 10:77–85. https://doi.org/10.5897/ajest2015.2014

Beesley L, Dickinson N (2011) Carbon and trace element fluxes in the pore water of an urban soil following greenwaste compost, woody and biochar amendments, inoculated with the earthworm *Lumbricus terrestris*. Soil Biol Biochem 43:188–196. https://doi.org/10.1016/j.soilbio.2010.09.035

Beesley L, Marmiroli M (2011) The immobilisation and retention of soluble arsenic, cadmium and zinc by biochar. Environ Pollut 159:474–480. https://doi.org/10.1016/j.envpol.2010.10.016

Behazin E, Ogunsona E, Rodriguez-Uribe A et al (2016) Mechanical, chemical, and physical properties of wood and perennial grass biochars for possible composite application. BioResources 11:1334–1348. https://doi.org/10.15376/biores.11.1.1334-1348

Bhattacharya T, Pandey SK, Pandey VC, Kumar A (2021) Potential and safe utilization of Fly ash as fertilizer for Pisum sativum L. Grown in phytoremediated and non-phytoremediated amendments. Environ Sci Pollut Res 28(36):50153–50166. https://doi.org/10.1007/s11356-021-14179-9

Bista P, Ghimire R, Machado S, Pritchett L (2019) Biochar effects on soil properties and wheat biomass vary with fertility management. Agronomy 9. https://doi.org/10.3390/agronomy9100623

Blanco-Canqui H (2017) Biochar and soil physical properties. Soil Sci Soc Am J 81:687–711. https://doi.org/10.2136/sssaj2017.01.0017

Bonanomi G, Ippolito F, Cesarano G et al (2017) Biochar as plant growth promoter: better off alone or mixed with organic amendments? Front Plant Sci 8. https://doi.org/10.3389/fpls.2017.01570

Borselli L, Carnicelli S, Ferrari GA et al (1996) Effects of gypsum on hydrological, mechanical and porosity properties of a kaolinitic crusting soil. Soil Technol 9:39–54. https://doi.org/10.1016/0933-3630(95)00034-8

Boshoff M, De Jonge M, Dardenne F et al (2014) The impact of metal pollution on soil faunal and microbial activity in two grassland ecosystems. Environ Res 134:169–180. https://doi.org/10.1016/j.envres.2014.06.024

Brennan A, Jiménez EM, Puschenreiter M et al (2014) Effects of biochar amendment on root traits and contaminant availability of maize plants in a copper and arsenic impacted soil. Plant Soil 379:351–360. https://doi.org/10.1007/s11104-014-2074-0

Briggs C, Breiner JM, Graham RC (2012) Physical and chemical properties of *Pinus ponderosa* charcoal: implications for soil modification. Soil Sci 177:263–268. https://doi.org/10.1097/SS.0b013e3182482784

Bryden KM, Hagge MJ (2003) Modeling the combined impact of moisture and char shrinkage on the pyrolysis of a biomass particle. Fuel 82:1633–1644. https://doi.org/10.1016/S0016-2361(03)00108-X

Burrell LD, Zehetner F, Rampazzo N et al (2016) Long-term effects of biochar on soil physical properties. Geoderma 282:96–102. https://doi.org/10.1016/j.geoderma.2016.07.019

Busscher WJ, Novak JM, Ahmedna M (2011) Physical effects of organic matter amendment of a southeastern US coastal loamy sand. Soil Sci 176:661–667. https://doi.org/10.1097/SS.0b013e3182357ca9

Byrne CE, Nagle DC (1997) Carbonization of wood for advanced materials applications. Carbon N Y 35:259–266. https://doi.org/10.1016/S0008-6223(96)00136-4

Caillat S, Vakkilainen E (2013) Large-scale biomass combustion plants: an overview. In: Biomass combustion science, technology and engineering, pp 189–224

Cantrell KB, Hunt PG, Uchimiya M et al (2012) Impact of pyrolysis temperature and manure source on physicochemical characteristics of biochar. Bioresour Technol 107:419–428. https://doi.org/10.1016/j.biortech.2011.11.084

Cao X, Harris W (2010) Properties of dairy-manure-derived biochar pertinent to its potential use in remediation. Bioresour Technol 101:5222–5228. https://doi.org/10.1016/j.biortech.2010.02.052

Cao X, Ma L, Liang Y et al (2011) Simultaneous immobilization of lead and atrazine in contaminated soils using dairy-manure biochar. Environ Sci Technol 45:4884–4889. https://doi.org/10.1021/es103752u

Cao Y, Ma Y, Guo D et al (2017) Chemical properties and microbial responses to biochar and compost amendments in the soil under continuous watermelon cropping. Plant Soil Environ 63:1–7. https://doi.org/10.17221/141/2016-PSE

Carter S, Shackley S, Sohi S et al (2013) The Impact of biochar application on soil properties and plant growth of pot grown lettuce (*Lactuca sativa*) and cabbage (*Brassica chinensis*). Agronomy 3:404–418. https://doi.org/10.3390/agronomy3020404

Cetin E, Moghtaderi B, Gupta R, Wall TF (2004) Influence of pyrolysis conditions on the structure and gasification reactivity of biomass chars. Fuel 83:2139–2150. https://doi.org/10.1016/j.fuel.2004.05.008

Cha JS, Park SH, Jung SC et al (2016) Production and utilization of biochar: a review. J Ind Eng Chem 40:1–15. https://doi.org/10.1016/j.jiec.2016.06.002

Chan KY, Van Zwieten L, Meszaros I et al (2007) Agronomic values of greenwaste biochar as a soil amendment. Aust J Soil Res 45:629–634. https://doi.org/10.1071/SR07109

Chen J, Sun X, Li L et al (2016) Change in active microbial community structure, abundance and carbon cycling in an acid rice paddy soil with the addition of biochar. Eur J Soil Sci 67:857–867. https://doi.org/10.1111/ejss.12388

Chen Y, Ding Q, Chao Y et al (2018) Structural development and assembly patterns of the root-associated microbiomes during phytoremediation. Sci Total Environ 644:1591–1601. https://doi.org/10.1016/j.scitotenv.2018.07.095

Chen D, Wang X, Wang X et al (2020) The mechanism of cadmium sorption by sulphur-modified wheat straw biochar and its application cadmium-contaminated soil. Sci Total Environ 714:136550. https://doi.org/10.1016/j.scitotenv.2020.136550

Choppala GK, Bolan NS, Megharaj M et al (2012) The influence of biochar and black carbon on reduction and bioavailability of chromate in soils. J Environ Qual 41:1175–1184. https://doi.org/10.2134/jeq2011.0145

Chun Y, Sheng G, Chiou GT, Xing B (2004) Compositions and sorptive properties of crop residue-derived chars. Environ Sci Technol 38:4649–4655. https://doi.org/10.1021/es035034w

Conti R, Rombolà AG, Modelli A et al (2014) Evaluation of the thermal and environmental stability of switchgrass biochars by Py-GC-MS. J Anal Appl Pyrolysis 110:239–247. https://doi.org/10.1016/j.jaap.2014.09.010

Cristaldi A, Conti GO, Jho EH et al (2017) Phytoremediation of contaminated soils by heavy metals and PAHs. A brief review. Environ Technol Innov 8:309–326. https://doi.org/10.1016/j.eti.2017.08.002

Crouse D (2018) Soils and plant nutrients, Chapter 1. In: Moore KA, Bradley LK (eds) North Carolina extension gardener handbook. NC State Extension, Raleigh, NC

Cui X, Shen Y, Yang Q et al (2018) Simultaneous syngas and biochar production during heavy metal separation from Cd/Zn hyperaccumulator (*Sedum alfredii*) by gasification. Chem Eng J 347:543–551. https://doi.org/10.1016/j.cej.2018.04.133

Cui L, Pan G, Li L et al (2012) The reduction of wheat cd uptake in contaminated soil via biochar amendment: a two-year field experiment. BioResources 7:5666–5676. https://doi.org/10.15376/biores.7.4.5666-5676

Cushman R, Robertson-Palmer K (1998) Protecting our children. Can J Public Heal 89:221–223. https://doi.org/10.1007/BF03403920

Dai Z, Meng J, Muhammad N et al (2013) The potential feasibility for soil improvement, based on the properties of biochars pyrolyzed from different feedstocks. J Soils Sediments 13:989–1000. https://doi.org/10.1007/s11368-013-0698-y

Dai Z, Zhang X, Tang C et al (2017) Potential role of biochars in decreasing soil acidification—a critical review. Sci Total Environ 581–582:601–611. https://doi.org/10.1016/j.scitotenv.2016.12.169

De Capitani EM, Algranti E, Handar AMZ et al (2007) Wood charcoal and activated carbon dust pneumoconiosis in three workers. Am J Ind Med 50:191–196. https://doi.org/10.1002/ajim.20418

De Meyer A, Poesen J, Isabirye M et al (2011a) Soil erosion rates in tropical villages: a case study from Lake Victoria Basin, Uganda. CATENA 84:89–98. https://doi.org/10.1016/j.catena.2010.10.001

Debela F, Thring RW, Arocena JM (2012) Immobilization of heavy metals by co-pyrolysis of contaminated soil with woody biomass. Water Air Soil Pollut 223:1161–1170. https://doi.org/10.1007/s11270-011-0934-2

Devereux RC, Sturrock CJ, Mooney SJ (2013) The effects of biochar on soil physical properties and winter wheat growth. Earth Environ Sci Trans R Soc Edinburgh 103:13–18. https://doi.org/10.1017/S1755691012000011

Díaz RA, Magrín GO, Travasso MI, Rodríguez RO (1997) Climate change and its impact on the properties of agricultural soils in the Argentinean rolling pampas. Clim Res 9:25–30. https://doi.org/10.3354/cr009025

Ding Y, Liu Y, Liu S et al (2016) Biochar to improve soil fertility. A review. Agron Sustain Dev 36:36. https://doi.org/10.1007/s13593-016-0372-z

Dwibedi SK, Pandey VC, Divyasree D, Bajpai O (2022) Biochar-based land development. Land Degrad Dev. https://doi.org/10.1002/ldr.4185

Ekström C, Lindman N, Pettersson R (1985) Catalytic conversion of tars, carbon black and methane from pyrolysis/gasification of biomass. In: Overend RP, Milne TA, Mudge LK (eds) Fundamentals of thermochemical biomass conversion. Elsevier Applied Science Publishers, Springer, Netherlands, pp 601–618

Emmerich FG, Luengo CA (1994) Reduction of emissions from blast furnaces by using blends of coke and babassu charcoal. Fuel 73:1235–1236. https://doi.org/10.1016/0016-2361(94)90266-6

Eom IY, Kim JY, Lee SM et al (2013) Comparison of pyrolytic products produced from inorganic-rich and demineralized rice straw (*Oryza sativa* L.) by fluidized bed pyrolyzer for future biorefinery approach. Bioresour Technol 128:664–672. https://doi.org/10.1016/j.biortech.2012.09.082

Fang G, Gao J, Liu C et al (2014) Key role of persistent free radicals in hydrogen peroxide activation by biochar: Implications to organic contaminant degradation. Environ Sci Technol 48:1902–1910. https://doi.org/10.1021/es4048126

Fellet G, Marchiol L, Delle Vedove G, Peressotti A (2011) Application of biochar on mine tailings: effects and perspectives for land reclamation. Chemosphere 83:1262–1267. https://doi.org/10.1016/j.chemosphere.2011.03.053

Fidel RB, Laird DA, Thompson ML, Lawrinenko M (2017) Characterization and quantification of biochar alkalinity. Chemosphere 167:367–373. https://doi.org/10.1016/j.chemosphere.2016.09.151

Forján R, Asensio V, Rodríguez-Vila A, Covelo EF (2016) Contribution of waste and biochar amendment to the sorption of metals in a copper mine tailing. CATENA 137:120–125. https://doi.org/10.1016/j.catena.2015.09.010

Forján R, Asensio V, Guedes RS et al (2017) Remediation of soils polluted with inorganic contaminants: role of organic amendments. Enhancing Cleanup Environ. Pollut. 2:313–338

Fresno T, Peñalosa JM, Flagmeier M, Moreno-Jiménez E (2020) Aided phytostabilisation over two years using iron sulphate and organic amendments: effects on soil quality and rye production. Chemosphere 240:124827. https://doi.org/10.1016/j.chemosphere.2019.124827

Fu P, Hu S, Xiang J et al (2012) Evaluation of the porous structure development of chars from pyrolysis of rice straw: effects of pyrolysis temperature and heating rate. J Anal Appl Pyrolysis 98:177–183. https://doi.org/10.1016/j.jaap.2012.08.005

Gai X, Wang H, Liu J et al (2014) Effects of feedstock and pyrolysis temperature on biochar adsorption of ammonium and nitrate. PLoS ONE 9:e113888. https://doi.org/10.1371/journal.pone.0113888

Gao F, Xue Y, Deng P et al (2015) Removal of aqueous ammonium by biochars derived from agricultural residuals at different pyrolysis temperatures. Chem Speciat Bioavailab 27:92–97. https://doi.org/10.1080/09542299.2015.1087162

Garau G, Silvetti M, Vasileiadis S et al (2017) Use of municipal solid wastes for chemical and microbiological recovery of soils contaminated with metal(loid)s. Soil Biol Biochem 111:25–35. https://doi.org/10.1016/j.soilbio.2017.03.014

Ghani WAWAK, Mohd A, da Silva G et al (2013) Biochar production from waste rubber-wood-sawdust and its potential use in C sequestration: chemical and physical characterization. Ind Crops Prod 44:18–24. https://doi.org/10.1016/j.indcrop.2012.10.017

Githinji L (2014) Effect of biochar application rate on soil physical and hydraulic properties of a sandy loam. Arch Agron Soil Sci 60:457–470. https://doi.org/10.1080/03650340.2013.821698

Giudicianni P, Cardone G, Ragucci R (2013) Cellulose, hemicellulose and lignin slow steam pyrolysis: thermal decomposition of biomass components mixtures. J Anal Appl Pyrolysis 100:213–222. https://doi.org/10.1016/j.jaap.2012.12.026

Glaser B, Lehmann J, Zech W (2002) Ameliorating physical and chemical properties of highly weathered soils in the tropics with charcoal—a review. Biol Fertil Soils 35:219–230. https://doi.org/10.1007/s00374-002-0466-4

Gomes MA da C, Hauser-Davis RA, de Souza AN, Vitória AP (2016) Metal phytoremediation: general strategies, genetically modified plants and applications in metal nanoparticle contamination. Ecotoxicol Environ Saf 134:133–147. https://doi.org/10.1016/j.ecoenv.2016.08.024

Gong X, Huang D, Liu Y et al (2019) Biochar facilitated the phytoremediation of cadmium contaminated sediments: metal behavior, plant toxicity, and microbial activity. Sci Total Environ 666:1126–1133. https://doi.org/10.1016/j.scitotenv.2019.02.215

Goodman AM, Ennos AR (1999) The effects of soil bulk density on the morphology and anchorage mechanics of the root systems of sunflower and maize. Ann Bot 83:293–302. https://doi.org/10.1006/anbo.1998.0822

Gray M, Johnson MG, Dragila MI, Kleber M (2014) Water uptake in biochars: the roles of porosity and hydrophobicity. Biomass Bioenerg 61:196–205. https://doi.org/10.1016/j.biombioe.2013.12.010

Gregory SJ, Anderson CWN, Camps-Arbestain M et al (2015) Biochar in co-contaminated soil manipulates arsenic solubility and microbiological community structure, and promotes organochlorine degradation. PLoS ONE 10:e0125393. https://doi.org/10.1371/journal.pone.0125393

Gul S, Whalen JK, Thomas BW et al (2015) Physico-chemical properties and microbial responses in biochar-amended soils: mechanisms and future directions. Agric Ecosyst Environ 206:46–59. https://doi.org/10.1016/j.agee.2015.03.015

Hariprasad VN, Dayananda SH (2013) Environmental impact due to agricultural runoff containing heavy metals—a review. Int J Sci Res Publ 3:1–6

Hass A, Gonzalez JM, Lima IM et al (2012) Chicken manure biochar as liming and nutrient source for acid Appalachian soil. J Environ Qual 41:1096–1106. https://doi.org/10.2134/jeq2011.0124

Herath HMSK, Camps-Arbestain M, Hedley M (2013) Effect of biochar on soil physical properties in two contrasting soils: an Alfisol and an Andisol. Geoderma 209–210:188–197. https://doi.org/10.1016/j.geoderma.2013.06.016

Hmid A, Al Chami Z, Sillen W et al (2014) Olive mill waste biochar: a promising soil amendment for metal immobilization in contaminated soils. Environ Sci Pollut Res 22:1444–1456. https://doi.org/10.1007/s11356-014-3467-6

Hossain MK, Strezov V, Yin Chan K, Nelson PF (2010) Agronomic properties of wastewater sludge biochar and bioavailability of metals in production of cherry tomato (*Lycopersicon esculentum*). Chemosphere 78:1167–1171. https://doi.org/10.1016/j.chemosphere.2010.01.009

Hossain MK, Strezov V, Nelson P (2007) Evaluation of agricultural char from sewage sludge. In: Proceedings international agrichar iniative 2007 conference

Hossain MM (2016) Recovery of valuable chemicals from agricultural waste through pyrolysis. Electronic thesis dissertation repository, University of Western Ontario, Ontario, Canada

Houben D, Evrard L, Sonnet P (2013) Beneficial effects of biochar application to contaminated soils on the bioavailability of Cd, Pb and Zn and the biomass production of rapeseed (*Brassica napus* L.). Biomass Bioenerg 57:196–204. https://doi.org/10.1016/j.biombioe.2013.07.019

Huang D, Liu L, Zeng G et al (2017) The effects of rice straw biochar on indigenous microbial community and enzymes activity in heavy metal-contaminated sediment. Chemosphere 174:545–553. https://doi.org/10.1016/j.chemosphere.2017.01.130

Huang M, Zhu Y, Li Z et al (2016) Compost as a soil amendment to remediate heavy metal-contaminated agricultural soil: mechanisms, efficacy, problems, and strategies. Water Air Soil Pollut 227. https://doi.org/10.1007/s11270-016-3068-8

Huang L, Li Y, Zhao M, et al (2018) Potential of *Cassia alata* L. coupled with biochar for heavy metal stabilization in multi-metal mine tailings. Int J Environ Res Public Health 15:494. https://doi.org/10.3390/ijerph15030494

IBI (2015) Standardized product definition and product testing guidelines for biochar that is used in soil. Int Biochar Initiat 23

Igalavithana AD, Mandal S, Niazi NK et al (2017) Advances and future directions of biochar characterization methods and applications. Crit Rev Environ Sci Technol 47:2275–2330. https://doi.org/10.1080/10643389.2017.1421844

Janus A, Pelfrêne A, Heymans S et al (2015) Elaboration, characteristics and advantages of biochars for the management of contaminated soils with a specific overview on Miscanthus biochars. J Environ Manage 162:275–289. https://doi.org/10.1016/j.jenvman.2015.07.056

Janus A, Waterlot C, Heymans S, et al (2018) Do biochars influence the availability and human oral bioaccessibility of Cd, Pb, and Zn in a contaminated slightly alkaline soil? Environ Monit Assess 190. https://doi.org/10.1007/s10661-018-6592-8

Jianping Z (1999) Soil erosion in Guizhou province of China: a case study in Bijie prefecture. Soil Use Manag 15:68–70. https://doi.org/10.1111/j.1475-2743.1999.tb00067.x

Jones DL, Rousk J, Edwards-Jones G et al (2012) Biochar-mediated changes in soil quality and plant growth in a three year field trial. Soil Biol Biochem 45:113–124. https://doi.org/10.1016/j.soilbio.2011.10.012

Junna S, Bingchen W, Gang X, Hongbo S (2014) Effects of wheat straw biochar on carbon mineralization and guidance for large-scale soil quality improvement in the coastal wetland. Ecol Eng 62:43–47. https://doi.org/10.1016/j.ecoleng.2013.10.014

Kammann CI, Schmidt HP, Messerschmidt N et al (2015) Plant growth improvement mediated by nitrate capture in co-composted biochar. Sci Rep 5. https://doi.org/10.1038/srep11080

Karhu K, Mattila T, Bergström I, Regina K (2011) Biochar addition to agricultural soil increased CH_4 uptake and water holding capacity—results from a short-term pilot field study. Agric Ecosyst Environ 140:309–313. https://doi.org/10.1016/j.agee.2010.12.005

Keiluweit M, Nico PS, Johnson M, Kleber M (2010) Dynamic molecular structure of plant biomass-derived black carbon (biochar). Environ Sci Technol 44:1247–1253. https://doi.org/10.1021/es9031419

Kelly CN, Peltz CD, Stanton M et al (2014) Biochar application to hardrock mine tailings: soil quality, microbial activity, and toxic element sorption. Appl Geochemistry 43:35–48. https://doi.org/10.1016/j.apgeochem.2014.02.003

Khadem A, Raiesi F (2017) Influence of biochar on potential enzyme activities in two calcareous soils of contrasting texture. Geoderma 308:149–158. https://doi.org/10.1016/j.geoderma.2017.08.004

Khalid S, Shahid M, Niazi NK et al (2017) A comparison of technologies for remediation of heavy metal contaminated soils. J Geochemical Explor 182:247–268. https://doi.org/10.1016/j.gexplo.2016.11.021

Khan S, Chao C, Waqas M et al (2013) Sewage sludge biochar influence upon rice (*Oryza sativa* L) yield, metal bioaccumulation and greenhouse gas emissions from acidic paddy soil. Environ Sci Technol 47:8624–8632. https://doi.org/10.1021/es400554x

Khan N, Clark I, Sánchez-Monedero MA et al (2014) Maturity indices in co-composting of chicken manure and sawdust with biochar. Bioresour Technol 168:245–251. https://doi.org/10.1016/j.biortech.2014.02.123

Khodadad CLM, Zimmerman AR, Green SJ et al (2011) Taxa-specific changes in soil microbial community composition induced by pyrogenic carbon amendments. Soil Biol Biochem 43:385–392. https://doi.org/10.1016/j.soilbio.2010.11.005

Kubier A, Wilkin RT, Pichler T (2019) Cadmium in soils and groundwater: a review. Appl Geochemistry 108:104388. https://doi.org/10.1016/j.apgeochem.2019.104388

Kumar A, Bhattacharya T (2020) Biochar: a sustainable solution. Environ Dev Sustain. https://doi.org/10.1007/s10668-020-00970-0

Kumar A, Bhattacharya T (2021) Biochar: a sustainable solution. Environ Develop Sustain 23(5):6642–6680. https://doi.org/10.1007/s10668-020-00970-0

Kumar A, Bhattacharya T (2022) Removal of arsenic by wheat straw biochar from soil. Bull Environ Contam Toxicol 108(3):415–422. https://doi.org/10.1007/s00128-020-03095-2

Kumar A, Bhattacharya T, Mozammil Hasnain SM, Kumar Nayak A, Hasnain Md S (2020) Applications of biomass-derived materials for energy production conversion and storage. Mater Sci Energy Technol 3:905–920 S2589299120300665. https://doi.org/10.1016/j.mset.2020.10.012

Kumar A, Bhattacharya T, Mukherjee S, Sarkar B (2022a) A perspective on biochar for repairing damages in the soil–plant system caused by climate change-driven extreme weather events. Biochar 4(1):22. https://doi.org/10.1007/s42773-022-00148-z

Kumar A, Bhattacharya T, Shaikh WA, Chakraborty S, Owens G, Naushad Mu (2022b) Valorization of fruit waste-based biochar for arsenic removal in soils. Environ Res 213113710-S0013935122010374 113710. https://doi.org/10.1016/j.envres.2022.113710

Kumar A, Bhattacharya T, Shaikh WA, Roy A, Mukherjee S, Kumar M (2021a) Performance evaluation of crop residue and kitchen waste-derived biochar for eco-efficient removal of arsenic from soils of the Indo-Gangetic plain: a step towards sustainable pollution management. Environ Res 200:111758-S0013935121010525 111758. https://doi.org/10.1016/j.envres.2021.111758

Kumar A, Nagar S, Anand S (2021b) Nanotechnology for sustainable crop production: recent development and strategies. In: Singh P, Singh R, Verma P, Bhadouria R, Kumar A, Kaushik M (eds) Plant-Microbes-Engineered Nano-particles (PM-ENPs) Nexus in Agro-Ecosystems. Advances in Science, Technology & Innovation. Springer, Cham. https://doi.org/10.1007/978-3-030-66956-0_3

Kumar A, Nagar S, Anand S (2021c) Climate change and existential threats. In: Suruchi Singh, Pardeep Singh, Rangabhashiyam S, Srivastava KK (eds) Global Climate Change,Elsevier, pp 1–31

Kuppusamy S, Thavamani P, Megharaj M et al (2016) Agronomic and remedial benefits and risks of applying biochar to soil: current knowledge and future research directions. Environ Int 87:1–12. https://doi.org/10.1016/j.envint.2015.10.018

Kwak JH, Islam MS, Wang S et al (2019) Biochar properties and lead(II) adsorption capacity depend on feedstock type, pyrolysis temperature, and steam activation. Chemosphere 231:393–404. https://doi.org/10.1016/j.chemosphere.2019.05.128

Lahori AH, Guo Z, Zhang Z et al (2017) Use of biochar as an amendment for remediation of heavy metal-contaminated soils: prospects and challenges. Pedosphere 27:991–1014. https://doi.org/10.1016/S1002-0160(17)60490-9

Laird D, Fleming P, Wang B et al (2010) Biochar impact on nutrient leaching from a Midwestern agricultural soil. Geoderma 158:436–442. https://doi.org/10.1016/j.geoderma.2010.05.012

Lebrun M, Macri C, Miard F et al (2017) Effect of biochar amendments on As and Pb mobility and phytoavailability in contaminated mine technosols phytoremediated by Salix. J Geochemical Explor 182:149–156. https://doi.org/10.1016/j.gexplo.2016.11.016

Lebrun M, Miard F, Hattab-Hambli N et al (2018a) Assisted phytoremediation of a multi-contaminated industrial soil using biochar and garden soil amendments associated with *Salix alba* or *Salix viminalis*: abilities to stabilize As, Pb, and Cu. Water Air Soil Pollut 229:. https://doi.org/10.1007/s11270-018-3816-z

Lebrun M, Miard F, Nandillon R et al (2018b) Assisted phytostabilization of a multicontaminated mine technosol using biochar amendment: early stage evaluation of biochar feedstock and particle size effects on As and Pb accumulation of two Salicaceae species (*Salix viminalis* and *Populus euramericana*). Chemosphere 194:316–326. https://doi.org/10.1016/j.chemosphere.2017.11.113

Lebrun M, Miard F, Nandillon R et al (2018c) Eco-restoration of a mine technosol according to biochar particle size and dose application: study of soil physico-chemical properties and phytostabilization capacities of *Salix viminalis*. J Soils Sediments 18:2188–2202. https://doi.org/10.1007/s11368-017-1763-8

Lebrun M, Miard F, Renouard S et al (2018d) Effect of Fe-functionalized biochar on toxicity of a technosol contaminated by Pb and As: sorption and phytotoxicity tests. Environ Sci Pollut Res 25:33678–33690. https://doi.org/10.1007/s11356-018-3247-9

Lebrun M, Miard F, Nandillon R et al (2019) Biochar effect associated with compost and iron to promote Pb and As soil stabilization and *Salix viminalis* L. growth. Chemosphere 222:810–822. https://doi.org/10.1016/j.chemosphere.2019.01.188

Lebrun M, De Zio E, Miard F et al (2020a) Amending an As/Pb contaminated soil with biochar, compost and iron grit: effect on *Salix viminalis* growth, root proteome profiles and metal(loid) accumulation indexes. Chemosphere 244:125397. https://doi.org/10.1016/j.chemosphere.2019.125397

Lebrun M, Miard F, Hattab-Hambli N et al (2020b) Effect of different tissue biochar amendments on As and Pb stabilization and phytoavailability in a contaminated mine technosol. Sci Total Environ 707:135657. https://doi.org/10.1016/j.scitotenv.2019.135657

Lebrun M, Miard F, Scippa GS et al (2020c) Effect of biochar and redmud amendment combinations on *Salix triandra* growth, metal(loid) accumulation and oxidative stress response. Ecotoxicol Environ Saf 195:110466. https://doi.org/10.1016/j.ecoenv.2020.110466

Lee HW, Kim YM, Kim S et al (2018) Review of the use of activated biochar for energy and environmental applications. Carbon Lett 26:1–10. https://doi.org/10.5714/CL.2018.26.001

Lehmann J, Gaunt J, Rondon M (2006) Bio-char sequestration in terrestrial ecosystems—a review. Mitig Adapt Strateg Glob Chang 11:403–427. https://doi.org/10.1007/s11027-005-9006-5

Lehmann J, Rillig MC, Thies J et al (2011) Biochar effects on soil biota—a review. Soil Biol Biochem 43:1812–1836. https://doi.org/10.1016/j.soilbio.2011.04.022

Li D, Hockaday WC, Masiello CA, Alvarez PJJ (2011) Earthworm avoidance of biochar can be mitigated by wetting. Soil Biol Biochem 43:1732–1737. https://doi.org/10.1016/j.soilbio.2011.04.019

Li H, Ye X, Geng Z et al (2016) The influence of biochar type on long-term stabilization for Cd and Cu in contaminated paddy soils. J Hazard Mater 304:40–48. https://doi.org/10.1016/j.jhazmat.2015.10.048

Li G, Khan S, Ibrahim M et al (2018a) Biochars induced modification of dissolved organic matter (DOM) in soil and its impact on mobility and bioaccumulation of arsenic and cadmium. J Hazard Mater 348:100–108. https://doi.org/10.1016/j.jhazmat.2018.01.031

Li J, Shen F, Yang G et al (2018b) Valorizing rice straw and its anaerobically digested residues for biochar to remove Pb(II) from aqueous solution. Int J Polym Sci 2018:1–11. https://doi.org/10.1155/2018/2684962

Liang B, Lehmann J, Solomon D et al (2006) Black carbon increases cation exchange capacity in soils. Soil Sci Soc Am J 70:1719–1730. https://doi.org/10.2136/sssaj2005.0383

Liang C, Gascó G, Fu S et al (2016) Biochar from pruning residues as a soil amendment: effects of pyrolysis temperature and particle size. Soil Tillage Res 164:3–10. https://doi.org/10.1016/j.still.2015.10.002

Liang J, Yang Z, Tang L et al (2017) Changes in heavy metal mobility and availability from contaminated wetland soil remediated with combined biochar-compost. Chemosphere 181:281–288. https://doi.org/10.1016/j.chemosphere.2017.04.081

Liao W, Thomas S (2019) Biochar particle size and post-pyrolysis mechanical processing affect soil pH, water retention capacity, and plant performance. Soil Syst 3:14. https://doi.org/10.3390/soilsystems3010014

Liu L, Li W, Song W, Guo M (2018) Remediation techniques for heavy metal-contaminated soils: principles and applicability. Sci Total Environ 633:206–219. https://doi.org/10.1016/j.scitotenv.2018.03.161

Lomaglio T, Hattab-Hambli N, Bret A et al (2017) Effect of biochar amendments on the mobility and (bio) availability of As, Sb and Pb in a contaminated mine technosol. J Geochemical Explor 182:138–148. https://doi.org/10.1016/j.gexplo.2016.08.007

Lomaglio T, Hattab-Hambli N, Miard F et al (2018) Cd, Pb, and Zn mobility and (bio)availability in contaminated soils from a former smelting site amended with biochar. Environ Sci Pollut Res 25:25744–25756. https://doi.org/10.1007/s11356-017-9521-4

Lu H, Zhang W, Yang Y et al (2012) Relative distribution of Pb2+ sorption mechanisms by sludge-derived biochar. Water Res 46:854–862. https://doi.org/10.1016/j.watres.2011.11.058

Lu K, Yang X, Shen J et al (2014) Effect of bamboo and rice straw biochars on the bioavailability of Cd, Cu, Pb and Zn to Sedum plumbizincicola. Agric Ecosyst Environ 191:124–132. https://doi.org/10.1016/j.agee.2014.04.010

Lu H, Li Z, Fu S et al (2015) Combining phytoextraction and biochar addition improves soil biochemical properties in a soil contaminated with Cd. Chemosphere 119:209–216. https://doi.org/10.1016/j.chemosphere.2014.06.024

Lv D, Xu M, Liu X et al (2010) Effect of cellulose, lignin, alkali and alkaline earth metallic species on biomass pyrolysis and gasification. Fuel Process Technol 91:903–909. https://doi.org/10.1016/j.fuproc.2009.09.014

Mahimairaja S, Shenbagavalli S (2012) Production and characterization of biochar from different biological wastes. Int J Plant, Anim Environ Sci 2:197–201

Manna S, Singh N, Purakayastha TJ, Berns AE (2020) Effect of deashing on physico-chemical properties of wheat and rice straw biochars and potential sorption of pyrazosulfuron-ethyl. Arab J Chem 13:1247–1258. https://doi.org/10.1016/j.arabjc.2017.10.005

Manyà JJ (2012) Pyrolysis for biochar purposes: a review to establish current knowledge gaps and research needs. Environ Sci Technol 46:7939–7954. https://doi.org/10.1021/es301029g

Manyà JJ, González B, Azuara M, Arner G (2018) Ultra-microporous adsorbents prepared from vine shoots-derived biochar with high CO_2 uptake and CO_2/N_2 selectivity. Chem Eng J 345:631–639. https://doi.org/10.1016/j.cej.2018.01.092

Marini Köpp M, Paixão Passos L, da Silva VR et al (2011) Effects of nutrient solution pH on growth parameters of alfalfa (*Medicago sativa* L.) genotypes. Comun Sci 2:135–141

Masulili A, Utomo WH, MS S (2010) Rice husk biochar for rice based cropping system in acid soil 1. The characteristics of rice husk biochar and its influence on the properties of acid sulfate soils and rice growth in West Kalimantan, Indonesia. J Agric Sci 2. https://doi.org/10.5539/jas.v2n1p39

Mazac R (2016) Assessing the use of food waste biochar as a biodynamic plant fertilizer. In: Dep. Honor. Proj. https://digitalcommons.hamline.edu/dhp/43

McKendry P (2002) Energy production from biomass (part 1): overview of biomass. Bioresour Technol 83:37–46. https://doi.org/10.1016/S0960-8524(01)00118-3

Melaku T, Ambaw G, Nigussie A et al (2020) Short-term application of biochar increases the amount of fertilizer required to obtain potential yield and reduces marginal agronomic efficiency in high phosphorus-fixing soils. Biochar. https://doi.org/10.1007/s42773-020-00059-x

Meng J, Tao M, Wang L et al (2018) Changes in heavy metal bioavailability and speciation from a Pb-Zn mining soil amended with biochars from co-pyrolysis of rice straw and swine manure. Sci Total Environ 633:300–307. https://doi.org/10.1016/j.scitotenv.2018.03.199

Meyer S, Glaser B, Quicker P (2011b) Technical, economical, and climate-related aspects of biochar production technologies: a literature review. Environ Sci Technol 45:9473–9483. https://doi.org/10.1021/es201792c

Middleton N (2004) Global desertification: do humans cause deserts? Environ Sci Policy 7:118–119. https://doi.org/10.1016/j.envsci.2003.12.005

Mohan D, Sarswat A, Ok YS, Pittman CU (2014) Organic and inorganic contaminants removal from water with biochar, a renewable, low cost and sustainable adsorbent—a critical review. Bioresour Technol 160:191–202. https://doi.org/10.1016/j.biortech.2014.01.120

Mokarram-Kashtiban S, Hosseini SM, Kouchaksaraei MT, Younesi H (2019) Biochar improves the morphological, physiological and biochemical properties of white willow seedlings in heavy metal-contaminated soil. Arch Biol Sci 71:281–291. https://doi.org/10.2298/ABS180918010M

Molnár M, Vaszita E, Farkas É et al (2016) Acidic sandy soil improvement with biochar—a microcosm study. Sci Total Environ 563–564:855–865. https://doi.org/10.1016/j.scitotenv.2016.01.091

Moore F, González ME, Khan N et al (2018) Copper immobilization by biochar and microbial community abundance in metal-contaminated soils. Sci Total Environ 616–617:960–969. https://doi.org/10.1016/j.scitotenv.2017.10.223

Muhammad N, Aziz R, Brookes PC, Xu J (2017) Impact of wheat straw biochar on yield of rice and some properties of Psammaquent and Plinthudult. J Soil Sci Plant Nutr 17:808–823. https://doi.org/10.4067/S0718-95162017000300019

Mukherjee A, Lal R (2013) Biochar impacts on soil physical properties and greenhouse gas emissions. Agronomy 3:313–339. https://doi.org/10.3390/agronomy3020313

Mukherjee A, Zimmerman AR, Harris W (2011) Surface chemistry variations among a series of laboratory-produced biochars. Geoderma 163:247–255. https://doi.org/10.1016/j.geoderma.2011.04.021

Mullen CA, Boateng AA, Goldberg NM et al (2010) Bio-oil and bio-char production from corn cobs and stover by fast pyrolysis. Biomass Bioenerg 34:67–74. https://doi.org/10.1016/j.biombioe.2009.09.012

Naeem MA, Khalid M, Aon M et al (2017) Effect of wheat and rice straw biochar produced at different temperatures on maize growth and nutrient dynamics of a calcareous soil. Arch Agron Soil Sci 63:2048–2061. https://doi.org/10.1080/03650340.2017.1325468

Naeem MA, Imran M, Amjad M et al (2019) Batch and column scale removal of cadmium from water using raw and acid activated wheat straw biochar. Water (Switzerland) 11:1–17. https://doi.org/10.3390/w11071438

Namgay T, Singh B, Singh BP (2010) Influence of biochar application to soil on the availability of As, Cd, Cu, Pb, and Zn to maize (*Zea mays* L.). Aust J Soil Res 48:638–647. https://doi.org/10.1071/SR10049

Nandillon R, Lebrun M, Miard F et al (2019a) Contrasted tolerance of Agrostis capillaris metallicolous and non-metallicolous ecotypes in the context of a mining technosol amended by biochar, compost and iron sulfate. Environ Geochem Health. https://doi.org/10.1007/s10653-019-00447-8

Nandillon R, Miard F, Lebrun M et al (2019b) Effect of biochar and amendments on Pb and As phytotoxicity and phytoavailability in a technosol. Clean Soil, Air, Water 47:1800220. https://doi.org/10.1002/clen.201800220

Neves D, Thunman H, Matos A et al (2011) Characterization and prediction of biomass pyrolysis products. Prog Energy Combust Sci 37:611–630. https://doi.org/10.1016/j.pecs.2011.01.001

Nie C, Yang X, Niazi NK et al (2018) Impact of sugarcane bagasse-derived biochar on heavy metal availability and microbial activity: a field study. Chemosphere 200:274–282. https://doi.org/10.1016/j.chemosphere.2018.02.134

Nigussie A, Kissi E, Misganaw M, Ambaw G (2012) Effect of biochar application on soil properties and nutrient uptake of lettuces (*Lactuca sativa*) grown in chromium polluted soils. Environ Sci 12:369376

Novak JM, Busscher WJ (2013) Selection and use of designer biochars to improve characteristics of southeastern USA coastal plain degraded soils. Springer, New York

Obia A, Mulder J, Martinsen V et al (2016) In situ effects of biochar on aggregation, water retention and porosity in light-textured tropical soils. Soil Tillage Res 155:35–44. https://doi.org/10.1016/j.still.2015.08.002

Olmo M, Alburquerque JA, Barrón V et al (2014) Wheat growth and yield responses to biochar addition under Mediterranean climate conditions. Biol Fertil Soils 50:1177–1187. https://doi.org/10.1007/s00374-014-0959-y

Onay O, Kockar OM (2003) Slow, fast and flash pyrolysis of rapeseed. Renew Energy 28:2417–2433. https://doi.org/10.1016/S0960-1481(03)00137-X

Overend RP (1999) Thermochemical conversion of biomass. Combustion 94–119

Panagos P, Van Liedekerke M, Yigini Y, Montanarella L (2013) Contaminated sites in Europe: review of the current situation based on data collected through a European network. J Environ Public Health 2013:1–11. https://doi.org/10.1155/2013/158764

Park JH, Choppala GK, Bolan NS et al (2011a) Biochar reduces the bioavailability and phytotoxicity of heavy metals. Plant Soil 348:439–451. https://doi.org/10.1007/s11104-011-0948-y

Park JH, Lamb D, Paneerselvam P et al (2011b) Role of organic amendments on enhanced bioremediation of heavy metal(loid) contaminated soils. J Hazard Mater 185:549–574. https://doi.org/10.1016/j.jhazmat.2010.09.082

Pathak H (2010) Trend of fertility status of Indian soils. Curr Adv Agric Sci 2:10–12

Paz-Ferreiro J, Lu H, Fu S et al (2014) Use of phytoremediation and biochar to remediate heavy metal polluted soils: a review. Solid Earth 5:65–75. https://doi.org/10.5194/se-5-65-2014

Peterson SC, Jackson MA, Kim S, Palmquist DE (2012) Increasing biochar surface area: optimization of ball milling parameters. Powder Technol 228:115–120. https://doi.org/10.1016/j.powtec.2012.05.005

Pettit RE (2014) Organic matter, humus, humate, humic acid, fulvic acid and humin: their importance in soil fertility and plant health. Igarss 2014:1–5. https://doi.org/10.1007/s13398-014-0173-7.2

Pimchuai A, Dutta A, Basu P (2010) Torrefaction of agriculture residue to enhance combustible properties. Energy Fuels 24:4638–4645. https://doi.org/10.1021/ef901168f

Placek A, Grobelak A, Kacprzak M (2016) Improving the phytoremediation of heavy metals contaminated soil by use of sewage sludge. Int J Phytoremediation 18:605–618. https://doi.org/10.1080/15226514.2015.1086308

Prashar P, Shah S (2016) Impact of fertilizers and pesticides on soil microflora in agriculture. Sustain Agric Rev 331–361. https://doi.org/10.1007/978-3-319-26777-7_8

Prendergast-Miller MT, Duvall M, Sohi SP (2011) Localisation of nitrate in the rhizosphere of biochar-amended soils. Soil Biol Biochem 43:2243–2246. https://doi.org/10.1016/j.soilbio.2011.07.019

Pröll T, Aichernig C, Rauch R, Hofbauer H (2007) Fluidized bed steam gasification of solid biomass—performance characteristics of an 8 MWth combined heat and power plant. Int J Chem React Eng 5. https://doi.org/10.2202/1542-6580.1398

Pulido-Novicio L, Hata T, Kurimoto Y et al (2001) Adsorption capacities and related characteristics of wood charcoals carbonized using a one-step or two-step process. J Wood Sci 47:48–57. https://doi.org/10.1007/BF00776645

Purakayastha TJ, Kumari S, Pathak H (2015) Characterisation, stability, and microbial effects of four biochars produced from crop residues. Geoderma 239–240:293–303. https://doi.org/10.1016/j.geoderma.2014.11.009

Qian L, Zhang W, Yan J et al (2016) Effective removal of heavy metal by biochar colloids under different pyrolysis temperatures. Bioresour Technol 206:217–224. https://doi.org/10.1016/j.biortech.2016.01.065

Rafael RBA, Fernandez-Marcos ML, Cocco S et al (2019) Benefits of biochars and NPK fertilizers for soil quality and growth of cowpea (*Vigna unguiculata* L. Walp.) in an acid arenosol. Pedosphere 29:311–333. https://doi.org/10.1016/S1002-0160(19)60805-2

Rajapaksha AU, Chen SS, Tsang DCW et al (2016) Engineered/designer biochar for contaminant removal/immobilization from soil and water: potential and implication of biochar modification. Chemosphere 148:276–291. https://doi.org/10.1016/j.chemosphere.2016.01.043

Rajkovich S, Enders A, Hanley K et al (2012) Corn growth and nitrogen nutrition after additions of biochars with varying properties to a temperate soil. Biol Fertil Soils 48:271–284. https://doi.org/10.1007/s00374-011-0624-7

Ramesh T, Bolan NS, Kirkham MB et al (2019) Soil organic carbon dynamics: Impact of land use changes and management practices: a review. Adv Agron 156:1–107. https://doi.org/10.1016/bs.agron.2019.02.001

Rehman MH, Holmes A, Saunders SJ, Islam KR (2011) Biochar workshop. NZ Biochar Research Centre, Massey University of Palmerst, North, New Zealand

Reynolds JF, Stafford DS (2002) Do human cause deserts? In: Reynolds JF, Stafford Smith DMS (eds) Global desertification: do human cause deserts? J. F. Dahlem University Press, Berlin. pp 1–22

Rouquerol F, Rouquerol J, Sing K (1999) Introduction. Adsorption by powders porous solids, 1–26

Sadaka S, Sharara MA, Ashworth A et al (2014) Characterization of biochar from switchgrass carbonization. Energies 7:548–567. https://doi.org/10.3390/en7020548

Sahota S, Vijay VK, Subbarao PMV et al (2018) Characterization of leaf waste based biochar for cost effective hydrogen sulphide removal from biogas. Bioresour Technol 250:635–641. https://doi.org/10.1016/j.biortech.2017.11.093

Saito T, Otani T, Seike N et al (2011) Suppressive effect of soil application of carbonaceous adsorbents on dieldrin uptake by cucumber fruits. Soil Sci Plant Nutr 57:157–166. https://doi.org/10.1080/00380768.2010.551281

Sarkhot DV, Ghezzehei TA, Berhe AA (2013) Effectiveness of biochar for sorption of ammonium and phosphate from dairy effluent. J Environ Qual 42:1545–1554. https://doi.org/10.2134/jeq2012.0482

Schmidt H, Pandit B, Martinsen V et al (2015) Fourfold increase in pumpkin yield in response to low-dosage root zone application of urine-enhanced biochar to a fertile tropical soil. Agriculture 5:723–741. https://doi.org/10.3390/agriculture5030723

Sevilla M, Maciá-Agulló JA, Fuertes AB (2011) Hydrothermal carbonization of biomass as a route for the sequestration of CO_2: chemical and structural properties of the carbonized products. Biomass Bioenerg 35:3152–3159. https://doi.org/10.1016/j.biombioe.2011.04.032

Shaikh WA, Islam R Ul, Chakraborty S (2021) Stable silver nanoparticle doped mesoporous biochar-based nanocomposite for efficient removal of toxic dyes. J Environ Chem Eng 9(1):104982-S2213343720313312 104982. https://doi.org/10.1016/j.jece.2020.104982

Shaikh WA, Chakraborty S, Islam R Ul, Ghfar AA, Naushad M, Bundschuh J, Maity JP, Mondal NK (2022a) Fabrication of biochar-based hybrid Ag nanocomposite from algal biomass waste for toxic dye-laden wastewater treatment. Chemosphere 289:133243-S0045653521037176 133243. https://doi.org/10.1016/j.chemosphere.2021.133243

Shaikh WA, Kumar A, Chakraborty S, Islam R Ul, Bhattacharya T, Biswas JK (2022b) Biochar-based nanocomposite from waste tea leaf for toxic dye removal: from facile fabrication to functional fitness. Chemosphere 291:132788-S0045653521032604 132788. https://doi.org/10.1016/j.chemosphere.2021.132788

Shivaram P, Leong YK, Yang H, Zhang DK (2013) Flow and yield stress behaviour of ultrafine Mallee biochar slurry fuels: the effect of particle size distribution and additives. Fuel 104:326–332. https://doi.org/10.1016/j.fuel.2012.09.015

Smebye AB, Sparrevik M, Schmidt HP, Cornelissen G (2017) Life-cycle assessment of biochar production systems in tropical rural areas: comparing flame curtain kilns to other production methods. Biomass Bioenerg 101:35–43. https://doi.org/10.1016/j.biombioe.2017.04.001

Smith J, Doran J (1996) Measurement and use of pH and electrical conductivity for soil quality analysis. In methods for assessing soil quality. Soil Sci Soc Am Spec Publ 49:169–182

Solaiman ZM, Anawar HM (2015) Application of biochars for soil constraints: challenges and solutions. Pedosphere 25:631–638. https://doi.org/10.1016/S1002-0160(15)30044-8

Spokas KA, Novak JM, Venterea RT (2012) Biochar's role as an alternative N-fertilizer: Ammonia capture. Plant Soil 350:35–42. https://doi.org/10.1007/s11104-011-0930-8

Stella Mary G, Sugumaran P, Niveditha S et al (2016) Production, characterization and evaluation of biochar from pod (*Pisum sativum*), leaf (*Brassica oleracea*) and peel (*Citrus sinensis*) wastes. Int J Recycl Org Waste Agric 5:43–53. https://doi.org/10.1007/s40093-016-0116-8

Sun H, Hockaday WC, Masiello CA, Zygourakis K (2012a) Multiple controls on the chemical and physical structure of biochars. Ind Eng Chem Res 51:1587–1597. https://doi.org/10.1021/ie201309r

Sun K, Gao B, Ro KS et al (2012b) Assessment of herbicide sorption by biochars and organic matter associated with soil and sediment. Environ Pollut 163:167–173. https://doi.org/10.1016/j.envpol.2011.12.015

Sun Y, Gao B, Yao Y et al (2014) Effects of feedstock type, production method, and pyrolysis temperature on biochar and hydrochar properties. Chem Eng J 240:574–578. https://doi.org/10.1016/j.cej.2013.10.081

Tan X, fei, Liu S bo, Liu Y guo, et al (2017) Biochar as potential sustainable precursors for activated carbon production: multiple applications in environmental protection and energy storage. Bioresour Technol 227:359–372. https://doi.org/10.1016/j.biortech.2016.12.083

Tang J, Zhu W, Kookana R, Katayama A (2013) Characteristics of biochar and its application in remediation of contaminated soil. J Biosci Bioeng 116:653–659. https://doi.org/10.1016/j.jbiosc.2013.05.035

Tchounwou PB, Yedjou CG, Patlolla AK, Sutton DJ (2012) Heavy metal toxicity and the environment. EXS 101:133–164

Tian J, Wang J, Dippold M et al (2016) Biochar affects soil organic matter cycling and microbial functions but does not alter microbial community structure in a paddy soil. Sci Total Environ 556:89–97. https://doi.org/10.1016/j.scitotenv.2016.03.010

Titirici MM, White RJ, Falco C, Sevilla M (2012) Black perspectives for a green future: hydrothermal carbons for environment protection and energy storage. Energy Environ Sci 5:6796–6822. https://doi.org/10.1039/c2ee21166a

Topoliantz S, Ponge JF (2003) Burrowing activity of the geophagous earthworm Pontoscolex corethrurus (Oligochaeta: Glossoscolecidae) in the presence of charcoal. Appl Soil Ecol 23:267–271. https://doi.org/10.1016/S0929-1393(03)00063-5

Tripathi M, Sahu JN, Ganesan P (2016) Effect of process parameters on production of biochar from biomass waste through pyrolysis: a review. Renew Sustain Energy Rev 55:467–481. https://doi.org/10.1016/j.rser.2015.10.122

Tryon EH (1948) Effect of charcoal on certain physical, chemical, and biological properties of forest soils. Ecol Monogr 18:81–115. https://doi.org/10.2307/1948629

Uchimiya M, Klasson KT, Wartelle LH, Lima IM (2011a) Influence of soil properties on heavy metal sequestration by biochar amendment: 1. Copper sorption isotherms and the release of cations. Chemosphere 82:1431–1437. https://doi.org/10.1016/j.chemosphere.2010.11.050

Uchimiya M, Wartelle LH, Klasson KT et al (2011b) Influence of pyrolysis temperature on biochar property and function as a heavy metal sorbent in soil. J Agric Food Chem 59:2501–2510. https://doi.org/10.1021/jf104206c

Usowicz B, Lipiec J, Marczewski W, Ferrero A (2006) Thermal conductivity modelling of terrestrial soil media—a comparative study. Planet Space Sci 54:1086–1095. https://doi.org/10.1016/j.pss.2006.05.018

Uzoma KC, Inoue M, Andry H et al (2011) Effect of cow manure biochar on maize productivity under sandy soil condition. Soil Use Manag 27:205–212. https://doi.org/10.1111/j.1475-2743.2011.00340.x

Valenzuela-Calahorro C, Bernalte-Garcia A, Gómez-Serrano V, Bernalte-García MJ (1987) Influence of particle size and pyrolysis conditions on yield, density and some textural parameters of chars prepared from holm-oak wood. J Anal Appl Pyrolysis 12:61–70. https://doi.org/10.1016/0165-2370(87)80015-3

Vassilev SV, Baxter D, Andersen LK, Vassileva CG (2010) An overview of the chemical composition of biomass. Fuel 89:913–933. https://doi.org/10.1016/j.fuel.2009.10.022

Vassilev SV, Baxter D, Andersen LK, Vassileva CG (2013) An overview of the composition and application of biomass ash. Part 1. Phase-mineral and chemical composition and classification. Fuel 105:40–76. https://doi.org/10.1016/j.fuel.2012.09.041

Verheijen F, Jeffery S, Bastos AC et al (2010) Biochar application to soils: a critical scientific review of effects on soil properties. Processes and Functions. Environment 8:144. https://doi.org/10.2788/472

Wang Y, Lin Y, Chiu PC et al (2015) Phosphorus release behaviors of poultry litter biochar as a soil amendment. Sci Total Environ 512–513:454–463. https://doi.org/10.1016/j.scitotenv.2015.01.093

Wang Z, Han L, Sun K et al (2016) Sorption of four hydrophobic organic contaminants by biochars derived from maize straw, wood dust and swine manure at different pyrolytic temperatures. Chemosphere 144:285–291. https://doi.org/10.1016/j.chemosphere.2015.08.042

Wang T, Zhai Y, Zhu Y et al (2018) A review of the hydrothermal carbonization of biomass waste for hydrochar formation: process conditions, fundamentals, and physicochemical properties. Renew Sustain Energy Rev 90:223–247. https://doi.org/10.1016/j.rser.2018.03.071

Wang R, Wei S, Jia P et al (2019a) Biochar significantly alters rhizobacterial communities and reduces Cd concentration in rice grains grown on Cd-contaminated soils. Sci Total Environ 676:627–638. https://doi.org/10.1016/j.scitotenv.2019.04.133

Wang Y, Wang H-S, Tang C-S et al (2019b) Remediation of heavy-metal-contaminated soils by biochar: a review. Environ Geotech 1–14. https://doi.org/10.1680/jenge.18.00091

Warnock DD, Lehmann J, Kuyper TW, Rillig MC (2007) Mycorrhizal responses to biochar in soil—concepts and mechanisms. Plant Soil 300:9–20. https://doi.org/10.1007/s11104-007-9391-5

Weber K, Quicker P (2018) Properties of biochar. Fuel 217:240–261. https://doi.org/10.1016/j.fuel.2017.12.054

Windeatt JH, Ross AB, Williams PT et al (2014) Characteristics of biochars from crop residues: potential for carbon sequestration and soil amendment. J Environ Manage 146:189–197. https://doi.org/10.1016/j.jenvman.2014.08.003

Wu F (2012) The community structure of microbial in arable soil under different long-term fertilization regimes in the Loess Plateau of China. African J Microbiol Res 6:. https://doi.org/10.5897/ajmr12.562

Wu W, Yang M, Feng Q et al (2012) Chemical characterization of rice straw-derived biochar for soil amendment. Biomass Bioenerg 47:268–276. https://doi.org/10.1016/j.biombioe.2012.09.034

Wu W, Li J, Niazi NK et al (2016) Influence of pyrolysis temperature on lead immobilization by chemically modified coconut fiber-derived biochars in aqueous environments. Environ Sci Pollut Res 23:22890–22896. https://doi.org/10.1007/s11356-016-7428-0

Xie T, Reddy KR, Wang C et al (2015) Characteristics and applications of biochar for environmental remediation: a review. Crit Rev Environ Sci Technol 45:939–969. https://doi.org/10.1080/10643389.2014.924180

Xie Y, Fan J, Zhu W, et al (2016) Effect of heavy metals pollution on soil microbial diversity and bermudagrass genetic variation. Front Plant Sci 7. https://doi.org/10.3389/fpls.2016.00755

Xu X, Cao X, Zhao L et al (2013) Removal of Cu, Zn, and Cd from aqueous solutions by the dairy manure-derived biochar. Environ Sci Pollut Res 20:358–368. https://doi.org/10.1007/s11356-012-0873-5

Xu M, Xia H, Wu J et al (2017) Shifts in the relative abundance of bacteria after wine-lees-derived biochar intervention in multi metal-contaminated paddy soil. Sci Total Environ 599–600:1297–1307. https://doi.org/10.1016/j.scitotenv.2017.05.086

Xu Y, Seshadri B, Sarkar B et al (2018) Biochar modulates heavy metal toxicity and improves microbial carbon use efficiency in soil. Sci Total Environ 621:148–159. https://doi.org/10.1016/j.scitotenv.2017.11.214

Yang J, Li H, Zhang D et al (2017) Limited role of biochars in nitrogen fixation through nitrate adsorption. Sci Total Environ 592:758–765. https://doi.org/10.1016/j.scitotenv.2016.10.182

Yargholi B, Azarneshan S (2014) Long-term effects of pesticides and chemical fertilizers usage on some soil properties and accumulation of heavy metals in the soil (case study of Moghan plain's (Iran) irrigation and drainage network). Int J Agric Crop Sci 7:518–523

Yeboah E, Ofori P, Quansah GW et al (2009) Improving soil productivity through biochar amendments to soils. African J Environ Sci Technol 3:34–41

Yin D, Wang X, Chen C et al (2016) Varying effect of biochar on Cd, Pb and As mobility in a multi-metal contaminated paddy soil. Chemosphere 152:196–206. https://doi.org/10.1016/j.chemosphere.2016.01.044

Yorgun S, Ensöz S, Koçkar ÖM (2001) Flash pyrolysis of sunflower oil cake for production of liquid fuels. J Anal Appl Pyrolysis 60:1–12. https://doi.org/10.1016/S0165-2370(00)00102-9

Young S (1995) Toxic metals in soil-plant systems. In: Ross SM (ed) xiv + 469 pp. Wiley, Chichester(1994). £55.00 (hardback). J Agric Sci 124:155–156. https://doi.org/10.1017/s00218596000 71422.ISBN 0471-94279-0

Yu Z, Qiu W, Wang F et al (2017) Effects of manganese oxide-modified biochar composites on arsenic speciation and accumulation in an indica rice (*Oryza sativa* L.) cultivar. Chemosphere 168:341–349. https://doi.org/10.1016/j.chemosphere.2016.10.069

Yuan JH, Xu RK, Zhang H (2011) The forms of alkalis in the biochar produced from crop residues at different temperatures. Bioresour Technol 102:3488–3497. https://doi.org/10.1016/j.biortech.2010.11.018

Yuan P, Wang J, Pan Y et al (2019) Review of biochar for the management of contaminated soil: preparation, application and prospect. Sci Total Environ 659:473–490. https://doi.org/10.1016/j.scitotenv.2018.12.400

Zeng Z, Da ZS, Li TQ et al (2013) Sorption of ammonium and phosphate from aqueous solution by biochar derived from phytoremediation plants. J Zhejiang Univ Sci B 14:1152–1161. https://doi.org/10.1631/jzus.B1300102

Zhang J, You C (2013) Water holding capacity and absorption properties of wood chars. Energy Fuels 27:2643–2648. https://doi.org/10.1021/ef4000769

Zhang X, Wang H, He L et al (2013) Using biochar for remediation of soils contaminated with heavy metals and organic pollutants. Environ Sci Pollut Res 20:8472–8483. https://doi.org/10.1007/s11356-013-1659-0

Zhang C, Liu L, Zhao M et al (2018a) The environmental characteristics and applications of biochar. Environ Sci Pollut Res 25:21525–21534. https://doi.org/10.1007/s11356-018-2521-1

Zhang S, Abdalla MAS, Luo Z, Xia S (2018b) The wheat straw biochar research on the adsorption/desorption behaviour of mercury in wastewater. Desalin Water Treat 112:147–160. https://doi.org/10.5004/dwt.2018.21850

Zhang M, Riaz M, Zhang L et al (2019) Biochar induces changes to basic soil properties and bacterial communities of different soils to varying degrees at 25 mm rainfall: more effective on acidic soils. Front Microbiol 10:1321

Zhao X, Ouyang W, Hao F et al (2013) Properties comparison of biochars from corn straw with different pretreatment and sorption behaviour of atrazine. Bioresour Technol 147:338–344. https://doi.org/10.1016/j.biortech.2013.08.042

Zhao H, Xue Y, Long L, Hu X (2018) Adsorption of nitrate onto biochar derived from agricultural residuals. Water Sci Technol 77:548–554. https://doi.org/10.2166/wst.2017.568

Zhou Y, Gao B, Zimmerman AR et al (2013) Sorption of heavy metals on chitosan-modified biochars and its biological effects. Chem Eng J 231:512–518. https://doi.org/10.1016/j.cej.2013.07.036

Zornoza R, Moreno-Barriga F, Acosta JA et al (2016) Stability, nutrient availability and hydrophobicity of biochars derived from manure, crop residues, and municipal solid waste for their use as soil amendments. Chemosphere 144:122–130. https://doi.org/10.1016/j.chemosphere.2015.08.046

Chapter 17
Biochar Production and Its Impact on Sustainable Agriculture

Sanat Kumar Dwibedi, Basudev Behera, and Farid Khawajazada

Abstract Biochar is a fine-grained, carbon-rich and porous organic derivative derived through pyrolytic combustion of biomass. Its use in agriculture since Amazonian *terra preta* civilization signifies its potential benefits in sustainable crop production and environmental remediation. It supports plant growth and yields through favourable soil physicochemical properties, enhanced water holding capacity, nutrient availability, heavy metal remediation and disease and pest suppression. It sequesters atmospheric carbon dioxide, pacifies the pace of global warming and contributes to quenching adverse effects of climate change in the long run. In this direction, large-scale biochar application in the agricultural production system is a holistic approach for socio-economic and ecological sustainability. Research results on biochar application, though miraculous, are mostly laboratory or greenhouse-based as the popularization of its wider field application in the agriculture sector is constrained by a higher rate of application incurring a high cost of production. This problem can be addressed through low-cost biochar generation from the locally available biowastes.

Keywords Biochar synthesis · Pyrolysis · Reclamation · Sustainable agriculture

17.1 Introduction

Healthy soil leads to a productive, profitable and sustainable agriculture production system (White and Barberchek 2017). Rhizospheric aeration, moisture, temperature, nutrients and microbial population are influenced by soil type and its physicochemical characteristics. However, continuous cropping over years depletes many essential plant nutrients from the soil that need to be replenished through judicious nutrient

---

S. K. Dwibedi (✉) · F. Khawajazada
Department of Agronomy, College of Agriculture, Odisha University of Agriculture and Technology, Bhubaneswar, India
e-mail: sanatdwibediouat@gmail.com

B. Behera
Institute of Agricultural Sciences, Siksha 'O' Anusandhan University (Deemed), Bhubaneswar, Odisha, India

© The Author(s), under exclusive license to Springer Nature Switzerland AG 2023
V. C. Pandey (ed.), *Bio-Inspired Land Remediation*, Environmental Contamination Remediation and Management, https://doi.org/10.1007/978-3-031-04931-6_17

management. Furthermore, excessive application of synthetic pesticides, hormones, probiotics and chemical fertilizers have perilous effects on the soil environment that ends up with pesticide resistance and pest resurgence (Wu et al. 2012). Intensive cropping and use of chemical fertilizers deteriorate the soil health and reduce crop yield (Rawat et al. 2017). Continuous cropping without manuring exhausts soil carbon pool that influences soil biota. Heavy metal accumulation, particularly nearby opencast mines need to be remediated for sustainable cropping and maintaining soil biodiversity (Rawat et al. 2017).

It is high time to feed the ever-growing population without degrading the environment. Selection of any soil ameliorant for land reclamation must be based on its compatibility, cost and availability. Hence, due care must be taken to maintain the soil carbon pool to facilitate soil biodiversity, natural cycles and to sequester atmospheric carbon. Among many options of soil fertility restoration and carbon sequestration, biochar application has been a well-proven, widely accepted and age-old practice dating back to Amazonian civilization (Lahori et al. 2017). Biochar is "a fine-grained, carbon-rich, and porous organic derivative derived through anaerobic thermo-chemical combustion of biomass" (Amonette and Joseph 2009). Pyrolytic burning of biomass produces oil and gas as co-products in addition to biochar depending on the substrate type and processing conditions (Gaunt and Rondon 2006).

17.2 History of Biochar Production and Use

As mentioned earlier, biochar production and application trail to the era of a fire-fallow system of cultivation during the Neolithic revolution when nomadic hunters and gatherers domesticated certain plants and animals for leading a settled life and getting more nutrition per unit area. These ancient nomads were clearing up the forests and grasslands, and burning biomass just before the rainiest part of the year to enrich the soil with valuable plant nutrients and to eliminate weeds and control the disease-pest infestation. However, after three to five years of cropping, the nomads were abandoning the land in search of new locations due to reduced soil fertility, and the resurgence of diseases, pests and weeds. After a gap of a few years, they were again returning to the same land on recovery. This cyclic process of burning and assorting is known as slash-burn or shifting cultivation. In India, it is known as *jhoom* or *jhum* cultivation (Singh 2018). As of 2004, an estimated area of 200 to 500 million hectares across the world was under this system of cultivation. As the slash and burn system of cultivation is not sustainable and scalable for the larger human population, an alternative system such as the *inga* alley cropping or slash and char system (Biederman 2012) with significantly less environmental repercussion had evolved (Elkan 2004).

17.2.1 Slash and Burn System Versus Slash and Char System

The slash and burn system of farming had evolved during the Neolithic era to expand crop area for feeding the growing human population by clearing thick vegetation. The burning of biomass was yielding ashes, that provided essential plant nutrients, but at the cost of devastating environmental pollution (Raison et al. 2009) by producing many toxic gases that polluted air in the near vicinity. The wood ash thus produced, being light in weight, was also getting washed away through natural drainage exposing the land to accelerated weathering and soil erosion. In long run, that eventually affected farming and large-scale ranching.

To mitigate the negative effects of burning, people started charring residues instead of burning after cutting. This alternate system of farming, known as the 'slash and char system', had tremendous environmental benefits over the slash and burn system as it significantly reduced toxic gases and improved the bio-physicochemical properties of soil. Slash and burn system with 1–3 years of cropping followed by 20 years of the fallow period could be sustainable but not practicable under growing food demands (Steiner et al. 2008).

In the slash and char system, biochar is produced which can be buried in the soil after mixing with biomass such as agricultural residues, manure and food waste for conditioning or *terra preta*. Terra preta is the most fertile black-coloured soil on the planet found in the Amazon basin, popularly known as Amazonian dark earth or Indian black earth (*Terra Preta de Indio*). It is known to regenerate on its own. It sequesters considerable quantities of atmospheric CO_2 into the soil as safe, stable but active form in contrast to the slash and burn system that increases carbon footprint opposite to it. Near about 50% of the carbon remains in stable form and remains active over hundreds of years (Lehmann et al. 2006).

17.2.2 Biochar in Traditional Agriculture

Charcoal, the precursor of biochar has been in use since the Paleolithic and Neolithic eras of slash and burn (Chen et al. 2019). Carbon dating of the charcoal paintings on the walls of the caves across the globe uncovers the story of charcoal use even more than 30,000 years ago (Zorich 2011). The International Biochar Initiative (IBI) defined biochar or pyrogenic carbon as "the solid material produced through thermochemical conversion of biomass in an oxygen-deprived environment". It is popular both in ancient and modern civilizations. The application of biochar in the ancient era is evident from the *Terra preta* in the Amazonian basin of South America (Glaser et al. 2001) for more than 2500 years (USBI News 2021a). Such a meaningful piece of ancient agricultural heritage was unveiled in 1966 by Wim Sombroek, a Dutch soil scientist who located a rich self-regenerating soil in the Amazon basin of Brazil (Wayne 2012). The nutrient and organic matter content of this Amazonian dark soil were extremely high (Harder 2006; Marris 2006; Tenenbaum 2009). Its chemical analyses indicated the presence of burned wood, crop and bone residues of animals and fishes (Sombroek et al. 2002). The productivity of *terra preta* is four times greater than the soil from similar parent material (Wayne 2012). Bruno Glaser of the University of Bayreuth in his article "the *Terra preta* phenomenon: a model for sustainable agriculture in the humid tropics" has estimated around 250 tons of carbon in *terra preta* compared to the maximum of 100 tons in unimproved soils from the same area (Glaser et al. 2001). The land size varied from 20 ha (Smith 1980; Zech et al. 1990; McCann et al. 2001) to 350 ha (Smith 1999) patches covering 50,000 ha in the central Amazonian region. Still today, 10% of the Amazonian basin is under terra preta soil (USBI News 2021a).

The porous structure of biochar facilitates nutrient accumulation, growth of beneficial microorganisms and helps in the slow release of nutrients in available form and a balanced ratio supporting vigorous plant growth (Shindo 1991; Cheng et al. 2008). The black carbon in charcoal exists in soil for over 1000 years or longer. This black soil from anthropogenic activity in the Amazonian basin of dense rainforest could be attributed to the sustenance of a large human population for thousands of years before it was exposed to the outer world by Christopher Columbus in 1498 (Petersen et al. 2001; Lehmann 2009).

China and India have a strong history of biochar production and application. Conversion of crop residues into biochar instead of burning in-situ has been an age-old practice in China, mostly in the southern region of the country (Yan et al. 2019). The use of charcoal in agriculture in the Himalayan hills of the Indian subcontinent is a traditional practice. People gather biomass in forests and fields, cover them under mud-coat and set fire to get biochar on subsequent cooling. *Terra preta* like soils have been identified in Peru, Ecuador, Benin and Liberia in West Africa also (USBI News 2021a). Archaeologists have claimed the fall of Mesopotamia civilization due to climate change leading to drought and depletion of soil carbon (Codur et al. 2017).

In some ancient civilizations, the production of biochar was not the only requirement. Rather, they were more acquainted with the liquid product recovery. Traces of wood-tar and pyroligneous acids on the embalmed body of the dead are widely observed in the remains of ancient Egyptian societies (Emrich 1985; Day et al. 2012). Macedonians obtained wood oil from burning biochar in pits (Klark and Rule 1925). Evidence dating 6000 years back shows the use of wood tar to attach arrowhead with the spear shaft (Klark and Rule 1925; Emrich 1985). However, few such practices of charcoal making in many developing countries are not completely anoxic and thus unhealthy for the environment but are better than open burning of residues (USBI News 2021b).

17.3 Benefits of Biochar Use

Pyrolysis of natural vegetation or farm residues generates biofuel without competition with crop production. Controlled burning of biomass with limited or no oxygen produces syngas and wood oil in addition to the biochar, while open burning generates greenhouse gases (GHGs) and deteriorates the environment. Biochar on incorporation into the soil enhances natural processes, improves soil physicochemical properties, promotes beneficial microbial growth (Ajema 2018) and facilitates plant growth, protects against moisture stress (Bera et al. 2018), induces disease-pest tolerance, provides anchorage, sequesters atmospheric CO_2 (Cornet and Escadafal 2009), reduces soil erosion (Jien and Wang 2013), remediates (Cheng et al. 2020) and rejuvenates the soil.

17.4 Procedure for Synthesis of Biochar

The carbonization of wood for heating or making biochar is as old as human civilization itself (Brown 1917; Emrich 1985). Although different methods of biochar making were employed by ancient civilizations, all of them were to generate heat without any intent to harness the released volatile gases during the combustion process releasing toxic gases and fumes into the surrounding environment. However, in some civilizations, wood tar was collected for embalming dead or inserting arrowheads.

The simple process of thermal decomposition of biomass for biochar production involves either pyrolysis or gasification. Pyrolysis is the temperature-mediated systematic chemical decomposition of organic substrates in an oxygen starved atmosphere without combustion (Demirbas 2004). The gasification system produces smaller quantities of biochar (10–20%) but the larger volume of syngas (80%) on direct heating at >700 °C or more (Nartey and Zhao 2014; Biochar International 2021). In pyrolysis kilns, retorts and other specialized equipment are used to bake the biomass at <600 °C in absence of oxygen. Pyrolytic gases, often called syngas, are allowed to escape or combusted to make the process self-sustaining (International Biochar Institute 2021). Broadly, two systems of pyrolysis are used today, viz. fast pyrolysis and slow pyrolysis. Fast pyrolysis produces 75% oils and 10–20% char while slow pyrolysis produces one-third each of oils, char and gases (Nartey and Zhao 2014; Biochar International 2021). Pyrolysis occurs in three basic steps: In the initial step, moisture and some volatiles are lost; in the middle step, organic residues are transformed into volatile gasses and biochar, and finally, chemical rearrangement of the biochar occurs slowly (Demirbas 2004).

17.4.1 Stages of Pyrolysis

Biomass constitutes five main components: water, cellulose, hemicelluloses, minerals (ash), and lignin at varying proportions depending on the biomass source. Seasoned wood contains 12–19% moisture and freshly cut crops or wood contain 40–80% water on a weight basis. On heating, most of the water escapes at 100 °C and biomass starts breaking down above 150 °C. At this temperature, biomass softens and chemically bound water is released with carbon dioxide (CO_2) and volatile organic compounds (VOCs). On 'torrefaction', means further heating into the range of 200–250 °C, chemical bonds start breaking. Acetic acid, methanol and other oxygenated volatile compounds along with carbon monoxide and CO_2 are released from cellulose and hemicelluloses. Torrefied biomass (e.g. boiler fuel) is brittle, easy to grind with less energy, resistant to microbial decomposition and water uptake. The liquid condensate, known as 'wood vinegar', 'smoke water' or 'pyrolignous acid', can be used as a fungicide, plant growth promoter, compost stimulant and to improve the effectiveness of biochar (International Biochar Institute 2021). The torrefaction process is endothermic—external heat is required for increasing the temperature of dry biomass. When the temperature reaches 250–300 °C, the thermal decomposition of biomass becomes more extreme with the release of a combustible mixture of H_2, CH_2, other hydrocarbons, CO, CO_2 and tars. At this stage, pyrolysis becomes exothermic with the release of heat due to break-up of large polymers of biomass and release of structural oxygen to support self-sustained combustion thereby increasing the temperature up to 400 °C till oxygen gets depleted completely leaving carbon-rich charcoal-like residues. As heat is released and lost outside the system, external heat is required for any further pyrolytic processes. At the end of this exothermic pyrolysis stage, the maximum yield is obtained but stable carbon is yet to be attained. The ash content, VOCs and fixed carbon of wood biochar may be around 1.5–5%, 25–35% and 60–70%, respectively (Biochar International 2021). The biochar at the end of the exothermic stage still contains a significant amount of VOCs. More heating is needed to enhance the fixed carbon content, surface area and porosity from the remaining VOCs. To elevate the fixed carbon content to 80–85% and reduce the VOCs below 12%, the biochar is heated further to a temperature range of 550–800 °C depending on the substrate and particle size (Biochar International 2021). At this stage, the biochar yield is 25–30% of the oven-dry weight of the feedstock.

Once the temperature goes above 600 °C, the addition of small quantities of steam and air can trigger up the temperature up to 700–800 °C which results in activation and gasification processes. Air and steam can activate the surface of biochar at high temperature and release more VOCs. Activation increases the surface area, porosity and CEC by adding acidic functional groups but at the cost of lowering the yield. If an excess of air and/or steam is added to the process then a relatively clean gas is produced that can be used for generation of electricity but the yield of biochar is reduced below 20% and the ash content increases significantly (Biochar International 2021).

17.4.2 Preprocessing of Feedstock

Preconditioning of the feedstock can alter the rate of pyrolysis and final properties of biochar. Pretreatment with phosphoric acid increases functional groups, reduces the pH of biochar and produces slow-release phosphatic fertilizer. Pretreatment with iron salts produces magnetic biochar that can remove heavy metals from water. Alkali (potassium hydroxide) pretreatment softens biomass and breaks-down lingo-cellulosic compounds. Mixing of clay, ferrous sulphate and rock phosphate with biomass slows down the rate of pyrolysis, captures nitrogen and increases the concentration of nutrient-rich nanoparticles (Biochar International 2021).

17.4.3 Post-processing of Biochar

Post-processing of biochar can alter its properties. Phosphoric acid can be treated to make slow-release phosphatic fertilizer, reduce pH and enhance functional groups. Urine is added to increase nitrogen content and alkali is added to increase pH and potassium content. Rock phosphate, dolomite, gypsum, iron oxides, lime are added to rectify soil constraints. Urea and diammonium phosphate (DAP) is added to biochar for making complex fertilizers (Mohiuddin et al. 2006).

17.4.4 Effect of Residential Time

The largest specific surface area ($155.77 \text{ m}^2 \text{ g}^{-1}$), a higher carbon content (67.45%) and a lower ash content (15.38%), and higher carboxylic and phenolic-hydroxyl group (1.74 and 0.86 mol kg^{-1}) were obtained in biochar from *Robinia pseudoacacia* biowaste with zero residential time (the gap between burning char falling on ground and cooling by the sprinkling of cold water). However, a longer exposure time (5.30 min) resulted in lower values of above parameters (Xiao et al. 2020).

17.5 Methods of Preparation

Biochar can be prepared in small quantities at the individual household level (Whitman and Lehmann 2009) and in large quantities in big industries (Amonette and Joseph 2009). A specific requirement driven procedure is adopted for the synthesis of biochar and other by-products (Srinivasarao et al. 2013). Various pyrolysis technologies are available for traditional and commercial production of biochar and other fractions.

The global biochar market in 2018 was US$1.3 billion while the demand was 395.3 kilotons in that year, which is expected to get doubled by 2025. Increased demand for organic food so also its application in waste treatment and water purification in emerging economies like India and China, are likely to trigger the biochar requirement in near future. Environment friendliness, cheaper cost and multifarious applicability render it indispensable to reorient government policies for wider market expansion (Grand View Research 2018). To popularize biochar among the farmers, low-cost biochar production technology with the least negative environmental impact needs to be developed at the community as well as individual farm-family level. A few traditional, as well as modern biochar making methods are discussed hereunder.

17.5.1 Heap Method

The heap or mound method of charcoal making is an oldest practice in many parts of the world where a heap or mound or pyramid-like structure is made up of dried wood, crop residues, weeds, sawdust, rice husks, etc. The heap is then covered with grasses, available agriculture waste or coir and moist earth to prevent the free flow of oxygen during burning (Fig. 17.1). Vents are opened at the top to downward to allow free out flow of the combustion gas and to facilitate uniform charring. The fire is set at the bottom hole or top hole of the heap which subsequently engulfs the entire heap within an hour or several days depending on the type and volume of substrates. The quantity of smoke during burning depends on the substrate type, oxygen supply and moisture content of the feedstock. When smoke production stops, the holes are plugged with mud for the final conditioning of the biochar. After several days of cooling, the earth cover is removed and water is sprinkled to wash away ash. Earth-mound kilns with adjustable chimneys at the top that regulate diameter and height controlling oxygen flow are the most advanced among earth kilns (Emrich 1985). This method is the cheapest, easiest, simplest, quickest and most popular way of making biochar.

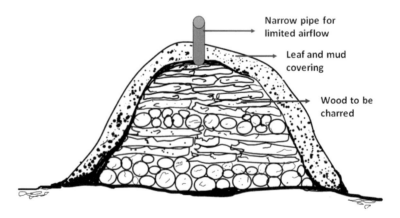

Fig. 17.1 Schematic diagram of the *heap* method of biochar making

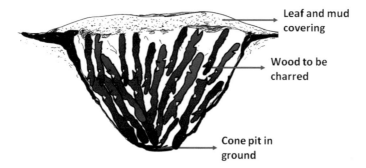

Fig. 17.2 Schematic diagram of *cone-pit* method of biochar making

17.5.2 Cone-Pit Method

Cone-pit method is also another traditional practice of producing charcoal. A pit of desired diameter and depth is dug in well-drained upland depending on the volume of the biomass (Fig. 17.2). A dried feedstock is put in it up to the ground level or below that at a time or in a phased manner after ignition of the fire. After completion of partial combustion, the pit is covered with fresh grasses or leaves followed by sealing with mud to restrict the inflow of oxygen into the pit. On cooling, the pit is opened and biochar is removed for further use in agricultural land or other purposes.

17.5.3 Drum Method

The drum method of biochar making is popular in areas where the transportation of biomass is cheaper *than in-situ* construction of kilns. Portable and handy metallic drums are easy to operate requiring less maintenance (Srinivasarao et al. 2013). Usually, cylindrical metal oil drums of about 200 L with both sides intact or of varying sizes depending on the volume of substrate capacity and are preferred for this purpose (Fig. 17.3). A square or round-shaped hole of 12–16 cm diameter or side length is made at the centre of the top lid to allow combustion syngas to escape through a chimney fitted to it. At the bottom of the drum, holes measuring about 4 cm^2 each are made covering 20% of the bottom area for uniform air flow from below. The pyrolytic temperature and quality of biochar depend on the inlet air volume and thus indirectly on the vent area at the bottom and side of the drum, if at all done in some designs. The entire drum is placed on 3–4 bricks to facilitate free airflow from the bottom. After putting feedstock systematically inside the drum, the fire is set by pouring some petroleum oil or using polythene pieces at the top or side-hole. Once the biomass catches fire, it is allowed to burn for about 15 min for partial combustion and then the top lid is covered. Initially, sooty smoke with luminous

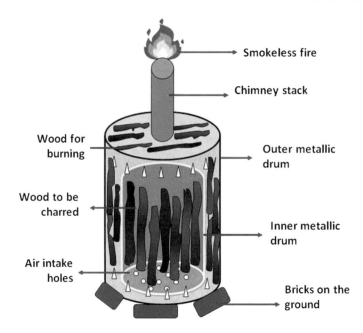

Fig. 17.3 Schematic diagram of *drum* method of biochar making

flame comes out of the chimney and subsequently bluish smokeless flame (non-luminous) come indicating completion of the heating phase of biochar making. The drum is then brought down from the top of bricks and placed on a muddy surface to prevent further entry of air. The top lid is also sealed by using mud to prevent airflow. After a few hours of cooling, the biochar is ready for direct use or grinding. The Central Research Institute for Dryland Agriculture (CRIDA), Hyderabad, India has developed a biochar kiln for the community as well as the individual level (Venkatesh et al. 2016).

17.5.4 Brick Kilns

Brick kilns are constructed at the place of origin of huge quantities of biomass. The size and quality of the kiln depend on the volume of feedstock and its expected longevity. Earthen bricks are mostly used but cemented bricks or fire bricks are also used in some designs. Earthen bricks are brittle and may break down if not specially baked and plastered thoroughly. Broken bricks allow free inflow of air resulting in vigorous burning and more ash production. Mud or cement mortar is used to plaster the bricks arranged in cylindrical or cubical shape (Fig. 17.4).

Fig. 17.4 Schematic diagram of *brick kiln* method of biochar making

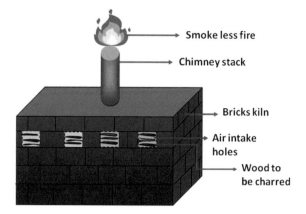

A simple biochar kiln, known as 'Holy Mother Biochar Kiln', has been made by the Sarada Matt (Holy Mother) at Almora, Uttarakhand, India by using clay mud-plaster and earthen bricks. Biomass is added continuously during combustion and the primary air vent at the bottom is kept open till biomass is added. Then further biomass addition is stopped and the primary vent is closed when the biomass reaches just below the secondary air vents. Thereafter, water is sprinkled over it to drop down its temperature and the biochar is collected and stored on drying.

17.5.5 Biochar Stoves

Biochar stoves are still widely used by more than two billion people across the globe, particularly in the developing and underdeveloped energy-starved countries to cook food or heat their homes with by burning wood, dried dung, crop residues or coal. Such inefficient traditional heating practices cause air pollution that can exacerbate global warming and bring health issues such as cardiac arrest and respiratory congestions. The UN Environment Programme (UNEP) has identified the Atmospheric Brown Clouds (ABCs) as a major contributing factor in climate change (UNEP 2008) resulting mostly from a forest fire and inefficient anthropogenic biomass combustion. Inefficient combustion of biomass produces black particles (soot) that absorb sunlight and heat up the air mass while suspended white particles reflect back the incident solar radiation. Black carbon significantly contributes to global warming, next only to CO_2 (Ramanathan and Carmichael 2008). Even non-biochar making cookstoves emit huge volume of black carbon. Black carbon from rocket stove equals to that of from an open fire (MacCarty et al. 2008). However, modern science-based technologies sequester carbon very efficiently through production of heat along with biochar without much gas release. Gasifier stoves such as Top-Lit Updraft Gasifier (TLUD) (Fig. 17.5) and the Anila stove are reported to have very low black carbon emissions. Four basic stratified zones viz. raw biomass, flaming pyrolysis, gas and

Fig. 17.5 Schematic diagram of top-lit updraft gassifier (TLUD)

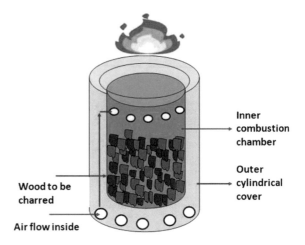

charcoal combustion are found in TLUD (Anderson and Reed 2004). If removed and quenched properly at right time then charcoal can also be obtained. During this process, the biomass is kept between two concentric cylindrical plates and a fire is ignited at the centre to pyrolyze the fuel in between the concentric rings. The gases from pyrolyzing fuel come out of the centre and they burn there to generate heat for cooking whereas the biomass becomes char (Srinivasarao et al. 2013). The modern *Anila stove* has been designed by U. N. Ravikumar of the Centre for Appropriate Rural Technology (CART) to take advantage of the huge biomass available in rural areas mostly in developing and underdeveloped countries and to minimize in-house air pollution that comes during cooking. The *Anila stove* works on the principle of top-lit updraft gasification. Hardwood fuel is lit at the top which burns downward and simultaneously combusts the released syngas. The stove is made from stainless steel and ordinarily weighs around 10 kg (Iliffe 2009). The IBI (Reddy 2011) has designed a fan-propelled biochar cooking stove that circulates air and liberates energy from the biomass for cooking and produces biochar in lesser quantity at the end of the process.

Three different pyrolysis reactors viz. kiln, retort and converter have been described by Emrich (1985) depending on the technology, size, purpose and the type of feedstock in use. In traditional biochar making process *kilns* are used solely to generate biochar. *Retorts* or reactors pyrolyzepile wood-log over 30 cm (length) × 18 cm (diameter) (Emrich 1985) whereas *converters* carbonize small biomass fragments like chipped or pelletized wood.

17.6 Economic Feasibility of Biochar Production

Application advantages of biochar for carbon sequestration, soil amendment, and bioremediation of heavy metals and organic pollutants are widely accepted however its large-scale use has been constrained by its high cost of production (Xiao et al. 2020). Slow pyrolysis of corn stover resulted in a higher yield of char (40% by weight) but with the lower gas release, while fast pyrolysis maximized bio-oil with lower biochar and gas yields as co-products (Brown et al. 2010). Anaerobic production of biochar from *Robinia pseudoacacia* biowaste demonstrated a low-cost of $20 t^{-1} (Xiao et al. 2020). As estimated in 2015, slow pyrolysis of corn stover was not profitable at offset value of biochar of $20 t^{-1} as feedstock cost was $83 t^{-1} in the USA while the fast pyrolysis resulted in 15% internal rate of return (IRR) as gasoline from bio-oil could value $2.96 per gallon gasoline-equivalent. By 2030, the carbon offset value of biochar is expected to rise to $60 t^{-1} and the gasoline price per gallon is presumed to reach $3.70 that could benefit investors with an IRR of 26% (Brown et al. 2010). A stochastic analysis of biochar production in Canada from spruce trees by slow pyrolysis mobile unit estimated fixed and variable cost of $505.14 and $499.13 t^{-1}. Its soil application @ 10 t ha^{-1} of carbon was reported to have increased the beet root yield from 2.9 to 11.4 t ha^{-1} with the maximum net profit of $11,288 ha^{-1} (Keske et al. 2019).

17.7 Effects of Biochar on Agriculture

17.7.1 Geomechanical Properties

Favourable soil tilth and an increase in root penetrability promote crop growth and yield (Jiang 2019). Although the literature on the biochar effects on soil tilth are rarely traceable but its ameliorative bio-physicochemical properties significantly improve soil tilth and tillage efficacy. Hseu et al. (2014) in a simulated rainfall experiment on biochar amendment in the degraded mudstone soil have observed increased macropores and reduced soil strength that invariably improved soil quality and physical properties for tilth. According to Snyder et al. (2009), reduced tillage requirements and residue retention due to biochar application significantly reduced GHG emission irrespective of the type of cultivation. Experiments conducted by Tim Crews (Cox 2013) in the Land Institute at Kansas revealed the importance of the 2000 year of the old practice of retaining soil nutrients that improves soil tilth too. Positive influence of biochar on soil tilth and soil aggregate stability has also been corroborated by Elad et al. (2010, 2011), Matt (2015), Yuniwati (2018) and Planet (2020).

Biochar can enhance the shear strength of clays and cyclic resistance of sand but can desaturate soil separates (Pardo et al. 2018). Sokolowska et al. (2020) in their experiment with wood waste and sunflower stick biochar experienced reduced tensile strength in all types of soils under test. Another experiment by Sadasivam and Reddy (2015) revealed a dramatic increase in cohesive strength of moist soil by almost thrice and an increase in shear strength of soil by incorporation of biochar at 10% (w/w) indicating induced stability to landfill covers. The above results were also corroborated by Reddy et al. (2015) with results showing positive relation between biochar amendments and geotechnical properties like hydraulic conductivity and shear strength of soil while compressibility had reverse relation. Looking at the paucity of information on the impact of biochar amendment on geomechanical properties, Renee (2019) has advocated for further intensive research for its effective geoenvironmental engineering applications.

17.7.2 Nutrient Dynamics

The role of biochar on nutrient dynamics in soil has already been touched upon earlier in this chapter. However, attempts are made in this section to review the research findings on the differential response of plants to varying levels of biochar applications only. Sukartono et al. (2011) have reported an increase in nutrient uptake in maize crop with the application of biochar. Olszyk et al. (2020) in their experiment reported variation in concentration of Ca, K, Mn, Mg, Zn and Fe in carrot taproot and lettuce leaf depending on the biochar type. The Ca, Mg and Zn were the most influenced and the concentration of K increased in the taproot system of carrot. The addition of corn stover biochar increased the uptake of macronutrients both in presence and absence of chemical fertilizers but switch-grass biochar had no effect on macronutrient uptake and pinewood biochar reduced the uptake (Chintala et al. 2013). The importance of P and K for the increase in crop productivity was revealed by Karer et al. (2013a) in an experiment on barley that resulted in reduced N uptake while P and K uptake improved with the biochar addition. In corn, omission of biochar from integrated chemical fertilizer application had at par effects on N, P and K uptake rates. However, the reduction in yield was severe under deficient N supply (Karer et al. 2013a). The uptake of nutrients in rice as studied by Ali et al. (2015) indicated a positive response of biochar on Ca, K, Mg, Cu and Mn uptake over control while the uptake of Zn, N and crude silica did not differ significantly. Moreover, the uptake of Fe was higher under normal fertilization than biochar supplementation in rice soil (Ali et al. 2015). In chickpea, application of maize stover biochar prepared by batch-wise hydrothermal carbonization (210 °C) had recorded better uptake of N, P, K and Mg than the biochar produced at 600 °C (Dilfuza et al. 2019). The utilization of biochar not only increased the growth of calendula (*Calendula officinalis* L.) but also increased the acquisition of macro and microelements from the soil (Karimi et al. 2020). A comparative report on the changes in chemical properties under biochar and cattle manure amendment in maize crop has been depicted at Table 17.1 for better understanding (Sukartono et al. 2011).

17.7.3 Disease Pest Infestation

Very few disease control methods are available to manage soil-borne pathogens whereas biochar has been successfully tested to fight against major diseases in fruit, vegetables, ornamental plants, trees, shrubs, etc. Elad et al. (2010) in their experiments with biochar amendments to the soil observed antagonistic effects against foliar fungal pathogens such as grey mould (*Botrytis cinerea*) and powdery mildew (*Leveillula taurica*) in pepper and tomato and to the broad mite pest (*Polyphagotarsonemus latus*) in pepper. In another experiment with biochar, they reported a shift in the bacterial community that could contribute to the resistance against bacterial wilt in tomato. The soil amendment with biochar altered microbial population and caused a shift towards beneficial microbial populations that promoted plant growth and induced resistance against soil-borne diseases (Lad et al. 2011). Graber et al. (2014) reported resistance of plants to pathogens in a U-shaped response curve depending on the dose of biochar, with a minimum disease outbreak at intermediate dose but severe effects at both the minimum and maximum doses. However, a relatively lower incidence of damping-off was seen in lower doses of biochar but at higher or moderate doses, the severity was similar to untreated control. Biochar has been affecting the progress of soil-borne diseases such as *Fusarium oxysporum* in asparagus (Elmer and Pignatello 2011), *Ralstonia solanaceaearum* in tomato (Nerome et al. 2005) and *Rhizoctonia solani* in cucumber (Jaiswal et al. 2014). Suppression of canker causing *Phytophthora* in woody plants was reported by Zwart and Kim (2012) under biochar addition to the soil.

17.7.4 Weed Dynamics

Study on weed dynamics is important, especially because, biochar can reduce the efficacy of herbicides. Many researchers have advocated for enhanced crop yield and ameliorative effects of biochar addition on bio-physicochemical properties of soil. Biochar has minimal effect on weed germination and emergence pattern as reported by Soni et al. (2015). Biochar mediated reduced germination and subsequent infestation of *Phelipanche aegyptiaca* (Egyptian broomrape), a weed in tomato has also been reported (Dilfuza et al. 2019). An increase in height and above-ground biomass of pig-weed and crabgrass was observed that might complicate the weed management strategy in biochar amended crop fields (Mitchell 2015). In a four year experiment with walnut shell biochar at 5 t ha^{-1}, 60–78% higher weed density was reported by Safaei et al. (2020) indicating more efficient utilization of macro and micronutrients by weeds compared with wheat and lentil crop. However, the reduced air-dry weight of weeds compared to the control plots in the rye crop grown with biochar has been reported in Poland (Kraska et al. 2016).

Preemergence herbicides are usually applied to the soil before the emergence of crop that might increase the adsorption of the applied herbicides by biochar thereby reducing efficacy. An experiment conducted by Soni et al. (2015) by incorporating biochar at 2 t ha^{-1} completely suppressed the herbicidal effects of atrazine and pendimethalin in corn crop due to the presence of organic carbon and higher surface area in biochar that resembled activated carbon thereby reducing the herbicidal efficacy (Soni et al. 2015). In another experiment, recommended dose of pendimethalin at 1 kg a.i. ha^{-1} along with biochar reduced grain yield of direct-seeded rice by 7.5% compared to pendimethalin without biochar. A higher dose of pendimethalin also reduced the biological yield of rice (Nath 2016). Hence, alternative weed management practices should be adopted for eradicating preemergence weeds in biochar amended soil (Sohi et al. 2010).

17.7.5 Water Use Efficiency

Biochar has the benefit of increasing water use efficiency (WUE) and water retention in soil (Monnie 2016; Dwibedi et al. 2022) at varying degrees depending on soil type, biochar characteristics and climatic parameters (Gao et al. 2020). Remarkable positive influence of biochar application on the WUE have also been observed by Benjamin et al. (2016), Lusiba et al. (2018) and Zhang et al. (2020). An experiment with corn cob biochar showed no remarkable effect on the water retention curve in sandy loam soil up to 20 t ha^{-1} but only at 80 t ha^{-1} the effect was significant (Monnie 2016). However, large application of biochar at 200 t ha^{-1} in sandy soil did not promote plant growth compared to 100 t ha^{-1} thereby fixing the upper limit of its beneficial effects (Kammann et al. 2011). This observation was corroborated by the result from low magnitude applications (1 and 2% of biochar in soil) that although slightly increased the water holding capacity but the effect was not sufficient to mitigate deficit moisture stress condition for which application with the higher rate was perhaps necessary (Afshar et al. 2016).

In the changing climatic scenario, it is imperative to develop a water balance agricultural method to improve resilience to climatic variability. A meta-analysis of observational data on biochar amendment revealed an increase in long-term evapotranspiration rates thereby increasing soil water retention capacity and water availability to crop (Benjamin et al. 2016). An increase in plant resistance to water stress (60% field capacity) was observed in biochar amended soil compared to the control (without biochar) (Aniqa et al. 2015). However, the negative effects of biochar on plant water availability are also cited by Fischer et al. (2019).

17.7.6 Crop Growth and Yield

Biochar has synergistic effects on crop growth and yield (Dwibedi et al. 2022). Its application in Chernozem soil significantly increased spinach (*Spinacia oleracea* L.) in terms of growth by 102 and 353% in spring and autumn, respectively (Zemanovai et al. 2017). In high drought-affected Chernozem soil, biochar application at 72 t ha^{-1} along with chemical N could increase barley crop yield by 10% compared to the control with N fertilizer but without biochar. However, reduction in maize and wheat grain yields by 46% and 70% at biochar application rate beyond 72 t ha^{-1} has been reported by Karer et al. (2013b). A single application of biochar at 20 t ha^{-1} to Colombian savanna soil increased maize yield by 28–140% compared to unamended control (Major et al. 2010). Perhaps the nutrient adsorptive capacity and antiallelopathic effects of biochar at 18 t ha^{-1} resulted in higher germination percentage, germination index and mean germination time of garden pea (*Pisum sativum* L.) seeds (Berihun et al. 2017) while biochar at 10 and 20 t ha^{-1} had positive influence on *Lepidium sativum* L. seed germination (Kraskal et al. 2016). Results of enhanced growth and yield parameters of bean (da Silva et al. 2017), wheat (Sial et al. 2019), maize (Zhu et al. 2015), rice (Muhammad et al. 2017), winter rye (Kraskal et al. 2016), sunflower (Qiang et al. 2020), and tomato (Yilangai et al. 2014) with positive effects on the plant height, root, shoot and grain dry mass, number of pods and/or grains due to application of biochar have also been reported. However, short-term application of biochar did not have any effect on grain yield or yield components of rice as reported by Yin et al. (2020). Rosenani et al. (2014) in their experiments with rice husk biochar reported higher biomass in *Amaranthus viridis* and *Ipomoea reptans* while no significant increase in yield was observed in sweet corn, except increase in total dry matter. The grain yield increase in cowpea with biochar amendment was irrespective of soil moisture regimes while the highest grain yield was reported under no-water deficit stress (Moosavi et al. 2020). However, significant interaction between biochar and maize productivity under limited water supply might prove a novel approach in enhancing yield as well as WUE (Faloye et al. 2019). Biochar amendment at 20 and 40 t ha^{-1} in rain fed region of North China although could significantly increase grain yield of maize by 23.9% and 25.3%, respectively with positive effects on root morphology and stalk biomass but its effects in the second year was not significant (Liu et al. 2020).

Application of biochar has also been influencing the cropping system as well (Dwibedi et al 2022). In rice–wheat system, Gupta et al. (2020) have reported higher grain yields for three consecutive years due to application of rice straw biochar and rice husk biochar at 5 t ha^{-1}. Significant positive correlation between N, P and K concentration in soil with total N, P and K in wheat indicated potential benefits of biochar application in supplementing plant nutrients in desired quantities (Gupta et al. 2020).

17.7.7 Climate Change

Carbon is an important basic constituent of all living organisms on this earth. Man is hunting for the traces of carbon in extraterrestrial bodies to explore any possible existence of life. On this earth, it cycles among the atmosphere, biosphere, hydrosphere and lithosphere in many forms. In the earth's atmosphere, carbon is present mostly as methane and carbon dioxide. The earth's largest carbon pool is found in the continental crusts and upper mantle, a large portion is present in form of sedimentary rocks. Oceanic carbon is the next largest stock, over 95% are present in inorganic dissolved carbon and only 5% (900 gigatons) of carbon (GtC) is available for exchange in the ocean surface (Kayler et al. 2017). The atmosphere contains only

839 GtC, a very small portion of total carbon but it plays a very significant role. Near about 19% of the carbon in earth's biosphere is stored in plants, and the rest remains in soil (FAO 2021). Soils contain 1325 GtC of top few feet and as much as 3000 GtC in total (Kayler et al., 2017). Oil and natural gases contain 270 and 260 GtC, respectively. Coal reserve accounts for 5000–8000 GtC and unconventional fossil fuels have whooping 15,000–40,000 GtC (Edmonds et al. 2004).

The concentration of CO_2 in the troposphere has elevated by 45%, from 280 ppm in 1750 to 415 ppm in 2019, due to the industrial revolution. The level of CO_2 has reached at this mark again after 3 million years, despite due absorption by various sinks involved in natural cycles. The earlier peak was natural and steady that had spread over many hundreds or even thousands of years allowing necessary adaption and adaptation by different species while the present rise is sudden and anthropogenic leading to mass extinctions of some life forms due to climate change.

Burning of fossil fuel, agricultural wastes and forest vegetation release fixed and structural carbons into the atmosphere elevating the CO_2 concentration of the atmosphere. Every year 30 GtC is fixed by crop plants, while on dying, it may return back to the atmosphere, resulting in little net change in soil carbon pool (Krounbi et al. 2019). Wildfires are estimated to add 8 billion tons of CO_2 every year for last 20 years and in 2017, the total CO_2 emission reached 32.5 billion tons as estimated by the International Energy Agency (Berwin 2018). In 2014, forest fires released 8.8 million tons of carbon compared to 104 million tons from all fires (Merzdorf 2019). Scientists have claimed that wildfires contribute less carbon than burning of fossil fuels, citing 15 years of carbon release from the wildfires in US at only 250 Gt as against fossil fuel contribution of 4800 GtC each year (Francovich 2019). However, their real worry began with the peatland fire in Indonesia in 1997–98 that released 3.7 billion tons of CO_2. Permafrost thaw due to global warming and climate change has increased the risks of uncontrolled fires in the northern peat that was previously not vulnerable to such hazard (Khadka 2018). Hence, issue of wildfires will be more challenging than mitigating the burning of fossil fuels in the future (Khadka 2018).

However, attempts to sequester significant amounts of free atmospheric carbon through afforestation and reforestation in forest fire affected areas are not successful in many cases due to global warming and related consequences. Restoration of the original wild biodiversity in such charred areas is quite difficult and time consuming. Many native species would be able to survive under changing climate due to mismatch with their physiological optima. Systematic planning and consistent efforts are required for altering the challenging and perilous effects of global warming and climate change.

Biochar can significantly smother climate change by reducing atmospheric GHG levels, and sequestering carbon dioxide. It can also increase productivity of marginal soils, reduce soil erodibility, recharge groundwater, reshapes soil biodiversity, regenerates natural vegetation and many more synergistic effects it can have in the line of sustainable agriculture and environment. Estimates reveal that application of biochar can reduce 12% of the global GHGs and doping of potassium can enhance carbon sequestration potential by 45% (Masek et al. 2019). Biochar in soil not only fixes atmospheric CO_2 but also ameliorates soil that facilitates plant growth. It induces dark colour in the topsoil, like *terra preta* of Amazon basin, which absorbs much incident solar radiation during daytime and reradiates it back as long-wave radiation during night thereby maintaining a steady range of diurnal temperature. Its presence in soil not only marginalizes diurnal air temperature but the soil temperature is maintained which protects vegetation against harmful effects of low temperature. Biochar is carbon negative and hence it can bring back the carbon from active cycle and sequester in an inactive native cycle that slows down the process of global warming and climate change (USBI News 2021b).

Studies on the application of biochar with poultry manure in maize (*Zea mays*) in rotation with soybean (*Glycine max*) in Canada showed a positive influence on carbon and nitrogen transformation in the soil–plant-atmosphere system (Mechler et al. 2018). In another experiment under soybean in Ohio, USA the cumulative N_2O emission over the growing period decreased by 92% in the biochar-amended soil compared to the control (without biochar) while the total cumulative CH_4 and CO_2 emissions did not get affected by any such amendment. Biochar amendment resulted in net soil carbon gain whereas humic acid and water treatment residual resulted in net soil carbon loss. However, all three amendments subsided the global warming potential (Mukherjee et al. 2014). A *meta*-analysis of Timmons et al. (2017) published papers with 552 paired comparisons conducted by He et al. (2017) indicated 22.14% increase in soil CO_2 fluxes, but 30.92% decrease in N_2O fluxes while CH_4 fluxes remained unaltered. However, under soil fertilization, the CO_2 fluxes were suppressed which implies that biochar is unlikely to stimulate CO_2 fluxes in the agriculture sector (He et al. 2017).

17.8 Future Prospects and Constraints in Biochar Systems

The significance of biochar in environmental remediation and agricultural production systems is now an undoubted fact. However, its in-depth study on ISO-based life cycle assessments in various systems has not yet been well attended. The potential of biochar and biochar systems is manifold. It can be potentially linked to many sectors for green-growth, development and climate resilience. Decision tools based on local environmental, agricultural, social constraints and opportunities requirement need to be designed and validated to select befitting biochar system technologies (Scholz et al. 2014).

17.8.1 Scaling up from Pilot to Programme

Biochar systems are nascent technologies in-spite of their wide adoption by many older civilizations. As of now, many researchers have intensively studied various biophysicochemical properties of biochar synthesized from different feedstock, ranging from wastes to wood under varying pyrolytic conditions. However, most of them are either laboratory- or GHG-based experimentations lacking wider replicability in the farmers' fields to adjudge their effects extensively. A deeper insight into the economic benefits of the carbon trading of biochar systems overweighs the return from crop growth. So also, due to lack of applicable methodologies and legislative yardstick to regulate the targeted source of feedstock, the engagement of private sector is unlikely to exist in larger scale, at least in the present scenario in most developing countries. Therefore, it is high time for the institutions such as World Bank, International Finance Corporation, Global Environment Facility and many other international and national institutions to test-demonstrate various sustainable biochar production systems across the globe prioritizing the economically deprived but resource stuffiest countries.

17.8.2 Further Research Needs

The quantum of funds pumped towards research and demonstration has not yet reached at its desired level to scale up biochar systems comfortably. Among the areas of further research, effective targeting of the 'true wastes' that degrade the environment in absence of judicious and alternative uses is of prime importance now. Furthermore, development of low cost pyrolysis units befitting to the socioeconomically deprived countries is an area for future research. Critical assessment of biochar application process and their bio-physicochemical effects on the soil and crop yields also deserve deeper attention. Characterization of biochar and their bio-physicochemical properties, depending on feedstocks, pyrolysis temperature and duration, would allow better prediction of soil fertility, target crops and soil types to which these biochars could be allowed. The farmers may be directly involved or the knowledge will be made available through intermediary extension service systems, preferably in the developing countries first. Moreover, social aspects of biochar system related technologies need further attention as certain biochar systems would increase drudgery that in turn would discourage the farmers, and farm women in particular, in adopting them. As biochar systems at higher rate of application, in many instances, may not be financially sustainable for small and marginal farmers, small-scale experimental use in limited areas could remediate soil and enhance crop productivity in a time series perspective. As biochar systems aim at 'triple win promise' viz. energy, climate and soil but no such evidence satisfies universal conditions without considering local conditions (Scholz et al. 2014). Hence, long-term applied research at scale of implementation could essentially resolve this problem.

17.8.3 Constraints and Risks

While considering the feasibility of biochar production and management systems certain key questions need to be addressed. Firstly, will the biomass be honestly sourced from the true waste materials? While answering this question a comparative analysis with alternative waste disposal systems should be performed giving importance to the energy capture and nutrient-recycling unlike open burning and land filling. Further question of safe-feedstock use could be addressed through incorporation of non-toxic rural and agricultural wastes in biochar systems and deliberately avoiding the industrial and urban wastes. However, the risk of rampant deforestation and cleaning of natural vegetation can never be set aside under lucrative government incentives to popularize biochar systems. Next challenge could be sufficient availability of suitable feedstock locally and its economic feasibility in long run. Such a challenge could be sorted out by indicators like sustainable availability of feedstock on-farm and its potential use in high value crops in intensive cropping system. Furthermore, the risk of methane, carbon oxide and other toxic volatile fume release must be addressed meticulously to safeguard global environment. A site specific biochar application repository could scientifically address variable soil and crop requirements. The constraint of non-adoption of technology in post demonstration phase could be ascribed to drudgery and valuable alternate energy services.

17.9 Conclusion

Application of biochar is an ancient practice of soil conditioning and sustainable yield enhancement. It ameliorates and improves physicochemical properties of soils, facilitates nutrient availability and enhances plant growth and yield, rendering it most suitable for organic, dryland and conservation agriculture and land reclamation. Its lower production cost from locally available biowastes could lead-support resource poor small and marginal farmers as an intriguing option in crop production. Although research results on biochar application are alluring but most of them are laboratory or greenhouse-based, lacking wider adaptability in open field conditions. Even today, large knowledge gaps on persistence, bio-geochemical cycles, GHG regulations, microbial behaviour and metal retention period are still lacking that need to be addressed in full-scale outdoor trials. Crop specific tailored biochar dose recommendations based on biochar feedstocks, pyrolytic conditions and soil type need to be designed.

Table 17.1 Soil characteristics of sandy loam at Lombok, Indonesia after application of biochar and cattle manure under maize cropping system (Sukartono et al. 2011)

Organic amendments	pH		CEC (mg kg^{-1})		C (mg kg^{-1})		N (mg kg^{-1})		P (mg kg^{-1})		K (cmol kg^{-1})		Ca (cmol kg^{-1})		Mg (cmol kg^{-1})	
	1st	2nd	1st	2nd	1st	2nd	1st	2nd	1st	2nd	1st	2nd	1st	2nd	1st	2nd
Coconut shell biochar	6.49^a	6.46^a	15.04^a	15.15^a	1.15^a	1.13^a	0.12^b	0.14^{ab}	26.48^a	22.39^{ab}	0.75^b	0.78^a	2.44^a	2.54^{ab}	1.42^b	1.54^b
Cattle dung biochar	6.45^a	6.46^a	15.1^a	15.14^a	1.14^a	1.11^a	0.16^a	0.15a^b	26.24^a	21.67^{ab}	0.89^a	0.78^a	2.6^b	2.78^b	1.5^a	1.53^b
Cattle manure once	6.39^b	6.36^b	15.02^a	14.67^{ab}	0.9^b	0.94^{ab}	0.14^{ab}	0.13^b	25.66^a	20.95^b	0.76^b	0.71^b	2.38^a	2.15^a	1.4^b	1.45^b
Without amendment	6.29^c	6.32^b	13.34^b	13.4^b	0.87^b	0.89^b	0.11^b	0.13^b	23.59^b	14.44^c	0.7^c	0.7b	2.22^c	2.08^a	1.37^b	1.32^c
Before expt.*	5.97	–	12.99	–	0.85	–	0.12	–	24.41	–	0.57	–	2.34	–	0.87	–

Mean with the same superscript letters within column do not differ significantly ($p = 0.05$); 1st and 2nd denote rainy season (2010–11) and dry season (2011) maize crops; * pre-treatment data

Acknowledgements Authors are highly thankful to the Dean, College of Agriculture, Odisha University of Agriculture and Technology, Bhubaneswar and the Dean, Institute of Agricultural Sciences, Siksha 'O' Anusandhan University (Deemed) for extending their kind support throughout preparation of this book chapter.

Funding No external funding was received for this review assignment.

Conflict of Interest There is no conflict of interest among the authors.

Data Availability Statement The authors confirm that the data supporting the findings of this study are available within the article.

References

Afshar RK, Hashemi M, DaCosta J, Spargo M, Sadeghpour A (2016) Biochar application and drought stress effects on physiological characteristics of *Silybum marianum*. Commun Soil Sci Plant Anal 47(6):743–752. https://doi.org/10.1080/00103624.2016.1146752

Ajema A (2018) Effects of biochar application on beneficial soil organism review. Int J Res Stud Sci Eng Tech 5(5):9–18. http://ijrsset.org/pdfs/v5-i5/2.pdf.

Ali M, Haruna OA, Charles W (2015) Coapplication of chicken litter biochar and urea only to improve nutrients use efficiency and yield of *Oryza sativa* L. cultivation on a tropical acid soil. Sci World J 943853. https://doi.org/10.1155/2015/943853

Amonette J, Joseph S (2009) Characteristics of biochar: micro-chemical properties. In: Lehmann J, Joseph S (eds) Biochar for environmental management: science and technology. Earth Scan, London. pp 33–52. https://www.osti.gov/biblio/985016

Anderson P, Reed TB (2004) Biomass gasification: clean residential stoves, commercial power generation, and global impacts, prepared for the LAMNET project international workshop on "bioenergy for a sustainable development," 8–10 Nov 2004, Viña del Mar, Chile. https://biochar-international.org/stoves/

Aniqa B, Samia T, Audil R, Azeem K, Samia Q, Aansa S, Muhammad G (2015) Potential of soil amendments (Biochar and Gypsum) in increasing water use efficiency of *Abelmoschus esculentus* L. Moench. Front Plant Sci 6. https://doi.org/10.3389/fpls.2015.00733

Benjamin MC, Fischer SM, Morillas L, Garcia M, Mark S, Johnson S, Lyon W (2016) Improving agricultural water use efficiency with biochar—a synthesis of biochar effects on water storage and fluxes across scales. Sci Total Environ 657:853–862. https://doi.org/10.1016/j.scitotenv.2018.11.312

Bera T, Purakayastha TJ, Patra AK, Datta SC (2018) Comparative analysis of physicochemical, nutrient, and spectral properties of agricultural residue biochars as influenced by pyrolysis temperatures. J Mater Cycles Waste 20:1115–1127. https://doi.org/10.1007/s10163-017-0675-4

Berihun T, Tolosa S, Tadele M, Kebede F (2017) Effect of biochar application on growth of garden pea (*Pisum sativum* L.) in acidic soils of Bule Woreda Gedeo Zone Southern Ethiopia. Int J Agron 6827323. https://doi.org/10.1155/2017/6827323

Berwin B (2018) How wildfires can affect climate change (and vice versa). *Inside Climate News*. https://insideclimatenews.org/news/23082018/extreme-wildfires-climate-change-global-warming-air-pollution-fire-management-black-carbon-co2. 23 Aug 2018

Biederman LA (2012) Biochar and its effects on plant productivity and nutrient cycling: a meta-analysis. GCB Bioenergy 5(2):202–214. https://doi.org/10.1111/gcbb.12037, 31 Dec 2012

Biochar International (2021) Basic principle of biochar making. https://biochar.international/guides/basic-principles-of-biochar-production/, 10 Mar 2021

Brown NC (1917) The hardwood distillation industry in New York. The New York State College of Forestry at Syracuse University, Jan 2017

Brown TR, Wright MM, Brown RC (2010) Estimating profitability of two biochar production scenarios: slow pyrolysis vs. fast pyrolysis. Biofuels Bioproducts Biorefining. https://www.abe.iastate.edu/wp-content/blogs.dir/7/files/2010/09/Profitability-of-Pyrolysis.pdf, 7 Dec 2010

Chen W, Meng J, Han X, Zhang W (2019) Past, present, and future of biochar. Biochar 1:75–87. https://doi.org/10.1007/s42773-019-00008-3

Cheng CH, Lehmann J, Thies JE, Burton SD (2008) Stability of black carbon in soils across a climatic gradient'. J Geophys Res 113:G02027. https://doi.org/10.1029/2007JG000642

Cheng S, Chen T, Xu W, Huang J, Jiang S, Yan B (2020) Application research of biochar for the remediation of soil heavy metals contamination: a review. Molecules (basel, Switzerland) 25(14):3167. https://doi.org/10.3390/molecules25143167

Chintala R, Gelderman RH, Schumacher TE, Malo DD (2013) Vegetative corn growth and nutrient uptake in biochar amended soils from an eroded landscape. Joint annual meeting of the association for the advancement of industrial crops and the USDA National Institute of Food and Agriculture, At: Washington D.C., October 2013. https://www.hort.purdue.edu/newcrop/proceedings2015/200-chintala.pdf

Codur AM, Itzkan S, Moomaw W, Thidemann K, Jonathan H (2017) Conserving and regenerating forests and soils to mitigate climate change. In: Climate policy brief no.4, Global Development and Environment Institute, Tufts University, https://sites.tufts.edu/gdae/files/2019/10/Codur_ConservingRegeneratingForestsSoils.pdf, 6 Dec 2017

Cornet A, Escadafal R (2009) Is biochar "green"? CSFD Viewpoint. Montpellier, France, 8 pp. http://www.csf-desertification.eu/combating-desertification/item/is-biochar-green#:~:text='Biochar'%2C%20which%20is%20short,sawmill%20waste%20or%20agricultural%20residue.&text=It%20is%20mostly%20composed%20of,and%20the%20pyrolysis%20process%20used

Cox S (2013) Biochar: Not all it's ground up to be? Mother Earth News. https://www.motherearthnews.com/nature-and-environment/biochar-not-all-its-ground-zb0z1307

da Silva ICBL, Fernandes A, Colen F, Sampaio RA (2017) Growth and production of common bean fertilized with biochar. Crop Production, Ciência Rural 47(11). https://doi.org/10.1590/0103-8478cr20170220

Day D, Evans RJ, Lee J, Reicosky D (2012) Methods of producing biochar and advanced biofuels in Washington state Part 1: literature review of the pyrolysis reactors pdf. http://pacificbiomass.org/documents/1207034.pdf

Dwibedi SK, Pandey VC, Divyasree D, Bajpai O (2022) Biochar-based land development. Land Degrad Dev. https://doi.org/10.1002/ldr.4185

Demirbas A (2004) Effects of temperature and particle size on biochar yield from pyrolysis of agricultural residues. J Anal Appl Pyrol 72:243–248. https://doi.org/10.1016/j.jaap.2004.07.003

Dilfuza E, Hua LL, M, Stephan W, Dorothea BKS, (2019) Soil amendment with different maize biochars improves chickpea growth under different moisture levels by improving symbiotic performance with mesorhizobium ciceri and soil biochemical properties to varying degrees. Front Microbiol 10:2423. https://doi.org/10.3389/fmicb.2019.02423

Edmonds J, Joos F, Nakicenovic N, Richels R, Sarmiento J (2004) Scenarios, targets, gaps and costs. In: Field BC, Raupach MR (eds) The global carbon cycle: integrating humans, climate and the natural world. Island Press, pp 77–102. ISBN 9-781-55963526-4

Elad Y, Cytryn E, Harel YM, Lew B, Graber ER (2011) The biochar effect: plant resistance to biotic stresses. Phytopatho Mediterr 50:335–349. https://doi.org/10.14601/Phytopathol_Mediterr-9807

Elad Y, David DR, Harel YM, Borenshtein M, Kalifa HB, Silber A, Graber ER (2010) Induction of systemic resistance in plants by biochar, a soil-applied carbon sequestering agent. Dis Control Pest Manag 100(9):913. https://doi.org/10.1094/PHYTO-100-9-0913

Elkan D (2004) Fired with ambition. The Guardian. https://www.theguardian.com/society/2004/apr/21/environment.environment, 21 Apr 2004

Elmer WH, Pignatello JJ (2011) Effect of biochar amendments on mycorrhizal associations and fusarium crown and root rot of asparagus in replant soils. Plant Dis 95(8):960–966. https://doi.org/10.1094/PDIS-10-10-0741

Emrich W (1985) Handbook of biochar making. The traditional and industrial methods. D. Reidel Publishing Company. https://www.springer.com/gp/book/9789048184118

FAO (2021, March 10) Forests and climate change. Retrieved from http://www.fao.org/3/ac836e/AC836E03.htm#:~:text=At%20the%20global%20level%2C%2019,69%20percent%20in%20the%20soil

Faloye OT, Alatise MO, Ajayi AE, Ewulo BS (2019) Effects of biochar and inorganic fertiliser applications on growth, yield and water use efficiency of maize under deficit irrigation. Agric Water Manage 217(C):165–178. https://doi.org/10.1016/j.agwat.2019.02.044

Fischer B, Manzoni S, Morillas L, Garcia M, Johnson MS, Lyon SW (2019) Can biochar improve agricultural water use efficiency? Geophys Res Abstr 21, EGU2019-7358. https://meetingorganizer.copernicus.org/EGU2019/EGU2019-7358.pdf

Francovich E (2019, June 7) Forest fires release less CO_2 than previously thought, challenging some forest management practices, study says. The Spokesman-Review. Retrieved from https://www.spokesman.com/stories/2019/jun/07/forest-fires-release-less-co2-than-previously-thou/

Gao Y, Shao G, Lu J, Zhang K, Wu S, Wang Z (2020) Effects of biochar application on crop water use efficiency depend on experimental conditions: a meta-analysis. Field Crops Res 249:107763. https://doi.org/10.1016/j.fcr.2020.107763

Glaser B, Haumaier L, Guggenberger G, Zech W (2001) The 'Terra Preta' phenomenon: a model for sustainable agriculture in the humid tropics. In: McEwan C, Barreto C, Neves EG (eds) Unknown Amazon: culture in nature in ancient Brazil. British Museum Press, London, pp 86–105. https://doi.org/10.1007/s001140000193

Graber ER, Frenkel O, Jaiswal AK, Elad Y (2014) How may biochar influence severity of diseases caused by soilborne pathogens? Carbon Management 5(2):169–183. https://doi.org/10.1080/17583004.2014.913360

Grand View Research (2018) Biochar market size, share & trends analysis report by technology (gasification, pyrolysis), by application (agriculture (farming, livestock)), by region, and segment forecasts, 2019–2025. https://www.grandviewresearch.com/industry-analysis/biochar-market

Gupta R, Hussain AY, Sooch S, Kang J, Sharma S, Dheri G (2020) Rice straw biochar improves soil fertility, growth, and yield of rice–wheat system on a sandy loam soil. Exp Agr 56(1):118–131. https://doi.org/10.1017/S0014479719000218

Harder B (2006) Smoldered-Earth policy: created by ancient Amazonia natives, fertile, dark soils retain abundant carbon. Sci News 169:133. https://doi.org/10.2307/3982299

He Y, Zhou X, Jiang L, Li M, Du Z, Zhou G, Shao J, Wang X, Xu Z, Bai SH, Wallace H, Xu C (2017) Effects of biochar application on soil greenhouse gas fluxes: a meta-analysis. Bioenergy 9:743–755. https://doi.org/10.1111/gcbb.12376

Hseu Z, Jien S, Chien W, Liou R (2014) Impacts of Biochar on Physical Properties and Erosion Potential of a Mudstone Slopeland Soil. Sci World J 2014:602197. https://doi.org/10.1155/2014/602197

Iliffe R (2009) Is the biochar produced by an Anila stove likely to be a beneficial soil additive? UKBRC Working Paper 4. https://www.biochar.ac.uk/abstract.php?id=15

International Biochar Institute (2021) Biochar Production Technologies. https://biochar-international.org/biochar-production-technologies/, 10 Mar 2021

Jaiswal AK, Elad Y, Graber ER, Frenkel O (2014) *Rhizoctonia solani* suppression and plant growth promotion in cucumber as affected by biochar pyrolysis temperature, feedstock and concentration. Soil Bio Biochem 69:110–118. https://doi.org/10.1016/j.soilbio.2013.10.051

Jiang R (2019) The effect of biochar amendment on the health, greenhouse gas emission, and climate change resilience of soil in a temperate agroecosystem. Master's thesis, Geography and Environmental Management, University of Waterloo. Retrieved from https://uwspace.uwaterloo.ca/bitstream/handle/10012/15169/Jiang_Runshan.pdf?isAllowed=y&sequence=7

Jien SH, Wang CS (2013) Effects of biochar on soil properties and erosion potential in a highly weathered soil. CATENA 110:225–233. https://doi.org/10.1016/j.catena.2013.06.021

Kammann C, Linsel S, Goessling JW, Koyro HW (2011) Influence of biochar on drought tolerance of Chenopodium quinoa Willd and on soil-plant relations. Plant Soil 345(1):195–210. https://doi.org/10.1007/s11104-011-0771-5

Karer J, Wimmer B, Zehetner F, Kloss S, Soja G (2013a) Biochar application to temperate soils: effects on nutrient uptake and crop yield under field conditions. Agric Food Sci 22:390–403. https://doi.org/10.23986/afsci.8155

Karer J, Zehetner F, Kloss S, Wimmer B, Soja G (2013b) Nutrient uptake by agricultural crops from biochar-amended soils: results from two field experiments in Austria. EGU General Assembly in Vienna, Austria, id. EGU2013b-4974. https://ui.adsabs.harvard.edu/abs/2013bEGUGA..15.4974K/abstract

Karimi E, Shirmardi M, Ardakani MD, Gholamnezhad J, Zarebanadkouki M (2020) The effect of humic acid and biochar on growth and nutrients uptake of Calendula (*Calendula officinalis* L.). Commun Soil Sci and Plant Anal 51(12):1–12. https://doi.org/10.1080/00103624.2020.1791157

Kayler Z, Janowiak M, Swanston C (2017) Global carbon. U.S. Department of Agriculture, Forest Service, Climate Change Resource Center. https://www.fs.usda.gov/ccrc/topics/global-carbon, June 2017

Keske C, Godfrey T, Hoag DLK, Abedin J (2019) Economic feasibility of biochar and agriculture coproduction from Canadian black spruce forest. Food Energy Secur 9(4). https://doi.org/10.1002/fes3.188

Khadka NS (2018) Climate change: worries over CO_2 emissions from intensifying wildfires. BBC News. https://www.bbc.com/news/science-environment-46212844, 15 Nov 2018

Klark M, Rule A (1925) The technology of wood distillation. Londin Chapman & Hill Ltd.

Kraska P, Oleszczuk P, Andruszczak S, Kwiecińska-Poppe1 E, Różyło1 K, Pałys1 E, Gierasimiuk P, Michałojć Z (2016) Effect of various biochar rates on winter rye yield and the concentration of available nutrients in the soil. Plant Soil Environ 62(11):483–489. https://doi.org/10.17221/94/2016-PSE

Krounbi L, Enders A, van Es H, Woolf D, von Herzen B, Lehmann J (2019) Biological and thermochemical conversion of human solid waste to soil amendments. Waste Manage 89:366–378. https://doi.org/10.1016/j.wasman.2019.04.010

Lad Y, Cytryn E, Harel Y, Lew B, Ggrbber E (2011, March 19) The biochar effect: plant resistance to biotic stresses. Phytopathol Mediterr 50(3):335–349. https://www.jstor.org/stable/26556455

Lahori AH, Zhanyu G, Zhang Z, Li R, Mahar A, Awasthi M, Shen F, Sial TA, Kumbhar F, Wang P, Jiang S (2017) Use of biochar as an amendment for remediation of heavy metal contaminated soils: prospects and challenges. Pedosphere 27:991–1014. https://doi.org/10.1016/1002-0160(17)60490-9

Lehmann J (2009) Terra preta Nova-where to from here? In: Woods WI, Teixeira WG, Lehmann J, Steiner C, WinklerPrins A (eds) Terra Preta Nova: a tribute to Wim Sombroek. Springer, Berlin, pp 473–486. https://doi.org/10.1007/978-1-4020-9031-8_28

Lehmann J, Gaunt J, Rondon M (2006) Bio-char sequestration in terrestrial ecosystems-a review. Mitig Adapt Strat Glob Change 11:403–427. https://doi.org/10.1007/s11027-005-9006-5

Liu X, Wang H, Liu C, Sun B, Zheng J, Bian R, Drosos M, Zhang X, Li L, Pan G (2020) Biochar increases maize yield by promoting root growth in the rainfed region. Arch Agron Soil Sci. https://doi.org/10.1080/03650340.2020.1796981

Lusiba S, Odhiambo J, Ogola J (2018) Growth, yield and water use efficiency of chickpea (*Cicer arietinum*): response to biochar and phosphorus fertilizer application. Arch Agron Soil Sci 64(6):819–833. https://doi.org/10.1080/03650340.2017.1407027

MacCarty N, Ogle D, Still D, Bond T, Roden C (2008) A laboratory comparison of the global warming impact of five major types of biomass cooking stoves. Energy Sustain Dev 12(2):56–65. https://doi.org/10.1016/S0973-0826(08)60429-9

Major J, Rondon M, Molina D, Riha SJ, Lehmann J (2010) Maize yield and nutrition during four years after biochar application to a Colombian savanna oxisol. Plant Soil 333:117–128. https://doi.org/10.1007/s11104-010-0327-0

Marris E (2006) Black is the new green. Nature 442:624–626. https://doi.org/10.1038/442624a

Masek O, Buss W, Brownsort P, Rovere M, Tagliaferro A, Zhao L, Cao X, Xu G (2019) Potassium doping increases biochar carbon sequestration potential by 45%, facilitating decoupling of carbon sequestration from soil improvement. Sci Reports 9:5514. https://doi.org/10.1038/s41598-019-41953-0

Matt CP (2015) An assessment of biochar amended soilless media for nursery propagation of northern Rocky Mountain native plants. Graduate student theses, dissertations, & professional papers, 4420. https://scholarworks.umt.edu/etd/4420

McCann JM, Woods WI, Meyer DW (2001) Organic matter and anthrosols in Amazonia: interpreting the Amerindian Legacy. In: Rees RM, Ball BC, Campbell D, Watson A (eds) Sustainable management of soil organic matter. CAB International, Wallingford, UK, pp 180–189. https://doi.org/10.1007/978-3-662-05683-7_4

Mechler AA, Jiang RW, Silverthorn TK, Oelbermann M (2018) Impact of biochar on soil characteristics and temporal greenhouse gas emissions: a field study from southern Canada. Biomass Bioener 118:154–162. https://doi.org/10.1016/j.biombioe.2018.08.019

Merzdorf J (2019) Boreal forest fires could release deep soil carbon. global climate change. https://climate.nasa.gov/news/2905/boreal-forest-fires-could-release-deep-soil-carbon/#:~:text=In%20total%2C%20about%2012%20percent,released%20by%20all%20the%20fires, 22 August 2019

Mitchell KA (2015) The effect of biochar on the growth of agricultural weed species. Master's thesis, Purdue University, West Lafayette, Indiana

Mohiuddin KM, Ali MS, Chowdhury MAH (2006) Transformation of nitrogen in urea and DAP amended soils. J Banglad Agricl Univ 4(2):201–210. https://doi.org/10.22004/ag.econ.276545

Monnie F (2016) Effect of biochar on soil physical properties, water use efficiency, and growth of maize in a sandy loam soil. Thesis (MPHIL)-University of Ghana. http://197.255.68.203/handle/123456789/21575

Moosavi SA, Shokuhfar A, Lak S, Mojaddam M, Alavifazel M (2020) Integrated application of biochar and bio-fertilizer improves yield and yield components of Cowpea under water-deficient stress. Ital J Agron 15(2). https://doi.org/10.4081/ija.2020.1581

Muhammad N, Aziz R, Brookes PC, Xu J (2017) Impact of wheat straw biochar on yield of rice and some properties of Psammaquent and Plinthudult. J Soil Sci Plant Nut 17(3). https://doi.org/10.4067/S0718-95162017000300019

Mukherjee A, Lal R, Zimmerman AR (2014) Effects of biochar and other amendments on the physical properties and greenhouse gas emissions of an artificially degraded soil. Sci Total Environ 487:26–36. https://doi.org/10.1016/j.scitotenv.2014.03.141

Nartey OD, Zhao B (2014) Biochar preparation, characterization, and adsorptive capacity and its effect on bioavailability of contaminants: an overview. Hindawi Publishing Corporation, 2014, Article ID 715398. https://doi.org/10.1155/2014/715398

Nath S (2016) Influence of biochar on weed control in dry direct seeded rice (*Oryza sativa* L.). Master thesis in G. B. Pant University of Agriculture and Technology, Pantnagar. Retrieved from https://krishikosh.egranth.ac.in/handle/1/5810115058

Nerome M, Toyota K, Islam TM (2005) Suppression of bacterial wilt of tomato by incorporation of municipal biowaste charcoal into soil. Soil Microorg 59(1):9–14 (in Japanese). https://doi.org/10.18946/jssm.59.1_9

Olszyk DM, Shiroyama T, Jeffrey M, Novak K, Cantrell B, Sigua G, Watts DW, Johnson MG (2020) Biochar affects essential nutrients of carrot taproots and lettuce leaves. J Am Soc Hortic Sci 55(2):261–271. https://doi.org/10.21273/HORTSCI14421-19

Pardo S, Orense RP, Sarmah AK (2018) Cyclic strength of sand mixed with biochar: some preliminary result. Soils Found 58(1):241–247. https://doi.org/10.1016/j.sandf.2017.11.004

Petersen JB, Neves E, Heckenberger MJ (2001) Gift from the past: Terra Preta and prehistoric Amerindian occupation in Amazonia. In: McEwan C, Barreto C, Neves E (eds) Unknown Amazonia. British Museum Press, London, pp 86–105. https://doi.org/10.1007/978-1-4020-9031-8_8

Planet S (2020) Using biochar to increase biodiversity in soil. Innovation News Network. https://www.innovationnewsnetwork.com/using-biochar-to-increase-biodiversity-in-soil/4090/, 9 Mar 2020

Qiang M, Jian-en G, Jian-qiao H, Haochen Z, Tingwu L, Long S (2020) How adding biochar improves Loessal soil fertility and sunflower yield on consolidation project land on the Chinese Loess plateau. Pol J Environ Stud 29:3759–3769. http://www.pjoes.com/How-Adding-Biochar-Improves-Loessal-Soil-nFertility-and-Sunflower-Yield-on-Consolidation,118204,0,2.html

Raison RJ, Khanna PK, Jacobsen K, Romanya J, Rarrasolses (2009) Effect of fire on forest nutrient cycles. In: Fire Effects on Soils and Restoration Strategies. https://doi.org/10.1201/9781431439843338-c8

Ramanathan V, Carmichael G (2008) Global and regional climate changes due to black carbon. Nat Geosci 1:221–227. https://doi.org/10.1038/ngeo156

Rawat J, Saxena J Sanwal P (2017) Biochar: a sustainable approach for improving plant growth and soil properties. In: Biochar an imperative amendment for soil and the environment. IntechOpen. https://doi.org/10.5772/intechopen,82151

Reddy B (2011) Biochar production and use. http://www.slideshare.net/saibhaskar/biocharproduction-and-uses-dr-reddy-5242206

Reddy KR, Yaghoubi P, Yukselen-Aksoy Y (2015) Effects of biochar amendment on geotechnical properties of landfill cover soil. Waste Manage Res 33(6):524–532. https://doi.org/10.1177/0734242X15580192

Renee L (2019) Evaluation of the geotechnical engineering properties of soil-biochar mixtures. Masters' thesis. University of Delaware. http://udspace.udel.edu/handle/19716/24921

Rosenani AB, Ahmad SH, Nurul AS, Loon TW (2014) Biochar as a soil amendment to improve crop yield and soil carbon sequestration. Acta Hortic 1018:203–209. https://doi.org/10.17660/ActaHortic.2014.1018.20

Sadasivam Y, Reddy KR (2015) Shear strength of waste-wood biochar and biochar-amended soil used for sustainable landfill cover systems. In: Fundamentals to applications in geotechnics. https://doi.org/10.3233/978-1-61499-603-3-745

Safaei Khorram M, Zhang G, Fatemi A, Kiefer R, Mahmood A, Jafarnia S, Zakaria MP, Li G (2020) Effect of walnut shell biochars on soil quality, crop yields, and weed dynamics in a 4-year field experiment. Environ Sci Pollut Res 27:18510–18520. https://doi.org/10.1007/s11356-020-08335-w

Scholz SM, Sembres T, Roberts K, Whitman T, Wilson K, Lehmann J (2014) Biochar systems for smallholders in developing countries: leveraging current knowledge and exploring future potential for climate-smart agriculture. World Bank Publications, The World Bank, number 18781, November. http://documents1.worldbank.org/curated/zh/188461468048530729/pdf/Biochar-systems-for-smallholders-in-developing-countries-leveraging-current-knowledge-and-exploring-future-potential-for-climate-smart-agriculture.pdf

Shindo H (1991) Elementary composition, humus composition, and decomposition in soil of charred grassland plants. J Soil Sci Plant Nut 37:651–657. https://doi.org/10.1080/00380768.1991.10416933

Sial TA, Lan Z, Wang L, Zhao Y, Zhang J, Kumbhar F, Memon M, Lashari MS, Shah AN (2019) Effects of different biochars on wheat growth parameters, yield and soil fertility status in a silty clay loam soil. Molecules 24(9):1798. https://doi.org/10.3390/molecules24091798

Singh SV (2018, September 18) NITI Aayog for clear policy on 'jhum' cultivation. The Hindu. https://www.thehindu.com/business/agri-business/niti-aayog-for-clear-policy-on-jhum-cultivation/article24970537.ece

Smith NJH (1980) Anthrosols and human carrying capacity in Amazonia. Ann Am Asso Geogr 70:553–566. https://doi.org/10.1111/j.1467-8306.1980.tb01332.x

Smith NJH (1999) The Amazon River forest: a natural history of plants, animals, and people. Oxford University Press, New York

Snyder CS, Bruulsema TW, Jensen TL, Fixen PE (2009) Review of greenhouse gas emissions from crop production systems and fertilizer management effects. Agric Ecosyst Environ 133:247–266. https://doi.org/10.1016/j.agee.2009.04.021

Sokołowska Z, Szewczuk-Karpisz K, Turski M, Tomczyk A, Skic CM, K, (2020) Effect of wood waste and sunflower husk biochar on tensile strength and porosity of dystric cambisol artificial aggregates. Agronomy 10(2):244. https://doi.org/10.3390/agronomy10020244

Sombroek W, Kern D, Rodriques T, da S Cravo M, Jarbas TC, Woods W, Glaser B (2002) 'Terra Preta and Terra Mulata: pre-Columbian Amazon kitchen middens and agricultural fields, their sustainability and their replication. In: Proceedings of the 17th World Congress of Soil Science, Thailand, Paper no. 1935. https://sswm.info/sites/default/files/reference_attachments/SOMBROEK%20et%20al%202002%20Terra%20Preta%20and%20Terra%20Mulata.pdf

Sohi S, Krull E, Lopez-Capel E, Bol R (2010) A review of biochar and its use and function in soil. Adv Agron 105:47–82. https://doi.org/10.1016/S0065-2113(10)05002-9

Soni N, Ramon GL, Erickson JE, Ferrell JA (2015) Biochar effects on weed management. UF/IFAS Extension. https://edis.ifas.ufl.edu/pdffiles/AG/AG39000.pdf

Srinivasarao C, Gopinath KA, Venkatesh G, Dubey AK, Wakudka, H, Purakayastha TJ, Pathak H, Jha P, Lakaria BL, Rajkhowa DJ, Mandal S, Jeyaraman S, Venkateswarlu B, Sikka AK (2013) Use of biochar for soil health management and greenhouse gas mitigation in India: potential and constraints, Central Research Institute for Dryland Agriculture, Hyderabad, Andhra Pradesh, 51p. http://www.nicra-icar.in/nicrarevised/images/Books/Biochor%20Bulletin.pdf

Steiner C, Glaser B, Teixeira WG, Lehmann L, Blum WEH, Zech W (2008) Nitrogen retention and plant uptake on a highly weathered central Amazonian ferrosol amended with compost and charcoal. J Plant Nutr Soil Sci 45:165–175. https://doi.org/10.1002/jpln.200625199

Sukartono W, Utomo H, Kusuma Z, Nugroho WH (2011) Soil fertility status, nutrient uptake, and maize (*Zea mays* L.) yield following biochar and cattle manure application on sandy soils of Lombok, Indonesia. J Trop Agric 49(1–2):47–52. http://jtropag.kau.in/index.php/ojs2/article/view/236

Tenenbaum D (2009) Biochar: carbon mitigation from the ground up. Environ Health Persp 117(2):70–73. https://doi.org/10.1289/ehp.117-a70

Timmons D, Lema-Driscoll A, Uddin G (2017) The economics of biochar carbon sequestration in Massachusetts. Report on Biochar prepared at University of Massachusetts Boston. https://ag.umass.edu/sites/ag.umass.edu/files/reports/timmons_-_biochar_report_10-16-17.pdf

UNEP (2008) United Nations Ministerial Conference of the Least Developed Countries. Energizing the least developed countries to achieve the Millennium Development Goals: the challenges and opportunities of globalization. Issues Paper. https://www.wto.org/english/thewto_e/coher_e/mdg_e/mdg_e.pdf

USBI News (2021a) Biochar slows climate change. https://biochar-us.org/biochar-slows-climate-change#:~:text=Biochar%20is%20Carbon%20Negative%E2%80%94%20Biochar,in%20the%20inactive%20carbon%20cycle

USBI News (2021b) Biochar then and now. https://biochar-us.org/biochar-then-now#:~:text=Biochar's%20History%20as%20an%20Ancient,a%20result%20of%20vegetation%20fires.&text=Terra%20preta%20was%20discovered%20in,Sombroek%20in%20the%20Amazon%20rainforest

Venkatesh G, Gopinath KA, Sammi Reddy K, Srinivasarao C (2016) Biochar production technology from forest biomass. In: Partiban KT, Seenivasan R (eds) Forestry technologies—complete value chain approach. Scientific Publisher, Jodhpur, pp 532–547. ISBN: 978-93-86102-60-7

Wayne E (2012) Conquistadors, cannibals and climate change A brief history of biochar. History of Biochar. Pro-Natura International, 1–5. http://www.pronatura.org/wp-content/uploads/2012/07/History-of-biochar.pdf

White C, Barberchek M (2017) Managing soil health: concepts and practices. Penn State University. https://extension.psu.edu/managing-soil-health-concepts-and-practices, 31 July 2017

Whitman T, Lehmann J (2009) Biochar—one way forward for soil carbon in offset mechanisms in Africa? Environ Sci Policy 12:1024–1027. https://doi.org/10.1016/j.envsci.2009.07.013

Wu F, Gai Y, Jiao Z, Liu Y, Ma X, An L, Wang W, Feng H (2012) The community structure of microbial in arable soil under different long-term fertilization regimes in the Loess Plateau of China. Afric J Microbiol Res 6:6152–6164. https://doi.org/10.5897/AMJR12.562

Xiao L, Feng L, Yuan G, Wei J (2020) Low-cost field production of biochars and their properties. Environ Geochem Health 42:1569–1578. https://doi.org/10.1007/s10653-019-00458-5

Yan S, Niu Z, Zhang A, Yan H, Zhang H, He K, Xiao X, Wang N, Guan C, Liu G (2019) Biochar application on paddy and purple soils in southern China: soil carbon and biotic activity. Royal Soc Open Sci 6:181499. https://doi.org/10.1098/rsos.181499

Yilangai MR, Manu AS, Pineau W, Mailumo SS, Okeke-Agulu KI (2014) The effect of biochar and crop veil on growth and yield of Tomato (Lycopersicum esculentus Mill) in Jos, North central Nigeria. Curr Agric Res J 2(1). https://doi.org/10.12944/CARJ.2.1.05

Yin X, Chen J, Cao F, Tao Z, Huang M (2020) Short-term application of biochar improves post-heading crop growth but reduces pre-heading biomass translocation in rice. Plant Prod Sci 23(4):522–528. https://doi.org/10.1080/1343943X.2020.1777879

Yuniwati ED (2018) The effect of chicken manure and corn cob biochar on soil fertility and crop yield on intercropping planting pattern of cassava and corn. Int J Environ Sci Natur Resour 54–55. https://doi.org/10.19080/IJESNR.2018.15.555910

Zech W, Haumaier L, Hempfling R (1990) Ecological aspects of soil organic matter in tropical land use. In: McCarthy P, Clapp CE, Malcolm RL, Bloom PR (eds) Humic substances in soil and crop sciences: selected readings. American Society of Agronomy and Soil Science Society of America, Madison, Wis., pp 187–202. https://doi.org/10.2136/1990.humicsubstances.c8

Zemanovai V, Brendova K, Pavlikova D, Kubatova P, Tlustos P (2017) Effect of biochar application on the content of nutrients (Ca, Fe, K, Mg, Na, P) and amino acids in subsequently growing spinach and mustard. Plant Soil Environ 63(7):322–327. https://doi.org/10.17221/318/2017-PSE

Zhang Y, Ding J, Wang H, Su L, Zhao C (2020) Biochar addition alleviate the negative effects of drought and salinity stress on soybean productivity and water use efficiency. BMC Plant Biol 20:288. https://doi.org/10.1186/s12870-020-02493-2

Zhu Q, Peng X, Huang T (2015) Contrasted effects of biochar on maize growth and N use efficiency depending on soil conditions. Int Agrophys 29(2):257–266. https://doi.org/10.1515/intag-2015-0023

Zorich Z (2011) March-April) "A Chauvet Primer. Archaeology 64(2):39

Zwart DC, Kim SH (2012) Biochar amendment increases resistance to stem lesions caused by *Phytophthora* spp. in tree seedlings. J Am Soc Hortic Sci 47(12):1736–1740. https://doi.org/10.21273/HORTSCI.47.12.1736

Index

A

Agents of vermicompost
   C/N ratio, 263
   heavy metals, 264
   microorganisms, 263
Agricultural biomass, 266
Agrochemical and pharmaceutical refuges, 173–174
Agrochemicals, 289–291
Amplified Ribosomal DNA Restriction Analysis (ARDRA), 117
Animal waste biomass, 269
Annual Ruderal plants, 36–39
Advantages of vermicompost, 279–281
Atmospheric contamination, 193

B

Bioavailability of heavy metals, 97–99, 99f
Biochar, 318–319
Biochar modification methods, 322
Biochar properties, 392–393, 408–409
Bioenergy crops, 4–7
Bioenergy sources, 8–10
Biofertilizers, 137
Biofortification, 70
Biology of earthworm, 250–252
Bioremediation, 92–95
Bioremediation through microbial biomass and their enzymes, 209, 210
Brick kilns, 454–455

C

Carbon sequestration, 364–368
Care of vermicomposting earthworms, 265

Climate change, 461–463
Connecting phytoremediation and food production, 76
Contaminated products, 179–180
Cover crop and crop rotation, 361
CPFP projects, 77t
Cropland soil carbon sequestration, 359–360
Crude oil polluted land, 293
Culture–dependent techniques, 116
Culture–independent techniques, 118
Cytochrome P450, 200

D

Degraded land, 32–33
Degraded land restoration, 10
Denaturing Gradient Gel Electrophoresis (DGGE), 119
Detergents, dyes and phthalates, 174–175
Detoxification of soil environment, 229–232
Disease pest infestation, 458–459

E

Edible crops, 70–71
Edible parts, 72–75
Endophyte-assisted land remediation, 136f
Endophytes in heavy metal remediation, 150, 151t
Endophytic microbes, 134–135
Energy crops
   Jatropha spp., 17f
   Miscanthus × *giganteus*, 13, 14–15f
   Panicum virgatum, 12t

Phalaris arundinacea, 12t
Populus nigra, 13t
Saccharum spp., 16f
Salix alba, 13t
Ex-situ bioremediation
  composting, 297–298
  land-farming, 297
  phytoremediation, 298

F

Fate and occurrence of PPCPs
  bio-adsorption and accumulation, 382
  degradation, 383
  transformation, 381
  translocation, 382–383
Fluorescence in Situ Hybridization (FISH), 119–120
Fly ash impact on soil characteristics, 245–246
Fly ash in agriculture, 246–247
Fly ash management, 241
Food security pillars, 70f
Forest biomass, 269
Forest soil carbon sequestration, 362
Fungal organisms and bioremediation, 167–168

G

Genetic engineering in pesticide degradation, 202–203
Global Carbon Cycle (GCC), 356–357t
Global land degradation, 2
Global regulations, 294
Global warming, 355
Grasslands soil carbon sequestration, 362

H

Haloalkane dehalogenases, 202
Health risk of PPCPs, 379–380
Heavy metal pollution
  Indian status, 88
  World status, 87
Heavy metal remediation mechanisms
  complexation, 330
  diffusion, 329
  electrostatic attraction/repulsion, 329–330
  hydrogen bond formation, 330–331
  ion exchange, 329
  physical adsorption, 328, 329f
  precipitation, 330

Human habitation waste
  vermicomposting of agro-based/sugar industry sludge, 272
  vermicomposting of carbide sludge, 274
  vermicomposting of contaminated groundwater, 274–275
  vermicomposting of distillery industry waste, 274
  vermicomposting of industrial sludge, 270–271
  vermicomposting of paper and pulp industry sludge, 271–272
  vermicomposting of sludge from food industry, 272
  vermicomposting of sludge from tanning industry, 273, 301
  vermicomposting of sludge from textile industries, 274
  vermicomposting of wastewater sludge from milk processing industry, 273
Hydrolases, 201
Hyper accumulation and biosorption, 152–153

I

Impact of heavy metal on land quality, 203–204
Invasive ruderal plants
  Ailanthus altissima, 55–56
  Ambrosia artemisiifolia, 51, 52f
  Amorpha fruticosa, 53, 54f
  Artemisia annua, 50
  Bromus tectorum, 50
  Cirsium arvense, 51
  Impatiens glandulifera, 50
  Leucaena leucocephala, 55
  Reynoutria japonica, 52
  Senecio jacobaea, 51
Irrigation, 362

L

Laccases, 201
Land pollution, 189–191
Land surface contamination, 194
Livelihood, 18–19

M

Major sources of heavy metals, 89t
Manure and nutrient management, 361
Mechanism of fungal bioremediation process, 175, 176f

Index

Mechanism of plant assisted bioremediation, 95–97
Mechanism of PPCPs removal from soil with biochar, 394, 395f
Metabolomics, 126
Metagenomics, 121–123
Metal transporters, 102t
Metaproteomics, 125
Metatranscriptomics, 124–125
Methods of biochar preparation
    Biochar stoves, 455–456
    Brick kilns, 454–455
    Cone-pit method, 453
    Drum method, 453, 454f
    Heap method, 452
Microbe–metabolite relationship, 127f
Microbial degradation, 197–198
Microbial enzymes in pesticide bioremediation, 198
Microorganisms mediated heavy metal bioremediation
    bioaugmentation, 206, 296
    biosparging, 205
    biostimulation, 205–206, 296
    bioventing, 205, 297
Municipal waste
    palm oil mill waste, 275–276
    vermicomposting of fly ash, 276
    vermicomposting of human excreta, 276
    vermiremediation of contaminated soils, 276–277
Mycoremediation, 166–167, 169f

N

N_2 fixing endophytic isolates, 146t
Nitrogen fixation, 135–136
Nutrient cycling, 138
Nutrient dynamics, 458

O

Oxidoreductase, 199
Oxygenase, 199–200

P

Pathogen antagonism, 137
Peatland soil carbon sequestration, 363
Peroxidase, 200–201
Persistent Organic Pollutants (POPs), 142, 147t, 195
Pesticides, 191, 192t, 193t
Pesticide toxicity and its degradation, 195, 196t, 197t
Petroleum, 175
Pharmaceuticals, 376
Pharmaceuticals and Personal Care Products (PPCPs), 376–378
Phytoavailability, 151–152
Phytoextraction, 90
Phytoremediation, 34
Phytostabilization, 90
Phytovolatilization, 90
Plant-endophytic interactions, 139–141
Plant growth promotion, 141–142
Plant uptake
    bioactivation of metals by root-microbe interaction, 100
    root absorption and compartmentalization, 100
Policy implementation, 78
Pollutant free edible part, 72–73
Polychlorinated Biphenyls (PCBs), 170–173
Polycyclic Aromatic Hydrocarbons (PAH), 170–171, 291–292
Potentially toxic metals and metalloids, 172–173
Potential microbes in bioremediation, 92–93, 94–95t
Potential plants in bioremediation, 95
PPCPs properties and behaviour, 393–394
Production of biochar, 406–408
Properties of biochar
    cation exchange capacity, 324
    composition, 323
    energy content and thermal conductivity, 326
    interaction with water, 326
    mechanical stability and grindability, 325–326
    pH and ash content, 323–324
    surface area, porosity and pore volume, 324–325
Properties of vermicompost
    chemical and biochemical properties, 277
    humus, 278
    microbial populations, 278
    physical properties, the, 277

Q

Quality of vermicompost, 278–279

R

Random Amplified Polymorphic DNA (RAPD), 118
Rhizoremediation, 299
Ribosomal Intergenic Spacer Analysis (RISA), 118
Ruderal perennials
 Plantago major, 43f
 Tussilago farfara, 41–42f
 Verbascum Thapsus, 44, 45f
Ruderal plants, 34–35

S

Sequestration and detoxification of heavy metals, 103–104
Siderophore production, 138
Soil Organic Carbon (SOC), 358–359
Soil systems, 177
Solid waste management, 225–226
Strategies for reducing pollutants in edible parts
 using low-accumulating cultivars, 74
 reducing the bioavailability of pollutants in the soil, 74
 minimizing the entry of toxic pollutants into the plant parts, 75
Strategies for remediation of PPCPs
 biochar for PPCPs removal, 385–386, 387–391t
Surface and ground water contamination, 194

T

Terminal Restriction Fragment Length Polymorphism (T-RFLP), 119
Tillage, 360–361
Toxicity Reduction, 153–154
Translocation
 root apoplast to aerial (stem and leaves) tissues, 103
 root symplast to apoplast through xylem tissues, 101
Transporter proteins, 101f
Types of earthworms in vermicomposting, 264–265

U

UN-SDGs, 20–21
Urban soil carbon sequestration, 364

V

Vermicompost, 302–303
Vermicompost and soil health management
 impact on nutrient cycling, 226–227
 mechanism of heavy metal detoxification, 227–229f
Vermicomposting, 223–224, 263
Vermi-reactor, 224, 243
Vermiremediation, 221–222, 247–249

W

Water bodies, 178–179
Water use efficiency, 460
Weed dynamics, 459
Woody Ruderals
 Betula pendula, 49f
 Pistacia lentiscus, 46
 Populus nigra, 47
 Populus spp., 47f

Printed in the United States
by Baker & Taylor Publisher Services